AF389218

Novel Applications of Optical Sensors and Machine Learning in Agricultural Monitoring

Novel Applications of Optical Sensors and Machine Learning in Agricultural Monitoring

Editors

Jibo Yue
Chengquan Zhou
Haikuan Feng
Yanjun Yang
Ning Zhang

Basel • Beijing • Wuhan • Barcelona • Belgrade • Novi Sad • Cluj • Manchester

Editors
Jibo Yue
Henan Agricultural University
Zhengzhou, China

Chengquan Zhou
Zhejiang Academy of Agricultural Sciences
Hangzhou, China

Haikuan Feng
Beijing Research Center for Information Technology in Agriculture
Beijing, China

Yanjun Yang
University of Kentucky
Lexington, KY, USA

Ning Zhang
Chinese Academy of Agricultural Sciences
Beijing, China

Editorial Office
MDPI
St. Alban-Anlage 66
4052 Basel, Switzerland

This is a reprint of articles from the Special Issue published online in the open access journal *Agriculture* (ISSN 2077-0472) (available at: https://www.mdpi.com/journal/agriculture/special_issues/3XU830GWF3).

For citation purposes, cite each article independently as indicated on the article page online and as indicated below:

Lastname, A.A.; Lastname, B.B. Article Title. *Journal Name* **Year**, *Volume Number*, Page Range.

ISBN 978-3-0365-9798-0 (Hbk)
ISBN 978-3-0365-9799-7 (PDF)
doi.org/10.3390/books978-3-0365-9799-7

Contents

Editorial

Novel Applications of Optical Sensors and Machine Learning in Agricultural Monitoring

Jibo Yue [1,*], Chengquan Zhou [2], Haikuan Feng [3], Yanjun Yang [4] and Ning Zhang [5]

[1] College of Information and Management Science, Henan Agricultural University, Zhengzhou 450002, China
[2] Institute of Agricultural Equipment, Zhejiang Academy of Agricultural Sciences, Hangzhou 310021, China; zhoucq@zaas.ac.cn
[3] Key Laboratory of Quantitative Remote Sensing in Agriculture, Ministry of Agriculture, Beijing Research Center for Information Technology in Agriculture, Beijing 100097, China; fenghaikuan@nercita.org.cn
[4] Department of Plant and Soil Sciences, College of Agriculture, Food and Environment, University of Kentucky, Lexington, KY 40546, USA; yanjun.yang@uky.edu
[5] Agricultural Information Institute, Chinese Academy of Agricultural Sciences, Beijing 100081, China; zhangning@caas.cn
* Correspondence: yuejibo@henau.edu.cn

Citation: Yue, J.; Zhou, C.; Feng, H.; Yang, Y.; Zhang, N. Novel Applications of Optical Sensors and Machine Learning in Agricultural Monitoring. *Agriculture* **2023**, *13*, 1970. https://doi.org/10.3390/agriculture13101970

Received: 25 September 2023
Revised: 5 October 2023
Accepted: 9 October 2023
Published: 10 October 2023

The rapid development of intelligence and automated technologies has provided new management opportunities for agricultural production. In particular, the progress of remote sensing equipment has allowed for vast improvements in the spatial, temporal, and spectral resolutions of optical sensors. Such sensors are key in current agricultural production management practices, with applications in areas that were previously explored using field observations, including the monitoring of plant health, growth conditions, and pest infestations.

The papers published in this Special Issue, "Novel Applications of Optical Sensors and Machine Learning in Agricultural Monitoring", present some of the most current and novel results of scholars' investigations on the applications of optical sensors and machine learning in the field of agriculture. Table 1 summarizes the 16 peer-reviewed articles included in this Special Issue. We found the guest editing for this exercise to be very inspiring, with contents including:

(1) The application of machine learning techniques to examine the key physiological development and production variables of crops.

(2) The use of datasets obtained from multiple sources and sensors to enhance crop mapping.

(3) Advanced target recognition algorithm techniques for weed and disease identification.

The optical sensors used in the presented research include a digital RGB camera, spectrometers, a 3D TOF sensor, a multispectral imaging sensor, and a satellite-based multispectral sensor. The machine learning methods include conventional machine learning techniques such as KNN, RF, SVM, and ANN, and deep learning techniques such as LSTM, VGG, YOLO, and SSD.

The contributions to this Special Issue are summarized in the following.

Wang et al. [1] employed LAI as the input to four machine learning models (RF, SVR, PLSR, and XGBOOST) and one deep learning model (LSTM) for winter wheat production estimates in Henan Province, China, during 2016. The results indicated that the LSTM performed better than the four traditional machine learning models, exhibiting the optimal R^2 and RMSE values. Kumar et al. [9] investigated the canopy cover of sugarcane and its relationship with dry matter and yield, and analyzed the relationship between (a) canopy temperature, chlorophyll fluorescence, SPAD index, and (b) yield. Luo et al. [13] fused vegetation indices determined using a UAV with brightness, greenness, and moisture indices estimated using tasseled cap transformation (TCT). The proposed approach was observed to enhance the accuracy of rice yield predictions and was able to avoid the saturation phenomenon.

Table 1. Summary of publications featured in this Special Issue.

Article	Agricultural Activities/Variables	Optical Sensors	Platforms	Machine Learning Methods
[1]	Winter wheat yield prediction	MODIS	Satellite	LSTM, RF, SVR, PLSR, and XGBoost
[2]	Land use/cover classification	Sentinel-2 MSI	Satellite	RF
[3]	Wheat fusarium head blight	Multispectral imaging sensor	UAV	KNN, SVM, XGBoost
[4]	Cropland spatial distribution	Landsat 8 OLI	Satellite	Blanket covering method
[5]	Soybean FVC, LCC, and maturity	SONY DSC-QX100	UAV	RF, PLSR, GPR, MSR
[6]	Apple leaf diseases	Canon Rebel T5i DSLR	Field	BTC-YOLOv5s, YOLOv5, SSD, R-CNN, Faster R-CNN, YOLOv4-tiny, and YOLOx, YOLOx-s
[7]	Crop classification	Sentinel-2	Satellite	1D-CNNs, LSTM, 2D-CNNs, 3D-CNNs, and ConvLSTM2D
[8]	Dairy herd fatness	3D TOF sensor	Field	BCS
[9]	Sugarcane dry matter and cane yield	Mobile phone camera	Field	Two-Way cluster
[10]	Peanut southern blight severity	ASD Field Spec3 VNIR-SWIR sensor	Field	SVM, DT, and KNN
[11]	Corn diseases	digital camera	Field	VGNet, VGG16
[12]	Soil moisture content	ASD Field Spec3 VNIR-SWIR sensor	Field	PCA, PCR, PLSR, and BP-ANN
[13]	Rice yield	Mini-MCA 1000	UAV	TCT
[14]	Weed detection in peanut fields	Fuji Finepixs4500	Field	YOLOv4-Tiny, YOLOv5s, Swin-Transformer, Faster-RCNN, YOLOv6-Tiny, and EM-YOLOv4-Tiny
[15]	Vegetation canopy reflectance angle normalization	GOCI	Satellite	SANM
[16]	Soybean maturity	SONY DSC-QX100	UAV	SVM, RF, InceptionResNetV2, MobileNetV2, Alexnet, ResNet50, and DS-SoybeanNet

Note: UAV, unmanned aerial vehicle; RF, random forest; TCT, tasseled cap transformation; SANM, synthetic angle normalization model; PCA, principal component analysis; LSTM, long short-term memory; SVR, support vector regression; PLSR, partial least squares regression; XGBoost, eXtreme gradient boosting; DT, decision tree; KNN, K-nearest neighbor; SVM, support vector machine; GPR, Gaussian process regression; MSR, stepwise multiple linear regression; YOLO, you only look once; SSD, single shot multi-box detector; CNN, convolutional neural network; R-CNN, regions-convolutional neural network; BCS, body condition scoring; PCA, principal component analysis; and BP-ANN, back propagation-artificial neural network.

In order to enhance the estimation accuracy of LULC models, Ibrahim [2] performed RF-based feature selection using data obtained from Sentinel-1, -2, and the Shuttle Radar Topographic Mission. The author revealed that integrating optical, radar, and elevation information is key to increasing the precision of LULC models for agriculturally dominated landscapes. Wang et al. [4] developed an information extraction method for the accurate determination of the spatial distribution of crops by integrating spatiotemporal image information using a fractal model. The authors demonstrated the ability of their approach to determine key cropland variables for the effective monitoring, conservation, and development of black soil. Li et al. [7] developed a 3D-CNN and ConvLSTM2D method for the classification of crops across time. Five deep learning models were tested, namely 1D-CNNs, LSTM, 2D-CNNs, 3D-CNNs, and ConvLSTM2D. 3D-CNN and ConvLSTM2D, which combine temporal, spectral, and spatial information, outperformed the other models in terms of crop classification using time series images.

Gao et al. [3] developed an approach based on UAV and multispectral imagery that integrated the spectral and textural features of images to examine wheat fusarium head blight (FHB) and estimate several Vis and Tis. The VIs, TIs, and combined VIs and TIs were adopted as the inputs to KNN, PSO-SVM, and XGBoost to develop wheat FHB monitoring models. The proposed approach was revealed to have potential for fast and nonintrusive observations of wheat FHB. Guo et al. [10] proposed the Peanut Southern Blight Severity method by combining hyperspectral data, continuous wavelet transform, and machine learning. The machine learning methods SVM, DT, and KNN were tested and compared. Fan et al. [11] developed a VGNet with the backbone set as VGG16, with the ability to improve the recognition of corn with poor health in fields. In particular, there was a 3.5% enhancement in the accuracy of the proposed VGNet compared to its predecessor VGG16.

Hu et al. [5] developed a soybean maturity recognition approach that combined UAV-based LCC and FVC maps with an anomaly detection method, exhibiting total monitoring accuracies greater than 98%. Zhang et al. [16] designed the novel CNN DS-SoybeanNet to enhance UAV-based soybean maturity observations, with the ability to extract and employ shallow and deep image features. The authors compared it with the widely used Alexnet, InceptionResNetV2, MobileNetV2, ResNet50, SVM, and RF, revealing the high accuracy of DS-SoybeanNet in soybean maturity classification.

Yurochka et al. [8] developed an approach for the automatic evaluation of dairy herd fatness using a 3D TOF sensor and the body condition score (BCS). The proposed approach was able to perform nonintrusive BCS evaluations of dairy herds throughout the lifetime of the herd while meeting the requirements of the farm. The overall accuracy of the system was estimated at 93.4%.

Jiang et al. [12] proposed an SMC estimation approach for mixed soil types based on PCA and machine learning, with hyperspectral data as the input. The R^2 and RMSE of the optimal model were determined as 0.932 and <2%, respectively. This approach proved to be valuable in extracting data on farm entropy prior to the sowing of crops on agricultural land, and provides a basis for the use of hyperspectral imagery to calculate SMC.

Geostationary satellites are able to extract information on the daily variations in crop canopy reflectance based on high-temporal-resolution imagery. Lin et al. [15] proposed the synthetic angle normalization model (SANM), which uses vegetation canopy reflectance as its input. The SANM makes use of the advantages of GSS imaging and is able to quantitatively compare spatiotemporal remote sensing data.

Advanced target recognition algorithm techniques, such as YOLO-, Swin-Transformer-, and Faster-RCNN-based models, have also been developed to identify weeds and diseases for farmland management.

For example, Zhang et al. [14] introduced EM-YOLOv4-Tiny to identify weeds and compared it with six other weed recognition deep learning models, namely YOLOv4-Tiny, YOLOv4, YOLOv5s, Swin-Transformer, and Faster-RCNN. The proposed approach was observed to outperform the majority of the models, with an mAP of 94.54%.

Li et al. [6] developed BTC-YOLOv5s based on YOLOv5s for the detection of apple leaf disease. In particular, the inclusion of the transformer and convolutional block attention modules decreased the background noise.

Intelligent agriculture can achieve information perception, quantitative decision-making, and intelligent control throughout agricultural production by integrating information technologies such as the Internet of Things, big data, artificial intelligence, and intelligent equipment with agriculture. Therefore, interdisciplinary cooperation is necessary for deepening the application of deep learning in intelligent agriculture. These collaborations include expert-assisted data annotation, machine learning methods, the design of agricultural-specific sensors, intelligent drones, intelligent robots, and more. Optical sensors and deep learning are fundamental in data collection, information perception, and decision analyses. Research on their combinations is crucial for promoting the development of intelligent agriculture. Therefore, we hope this work can attract the attention of the agricultural, electronic, and computer communities and promote more research on optical sensors and machine learning. The research published in this Special Issue focus on a variety of machine learning methods, optical sensors, and platforms for agricultural monitoring. The novel results and progress made by the papers will hopefully stimulate further research in these areas.

Funding: This study was supported by the Henan Province Science and Technology Research Project (232102111123) and the National Natural Science Foundation of China (grant number 42101362).

Institutional Review Board Statement: Not applicable.

Data Availability Statement: No new data were created or analyzed in this study.

Conflicts of Interest: The authors declare no conflict of interest.

References

1. Wang, J.; Si, H.; Gao, Z.; Shi, L. Winter Wheat Yield Prediction Using an LSTM Model from MODIS LAI Products. *Agriculture* **2022**, *12*, 1707. [CrossRef]
2. Ibrahim, S. Improving Land Use/Cover Classification Accuracy from Random Forest Feature Importance Selection Based on Synergistic Use of Sentinel Data and Digital Elevation Model in Agriculturally Dominated Landscape. *Agriculture* **2023**, *13*, 98. [CrossRef]
3. Gao, C.; Ji, X.; He, Q.; Gong, Z.; Sun, H.; Wen, T.; Guo, W. Monitoring of Wheat Fusarium Head Blight on Spectral and Textural Analysis of UAV Multispectral Imagery. *Agriculture* **2023**, *13*, 293. [CrossRef]
4. Wang, Q.; Guo, P.; Dong, S.; Liu, Y.; Pan, Y.; Li, C. Extraction of Cropland Spatial Distribution Information Using Multi-Seasonal Fractal Features: A Case Study of Black Soil in Lishu County, China. *Agriculture* **2023**, *13*, 486. [CrossRef]
5. Hu, J.; Yue, J.; Xu, X.; Han, S.; Sun, T.; Liu, Y.; Feng, H.; Qiao, H. UAV-Based Remote Sensing for Soybean FVC, LCC, and Maturity Monitoring. *Agriculture* **2023**, *13*, 692. [CrossRef]
6. Li, H.; Shi, L.; Fang, S.; Yin, F. Real-Time Detection of Apple Leaf Diseases in Natural Scenes Based on YOLOv5. *Agriculture* **2023**, *13*, 878. [CrossRef]
7. Li, Q.; Tian, J.; Tian, Q. Deep Learning Application for Crop Classification via Multi-Temporal Remote Sensing Images. *Agriculture* **2023**, *13*, 906. [CrossRef]
8. Yurochka, S.S.; Dovlatov, I.M.; Pavkin, D.Y.; Panchenko, V.A.; Smirnov, A.A.; Proshkin, Y.A.; Yudaev, I. Technology of Automatic Evaluation of Dairy Herd Fatness. *Agriculture* **2023**, *13*, 1363. [CrossRef]
9. Kumar, R.A.; Vasantha, S.; Gomathi, R.; Hemaprabha, G.; Alarmelu, S.; Srinivasa, V.; Vengavasi, K.; Alagupalamuthirsolai, M.; Hari, K.; Palaniswami, C.; et al. Rapid and Non-Destructive Methodology for Measuring Canopy Coverage at an Early Stage and Its Correlation with Physiological and Morphological Traits and Yield in Sugarcane. *Agriculture* **2023**, *13*, 1481. [CrossRef]
10. Guo, W.; Sun, H.; Qiao, H.; Zhang, H.; Zhou, L.; Dong, P.; Song, X. Spectral Detection of Peanut Southern Blight Severity Based on Continuous Wavelet Transform and Machine Learning. *Agriculture* **2023**, *13*, 1504. [CrossRef]
11. Fan, X.; Guan, Z. VGNet: A Lightweight Intelligent Learning Method for Corn Diseases Recognition. *Agriculture* **2023**, *13*, 1606. [CrossRef]
12. Jiang, X.; Luo, S.; Ye, Q.; Li, X.; Jiao, W. Hyperspectral Estimates of Soil Moisture Content Incorporating Harmonic Indicators and Machine Learning. *Agriculture* **2022**, *12*, 1188. [CrossRef]
13. Luo, S.; Jiang, X.; Jiao, W.; Yang, K.; Li, Y.; Fang, S. Remotely Sensed Prediction of Rice Yield at Different Growth Durations Using UAV Multispectral Imagery. *Agriculture* **2022**, *12*, 1447. [CrossRef]
14. Zhang, H.; Wang, Z.; Guo, Y.; Ma, Y.; Cao, W.; Chen, D.; Yang, S.; Gao, R. Weed Detection in Peanut Fields Based on Machine Vision. *Agriculture* **2022**, *12*, 1541. [CrossRef]
15. Lin, Y.; Tian, Q.; Qiao, B.; Wu, Y.; Zuo, X.; Xie, Y.; Lian, Y. A Synthetic Angle Normalization Model of Vegetation Canopy Reflectance for Geostationary Satellite Remote Sensing Data. *Agriculture* **2022**, *12*, 1658. [CrossRef]
16. Zhang, S.; Feng, H.; Han, S.; Shi, Z.; Xu, H.; Liu, Y.; Feng, H.; Zhou, C.; Yue, J. Monitoring of Soybean Maturity Using UAV Remote Sensing and Deep Learning. *Agriculture* **2022**, *13*, 110. [CrossRef]

Article

VGNet: A Lightweight Intelligent Learning Method for Corn Diseases Recognition

Xiangpeng Fan [1,2] and Zhibin Guan [1,2,*]

1   Agricultural Information Institute, Chinese Academy of Agricultural Sciences, Beijing 100081, China; fanxiangpeng@caas.cn
2   National Nanfan Research Institute (Sanya), Chinese Academy of Agricultural Sciences, Sanya 572024, China
*   Correspondence: guanzhibin@caas.cn

Abstract: The automatic recognition of crop diseases based on visual perception algorithms is one of the important research directions in the current prevention and control of crop diseases. However, there are two issues to be addressed in corn disease identification: (1) A lack of multicategory corn disease image datasets that can be used for disease recognition model training. (2) The existing methods for identifying corn diseases have difficulty satisfying the dual requirements of disease recognition speed and accuracy in actual corn planting scenarios. Therefore, a corn diseases recognition system based on pretrained VGG16 is investigated and devised, termed as VGNet, which consists of batch normalization (BN), global average pooling (GAP) and L2 normalization. The performance of the proposed method is improved by using transfer learning for the task of corn disease classification. Experiment results show that the Adam optimizer is more suitable for crop disease recognition than the stochastic gradient descent (SGD) algorithm. When the learning rate is 0.001, the model performance reaches a highest accuracy of 98.3% and a lowest loss of 0.035. After data augmentation, the precision of nine corn diseases is between 98.1% and 100%, and the recall value ranges from 98.6% to 100%. What is more, the designed lightweight VGNet only occupies 79.5 MB of space, and the testing time for 230 images is 75.21 s, which demonstrates better transferability and accuracy in crop disease image recognition.

Keywords: VGNet; corn diseases; leaf detection; lightweight; transfer learning; agriculture

Citation: Fan, X.; Guan, Z. VGNet: A Lightweight Intelligent Learning Method for Corn Diseases Recognition. *Agriculture* **2023**, *13*, 1606. https://doi.org/10.3390/agriculture13081606

Academic Editor: Roberto Alves Braga Júnior

Received: 19 June 2023
Revised: 8 August 2023
Accepted: 9 August 2023
Published: 14 August 2023

1. Introduction

Crop diseases can cause irreversible damage to crop growth and are considered one of the main limiting factors for crop cultivation, and spraying pesticides is the main measure to address crop diseases. Appropriate pesticide category selection and dosage regulation can ensure effective crop disease resolution and avoid pesticide residues' ecological impact. Therefore, accurately identifying the types and degrees of crop diseases is a prerequisite for achieving precise agricultural spraying [1–8]. In traditional methods, professionals mainly detect and identify crop diseases based on their naked eyes and experience, but it is time-consuming, laborious, and subjective. With the development of deep learning (DL) and visual perception technology, visual feature learning methods based on deep learning have become the mainstream of crop disease recognition, which realizes automatic recognition of crop diseases by extracting and learning the pest and disease features of crop images [9,10].

Deep learning is a branch of machine learning that mainly utilizes deep artificial neural networks to extract multilayer visual features and fuse multigranularity features of input images, thereby achieving high-level semantic learning of images [11]. Unlike traditional machine learning methods, deep learning methods require significant computational resources, because deep artificial neural network models optimize model parameters through a large number of parameter calculations in the high-level semantic learning of images.

With the rapid development of high-performance computing and image processing units, deep learning methods have been successfully applied in various fields, which has turned out to be very excellent in discovering intricate structures in high-dimensional data and is therefore applicable to many domains of science, engineering [12–14], industries[15–17], bioinformatics [18–20], and agriculture [21–26]. Concretely, deep learning has provided many significant works in the field of plant stress phenotyping and image analysis for detection [27–30], recognition [31–34], classification [35–38], quantification [39], and prediction [40] in agriculture to tackle the challenges of agricultural production [41]. And the convolutional neural network (CNN)-based approaches are arguably the most commonly used [42].

Ferentinos developed a plant diseases detection model with a best performance of 99.5% using 87,848 images under controlled conditions [43]. Liang et al. designed a deep plant diseases diagnosis and severity estimation network (PD2-SE-Net) model to identify plant species, diseases, and their severities with a final accuracy of 99% [44]. They utilized the artificial intelligence (AI) Challenger [45] images for experiment data. The approach they proposed reached an accuracy of 99.4%. Zhong et al. proposed an apple diseases classification method based on dense networks with 121 layers (DenseNet-121) and 2462 apple leaf images from AI Challenger, which achieved an accuracy of 93.71% [46]. He et al. proposed an approach to detect oilseed rape pests based on SSD with an Inception module, which was helpful for integrated pest management [47]. Zeng et al. introduced a self-attention mechanism to a convolutional neural network, and the accuracy of the proposed model reached 98% using 9244 diseased cucumber images [48].

Deep convolutional neural networks have a strong ability for feature learning and expression. The above crop disease recognition methods based on CNNs have achieved good accuracies or success rates. However, the accuracy and robustness of deep learning models require training on a large amount of image data. There are two issues that need to be addressed in crop disease identification. On the one hand, there is a lack of diverse maize disease training datasets, as most of the crop disease images used in the existing methods are created under controlled or laboratory conditions. On the other hand, the complexity of existing corp disease models is high, making it difficult to meet the actual detection needs of field scenarios, and their performance in identifying fine-grained corn diseases is insufficient. Therefore, we introduced transfer learning and designed VGNet to solve the above problems. Specifically, we first collected corn disease image data from real field scenarios, covering nine types of corn diseases, which can be used for parameter optimization of fine-grained corn disease recognition models. Afterwards, we designed a relatively simple VGNet model based on the VGG16 model but with relatively high accuracy in identifying crop diseases, which can meet the disease detection needs of actual corn planting scenarios.

The reason why the VGG16 model is selected as the backbone network is that the VGG network is a straight cylinder network structure, and its computing resource consumption is significantly less than the residual network structure, which can satisfy the dual needs of speed and accuracy in real-time crop disease detection. In the VGNet method, the structure of VGG16 is modified by adding the BN, replacing two hidden fully connected layers with a GAP layer, and adding L2 normalization. Through the comparative experiment of different training methods, parameters, and datasets, the redesigned VGNet after fine-tuning achieves an accuracy of 98.3%, which can achieve a 66.8% reduction in testing time compared with the original VGG16 model. The following summary provides the main contributions of this paper:

- A lightweight intelligent learning method, termed as VGNet, is proposed for multiple categories of corn disease detection.
- Fine-grained corn disease images are collected and can be used for the parameter optimization of corn disease recognition models.
- Evaluation results show that the accuracy of the proposed method in disease detection reaches 98.3%, which can satisfy the detection requirements of practical scenarios.

The remainder of this paper is organized as follows. Section 2 describes the materials and methods. The experiment results of VGNet are detailed in Section 3. In Section 4, the discussion of VGNet for fine-grained corn disease recognition is given. Finally, the conclusions are drawn in Section 5. Further research directions are also proposed.

2. Materials and Methods

2.1. Image Samples

2.1.1. Images for Pretraining

In the field of crop disease recognition, many crop disease datasets have appeared, among which the most commonly used ones are PlantVillage [49] and AI Challenger datasets. PlantVillage contains open and free datasets with 54,306 annotated images and 26 diseases for 14 crop plants, and it was created by Mohanty et al. under controlled conditions [50]. AI Challenger is provided by the Shanghai Science and Technology Innovation Center as a new guest competition crop leaf image datasets, with 45,285 marked images, containing 10 kinds of plants (apple, cherry, grape, orange, peach, strawberry, tomatoes, peppers, corn, and potato), 27 kinds of diseases, and a total 61 categories. Both of these datasets are open-source image datasets containing healthy plant leaves and diseased leaves and have great similarity with the target disease image dataset in this research area. ImageNet dataset [51] contains a large number of images from all aspects of life, and the initial training of VGG16 was obtained through the ImageNet dataset, which has achieved excellent results. These three different large open datasets were used for pretraining the selected CNN structure. The properties of the three pretrained experimented datasets are shown in Table 1.

Table 1. Properties of the pretrained experimented datasets.

Dataset	Classes	Samples	Features Type	Image Type
ImageNet	1000	14,197,122	coarse-grained	RGB
PlantVillage	38	54,306	fine-grained	RGB
AI Challenger	61	45,285	fine-grained	RGB

2.1.2. Images for Parameter Optimization

In this experiment, the images used for recognition and fine-tuning training were composed of symptom pictures of nine corn diseases caused by fungus. They were Anthracnose (ANTH), Tropical Rust (TR), Southern Corn Rust (SCR), Common Rust (CR), Southern Leaf Blight (SLB), Phaeosphaeria Leaf Blight (PHLB), Diplodia Leaf Streak (DLS), Physoderma Brown Spot (PHBS), and Northern Leaf Blight (NLB) of corn. The images were captured using a digital camera (Nikon D750) under natural field conditions at the Western Corn Farm of Urumqi, Xinjiang, China. In order to make the collected images be more representative, symptom images were obtained, respectively, in sunny, cloudy, and windy weather conditions from different times in the morning, noon, and evening with multiangle shooting. The shooting background was complicated, containing corn stalks, soil, weeds, and blades covering each other, etc., to reflect the practical growth situation of corn. There is a total of 1150 images obtained in a 3096 × 3096 pixel spatial resolution. The sample numbers of various diseases are kept balanced relatively. The quantity distribution of corn disease images is shown in Figure 1. Some image examples are shown in Figure 2.

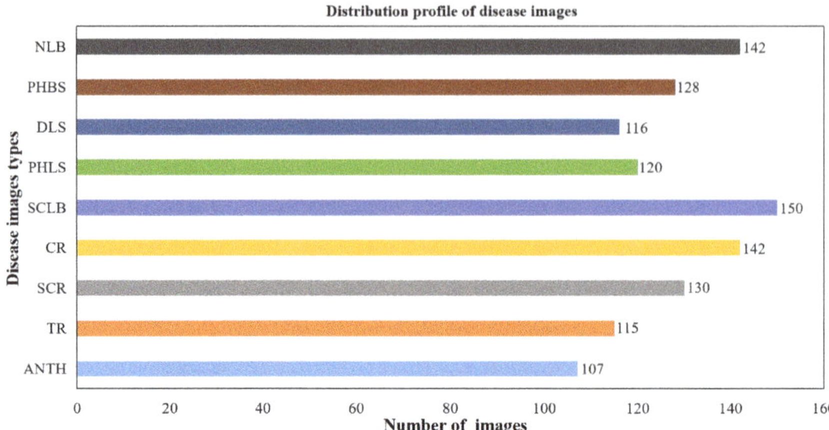

Figure 1. The quantity distribution of maize disease images with complex background.

Figure 2. Some examples of corn disease images with complicated backgrounds from a field: (**a**) Northern leaf blight. (**b**) Common rust. (**c**) Anthracnose. (**d**) Diplodia leaf streak. (**e**) Phaeosphaeria spot. (**f**) Physoderma brown spot. (**g**) Sourthern corn rust. (**h**) Sourthern corn leaf blight. (**i**) Tropical rust.

2.1.3. Data Preprocessing

Data preprocessing includes annotation, cropping, or zooming. Firstly, the CNN model needs supervised training and learning; so, it is necessary to manually annotate the disease images acquired in the field. After the images were confirmed by corn pathologists, the LabelMe tool was used for annotation, and the annotated images were saved as PASCAL VOC2007 format. Secondly, because the images from the corn field and public dataset websites have different resolution and sizes, the size of each image is uniformly cropped and resized to $(224, 224, 3)$ channels.

2.2. Backbone Network

2.2.1. CNN and VGG16 Network

The CNN is one of the classical network algorithms of deep learning. A CNN consists of input layers, convolutional layer, activation function, pooling layers (sampling layer), fully connected layers, and classification layers. Several baseline architectures of CNN have been developed for image recognition, including AlexNet, GoogLeNet, VGGNet, XceptionNet, and ResNet et al. [52]. VGG Net was first devised by Simonyan and Zisserman (2015) for the ILSVRC-2014 challenge. It has been proven to have excellent performance for image classification. The most significant superiority of VGG Net is the utilization of a smaller convolution kernel and pooling window in the feature extractor, which can extract fine-grained features from the input data. Figure 3 shows the basic structure diagram of VGG16. VGG16 contains thirteen convolutional layers and three fully connected layers with 4096, 4096, and 1000 dimensions, respectively. There are five maximum pooling layers between the convolutional layers. During training, the input to VGG16 is a fixed $(224, 224, 3)$-channel RGB image. Large receptive fields in VGG16 were substituted with consecutive layers of 3×3 convolution filters. The convolutional stride was fixed to 1 pixel. The padding of the convolution layer input was maintained as 1 pixel and max-pooling was performed with a stride of 2 over a 2×2 pixel pooling window. The neuron activation function used in VGG16 is the rectified linear unit (ReLU) function.

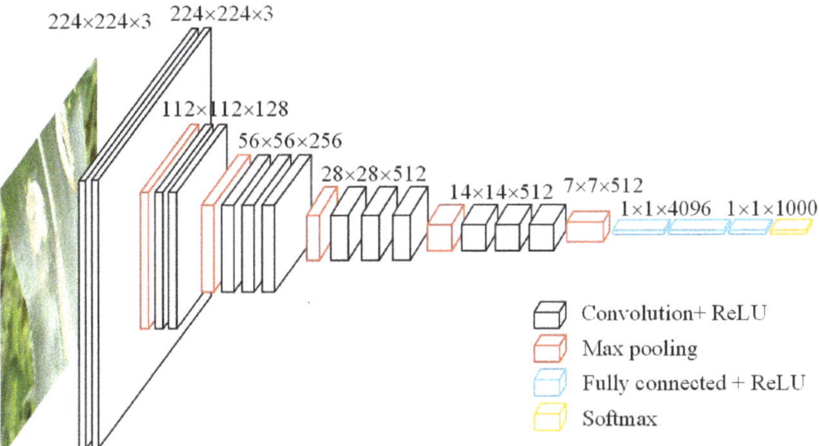

Figure 3. Structure diagram of original VGG16 convolutional neural network.

2.2.2. Proposed Approach and Processes

Figure 4 describes the main process of the VGNet with transfer learning for corn disease recognition. The whole recognition process includes three parts. Part one is the pretraining and parameters transfer process of original VGG16 using three different large datasets, the aim of transfer learning is to shift the general knowledge of image classification acquired by VGG16 from a large image dataset to the new corn leaf disease recognition model. Part two is the establishment of VGNet, the remaining part is fine-tuning the updated VGNet with a new image dataset. After acquiring the new images, they were preprocessed and divided into training set and test set. The modification of the VGG16 network included adding a batch normalization layer to speed up fine-tuning training, replacing the two hidden dense layers by a global average pooling layer to reduce feature dimension, and integrating the L2 regularization algorithm to improve the ability of the model to extract effective features from complex backgrounds. The last layer of the VGG Net was changed by a 9-tag softmax classifier instead of the original softmax classifier with 1000 tags. Three large open datasets were used to obtain the model parameters and feature extraction abilities in the pretraining process, and different training tactics in the parameter tuning were utilized to optimize the VGNet model. After pretraining, the convolutional layers and pooling layers remained unchanged. Their parameters were loaded to the newly designed VGG16 Net and then they were frozen. The VGNet was fine-tuned through the iteration of loss function to reoptimize the parameters of the remaining fully connected layer and softmax function. Finally, the test process was executed by the designed model.

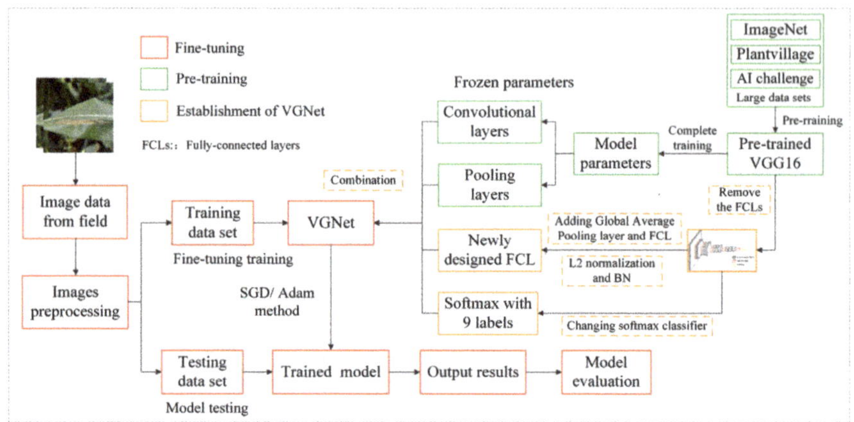

Figure 4. Flowchart of corn disease image recognition method based on transfer learning and VGNet.

2.3. VGNet

As described in Section 2.2.1, the original VGG16 network has 13 convolutional layers, 5 pooling layers, and 3 fully connected layers, and it has 138 million parameters and large amounts of computation, leading to the consumption of both memory and time. The model will easily fall into an overfitting state and lower convergence. Thus, we redesigned VGNet to improve the accuracy and real-time performance of the VGG-based network. Normalization strategies were also adopted, including adding batch normalization (BN) processing and the L2 normalization algorithm. The number of our class labels in the softmax layer of VGNet is 9.

2.3.1. Batch Normalization

For the convolutional neural network, the normalization of datasets is required in the gradient descent process, which can prevent gradient explosion and accelerate the convergence of the network. Thus, batch normalization (BN) processing was applied to normalize the feature map of each sample after the convolutional layers. The mean (μ) and variance (σ) of the total number of pixels in the feature graph were obtained firstly; then, the normalization equation was utilized to calculate the sample normalization values, and the optimal value search data are converted into the standard normal distribution. The BN layer can effectively solve the problem of the data distribution changes in the middle layer during the training process of the model. BN can also accelerate convergence, improve accuracy, and reduce the overfitting phenomenon. The calculation equations of mean (μ) and variance (σ) of the feature maps are described as Equations (1) and (2).

$$\mu = \frac{1}{n} \sum_{i=1}^{n} x_i.$$

(1)

$$\sigma = \frac{1}{n} \sum_{i=1}^{n} (x_i - \mu)^2.$$

(2)

where x_i represent the value of the ith pixel in the image sample. n represents the total number of pixels in the sample. The normalization equation is shown in Formula (3).

$$\bar{x} = \frac{x_i - \mu}{\sqrt{\sigma^2 + \varepsilon}}. \tag{3}$$

where x represents the normalized pixel value of the ith pixel of the sample. ε is a small constant value greater than 0 to ensure that the denominator in Equation (3) is greater than 0. According to the batch normalization algorithm in the training process, the average value and variance of the data estimated based on each batch will be used to replace the actual average value and variance, and the data will be converted to the standard normal distribution according to the estimated average value and variance. The data of the standard normal distribution will be restored by constantly updating the values of x_i and u during the training process. And then they are output by the model.

2.3.2. Replacing Fully Connected Layers by GAP Layer

Although the original VGG16 network structure has 16 weight layers, there is a large number of parameters in the fully connected layer, which leads to excessive computation in the training and testing process. Thus, we decided to compress its weight matrix using a global average pooling (GAP) layer after the last convolutional layer, which outputs a series of feature maps with a depth the same as the number of classes in the classification problems. A GAP layer could enhance the relationship between feature map and category. It has been proven that GAP layers can replace fully connected layers in a conventional structure and thus reduce the storage required by the large weight matrices of the fully connected layers [53]. Performing GAP on a feature map involves computing the average value of all the elements in the feature map.

The principle of GAP is to shrink the parameter space to avoid overfitting and enable precise adjustment of the dropout ratio, which can be treated as the process of dimension reduction in a feature matrix. As shown in Figure 5, the output feature maps from C_l, which is the last convolutional layer, are downsampled into fm_{GAP}, which has a size of $1 \times 1 \times size_{fm}$ after global average pooling. In GAP, the weight matrices of f_1, W can be adjusted as Equation (4) as follows:

$$W' = \sum_{l=(j-1)size_{fm}^2+1}^{j*size_{fm}^2} W_{i,j}. \tag{4}$$

where $size_{fm}$ is the size of the input feature map, i, j is the index of the output neurons and input feature maps, and W' is the modified weight matrix. As shown in Figure 5, the corresponding weights of each feature map are summed up, and each matrix in W is modified and reduced to a column vector composed of $1 \times 1 \times$ depth of fm_{GAP}. Thus, the dimension reduction in the feature matrix is realized. Instead of adding fully connected layers on top of the feature maps, we take the average of each feature map, and the resulting vector is fed directly into the softmax layer. One advantage of the GAP layer over the fully connected layers is that it is more native to the convolution structure by enforcing correspondences between feature maps and categories. Another advantage is that there is no parameter to optimize in the GAP layer, thus overfitting is avoided at this layer. Furthermore, the GAP layer sums out the spatial information, thus it is more robust to spatial translations of the input.

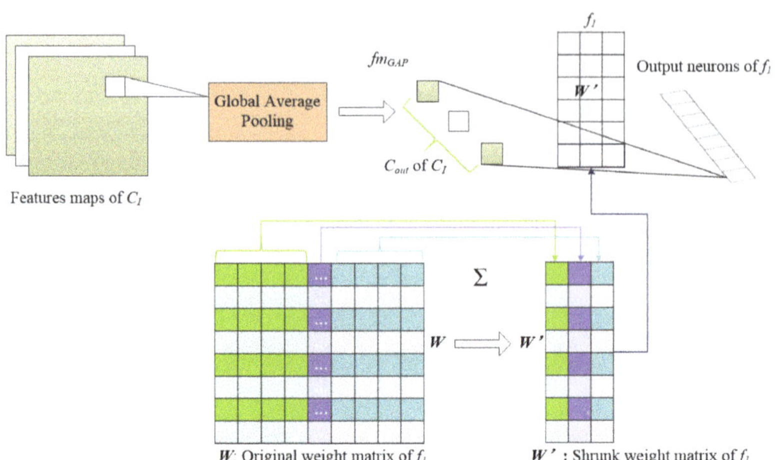

Figure 5. Flowchart of matrix dimension reduction by GAP layer feature.

2.3.3. L2 Normalization

The idea of L2 normalization is to add the regularization term (penalty term) to the loss function, which prevents the model from arbitrarily fitting the complex background and other noise information in the training set by restricting the most weight value ω in the model. Suppose the original loss function in the training process is $J_0(\omega, b)$, the utilization of L2 normalization is to optimize $J_0(\omega, b) + c\lambda R(\omega)$, and $R(\omega)$ is the regularization term or penalty term, which describes the complexity of the model. Relative equations above are illustrated in Equations (5)–(7).

$$J_0(\omega, b) = \frac{1}{m} \sum_{i=1}^{m} L(y'^{(i)}, y^{(i)}). \tag{5}$$

$$R(\omega) = \|W\|^2 = \sum_{j=1}^{l} \omega_j^2. \tag{6}$$

$$J(\omega, b) = \frac{1}{m} \sum_{i=1}^{m} L(y'^{(i)}, y^{(i)}) + \frac{\lambda}{2m} \sum_{j=1}^{l} \omega_j^2. \tag{7}$$

where, $J_0(\omega, b)$ is the original loss function; ω is the weight in the neuronal transmission process; relatively, ωj stands for the weight of the jth neuron and b represents the bias of neuronal transmission process; m represents the size of the sample dataset; $y'(i)$ represents the actual output value; $y(i)$ represents the expected output of a neuron; l is the number of dense; k is the number of neurons; $J_0(\omega, b)$ represents the new updated loss function; and λ is the parameter of L2 normalization. From Equation (9), it can be seen that the realization of L2 normalization is adding the sum of squares of the weight coefficients to the original loss function. In this experiment, the λ parameter was set to 0.12.

2.4. Transfer Learning and Fine-Tuning

In the field of deep learning, it is often necessary to train the model with a large number of datasets. However, in practical application, it is often difficult to obtain a large-scale dataset in the target field. Therefore, the idea of transfer learning can be adopted, and the image classification and recognition ability acquired by the deep convolutional neural network model trained on a large dataset after full training can be used to transfer the useful knowledge from the source domain to the new target domain. This makes the utility and inference scope from learned models much wider than an isolated model specific to individual plant species. Transfer learning also enables rapid progress and improved performance in modeling subsequent tasks by fine-tuning training. The most commonly used transfer learning approach is parameter-based transfer learning, which uses a model but, after fine-tuning, the partial parameters are based on the new dataset. This process is often referred to as domain adaption. Thus, in the experiment, VGG16 was pretrained, and the parameters of the convolutional layers and pooling layers were transferred to the newly designed VGNet. The internal weights of the newly designed model are automatically updated by fine-tuning training. To obtain a preferable model for this research, external factors containing training methods, regularization techniques, and the value of the hyperparameters are considered in the fine-tuning process.

2.4.1. Parameter Fine-Tuning

In deep learning networks, making each network parameter learn automatically and effectively with the input of training data is the key procedure to let the network training converge towards the required direction. The learning rate defines the learning progress of the proposed model and updates the weight parameters to reduce the loss function of the network. Thus, learning rate is an important parameter in the training algorithm. Some optimization strategies for network training parameters have been put forward [54], such as SGD, AdaGrad, AdaDelta, RMSProp, Adam [55], etc. The SGD and Adam optimizer are the most commonly used in image classification applications. In this experiment, we compared performance with the fine-tuning training algorithm involving the SGD and Adam optimizer to obtain better performance of the VGNet model.

2.4.2. Experimental Environment

All of the experiments were performed on Windows 7 (64-bit) operation system. The RAM of the computer is 16 GB, with Intel(R) Xeon(R) CPU E5-2630 v4 @2.20GHz CPU. The program platform was Anaconda 3.5.0, CUDA 8.0. CuDNN was the library for CUDA, developed by NVIDIA, which provided highly tuned implementations of primitives for deep neural networks. Python 3.5.6 was applied based on TensorFlow environment. The image dataset of the fine-tuning process was divided into two parts: 80% of image data were for training and the remaining 20% were for testing. Table 2 presents the hyperparameters of the fine-tuning training process of VGNet.

Table 2. Specification of hyperparameters in the experiment.

Parameters	Setting Values
Initial learning rate (SGD, Adam)	0.001, 0.005, 0.01
Momentum (SGD)	0.9
Small constant τ (Adam)	10^{-8}
Weight decay (SGD, Adam)	0.00005
L2 normalization parameter λ	0.12
Iteration	5000

2.5. Evaluation of Proposed Method

The performances are graphically depicted for each model with accuracy and loss. An overall loss score and accuracy based on the test dataset are computed and used to determine the performance of the models. The accuracy is calculated on the testing dataset in a regular interval with validation frequency of 25 iterations, and it is given as Equation (8).

$$Acc = \frac{\text{Predicted samples}}{\text{Disease samples}}. \tag{8}$$

Meanwhile, categorical cross-entropy is used as the loss function, which has softmax activations in the output layer, which is illustrated as Equation (9)

$$Loss = \sum_{i=1}^{N} \sum_{j=1}^{K} t_{ij} \ln y_{ij}. \tag{9}$$

where N represents the number of corn disease images, K is the number of diseases classes, t_{ij} indicates that the ith disease image belongs to the jth disease class, and y_{ij} stands for the output for sample i for disease class j. To evaluate the results of the disease recognition and classification experiment in the confusion matrix intuitively, P_{re} (Precision) and R_{ec} (Recall) are calculated after testing the samples. They are used to measure how accurately the results for each category are with respect to the corresponding ground-truth data. A comprehensive evaluation index, the *F1* score, is used as the evaluation value of P_{re} and R_{ec}. Equations for P_{re}, R_{ec}, and *F1* score are as follows in Equations (10)–(12).

$$P_{re} = \frac{TP}{TP + FP}. \tag{10}$$

$$R_{ec} = \frac{TP}{TP + FN}. \tag{11}$$

$$F1 = \frac{2P_{re}R_{ec}}{P_{re} + R_{ec}}. \tag{12}$$

where, the TP (true positive) is the amount of positive data that are correctly predicted as positive. The FP (false positive) represents the amount of negative data points that are wrongly predicted as positive. The FN (false negative) is the amount of negative data that are misclassified as negative. *Pre* (Precision) is used to find the proportion of positive identifications that are true. *Rec* is used to determine the proportion of actual positives that were correctly identified. The *F1* score reflects the number of instances that are correctly classified by the learning models.

3. Results

In this study, an assessment of the appropriateness of VGNet with transfer learning and fine-tuning training for the task of crop disease recognition was carried out. Our focus was to pretrain the VGG 16 Network with different public datasets and to fine-tune the newly designed VGNet model with different a training mechanism and parameters. Large open datasets like ImageNet, PlantVillage, and AI Challenger were utilized to pretrain the model; then, the weights and parameters of the convolutional layers and pooling layers were transferred to the new model and frozen. After updating the structure of VGNet, the parameters of the GAP layer, the remaining fully connected layers, and the softmax layer were retrained and fine-tuned by the new dataset obtained from corn fields. The performance of the proposed method was analyzed after five-fold cross-validation experiments to acquire convincing results. K-fold cross-validation is a common method used to test the accuracy of DL algorithms. To perform K-fold cross-validation on the overall data, the image dataset C is divided into K parts for disjoint subsets. In order to prevent data leakage, suppose the number of training samples in dataset C is M; then, the number of samples in each subset is M/K. When training the network model, one subset is selected each time as the verification set, and the other (K-1) subsets are selected as the training set, and the classification accuracy of the network model on the selected verification set can be obtained. After repeating the above process for K times, the average of classification ac-curacy is obtained as the true classification accuracy of the model. In our research, the K is set as 5, since the results of 5-fold validation and 10-fold validation are the same in the previous experimental experience.

3.1. Effects of Fine-Turning Training Mechanism

The following sections analyze the effects on model performance with a different training mechanism in the fine-tuning VGNet process, including different training methods and initial learning rates. Table 3 shows the testing loss and accuracy of the different training mechanism in the fine-tuning process.

Table 3. Testing loss and accuracy of the method based on SGD or Adam with different learning ranges.

Optimizer	Initial Learning Rate	Loss	Accuracy (%)
SGD	0.01	0.103	85.6
SGD	0.005	0.089	89.1
SGD	0.001	0.061	93.0
Adam	0.01	0.074	91.3
Adam	0.005	0.058	94.4
Adam	0.001	0.035	98.3

From Table 3, it can be seen that six different experiments were carried out; their final loss values and accuracies of testing vary with the training methods and initial learning rate. Figures 6 and 7 show the loss and accuracy curves of two training methods with initial learning rates of 0.01 and 0.001, respectively. As seen in Figures 6 and 7 and Table 3, training methods and initial learning rate have great influence on the performance of the model. By comparing experiment 1, 2, and 3 using the SGD method, it can be found that the loss value decreases as the learning rate declines, while the accuracy increases with the fall in learning rate. When the learning rate is set to 0.01, the loss value of the model test is 0.103, and the accuracy is only 85.65%. In this process, the performance is unstable, and the loss and accuracy shake violently, which can be seen by the green curves in Figure 6. When the initial learning rate drops to 0.001, the loss value of the model test decreases to 0.061, and the accuracy is improved to 93.04%. At this time, the testing process has fewer shocks, and the model can converge at about 4500 iterations, which is described by green curves in Figure 7. Rows 4, 5, and 6 in Table 3 were fine-tuning-trained with the Adam optimizer. Their variation in loss value and accuracy are consistent with former experiments 1, 2, and 3. The reason is that with the aid of transfer learning, all the front layers of the network obtained good training, and the weight parameters at the initial time of training are close to the optimal state. If the initial learning rate is not set properly, the training process will shock and even diverge. If a higher learning rate (0.01) is used in the fine-tuning training phase, the model is likely to skip the optimal solution, resulting in larger loss, lower accuracy, or severe oscillation. When the initial learning rate is 0.001, the model is more stable, and its performances are much better. Therefore, when the transfer learning mechanism is applied to the training of a convolutional neural network, the initial learning rate in the fine-tuning training stage needs to be lower than that of the model trained from scratch.

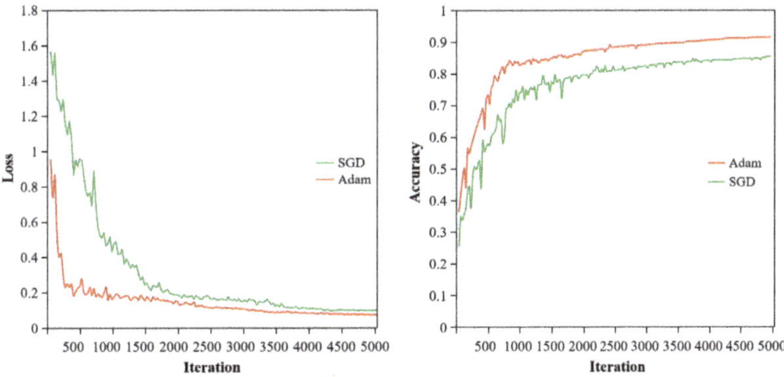

Figure 6. Comparison of loss and accuracy of two learning methods when the learning rate is 0.01.

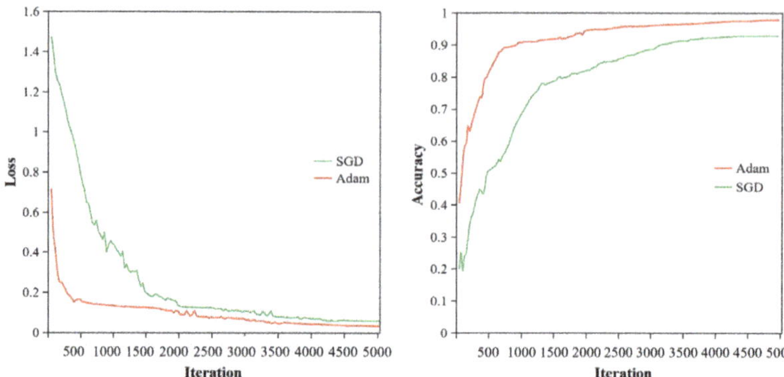

Figure 7. Comparison of loss and accuracy of two learning methods when the learning rate is 0.001.

Compare experiment 3 with experiment 6 in Table 3, where the initial learning rate was set as 0.001 with the SGD algorithm and Adam optimizer, respectively. At this point, the final performance of the model was different due to the different training methods. The loss value of the model trained by the Adam optimizer is lower than that of the model trained by SGD algorithm. Furthermore, the model trained by the Adam optimizer reaches convergence first and becomes stable after 3500 iterations, which is illustrated by the red curve in Figure 7. However, the model trained by the SGD method converges slowly, and the final loss value after convergence is 0.061, which is higher than the model trained by the Adam optimizer. Moreover, since the SGD training algorithm adjusts the weight for each data point, the network performance fluctuates up and down a lot more than the Adam optimizer during the learning process. The right part of Figure 7 shows the variation in the accuracy of the two training methods. It can be found that the model retrained by the Adam optimizer reached an accuracy of 98.26%, while the model retrained by the SGD algorithm did not perform as well. Apparently, when the model is fine-tuned by the SGD algorithm, it is always lower than when trained by the Adam optimizer. In general, the Adam optimizer algorithm has the advantage of faster model convergence than the SGD training algorithm and is more stable in the testing process. Therefore, the Adam optimizer in the fine-tuning training stage of the model is more in line with the corn disease recognition model.

3.2. Effects of Transfer Learning on Multiple Datasets

To explore the impact of training mechanisms and different datasets in the pretraining process, four completely selfsame VGNet models were utilized in the form of learning from scratch and transfer learning, respectively. The scratched learning model only adopted the image obtained from corn fields without pretraining. The other three models utilized three different large open datasets for pretraining and parameter transfer learning. The experimental results of applying four different learning types and datasets are listed in Table 4. From Table 4, it can be seen that the accuracy of learning from scratch is the lowest, reaching an accuracy of 69.57%. Under the condition of transfer learning and fine-tuning learning, the model pretrained using the PlantVillage dataset has the best performance, with an accuracy of 98.26%. Since training the VGNet model from scratch needs more images and time to optimize network parameters, and the training dataset only has 920 images, it is not enough for a deep convolutional neural network. This leads to the nonideal classification effect. Pretraining and transfer learning make the VGNet model acquire the ability of feature extraction and the knowledge of classification; thus, it is easier to achieve higher accuracy than with the scratched learning model. Therefore, transfer learning seems to be a better approach than learning from scratch when the dataset is not big enough. Though the original VGG16 Net is a model with excellent performance trained on ImageNet, a large public dataset, in general, the filter at the bottom of the model can acquire different local edge and texture information through training, which has good universality for any image. However, the feature gaps between the ImageNet dataset from source area and the corn disease images in this new area are too large, while the other two datasets have much more similar features in color, texture, and shape to the corn disease images. Thus, the accuracies of the models pretrained with PlantVillage and AI Challenger are higher than the model pretrained with ImageNet. Images from PlantVillage are very similar to those from AI Challenger, but the number of PlantVillage is bigger than that of AI Challenger. Thus, the model pretrained with PlantVillage obtains a better learning effect, and PlantVillage is more suitable for the pretraining in this research. This indicates that in transfer learning, the source domain and target domain should have a high fitting degree for better performance.

Table 4. Experiment results of different learning types and datasets for pretraining and fine-tuning.

Learning Types	Pretrained Images	Accuracy on Original Images (%)	Accuracy on Augmented Images (%)
Learning from Scratch	—	69.6	89.5
Transfer learning	ImageNet	93.5	94.6
Transfer learning	PlantVillage	**98.3**	**99.4**
Transfer learning	AI Challenger	97.3	91.3

3.3. Effects of Augmentation

Data augmentation was applied here based on image transformations, such as geometric transformation, color changing, and noise adding, to generate new training images from the original ones by applying such random image transformations. The size of the dataset was enlarged from 1150 to 11,500. The ratio of the training dataset and testing dataset was also 8:2. The effects of image augmentation for fine-tuning learning are also illustrated in Table 4. It can be concluded that the effects of image data augmentation on different training models are different. In the mode of learning from scratch, data augmentation improves the accuracy by nearly 20%. Because the original dataset is too small, and the structure of the network structure is deep, the overfitting phenomenon reduces the performance of the network. When the image data are enlarged by data augmentation, the number and diversity of the data are increased. Thus, data augmentation has a larger role in avoiding overfitting and increasing accuracy when the model is learning from scratch. In the transfer learning mode, the accuracy of the fine-tuned model trained with augmentation is at least 2% higher than that of the model fine-tune-trained by original image data. This is because the pretraining model has learned a lot of knowledge from the large image dataset, which weakens the role of data augmentation. Hence, enlarging data plays a slight role in improving the performance of model classification in transfer learning.

4. Discussion

4.1. Obfuscation Matrix Analysis and Quantitative Statistics

To clearly show the recognition precision and classification results based on the fine-tuning training of the designed VGNet with augmented datasets, the confusion matrix drawn on the basis of the model classification results is shown in Figure 8. ANTH, TR, SCR, CR, SLB, PHLS, DLS, PHBS, and NLB, respectively, represent the abbreviations of nine types of corn diseases. The values in darker diagonal lines in Figure 8 (left) illustrate the number of correct classifications for each disease category, while the results of darker diagonal lines in Figure 8 (right) represent the recognition accuracies of correct classifications. It can be found that the recognition accuracies of nine corn diseases present some differences. Relatively, the accuracy of ANTH (Anthracnose) is lower than others; this probably because the sample number is fewer than other types. And the accuracy of SCR (Southern corn rust) reaches 100%. On the whole, the accuracies are kept in the range of 98.6% and 100%, which can be treated as a balanced result.

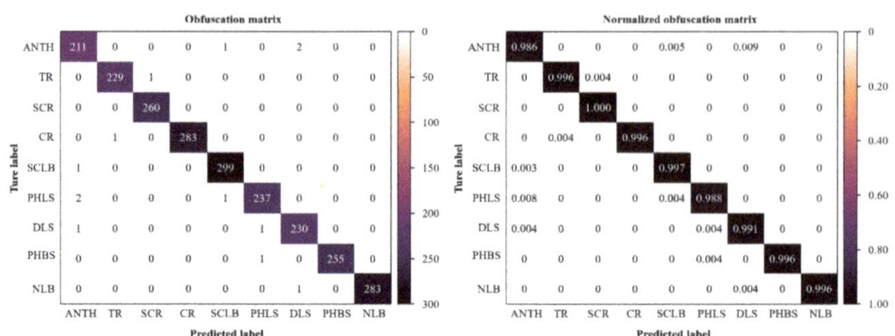

Figure 8. Obfuscation matrix analysis of classification based on transfer learning and data augmentation. The left is the obfuscation matrix, and the right is the normalized obfuscation matrix.

After the analysis and statistics of the confounding matrix, each parameter reflecting the model performance is obtained, as shown in Table 5, which describes the more detailed original and testing classification information of the proposed VGNet. It can be found in Table 5 that the precision and recall values of each disease type are different, which is related to the characteristic types and image numbers of each disease. The precision value in Table 5 is between 98.1% and 100%. The recall value ranges from 98.6% to 100%. The F1 value ranges from 98.4% to 99.8%, with an average accuracy of 99.4%. This indicates that the proposed method performs well in the established dataset after transfer learning and fine-tuning training, which could be applied to the actual detection of crop diseases in the field environment.

Table 5. Obfuscation matrix statistics for nine types of corn diseases with transfer learning and augmentation.

Types	ANTH	TR	SCR	CR	SLB	PHLS	DLS	PHBS	NLB
Samples	1070	1150	1300	1420	1500	1200	1160	1280	1420
Positive	214	230	260	284	300	240	232	256	284
Negative	2086	2070	2040	2016	2000	2060	2068	2044	2016
TP	211	229	260	283	299	237	230	255	283
FN	3	1	0	1	1	3	2	1	1
TN	2076	2058	2027	2004	1988	2050	2057	2032	2004
FP	4	1	1	0	2	2	3	0	0
Pre (%)	98.1	99.6	99.6	100.0	99.3	99.2	98.7	100.0	100.0
Rec (%)	98.6	99.6	100.0	99.7	99.7	98.8	99.1	99.6	99.7
F1 (%)	98.4	99.6	99.8	99.8	99.5	99.0	98.9	99.8	99.8
Acc (%)					99.4				

4.2. Comparison with State-of-the-Art Methods

To further validate the effect of our method based on fine-tuning training and VGNet, we compared the proposed method with the traditional machine learning classifiers and state-of-the-art models (deep learning methods), respectively, under the same experiment conditions as well as the same dataset. The total number of images was 1150. Traditional machine learning methods include random forest (RF) classification algorithm, support vector machine (SVM), and BP neural network. AlexNet, ResNet50, Inception v3, and the original VGG16 Net are the selected deep convolutional neural networks for the comparative experiment. For conventional machine learning methods, we preprocessed the corn disease images, including image enhancement, segmentation, and feature extraction. After removing background information, the disease spots with clear boundaries were obtained. Then color histogram feature in HSV color space and the matrix characteristics in RGB color space were extracted, respectively. The gray-level co-occurrence matrix was used for texture features and a seven-hue invariant matrix was used for shape feature extraction. Then, the extracted features were fused as input vectors of the BP, SVM, and RF classifiers. The learning experiments of AlexNet, ResNet50, Inception v3, the original VGG16, and VGNet models adopt the method of transfer learning and fine-tuning mechanism. The experiment parameters were consistent with the proposed method. After training, the models were test tested and identification results were output. The accuracies obtained from different traditional machine learning classifiers and deep learning methods are shown in Figure 9. It can be seen in Figure 9 that the accuracies of traditional methods are generally lower than 87%. In addition, conventional classifiers often require tedious preprocesses involving image enhancement, segmentation, and extraction of features manually. In deep learning methods, the accuracies are greater than 92%, and they vary because of the different deep structures and abilities of feature extraction. The accuracy of AlexNet is the lowest among the five deep architectures, because the structure of AlexNet is shallower than others, which leads to the insufficient ability to extract the features of corn disease images. The accuracy of the original VGG16 Net is 94.78%, the ResNet50 is 95.22%, and Inception v3 achieves an accuracy of 96.96%. Experimental results indicate that deep learning methods are superior to conventional machine learning. It can also be seen that our model reaches a highest accuracy of 98.26%, which is improved by 3.48% compared with the original VGG16 Net. The addition of BN, a GAP layer, and L2 normalization makes the VGG16 Net more robust with higher accuracy. The improvement of our method based on the classical VGG16 Net has the capability to learn more complex features, as more convolutional layers are in the stack with smaller filter sizes compared with other deep learning models.

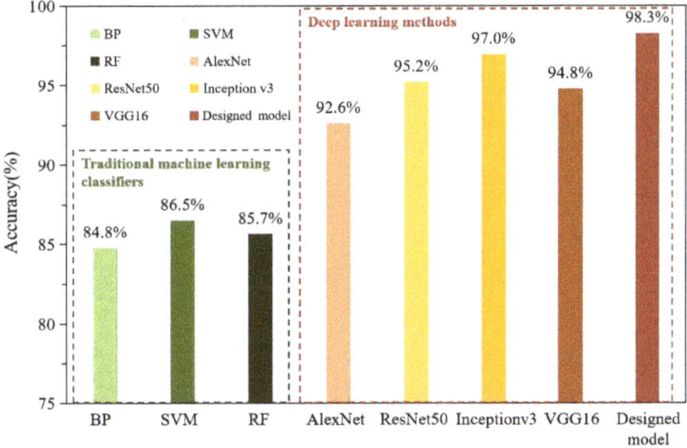

Figure 9. Comparison of accuracy between different models based on the same dataset.

Table 6 shows the comparative parameters and testing time of different deep learning methods. From Table 6, we can see that the original VGG16 Net has the most parameters and the longest testing time. AlexNet has eight weight layers and 58.3 million parameters; the testing time of AlexNet is the shortest, only 50.14 s for 230 images. However, the accuracy of AlexNet is the lowest (Figure 9). The parameters and testing time of ResNet50 and Inception v3 are slightly different. Our VGNet has 14 weight layers and 22.9 million parameters after replacing huge hidden fully connected layers by a GAP layer, and it only occupies 79.5 MB of memory space. The testing time of our model is only 75.21 s for 230 images, which improves by 151.11 s compared with the original VGG16 Net. In addition, the loss value of the designed VGNet is only 0.035, which is significantly smaller than other models, such as VGG16 and ResNet50. The proposed method can achieve real-time detection of corn diseases. In general, our proposed method has the best recognition effect after transfer learning and fine-tuning. The utilization of the GAP layer realized the feature dimension reduction. The parameters of the network were greatly reduced, as well as the calculation amount. This means the network regularization in the structure to prevent overfitting. The connections between each category in the feature map are more intuitive (compared with the fully connected layers), and it is easier for the feature map to be converted into classification probability. Thus, the proposed VGNet is lightweight and robust, which could obtain the best performance among the state-of-the-art models.

Table 6. Comparison of the classic convolutional neural networks and corresponding parameters.

Methods	Network Layers	Parameters (Millions)	Weights (MB)	Times (s)	Loss Value
AlexNet	8	60.9	224	50.14	0.912
ResNet50	50	25.5	102	88.78	0.587
InceptionV3	46	24.7	96	86.02	0.271
VGG16	16	138	533	226.32	0.196
VGNet	14	22.9	79.5	75.21	0.035

Actually, our method utilizes 1150 corn disease images from field conditions, and the recognition accuracy reaches 98.3%, which is better than the models learning from scratch. After data augmentation, the accuracy of the model improves slightly by 1.2%. The dataset in this research is small compared with many deep convolutional models. Actually, Ferentinos et al. collected 87,848 images of plant diseases to train a convolutional neural network model, whose performance finally reached 99.5% accuracy [43]. In our experiment, when the dataset is enlarged to 11,500, the accuracy of VGNet increases to 99.4%. Compared with the study of Ferentinos, our success rate is only 0.1% lower than that of the model using 86,000 images. Thus, transfer learning seems to be an ideal method for the CNN model to achieve better performance. With the aid of the parameters transfer of the pretrained model, a more accurate model can be generated when fine-tuning several layers for disease image classification.

Three types of open large datasets, including ImageNet, PlantVillage, and AI Challenger, were used, and the results show that the models pretrained with PlantVillage or AI Challenger were better than that pretrained ones with ImageNet. The similarity of the training data to the experimental data results in easier transferability. The SGD algorithm and Adam optimizer are compared and analyzed in the fine-tuning phase. The experiments prove that the Adam optimizer for training the VGG16 Net is more accurate and more stable than the SGD algorithm. The initial learning rate is also an important parameter in model training. In regard to the pretrained model, smaller learning rates for convolutional nets are common, as network parameters should not be changed dramatically.

4.3. Feature Visualization

The ability of automatic feature extraction is an important factor to reflect the performance of the model. To examine the effect of feature extraction on the proposed model, feature map visualization was carried out. Figure 10 illustrates the original input image and the feature maps derived from the pooling layer of the model. From the right of Figure 10, we find that the disease spots were abstracted high-dimensional features; the VGNet obviously had high-quality feature extraction, which was beneficial for recognition and classification.

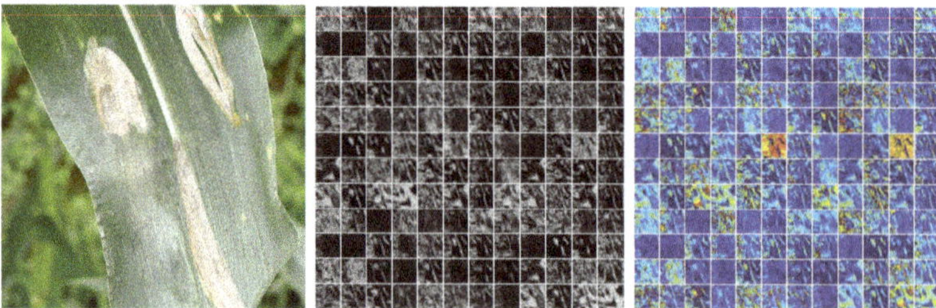

Figure 10. Obfuscation matrix analysis of classification based on transfer learning and data augmentation. The left is the original image; the middle is the grey feature map; and the right is the color feature map.

5. Conclusions

Data diversity and representativeness are the key elements to ensure the generalization of the model. In this paper, we devised a VGNet which takes VGG16 as the backbone and adds batch normalization, as well as replacing two fully connected layers with a GPA layer and adding L2 normalization. The parameters of the convolutional layers and pooling layers are transferred to the newly designed VGNet; then, the fine-tuning learning for VGNet is studied to enhance the ability of recognizing corn disease images from real field conditions.

Data augmentation has greater promotion of model learning from scratch than on pretrained model, because the parameters of pretrained models are trained enough by open large datasets. Compared with traditional machine learning methods and state-of-the-art deep learning methods, the proposed VGNet has a stronger ability to identify a hierarchy of features of corn diseases. The accuracy of VGNet is improved by 3.5% compared with the original VGG16 Net, and the testing time for 230 images is reduced by 66.8%, with balanced precision, recall, and F1 indexes. The parameters and memory occupation of the proposed VGNet are reduced by 83.4% and 85.1%, respectively. The comparative experiments and performance analysis illustrated the wide adaptability of the proposed method. In addition, the proposed method could provide baseline architecture for other types of phenotypic information recognition or interpretation with much fewer parameters and computation time. In future work, we will focus on collecting multiple crop disease images from real scenes and developing fine-grained disease detection methods that can be used for multiple categories of crops.

Author Contributions: Conceptualization, Z.G.; methodology, X.F. and Z.G.; software, X.F.; validation, X.F.; visualization, Z.G.; writing—original draft, X.F.; writing—review and editing, Z.G. All authors have read and agreed to the published version of the manuscript.

Funding: This research was funded in part by the National Major Science and Technology Projects under grant number 2022YFD2022205; in part by the Nanfan special project, CAAS under grant number YBXM10; in part by the Science and Technology Innovation Program of the Chinese Academy of Agricultural Sciences under grant number CAAS-ZDRW202107 and CAAS-ASTIP-2023-AII.

Data Availability Statement: The data presented in this study are available on request from the corresponding author or the first author.

Acknowledgments: The authors are grateful for the constructive advice on the revision of this manuscript from anonymous reviewers.

Conflicts of Interest: The authors declare that they have no conflict of interest.

References

1. Xu, Y.; Wang, X.; Zhai, Y.; Li, C.; Gao, Z. Precise variable spraying system based on improved genetic proportional-integral-derivative control algorithm. *Trans. Inst. Meas. Control* **2021**, *43*, 3255–3266. [CrossRef]
2. Roshan, S.H.; Kazemitabar, S.J.; Kheradmandian, G. Artificial Intelligence Aided Agricultural Sensors for Plant Frostbite Protection. *Appl. Artif. Intell.* **2022**, *36*, 2031814. [CrossRef]
3. Xu, Y.; Xue, X.; Sun, Z.; Gu, W.; Cui, L.; Jin, Y.; Lan, Y. Joint path planning and scheduling for vehicle-assisted multiple Unmanned Aerial Systems plant protection operation. *Comput. Electron. Agric.* **2022**, *200*, 107221. [CrossRef]
4. Godara, S.; Toshniwal, D.; Bana, R.S.; Singh, D.; Bedi, J.; Parsad, R.; Dabas, J.P.S.; Jhajhria, A.; Godara, S.; Kumar, R.; et al. AgrIntel: Spatio-temporal profiling of nationwide plant-protection problems using helpline data. *Eng. Appl. Artif. Intell.* **2023**, *117*, 105555. [CrossRef]
5. Li, Z.; Wang, W.; Zhang, C.; Zheng, Q.; Liu, L. Fault-tolerant control based on fractional sliding mode: Crawler plant protection robot. *Comput. Electr. Eng.* **2023**, *105*, 108527. [CrossRef]
6. Tang, Y.; Fu, Y.; Guo, Q.; Huang, H.; Tan, Z.; Luo, S. Numerical simulation of the spatial and temporal distributions of the downwash airflow and spray field of a co-axial eight-rotor plant protection UAV in hover. *Comput. Electron. Agric.* **2023**, *206*, 107634. [CrossRef]
7. Liu, Y.; Gao, G.; Zhang, Z. Crop Disease Recognition Based on Modified Light-Weight CNN With Attention Mechanism. *IEEE Access* **2022**, *10*, 112066–112075. [CrossRef]
8. Haque, M.A.; Marwaha, S.; Deb, C.K.; Nigam, S.; Arora, A. Recognition of diseases of maize crop using deep learning models. *Neural Comput. Appl.* **2023**, *35*, 7407–7421. [CrossRef]

9. Hua, J.; Zhu, T.; Liu, J. Leaf Classification for Crop Pests and Diseases in the Compressed Domain. *Sensors* **2023**, *23*, 48. [CrossRef]
10. Kurmi, Y.; Gangwar, S.; Chaurasia, V.; Goel, A. Leaf images classification for the crops diseases detection. *Multimed. Tools Appl.* **2022**, *81*, 8155–8178. [CrossRef]
11. Youk, G.; Kim, M. Transformer-Based Synthetic-to-Measured SAR Image Translation via Learning of Representational Features. *IEEE Trans. Geosci. Remote Sens.* **2023**, *61*, 1–18. [CrossRef]
12. Naseem, A.; Rehman, M.A.; Qureshi, S.; Ide, N. Graphical and Numerical Study of a Newly Developed Root-Finding Algorithm and Its Engineering Applications. *IEEE Access* **2023**, *11*, 2375–2383. [CrossRef]
13. Chakraborty, S.K.; Chandel, N.S.; Jat, D.; Tiwari, M.K.; Rajwade, Y.A.; Subeesh, A. Deep learning approaches and interventions for futuristic engineering in agriculture. *Neural Comput. Appl.* **2022**, *34*, 20539–20573. [CrossRef]
14. Liang, D.; Tang, W.; Fu, Y. Sustainable Modern Agricultural Technology Assessment by a Multistakeholder Transdisciplinary Approach. *IEEE Trans. Eng. Manag.* **2023**, *70*, 1061–1075. [CrossRef]
15. Tu, J.; Aznoli, F.; Navimipour, N.J.; Yalçin, S. A new service recommendation method for agricultural industries in the fog-based Internet of Things environment using a hybrid meta-heuristic algorithm. *Comput. Ind. Eng.* **2022**, *172*, 108605. [CrossRef]
16. Almadani, B.; Mostafa, S.M. IIoT Based Multimodal Communication Model for Agriculture and Agro-Industries. *IEEE Access* **2021**, *9*, 10070–10088. [CrossRef]
17. Wang, F.; Yang, J.; Wang, X.; Li, J.; Han, Q. Chat with ChatGPT on Industry 5.0: Learning and Decision-Making for Intelligent Industries. *IEEE CAA J. Autom. Sinica* **2023**, *10*, 831–834. [CrossRef]
18. Jia, M.; Li, J.; Zhang, J.; Wei, N.; Yin, Y.; Chen, H.; Yan, S.; Wang, Y. Identification and validation of cuproptosis related genes and signature markers in bronchopulmonary dysplasia disease using bioinformatics analysis and machine learning. *BMC Med. Inform. Decis. Mak.* **2023**, *23*, 69. [CrossRef]
19. Liu, C.; Zhou, Y.; Zhou, Y.; Tang, X.; Tang, L.; Wang, J. Identification of crucial genes for predicting the risk of atherosclerosis with system lupus erythematosus based on comprehensive bioinformatics analysis and machine learning. *Comput. Biol. Med.* **2023**, *152*, 106388. [CrossRef]
20. Bacon, W.; Holinski, A.; Pujol, M.; Wilmott, M.; Morgan, S.L. Correction: Ten simple rules for leveraging virtual interaction to build higher-level learning into bioinformatics short courses. *PLoS Comput. Biol.* **2023**, *19*, e1010964. [CrossRef]
21. Ramana, K.; Aluvala, R.; Kumar, M.R.; Nagaraja, G.; Krishna, A.V.; Nagendra, P. Leaf Disease Classification in Smart Agriculture Using Deep Neural Network Architecture and IoT. *J. Circuits Syst. Comput.* **2022**, *31*, 2240004:1–2240004:27. [CrossRef]
22. Bajpai, C.; Sahu, R.; Naik, K.J. Deep learning model for plant-leaf disease detection in precision agriculture. *Int. J. Intell. Syst. Technol. Appl.* **2023**, *21*, 72–91. [CrossRef]
23. Pal, A.; Kumar, V. AgriDet: Plant Leaf Disease severity classification using agriculture detection framework. *Eng. Appl. Artif. Intell.* **2023**, *119*, 105754. [CrossRef]
24. Kwaghtyo, D.K.; Eke, C.I. Smart farming prediction models for precision agriculture: A comprehensive survey. *Artif. Intell. Rev.* **2023**, *56*, 5729–5772. [CrossRef]
25. Surampudi, S.; Kumar, V. Flood Depth Estimation in Agricultural Lands From L and C-Band Synthetic Aperture Radar Images and Digital Elevation Model. *IEEE Access* **2023**, *11*, 3241–3256. [CrossRef]
26. Tong, K.; Wu, Y.; Zhou, F. Recent advances in small object detection based on deep learning: A review. *Image Vis. Comput.* **2020**, *97*, 103910. [CrossRef]
27. Zhao, X.; Zhang, J.; Huang, Y.; Tian, Y.; Yuan, L. Detection and discrimination of disease and insect stress of tea plants using hyperspectral imaging combined with wavelet analysis. *Comput. Electron. Agric.* **2022**, *193*, 106717. [CrossRef]
28. Vishnoi, V.K.; Kumar, K.; Kumar, B.; Mohan, S.; Khan, A.A. Detection of Apple Plant Diseases Using Leaf Images Through Convolutional Neural Network. *IEEE Access* **2023**, *11*, 6594–6609. [CrossRef]
29. Prabhakar, M.L.C.; Merina, R.D.; Mani, V. IoT Based Air Quality Monitoring and Plant Disease Detection for Agriculture. *Autom. Control Comput. Sci.* **2023**, *57*, 115–122. [CrossRef]
30. Amrani, A.; Sohel, F.; Diepeveen, D.; Murray, D.; Jones, M.G.K. Deep learning-based detection of aphid colonies on plants from a reconstructed *Brassica* image dataset. *Comput. Electron. Agric.* **2023**, *205*, 107587. [CrossRef]
31. Quach, B.M.; Cuong, D.V.; Pham, N.; Huynh, D.; Nguyen, B.T. Leaf recognition using convolutional neural networks based features. *Multimed. Tools Appl.* **2023**, *82*, 777–801. [CrossRef]
32. Lv, Z.; Zhang, Z. Research on plant leaf recognition method based on multi-feature fusion in different partition blocks. *Digit. Signal Process.* **2023**, *134*, 103907. [CrossRef]
33. Jin, H.; Li, Y.; Qi, J.; Feng, J.; Tian, D.; Mu, W. GrapeGAN: Unsupervised image enhancement for improved grape leaf disease recognition. *Comput. Electron. Agric.* **2022**, *198*, 107055. [CrossRef]
34. Laxmi, S.; Gupta, S.K. Multi-category intuitionistic fuzzy twin support vector machines with an application to plant leaf recognition. *Eng. Appl. Artif. Intell.* **2022**, *110*, 104687. [CrossRef]
35. Reddy, S.R.G.; Varma, G.P.S.; Davuluri, R.L. Resnet-based modified red deer optimization with DLCNN classifier for plant disease identification and classification. *Comput. Electr. Eng.* **2023**, *105*, 108492. [CrossRef]
36. Janani, M.; Jebakumar, R. Detection and classification of groundnut leaf nutrient level extraction in RGB images. *Adv. Eng. Softw.* **2023**, *175*, 103320. [CrossRef]
37. Kumar, R.R.; Athimoolam, J.; Appathurai, A.; Rajendiran, S. Novel segmentation and classification algorithm for detection of tomato leaf disease. *Concurr. Comput. Pract. Exp.* **2023**, *35*, e7674. [CrossRef]

38. Cui, S.; Su, Y.L.; Duan, K.; Liu, Y. Maize leaf disease classification using CBAM and lightweight Autoencoder network. *J. Ambient Intell. Humaniz. Comput.* **2023**, *14*, 7297–7307. [CrossRef]

39. Charak, A.S.; Sinha, A.; Jain, T. Novel approach for quantification for severity estimation of blight diseases on leaves of tomato plant. *Expert Syst. J. Knowl. Eng.* **2023**, *40*, e13174. [CrossRef]

40. Ashwini, C.; Sellam, V. EOS-3D-DCNN: Ebola optimization search-based 3D-dense convolutional neural network for corn leaf disease prediction. *Neural Comput. Appl.* **2023**, *35*, 11125–11139. [CrossRef]

41. Cen, H.; Zhu, Y.; Sun, D.; Zhai, L.; He, Y. Current status and future perspective of the application of deep learning in plant phenotype research. *Trans. Chin. Soc. Agric. Eng.* **2020**, *36*, 1–16.

42. Zhang, J.; Rao, Y.; Man, C.; Jiang, Z.; Li, S. Identification of cucumber leaf diseases using deep learning and small sample size for agricultural Internet of Things. *Int. J. Distrib. Sens. Netw.* **2021**, *17*, 155014772110074. [CrossRef]

43. Ferentinos, K.P. Deep learning models for plant disease detection and diagnosis. *Comput. Electron. Agric.* **2018**, *145*, 311–318. [CrossRef]

44. Liang, Q.; Xiang, S.; Hu, Y.; Coppola, G.; Zhang, D.; Sun, W. PD^{2}SE-Net: Computer-assisted plant disease diagnosis and severity estimation network. *Comput. Electron. Agric.* **2019**, *157*, 518–529. [CrossRef]

45. AI Challenger 2018. Available online: https://github.com/AIChallenger/AI_Challenger_2018 (accessed on 6 December 2022)

46. Zhong, Y.; Zhao, M. Research on deep learning in apple leaf disease recognition. *Comput. Electron. Agric.* **2020**, *168*. [CrossRef]

47. He, Y.; Zeng, H.; Fan, Y.; Ji, S.; Wu, J. Application of Deep Learning in Integrated Pest Management: A Real-Time System for Detection and Diagnosis of Oilseed Rape Pests. *Mob. Inf. Syst.* **2019**, *2019*, 4570808:1–4570808:14. [CrossRef]

48. Zeng, W.; Li, M. Crop leaf disease recognition based on Self-Attention convolutional neural network. *Comput. Electron. Agric.* **2020**, *172*, 105341. [CrossRef]

49. Hughes, D.P.; Salathé, M. An open access repository of images on plant health to enable the development of mobile disease diagnostics through machine learning and crowdsourcing. *arXiv* **2015**, arXiv:1511.08060.

50. Mohanty, S.P.; Hughes, D.P.; Salathé, M. Using Deep Learning for Image-Based Plant Disease Detection. *Front Plant Sci.* **2016**, *7*, 1419. [CrossRef]

51. Deng, J.; Dong, W.; Socher, R.; Li, L.; Li, K.; Fei-Fei, L. ImageNet: A large-scale hierarchical image database. In Proceedings of the 2009 IEEE Computer Society Conference on Computer Vision and Pattern Recognition (CVPR 2009), Miami, FL, USA, 20–25 June 2009; pp. 248–255. [CrossRef]

52. Waheed, A.; Goyal, M.; Gupta, D.; Khanna, A.; Hassanien, A.E.; Pandey, H.M. An optimized dense convolutional neural network model for disease recognition and classification in corn leaf. *Comput. Electron. Agric.* **2020**, *175*, 105456. [CrossRef]

53. Hsiao, T.; Chang, Y.; Chou, H.; Chiu, C. Filter-based deep-compression with global average pooling for convolutional networks. *J. Syst. Archit.* **2019**, *95*, 9–18. [CrossRef]

54. Kaya, A.; Keçeli, A.S.; Catal, C.; Yalic, H.Y.; Temuçin, H.; Tekinerdogan, B. Analysis of transfer learning for deep neural network based plant classification models. *Comput. Electron. Agric.* **2019**, *158*, 20–29. [CrossRef]

55. Kingma, D.P.; Ba, J. Adam: A Method for Stochastic Optimization. In Proceedings of the 3rd International Conference on Learning Representations, ICLR 2015, San Diego, CA, USA, 7–9 May 2015.

Article

Spectral Detection of Peanut Southern Blight Severity Based on Continuous Wavelet Transform and Machine Learning

Wei Guo [1], Heguang Sun [1,2], Hongbo Qiao [1], Hui Zhang [1], Lin Zhou [3], Ping Dong [1,*] and Xiaoyu Song [2,*]

[1] College of Information and Management Science, Henan Agricultural University, Zhengzhou 450046, China; guowei@henau.edu.cn (W.G.); sunheguang@henau.edu.cn (H.S.); qiaohb@henau.edu.cn (H.Q.); huizi@henau.edu.cn (H.Z.)

[2] Information Technology Research Center, Beijing Academy of Agriculture and Forestry Sciences, Beijing 100094, China

[3] College of Plant Protection, Henan Agricultural University, Zhengzhou 450002, China; zhoulinhenau@163.com

* Correspondence: dongping@henau.edu.cn (P.D.); songxy@nercita.org.cn (X.S.)

Abstract: Peanut southern blight has a severe impact on peanut production and is one of the most devastating soil-borne fungal diseases. We conducted a hyperspectral analysis of the spectral responses of plants to peanut southern blight to provide theoretical support for detecting the severity of the disease via remote sensing. In this study, we collected leaf-level spectral data during the winter of 2021 and the spring of 2022 in a greenhouse laboratory. We explored the spectral response mechanisms of diseased peanut leaves and developed a method for assessing the severity of peanut southern blight disease by comparing the continuous wavelet transform (CWT) with traditional spectral indices and incorporating machine learning techniques. The results showed that the SVM model performed best and was able to effectively detect the severity of peanut southern blight when using CWT ($WF_{770 \sim 780}$, 5) as an input feature. The overall accuracy (OA) of the modeling dataset was 91.8% and the kappa coefficient was 0.88. For the validation dataset, the OA was 90.5% and the kappa coefficient was 0.87. These findings highlight the potential of this CWT-based method for accurately assessing the severity of peanut southern blight.

Keywords: peanut southern blight; reflection spectrum; spectral index; continuous wavelet transform; machine learning

Citation: Guo, W.; Sun, H.; Qiao, H.; Zhang, H.; Zhou, L.; Dong, P.; Song, X. Spectral Detection of Peanut Southern Blight Severity Based on Continuous Wavelet Transform and Machine Learning. *Agriculture* **2023**, *13*, 1504. https://doi.org/10.3390/agriculture13081504

Academic Editor: Hyeon Tae Kim

Received: 14 June 2023
Revised: 18 July 2023
Accepted: 20 July 2023
Published: 27 July 2023

1. Introduction

Peanut southern blight, which is caused by the soil-borne fungus Sclerotium rolfsii Sacc, is a fungal pathogen that significantly impacts global peanut production [1,2]. This pathogen gradually turns peanut leaves brown or yellow, eventually leading to their detachment. The fungus destroys the fleshy tissues within the stems. Noticeable white mycelia appear on the roots, and at high temperatures, light brown spherical sclerotia develop within the infected tissues. Ultimately, this can lead to complete crop failure [3]. Due to the rapid onset of peanut southern blight, current field surveys and control measures are insufficient. Therefore, it is essential to explore the spectral response mechanism of peanut southern blight in order to achieve precise prevention and control strategies [4].

In recent years, the majority of research efforts have focused on viruses, bacteria, fungi, and nematodes, which have long been recognized as the main culprits behind infectious diseases. The changes in a pathogen and in the interactions between plants and pathogens can be reflected through variations in plant tissue color [5], leaf shape [6], transpiration rate, and plant density. The physiological and biochemical changes that occur during this process are inevitably reflected in certain spectral bands. Typically, healthy green plants exhibit low reflectance in the visible (VIS) spectrum, high reflectance in the near-infrared (NIR) spectrum, and low wide-band reflectance in the shortwave infrared (SWIR) spectrum [7]. In recent years, there has been an increasing number of reports regarding pests and diseases affecting plant leaves [8–10]. With leaf infection, various spots or necrotic areas often appear [11]. This leads to a reduction in leaf pigmentation and photosynthesis [12,13]. The result is a typical red-edge "blue-shifting" phenomenon that can be observed in the visible-light range [14]. Ray et al. pointed out that red-edge information becomes particularly important when subtle structural changes occur [15]. At this point, the importance of spectral resolution becomes clear, as a higher spectral resolution enables the more detailed observation of spectral responses. Hyperspectral sensors contain hundreds to thousands of useful narrow-band data [16], and they have been proven to detect the spectral response mechanisms of plants under stress, such as wheat stripe rust [17,18] and rice blast [19]. However, few studies have used hyperspectral technology to investigate the spectral response mechanisms of plants with peanut southern blight.

Currently, there are two main categories of methods for monitoring plants under stress: empirical methods and physical methods [20]. Physical methods based on Radiative Transfer Models (RTMs) have consistently attracted attention in the field of pest and disease monitoring [21,22]. The main advantage of this approach is that it does not require parameterization [23]. Rather, it uses existing leaf or canopy spectra to simulate changes in plant growth and developmental traits. For example, Saddik et al. [24] combined RGB images and hyperspectral reflectance data with an RTM to differentiate spectra affected by yellowness and esca infections. Although RTMs have model interpretability and mechanistic modeling advantages, they rely on the calibration of the input feature set, and this may limit their applicability in real-world scenarios [25]. Empirical methods can effectively characterize spectral changes [26]. Some studies focus on developing crop-specific spectral indices [27,28]. In addition, some studies have analyzed different spectral transformation forms, such as logarithms, derivatives, and continuous wavelet transforms, to enhance the separability of spectra under different severity levels [29,30]. In order to ascertain the spectral response mechanism of plants with peanut southern blight, we have employed the Continuous Wavelet Transform (CWT) technique. This method decomposes the reflectance spectra of leaves into multiple scale components, amplifying the underlying spectral differences [31]. Previous studies have used the CWT technique in various domains, such as the study of vegetation [32], minerals [33], and inland water bodies [34]. Specifically, wavelet analysis has been applied in the detection and assessment of plant physiological stress [35]. Some authors have also utilized wavelet analysis in the study of airborne imaging spectroscopy data to quantify forest structural parameters [36] and identify plant species [37]. The use of the Standard Normal Variate (SNV) method and some previously reported spectral indices has also been evaluated in detail and has been compared with the CWT technique [38].

In recent years, the combination of Feature Selection (FS) methods and Machine Learning (ML) algorithms has been widely applied in the field of remote sensing [39,40]. By utilizing selected features as input, it is possible to significantly reduce model running time and enhance model accuracy [41]. Wang et al. utilized Principal Component Analysis (PCA) to reduce the dimensionality of features and combined it with the Backpropagation Neural Network (BPNN) machine learning algorithm to analyze grape and wheat diseases [42]. Huang et al. employed the relief algorithm to extract wavelength information concerning different diseases from wheat leaf spectral data and used machine learning modeling to monitor various wheat diseases [10]. To evaluate the severity of peanut southern blight, we applied the relief algorithm to determine the feature weights of the vegetation indices [43]. Through feature stacking, we identified the most sensitive features for the classification task. In addition, we evaluated three classification models (Support Vector Machine (SVM), decision tree, and K-Nearest Neighbors (KNN)) in combination with the selected features.

Different stress conditions caused by various pathogens have different effects on crop growth and development [20]. Currently, hyperspectral remote sensing research mainly focuses on aspects such as chlorophyll content, nitrogen content, and pest and disease detection. For peanuts, most studies concentrate on diseases with obvious pathogenic characteristics, such as leaf spot and stem rot. For instance, Guan et al. used portable spectroradiometers and spectrophotometers to study the spectral characteristics of peanut leaf spot disease [44]. Wei et al. used hyperspectral sensors and machine learning techniques to identify the optimal wavelength features for detecting peanut stem rot [45]. However, to the best of our knowledge, there have been no research reports on the remote sensing monitoring mechanism of peanut southern blight. Whether the progress of previous work is applicable to our research presents new challenges.

The overall objective of this study is to investigate the spectral response mechanism of peanut southern blight and distinguish peanuts with different levels of severity. Specifically, our goals are to address the following questions: (1) Can we extract the spectral response mechanism of peanut southern blight from the hyperspectral remote sensing data? (2) Can CWT be applied to our hyperspectral data to differentiate the severity of peanut southern blight at the leaf level? (3) Is the combination of CWT and ML models more effective than traditional spectral indices and spectral preprocessing methods?

2. Materials and Methods

2.1. Experimental Design

The peanut trial was conducted in 2021 and 2022 at the Wenhua Road Campus of Henan Agricultural University. A laboratory pot was used to control the publication-grade experiment manually. The experimental peanut variety was Yuhua 37, sown in the greenhouse laboratory and managed regularly. The soil for peanut culture was a mixture of matrix and vermiculite with a volume ratio of 3:1 after autoclaving. Peanut plants with uniform size and healthy growth in the greenhouse for ten days were inoculated with different concentration doses (namely benzovindiflupyr and thifluzamide). The experimental concentrations of benzovindiflupyr were 50, 100, and 200 mg L^{-1}, respectively, and thifluzamide was used as the control agent with a concentration of 100 mg/L. Blank control peanut plants were treated with distilled water, and 10 mL of each concentration of fungicide was applied with a 5 mL pipettor to the stem base of the plant 48 h before inoculation (preventive activity) or after inoculation (therapeutic exercise). This study selected the inoculation strain for highly virulent Sclerotium rolfsii Sacc (ZMGD-2). Four agar disks containing mycelium (5 mm in diameter) were placed around the root and stem of each peanut plant and buried with the matrix. The inoculated plants were kept at 30 °C and 80% relative humidity for seven days as far as possible. The data collection is shown in Table 1.

Table 1. Sample inoculation and acquisition time.

Sample Inoculation Time	Sample Acquisition Time	Quantity
02 Nov. 2021	05 Dec. 2021	76
04 Dec. 2021	05 Jan. 2022	46
20 Mar. 2022	26 Apr.2022	53

2.2. Data Collection

2.2.1. Classification and Analysis of Disease Severity

The samples for the pot experiment were obtained through investigation conducted by plant protection experts from Henan Agricultural University. The surveyed peanut plants had an average height of approximately 10 cm. Samples were selected from peanut plants treated with different chemicals and concentrations to assess the disease grade of southern blight. Based on previous research on the genetic and phenotypic diversity of peanuts, the severity of southern blight was defined as Grade 0 = healthy plants, Grade 1 = mild, and Grade 2 = severe, as shown in Table 2 and Figure 1.

2.2.2. Reflectance Spectral Measurement

The spectral measurement instrument of this experiment adopted the ASD Field Spec3 spectrometer and the matching plant probe to collect the spectral data of peanut leaves.

The dimensions of the equipment are 12.7 cm × 36.8 cm × 29.2 cm and its weight is 5.44 kg. The wavelength range is 350–2500 nm, the sampling intervals are 1.4 nm (350–1000 nm) and 2 nm (1001–2500 nm), and the resampling interval is 1 nm. To avoid signal loss due to light absorption by atmospheric water vapor at wavelengths between 1400 nm and 1800 nm, the handheld Leaf Clip (ASD Leaf Clip) of the matching spectrometer was used to measure the spectrum of peanut leaves in this experiment. The built-in standard whiteboard was calibrated every 3 min to obtain a baseline close to 100% to ensure the accuracy of spectral data during the experiment.

Table 2. Grading standard of peanut southern blight disease.

Disease Severity	Symptom
Health (Grade 0)	No apparent symptoms
Mild (Grade 1)	Most of the leaves exhibit yellowing and wilting, while a significant amount of white mycelium is observed at the plant's root base.
Severe (Grade 2)	The entire plant exhibits complete wilting of leaves, while brown spherical sclerotia are present at the plant's root base.

Figure 1. Experimental potted plant for peanut southern blight. (**a**,**b**) Healthy, (**c**) mild, (**d**) severe.

2.3. Data Analysis Methods

2.3.1. Continuous Wavelet Transform

CWT is a linear operation that transforms a reflectance spectrum $f(\lambda)$ ($\lambda = 1, 2, \ldots, n$, where n is the number of spectral bands) into sets of coefficients at various scales by using a mother wavelet function. The mother wavelet $\psi(\lambda)$ is a small wave and has an average value of zero, which can be shifted (translated) and scaled (stretched or compressed) to produce a series of continuous wavelets $\psi_{a,b}(\lambda)$ as follows (dyadic numbers $2^1, 2^2, 2^3, \ldots, 2^8$ are denoted as Scale 1, Scale 2, Scale 3, ..., Scale 8 for simplicity, respectively) [44]. In Formula (1), a represents the wavelength and b represents the phase. After spectrum decomposition, the complete wavelet coefficient matrix of different bands and decomposition scales can be obtained:

$$\psi_{a,b}(\lambda) = \frac{1}{\sqrt{a}}\psi\left(\frac{\lambda - b}{a}\right) \tag{1}$$

$$W_f(a,b) = \int_{-\infty}^{\infty} f(\lambda)\psi_{a,b}(\lambda)d\lambda \tag{2}$$

where $\psi_{a,b}(\lambda)$ denotes the inner products of wavelets and the input spectrum. The output $W_f(a, b)$ of a one-dimentional input spectrum comprises a two-dimentional wavelet power scalogram. Each element of the scalogram is a wavelet feature or wavelet coefficient that characterizes the correlation between a subset of the input spectrum and a scaled, shifted version of the mother wavelet [45].

The wavelet transform has proven to be an effective technique for extracting spectral information related to foliar chemistry and species composition from vegetation reflectance spectra when applied to spectroscopic data in remote sensing [46,47]. The Continuous Wavelet Transform (CWT) was utilized instead of the Discrete Wavelet Transform (DWT), because CWT provides scale components that are directly comparable to the input reflectance spectrum on a band-by-band basis, making the results easier to interpret.

2.3.2. Standard Normal Variable Transformation Processing

The SNV transformation was used to eliminate the influence of diffuse reflectance spectra caused by surface scattering and solid particle sizes during data collection. The average value of the spectral data was subtracted from the initial spectral reflectance data and then divided by its standard deviation [48]. The formula is as follows:

$$X_{snv} = \frac{X - \bar{x}}{\sqrt{\frac{\sum_{k=1}^{m}(X_{k-\bar{x}})}{(m-1)}}} \tag{3}$$

where $\bar{x} = \frac{\sum_{k=1}^{m} x^k}{m}$, m is the total number of wavelengths, and $k = 1, 2 \ldots, m$.

2.3.3. Spectral Index

After reviewing the previous research on spectral indices, 12 spectral indices related to pest and disease stress were selected from the highly cited literature (Table 3). We then analyzed their weights using the relief algorithm to retain the most sensitive features for assessing their transferability.

Table 3. The spectral indices included in this study.

Index	Formulation	Reference
SIPI	(R800 − R445)/(R800 + R680)	[49]
R	R700/R670	[50]
G	R570/R670	[51]
B	R450/R490	[51]
NRI	(R570 − R670)/(R570 + R670)	[52]
WI	R900/R970	[53]
mNDI	(R750 − R705)/(R750 − R705 − 2R445)	[54]
HI	(R739 − R402)/(R739 + R402) − 0.5R403	[10]
NSRI	R890/R780	[55]
PSRI	(R680 − R500)/R750	[56]
MSR	(R750 − R445)/(R705 − R445)	[54]
PSSRa	R800/R675	[57]

2.3.4. Relief

The relief algorithm is a classic feature weight selection method that assigns weights to different features based on their relevance to the target variable.

In the initial feature set, the relief algorithm randomly selects a sample, denoted as "a", and then searches for the nearest neighbor sample within the same class, known as the "Near Hit". It also searches for the nearest neighbor sample outside the same class, referred to as the "Near Miss". Feature weights are defined as follows: if the distance between the feature of interest and the Near Hit (H) is smaller than the distance between the same feature and the Near Miss (M), the weight is increased, which indicates that the feature effectively distinguishes different classes. Conversely, the weight is decreased for the reverse case [58].

$$w = diff(l, a, M) - diff(l, a, H) \qquad (4)$$

$$diff(l, a, b) = \frac{|a - b|}{\max(l) - \min(l)} \qquad (5)$$

where $diff(l, a, b)$ represents the distance between samples a and b for feature l, and $\max(l)$ and $\min(l)$ represent the upper and lower bounds of feature l, respectively.

2.3.5. Machine Learning

In this study, three non-parametric machine learning algorithms, namely SVM, decision tree, and KNN, were employed to detect the severity of peanut southern blight.

The working principle of SVM is to create an optimal classification hyperplane using the training dataset and achieve different sample classifications based on minimal errors. In this study, we employed grid search to determine the best parameters, including the Radial Basis Function (RBF) kernel and polynomial kernel functions, for SVM classification [59]. Decision tree is a supervised learning algorithm that learns from a labeled training dataset to construct a root node and selects the best feature to further partition the data, aiming to achieve the best classification for each data point at each step [60]. KNN is a non-parametric classification method that assigns labels to data points based on the classification of K similar training samples. It does not assume any specific distribution for the data [61].

2.3.6. Evaluation of Accuracy

In this study, a 5-fold cross-validation with 100 repetitions was performed to evaluate the accuracy and robustness of all models. The first two sets of data (n = 122) were used to build and validate the models, while the third set of data (n = 53) was used for independent validation. The sample sizes for each severity level were approximately balanced across the three sets. The OA and kappa coefficient were used to assess the performance of the models. The formulas for calculating these two metrics are shown as Equations (6) and (7), respectively. In the equations, N represents the total number of classes; n represents the number of samples; akk represents the number of correctly classified samples; x_{ii} represents the diagonal elements of the confusion matrix; and x_{ij} represents each element of the confusion matrix.

$$OA = \frac{\left(\sum_{k=1}^{N} akk \right)}{n} \qquad (6)$$

$$kappa = \frac{N\sum_{i=1}^{m} x_{ii} - \sum_{k=1}^{m} \left(\sum_{i=1}^{m} x_{ij} \sum_{j=1}^{m} x_{ij} \right)}{N^2 - \sum_{k=1}^{m} \left(\sum_{i=1}^{m} x_{ij} \sum_{j=1}^{m} x_{ij} \right)} \qquad (7)$$

3. Results

3.1. Spectral Response of Peanut Southern Blight

The sample contained 175 healthy, mild, moderate, and severe peanut leaves. The average spectral responses of each leaf type in different wavelength bands are shown in Figure 2. The findings indicate that in the green-light wavelength band (530–580 nm), healthy leaves exhibited the highest reflectance, while severely affected leaves showed the lowest reflectance. In the red-light wavelength band (620–670 nm), although the differences were not significant, some features were observed. Specifically, the spectral reflectance followed the pattern of healthy leaves > mild leaves > severe leaves, with a relatively small peak at 640 nm. In the red-edge wavelength band (700–780 nm), there were significant differences between healthy and severely affected leaves. The reason for this difference is attributed to the destruction of photosynthetic pigments, including chlorophyll, in infected leaves. The absorption capacity in the blue-light wavelength band (centered at 450 nm) and red-light wavelength band (centered at 660 nm) weakened, resulting in relatively small peaks. As chlorophyll continued to be destroyed and the photosynthetic ability weakened, the reflectance in the green-light wavelength band (centered at 550 nm) decreased, with a noticeable difference at 560 nm, due to changes in cell structure, loss of water content, and a decrease in chlorophyll and photosynthetic intensity in leaf cells caused by the continuous invasion of Sclerotium rolfsii in the intercellular space of the leaves. The hyperspectral reflectance of southern blight was relatively low in the visible band (400–760 nm) and relatively high in the near-infrared band (760–1350 nm). In comparison with healthy plants, the red edge of the infected southern blight largely shifted toward shorter wavelengths, indicating a "blue shift" phenomenon. In the overall analysis, the spectral reflectance of infected leaves in the visible light and near-infrared wavelength bands showed a decreasing trend with the increasing severity of the disease.

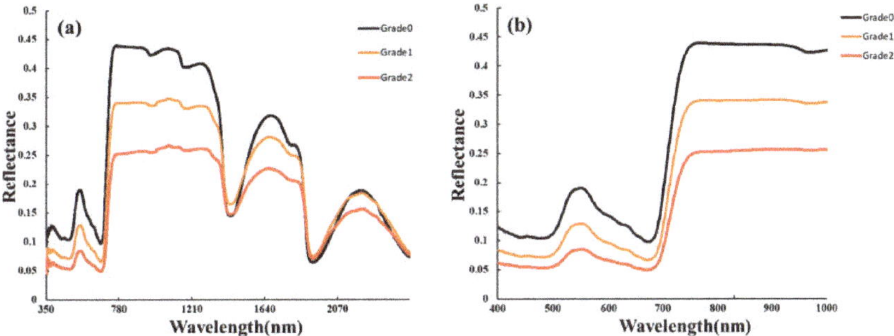

Figure 2. Spectral characteristics of leaves of southern blight with different disease degrees: (**a**) 350–2500 nm; (**b**) 400–1000 nm.

3.2. Continuous-Wavelet-Transform-Sensitive Spectral Characterization

The spectral reflectance data of different severity levels collected in 2021 and 2022 were applied to CWT, and then the spectral results for each severity level were averaged, as shown in Figure 3b. Compared to the original spectra, CWT amplified the spectral differences in the red-edge range (700–790 nm) for different severity levels. Furthermore, we generated a classification scale based on the data, as shown in Figure 3a. We evaluated the accuracy of the machine learning models by incorporating the most sensitive wavelet features for each scale separately. Ultimately, we found that the SVM model using CWT (WF$_{770\sim780}$, 5) performed best. The validation set OA was 90.5% and the kappa coefficient was 0.87 (Table 4).

3.3. Standard Normal Variable Transformation Processing

Figure 4 shows the spectral curves after SNV preprocessing, which altered the shape of the spectra compared to the original spectra (OR). SNV increased the separability of the spectral curves for different severity levels in the range of 350–1200 nm. We employed simple Linear Discriminant Analysis (LDA) to explore the sensitive bands of OR and SNV. For OR, the 940–1300 nm range exhibited the best performance, with the 942 nm band having the highest accuracy and an OA of 87.4%. For SNV, the 780–1300 nm range demonstrated the best performance, with the 903 nm band having the highest accuracy and an OA of 88.7%. Overall, both OR and SNV showed their highest accuracies in the near-infrared (NIR) range, and SNV enhanced NIR spectral differences among different severity levels (Figure 5).

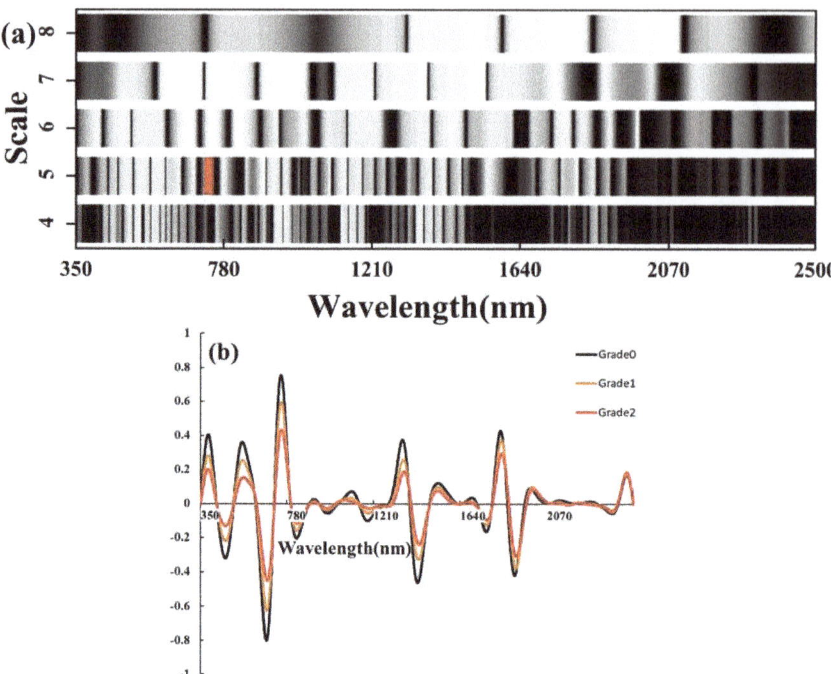

Figure 3. (**a**) The X-axis represents the spectral wavelength range from 350 to 2500 nm, and the Y-axis represents the fourth to eighth wavelet scales. The grayscale brightness in the scale chart indicates the magnitude of classification accuracy (brighter indicates higher accuracy). The red region corresponds to the top 1% of the highest accuracy achieved. (**b**) CWT spectral curve.

Table 4. Accuracy evaluation using CWT machine learning models.

Features	Model	Calibration		Validation	
		OA (%)	Kappa	OA (%)	Kappa
	SVM	91.8%	0.88	90.5%	0.87
$WF_{770\sim780}$, 5	KNN	85.2%	0.78	86.6%	0.79
	Decision Trees	89.3%	0.84	86.8%	0.79

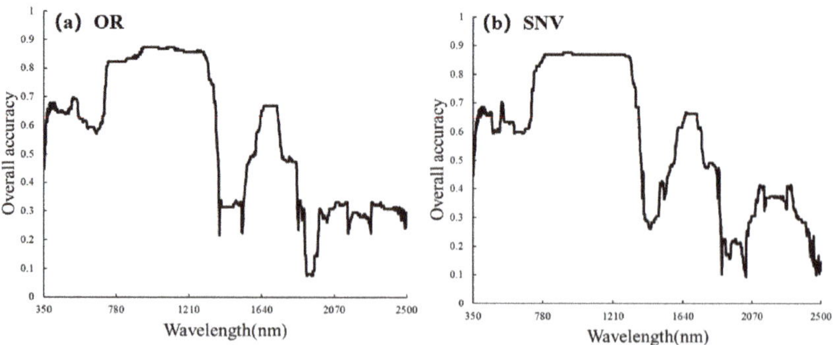

Figure 4. Analysis of spectral separability based on Linear Discriminant Analysis. (**a**) Original spectrum, (**b**) SNV spectrum.

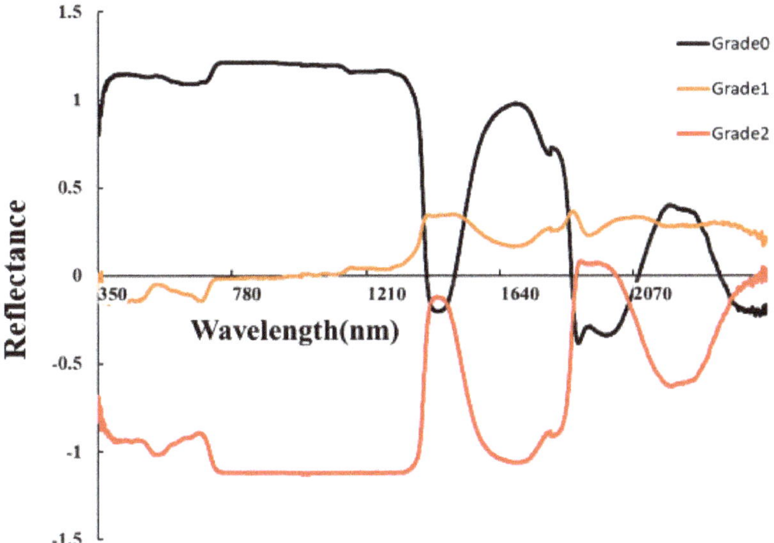

Figure 5. SNV spectral curve.

3.4. Assessing the Transferability of Spectral Indices

To explore the spectral indices characterizing the severity of peanut southern blight, we analyzed the weights of 12 spectral indices using the relief algorithm (Figure 6a). To further evaluate the accuracy of these features, we performed the SVM modeling with different combinations of the features and evaluated their performance, as shown in Figure 6b. The highest OA was 74.9%, which was achieved by using 11 features. Notably, these features exhibited complementarity in the model. Removing the NSRI feature resulted in a 2.9% reduction in OA when the remaining 10 features were used. Likewise, removing the HI feature resulted in a 5.7% drop in OA when using the remaining nine features. Removing the G feature resulted in a 2.9% reduction in OA when using the remaining four features. Finally, removing the SIPI features led to a 2.9% decrease in OA. Based on these findings, we can conclude that the NSRI, HI, G, and SIPI features had a particular impact on the performance of the model.

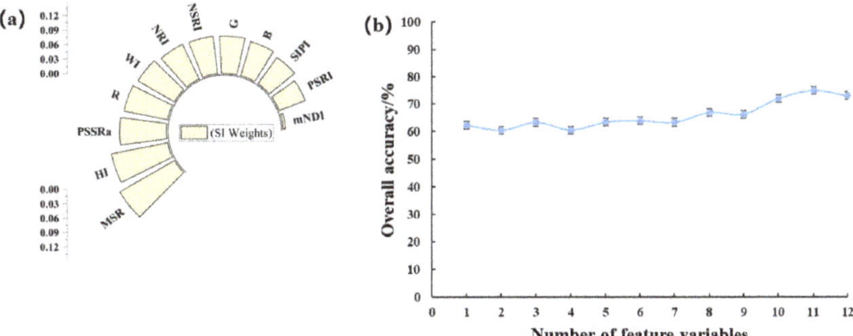

Figure 6. (a) Weights of VIS; (b) plot of OA of VIS-superimposed SVM model.

We further conducted an autocorrelation analysis on the four selected features (Table 5). The correlation between each feature was found to be very low, indicating the absence of multicollinearity among the features. We proceeded to evaluate the accuracy of different machine learning models (Table 6). Among them, the SVM model achieved the highest performance on the training set, with an OA of 92.6% and kappa coefficient of 0.89. However, when applied to the independent validation data, the SVM model showed lower performance, with an OA of only 62.3% and kappa coefficient of 0.43, indicating poor robustness of the model.

Table 5. Correlation analysis of NSRI, HI, G, and SIPI.

	NSRI	**HI**	**G**	**SIPI**
NSRI	1			
HI	0.330943	1		
G	−0.36627	0.340911	1	
SIPI	−0.05788	0.539683	0.202275	1

Table 6. Accuracy of different machine learning models.

Features	**Model**	**Calibration**		**Validation**	
		OA (%)	**Kappa**	**OA (%)**	**Kappa**
	SVM	92.6%	0.89	62.3.%	0.43
NSRI, HI, G, SIPI	KNN	86.9%	0.8	67.9%	0.51
	Decision Trees	74.6.%	0.61	64.2%	0.47

4. Discussion

4.1. Spectral Response Mechanism of Peanut Southern Blight

Peanut southern blight is a highly contagious and extremely destructive soil-borne fungal disease that occurs in most countries. It has become a key factor limiting peanut yield and quality [62,63]. Currently, there are few reports on the spectral response mechanism of peanut southern blight. In this study, we obtained spectral curves of samples with different severity levels through variable-controlled experiments. Machine learning techniques have been applied to hyperspectral data to enhance the detection capability of peanut southern blight severity. The main focus in this regard is to explore the spectral response of southern blight and obtain the optimal spectral features. These methods are consistent with previous advancements in the field [64].

Previous research has indicated that when plants are under stress from pests and diseases, their spectra tend to shift toward shorter wavelengths, and the amplitude of the red edge decreases [65]. When peanut plants are infected with southern blight disease, their photosynthesis is disrupted, resulting in a decrease in the absorption capacity of blue- and red-light wavelengths and a decrease in reflectance. As the disease progresses over time, it further damages the leaf structure, leading to the loss of chlorophyll and water content. Therefore, samples with different severity levels exhibit significant differences in the red-edge and near-infrared range (725–1200 nm). Infected samples generally show a decreasing trend in spectral reflectance, accompanied by a red-edge shift toward shorter wavelengths.

4.2. Advantages of Wavelet Analysis in Pest and Disease Detection

CWT can perform spectral decomposition at continuous wavelengths and scales. It effectively reduces noise interference, amplifies implicit weak spectral information, and plays a significant role in eliminating spectral background differences. Moreover, it enhances the sensitivity of spectra to the severity of peanut southern blight disease [66,67].

In this study, we conducted a comparative analysis using CWT at different scales, and the results showed that CWT at five scales achieved the best performance. Additionally, we found that the SVM model constructed using CWT ($WF_{770\sim780}$, 5) outperformed the models based on the SNV and the original spectra. The main reason for this improvement is that CWT enhances the spectral response in the red-edge region, enabling effective differentiation of different severity levels of peanut southern blight disease [68].

4.3. Application of Spectral Index in Pest and Disease Detection

To assess the transferability of spectral indices under previous disease and pest stress, this study selected 12 spectral indices that contain information in the red-edge region. These indices were chosen based on current reports on the remote sensing of diseases and pests, such as wheat stripe rust [69] and apple fire blight [70]. Furthermore, relief analysis was employed to determine feature weights, and SVM models were evaluated by individually incorporating each feature. We observed that the model's accuracy significantly changed when certain features were removed (Figure 6). This variation can be attributed to the complementary nature of different features in the model. As a result, we identified and confirmed four features as the final inputs for the model.

Although the SVM model achieved the highest accuracy on the training dataset, we observed poor robustness when validating the model using independent data. This may be attributed to the fact that the data from the final period were collected in spring, while the training dataset consisted of data collected in winter (Table 1). The different growth stages could have led to suboptimal model performance. However, we also identified several spectral indices that are related to the severity of peanut southern blight, indicating that spectral indices can rapidly detect the stress of plant diseases and pests, which is consistent with previous findings. Moving forward, our future work will likely focus on exploring additional vegetation indices that can accurately detect the severity of peanut southern blight, thus providing feasibility analysis for large-scale remote sensing of this disease.

4.4. Implications for Future Applications

The study also found some complex challenges in the early monitoring of peanut southern blight. The first problem is that the physiological interaction between fungal pathogens and host plants depends on pathogenic fungi. So, more in-depth investigation is needed to explore the interaction between different pathogens. Previously unconsidered variations can be revealed as the original source of reflectance data. A non-imaging sensor, to capture the average of healthy and diseased plant tissue parts, has been used to measure the reflectance curve, which causes many typical single-point measurement problems [71]. The second challenge lies in the complexity of field environments, where phenomena like spectral variations from the same object and the co-occurrence of multiple diseases can occur. Our severity classification model for peanut southern blight built at the leaf scale may be influenced by various factors. For example, in terms of spectral response, peanut leaf spot disease shows a significant negative correlation between the disease index and the spectral curve in the NIR range, which is very similar to the spectral response of peanut southern blight [11]. In terms of plant structure, peanut stem rot disease also exhibits yellowish-brown rotting signs at the base of the stem during the early stages of infection [72]. However, without the presence of white mycelium and brown sclerotia at the base, it can often lead to misinterpretation. Therefore, it requires the integration of field meteorological data, agronomic background, and other relevant data for comprehensive discrimination, which is a difficult task. The third challenge is to integrate multiple data sources and enable data sharing of peanut southern blight between different provinces, aiming to improve the model's transferability. Our future focus is on integrating multi-source remote sensing data to achieve data exchange between provinces and establishing a dynamic monitoring platform for peanut southern blight. This platform aims to provide technical support for disease prevention and control in peanuts.

5. Conclusions

This study analyzed the spectral response mechanism of different severity levels of peanut southern blight. For the severity classification problem, we compared the machine learning modeling using CWT with traditional spectral indices and spectral preprocessing methods. The results showed that CWT was more effective and amplified the spectral differences between different levels of severity. Furthermore, this study emphasized the potential of using hyperspectral sensors for monitoring peanut southern blight, which is an exciting tool for disease management and control in peanuts.

Author Contributions: W.G.: conceptualization, methodology, writing—original draft. H.S.: methodology, writing—review and editing. H.Q.: methodology, investigation, supervision. H.Z.: data curation, methodology. L.Z.: materials and methods. P.D.: data curation, methodology. X.S.: methodology, investigation, conceptualization. All authors have read and agreed to the published version of the manuscript.

Funding: This research was funded by The Henan Provincial Science and Technology Major Project (221100110100); National Natural Science Foundation of China, grant number (32271993); The Joint Fund of Science and Technology Research Development program (Cultivation project of preponderant discipline) of Henan Province, China (222301420114).

Data Availability Statement: Not applicable.

Conflicts of Interest: The authors declare no conflict of interest.

References

1. Avijit, T.; Swaroopa, R.T.; Sharath, C.U.S.; Raju, G.; Chobe, D.R.; Mamta, S. Exploring Combined Effect of Abiotic (Soil Moisture) and Biotic (Sclerotium rolfsii Sacc.) Stress on Collar Rot Development in Chickpea. *Front. Plant Ence* **2018**, *9*, 1154.
2. Thiessen, L.D.; Woodward, J.E. Diseases of peanut caused by soilborne pathogens in the Southwestern United States. *Int. Sch. Res. Not.* **2012**, *2012*, 1–9. [CrossRef]
3. Xu, M.; Zhang, X.; Yu, J.; Guo, Z.; Wan, S. Biological control of peanut southern blight (Sclerotium rolfsii) by the strain Bacillus pumilus LX11. *Biocontrol Sci. Technol.* **2020**, *30*, 485–489. [CrossRef]
4. Zhang, W.; Zhang, B.-W.; Deng, J.-F.; Li, L.; Yi, T.-Y.; Hong, Y.-Y. The resistance of peanut to soil-borne pathogens improved by rhizosphere probiotics under calcium treatment. *BMC Microbiol.* **2021**, *21*, 299. [CrossRef] [PubMed]
5. Zhang, J.; Huang, Y.; Pu, R.; Gonzalez-Moreno, P.; Yuan, L.; Wu, K.; Huang, W. Monitoring plant diseases and pests through remote sensing technology: A review. *Comput. Electron. Agric.* **2019**, *165*, 104943. [CrossRef]
6. Long, T.; Bowen, X.; Ziyi, W.; Dong, L.; Xia, Y.; Qiang, C.; Yan, Z.; Weixing, C.; Tao, C. Spectroscopic detection of rice leaf blast infection from asymptomatic to mild stages with integrated machine learning and feature selection. *Remote Sens. Environ.* **2021**, *257*, 112350.
7. Zhang, N.; Yang, G.; Pan, Y.; Yang, X.; Chen, L.; Zhao, C. A Review of Advanced Technologies and Development for Hyperspectral-Based Plant Disease Detection in the Past Three Decades. *Remote Sens.* **2020**, *12*, 3188. [CrossRef]
8. Feng, L.; WU, D.; HE, Y. Identification and classification of rice leaf blast based on multi-spectral imaging sensor. *Spectrosc. Spectr. Anal.* **2009**, *29*, 2730–2733.
9. Zhou, L.-N.; Yu, H.-Y.; Zhang, L.; Ren, S.; Sui, Y.-Y.; Yu, L.-J. Rice blast prediction model based on analysis of chlorophyll fluorescence spectrum. *Spectrosc. Spectr. Anal.* **2014**, *34*, 1003–1006.
10. Huang, W.; Guan, Q.; Luo, J.; Zhang, J.; Zhao, J.; Liang, D.; Huang, L.; Zhang, D. New optimized spectral indices for identifying and monitoring winter wheat diseases. *IEEE J. Sel. Top. Appl. Earth Obs. Remote Sens.* **2014**, *7*, 2516–2524. [CrossRef]
11. Chen, T.; Zhang, J.; Chen, Y.; Wan, S.; Zhang, L. Detection of peanut leaf spots disease using canopy hyperspectral reflectance. *Comput. Electron. Agric.* **2019**, *156*, 677–683. [CrossRef]
12. Jiang, X.; Zhen, J.; Miao, J.; Zhao, D.; Shen, Z.; Jiang, J.; Gao, C.; Wu, G.; Wang, J. Newly-developed three-band hyperspectral vegetation index for estimating leaf relative chlorophyll content of mangrove under different severities of pest and disease. *Ecol. Indic.* **2022**, *140*, 108978. [CrossRef]
13. Talbot, N.J. On the trail of a cereal killer: Exploring the biology of Magnaporthe grisea. *Annu. Rev. Microbiol.* **2003**, *57*, 177–202. [CrossRef]
14. Baranoski, G.; Rokne, J. A practical approach for estimating the red edge position of plant leaf reflectance. *Int. J. Remote Sens.* **2005**, *26*, 503–521. [CrossRef]
15. Ray, S.S.; Jain, N.; Arora, R.; Chavan, S.; Panigrahy, S. Utility of hyperspectral data for potato late blight disease detection. *J. Indian Soc. Remote Sens.* **2011**, *39*, 161–169. [CrossRef]
16. Sathish, C.; Nakhawa, A.; Bharti, V.S.; Jaiswar, A.; Deshmukhe, G. Estimation of extent of the mangrove defoliation caused by insect Hyblaea puera (Cramer, 1777) around Dharamtar creek, India using Sentinel 2 images. *Reg. Stud. Mar. Sci.* **2021**, *48*, 102054. [CrossRef]
17. Wang, H.-G.; Ma, Z.-H.; Wang, T.; Cai, C.-J.; An, H.; Zhang, L.-D. Application of hyperspectral data to the classification and identification of severity of wheat stripe rust. *Guang Pu Xue Yu Guang Pu Fen Xi = Guang Pu* **2007**, *27*, 1811–1814.
18. He, R.; Li, H.; Qiao, X.; Jiang, J. Using wavelet analysis of hyperspectral remote-sensing data to estimate canopy chlorophyll content of winter wheat under stripe rust stress. *Int. J. Remote Sens.* **2018**, *39*, 4059–4076. [CrossRef]
19. Yang, Y.; Chai, R.; He, Y. Early detection of rice blast (Pyricularia) at seedling stage in Nipponbare rice variety using near-infrared hyper-spectral image. *Afr. J. Biotechnol.* **2012**, *11*, 6809–6817. [CrossRef]

20. Feng, Z.; Guan, H.; Yang, T.; He, L.; Duan, J.; Song, L.; Wang, C.; Feng, W. Estimating the canopy chlorophyll content of winter wheat under nitrogen deficiency and powdery mildew stress using machine learning. *Comput. Electron. Agric.* **2023**, *211*, 107989. [CrossRef]

21. Hernández-Clemente, R.; Hornero, A.; Mottus, M.; Peñuelas, J.; González-Dugo, V.; Jiménez, J.; Suárez, L.; Alonso, L.; Zarco-Tejada, P.J. Early diagnosis of vegetation health from high-resolution hyperspectral and thermal imagery: Lessons learned from empirical relationships and radiative transfer modelling. *Curr. For. Rep.* **2019**, *5*, 169–183. [CrossRef]

22. Berger, K.; Machwitz, M.; Kycko, M.; Kefauver, S.C.; Van Wittenberghe, S.; Gerhards, M.; Verrelst, J.; Atzberger, C.; van der Tol, C.; Damm, A. Multi-sensor spectral synergies for crop stress detection and monitoring in the optical domain: A review. *Remote Sens. Environ.* **2022**, *280*, 113198. [CrossRef] [PubMed]

23. Feret, J.-B.; François, C.; Asner, G.P.; Gitelson, A.A.; Martin, R.E.; Bidel, L.P.; Ustin, S.L.; Le Maire, G.; Jacquemoud, S. PROSPECT-4 and 5: Advances in the leaf optical properties model separating photosynthetic pigments. *Remote Sens. Environ.* **2008**, *112*, 3030–3043. [CrossRef]

24. Al-Saddik, H.; Laybros, A.; Billiot, B.; Cointault, F. Using image texture and spectral reflectance analysis to detect Yellowness and Esca in grapevines at leaf-level. *Remote Sens.* **2018**, *10*, 618. [CrossRef]

25. Verrelst, J.; Camps-Valls, G.; Muñoz-Marí, J.; Rivera, J.P.; Veroustraete, F.; Clevers, J.G.; Moreno, J. Optical remote sensing and the retrieval of terrestrial vegetation bio-geophysical properties–A review. *ISPRS J. Photogramm. Remote Sens.* **2015**, *108*, 273–290. [CrossRef]

26. Calderón, R.; Navas-Cortés, J.A.; Lucena, C.; Zarco-Tejada, P.J. High-resolution airborne hyperspectral and thermal imagery for early detection of Verticillium wilt of olive using fluorescence, temperature and narrow-band spectral indices. *Remote Sens. Environ.* **2013**, *139*, 231–245. [CrossRef]

27. Meena, S.V.; Dhaka, V.S.; Sinwar, D. Exploring the Role of Vegetation Indices in Plant Diseases Identification. In Proceedings of the 2020 Sixth International Conference on Parallel, Distributed and Grid Computing (PDGC), Solan, India, 6–8 November 2020; pp. 372–377.

28. Zhao, H.; Yang, C.; Guo, W.; Zhang, L.; Zhang, D. Automatic estimation of crop disease severity levels based on vegetation index normalization. *Remote Sens.* **2020**, *12*, 1930. [CrossRef]

29. Fu, H.; Zhao, H.; Song, R.; Yang, Y.; Li, Z.; Zhang, S. Cotton aphid infestation monitoring using Sentinel-2 MSI imagery coupled with derivative of ratio spectroscopy and random forest algorithm. *Front. Plant Sci.* **2022**, *13*, 1029529. [CrossRef]

30. Huang, W.; Lu, J.; Ye, H.; Kong, W.; Mortimer, A.H.; Shi, Y. Quantitative identification of crop disease and nitrogen-water stress in winter wheat using continuous wavelet analysis. *Int. J. Agric. Biol. Eng.* **2018**, *11*, 145–152. [CrossRef]

31. Cheng, T.; Riaño, D.; Ustin, S.L. Detecting diurnal and seasonal variation in canopy water content of nut tree orchards from airborne imaging spectroscopy data using continuous wavelet analysis. *Remote Sens. Environ.* **2014**, *143*, 39–53. [CrossRef]

32. Blackburn, G.A.; Ferwerda, J.G. Retrieval of chlorophyll concentration from leaf reflectance spectra using wavelet analysis. *Remote Sens. Environ.* **2007**, *112*, 1614–1632. [CrossRef]

33. Rivard, B.; Feng, J.; Gallie, A.; Sanchez-Azofeifa, A. Continuous wavelets for the improved use of spectral libraries and hyperspectral data. *Remote Sens. Environ.* **2008**, *112*, 2850–2862. [CrossRef]

34. Ampe, E.M.; Hestir, E.L.; Bresciani, M.; Salvadore, E.; Brando, V.E.; Dekker, A.G.; Malthus, T.J.; Jansen, M.; Triest, L.; Batelaan, O. A Wavelet Approach for Estimating Chlorophyll-A From Inland Waters with Reflectance Spectroscopy. *IEEE Geosci. Remote Sens. Lett.* **2014**, *11*, 89–93. [CrossRef]

35. Juhua, L.; Wenjiang, H.; Jinling, Z.; Jingcheng, Z.; Chunjiang, Z.; Ronghua, M. Detecting Aphid Density of Winter Wheat Leaf Using Hyperspectral Measurements. *IEEE J. Sel. Top. Appl. Earth Obs. Remote Sens.* **2013**, *6*, 690–698.

36. Pu, R.; Gong, P. Wavelet transform applied to EO-1 hyperspectral data for forest LAI and crown closure mapping. *Remote Sens. Environ.* **2004**, *91*, 212–224. [CrossRef]

37. Banskota, A.; Wynne, R.H.; Kayastha, N. Improving within-genus tree species discrimination using the discrete wavelet transform applied to airborne hyperspectral data. *Int. J. Remote Sens.* **2011**, *32*, 3551–3563. [CrossRef]

38. Asaari, M.S.M.; Mishra, P.; Mertens, S.; Dhondt, S.; Wuyts, N.; Scheunders, P. Close-range hyperspectral image analysis for the early detection of plant stress responses in individual plants in a high-throughput phenotyping platform. *ISPRS J. Photogramm. Remote Sens.* **2018**, *138*, 121–138. [CrossRef]

39. Feng, Z.; Zhang, H.; Duan, J.; He, L.; Yuan, X.; Gao, Y.; Liu, W.; Li, X.; Feng, W. Improved Spectral Detection of Nitrogen Deficiency and Yellow Mosaic Disease Stresses in Wheat Using a Soil Effect Removal Algorithm and Machine Learning. *Remote Sens.* **2023**, *15*, 2513. [CrossRef]

40. Feng, Z.; Song, L.; Duan, J.; He, L.; Zhang, Y.; Wei, Y.; Feng, W. Monitoring wheat powdery mildew based on hyperspectral, thermal infrared, and RGB image data fusion. *Sensors* **2022**, *22*, 31. [CrossRef]

41. Hamed, A.M.S.; Mehdi, M.; Asghari, B.B. A feature extraction method based on spectral segmentation and integration of hyperspectral images. *Int. J. Appl. Earth Obs. Geoinf.* **2019**, *89*, 102097.

42. Wang, H.; Li, G.; Ma, Z.; Li, X. Image recognition of plant diseases based on backpropagation networks. In Proceedings of the 2012 5th International Congress on Image and Signal Processing, Agadir, Morocco, 28–30 June 2012; pp. 894–900.

43. Kononenko, I. Estimating attributes: Analysis and extensions of RELIEF. In Proceedings of the European conference on machine learning, Catania, Italy, 6–8 April 1994; pp. 171–182.

44. Bruce, L.M.; Li, J.; Huang, Y. Automated detection of subpixel hyperspectral targets with adaptive multichannel discrete wavelet transform. *IEEE Trans. Geosci. Remote Sens.* **2002**, *40*, 977–980. [CrossRef]
45. Bruce, L.M.; Li, J. Wavelets for computationally efficient hyperspectral derivative analysis. *IEEE Trans. Geosci. Remote Sens.* **2001**, *39*, 1540–1546. [CrossRef]
46. Blackburn, G.A. Wavelet decomposition of hyperspectral data: A novel approach to quantifying pigment concentrations in vegetation. *Int. J. Remote Sens.* **2007**, *28*, 2831–2855. [CrossRef]
47. Cheng, T.; Rivard, B.; Sánchez-Azofeifa, A. Spectroscopic determination of leaf water content using continuous wavelet analysis. *Remote Sens. Environ.* **2010**, *115*, 659–670. [CrossRef]
48. Dhanoa, M.; Lister, S.; Sanderson, R.; Barnes, R. The link between multiplicative scatter correction (MSC) and standard normal variate (SNV) transformations of NIR spectra. *J. Near Infrared Spectrosc.* **1994**, *2*, 43–47. [CrossRef]
49. Penuelas, J.; Baret, F.; Filella, I. Semi-empirical indices to assess carotenoids/chlorophyll a ratio from leaf spectral reflectance. *Photosynthetica* **1995**, *31*, 221–230.
50. Gitelson, A.; Yacobi, Y.; Schalles, J.; Rundquist, D.; Han, L.; Stark, R.; Etzion, D. Remote estimation of phytoplankton density in productive waters. *Adv. Limnol. Stuttg.* **2000**, *55*, 121–136.
51. Mahlein, A.-K.; Rumpf, T.; Welke, P.; Dehne, H.-W.; Plümer, L.; Steiner, U.; Oerke, E.-C. Development of spectral indices for detecting and identifying plant diseases. *Remote Sens. Environ.* **2013**, *128*, 21–30. [CrossRef]
52. Ferwerda, J.G.; Skidmore, A.K.; Mutanga, O. Nitrogen detection with hyperspectral normalized ratio indices across multiple plant species. *Int. J. Remote Sens.* **2005**, *26*, 4083–4095. [CrossRef]
53. Peñuelas, J.; Pinol, J.; Ogaya, R.; Filella, I. Estimation of plant water concentration by the reflectance water index WI (R900/R970). *Int. J. Remote Sens.* **1997**, *18*, 2869–2875. [CrossRef]
54. Sims, D.A.; Gamon, J.A. Relationships between leaf pigment content and spectral reflectance across a wide range of species, leaf structures and developmental stages. *Remote Sens. Environ.* **2002**, *81*, 337–354. [CrossRef]
55. Liu, L.-Y.; Huang, W.-J.; Pu, R.-L.; Wang, J.-H. Detection of internal leaf structure deterioration using a new spectral ratio index in the near-infrared shoulder region. *J. Integr. Agric.* **2014**, *13*, 760–769. [CrossRef]
56. Merzlyak, M.N.; Gitelson, A.A.; Chivkunova, O.B.; Rakitin, V.Y. Non-destructive optical detection of pigment changes during leaf senescence and fruit ripening. *Physiol. Plant.* **1999**, *106*, 135–141. [CrossRef]
57. Blackburn, G.A. Spectral indices for estimating photosynthetic pigment concentrations: A test using senescent tree leaves. *Int. J. Remote Sens.* **1998**, *19*, 657–675. [CrossRef]
58. Urbanowicz, R.J.; Meeker, M.; La Cava, W.; Olson, R.S.; Moore, J.H. Relief-based feature selection: Introduction and review. *J. Biomed. Inform.* **2018**, *85*, 189–203. [CrossRef]
59. Cortes, C.; Vapnik, V. Support-vector networks. *Mach. Learn.* **1995**, *20*, 273–297. [CrossRef]
60. Rokach, L.; Maimon, O. Decision trees. In *Data Mining and Knowledge Discovery Handbook*; Springer: New York, NY, USA, 2005; pp. 165–192.
61. Guo, G.; Wang, H.; Bell, D.; Bi, Y.; Greer, K. KNN model-based approach in classification. In Proceedings of the On the Move to Meaningful Internet Systems 2003: CoopIS, DOA, and ODBASE: OTM Confederated International Conferences, CoopIS, DOA, and ODBASE 2003, Catania, Italy, 3–7 November 2003; pp. 986–996.
62. Zhiyuan, H.; Kaidi, C.; Mengke, W.; Chaofan, J.; Te, Z.; Meizi, W.; Pengqiang, D.; Leiming, H.; Lin, Z. Bioactivity of the DMI fungicide mefentrifluconazole against Sclerotium rolfsii, the causal agent of peanut southern blight. *Pest Manag. Sci.* **2023**, *79*, 2126–2134.
63. Damicone, J.; Jackson, K. Factors affecting chemical control of southern blight of peanut in Oklahoma. *Plant Dis.* **1994**, *78*, 482–486. [CrossRef]
64. Rumpf, T.; Mahlein, A.-K.; Steiner, U.; Oerke, E.-C.; Dehne, H.-W.; Plümer, L. Early detection and classification of plant diseases with support vector machines based on hyperspectral reflectance. *Comput. Electron. Agric.* **2010**, *74*, 91–99. [CrossRef]
65. Jiang, J.-B.; Chen, Y.-H.; Huang, W.-J. Using the distance between hyperspectral red edge position and yellow edge position to identify wheat yellow rust disease. *Spectrosc. Spectr. Anal.* **2010**, *30*, 1614–1618.
66. Zhang, J.-C.; Lin, Y.; Wang, J.-H.; Huang, W.-J.; Chen, L.-P.; Zhang, D.-Y. Spectroscopic leaf level detection of powdery mildew for winter wheat using continuous wavelet analysis. *J. Integr. Agric.* **2012**, *11*, 1474–1484. [CrossRef]
67. Zhang, J.; Pu, R.; Loraamm, R.W.; Yang, G.; Wang, J. Comparison between wavelet spectral features and conventional spectral features in detecting yellow rust for winter wheat. *Comput. Electron. Agric.* **2014**, *100*, 79–87. [CrossRef]
68. Li, D.; Cheng, T.; Zhou, K.; Zheng, H.; Yao, X.; Tian, Y.; Zhu, Y.; Cao, W. WREP: A wavelet-based technique for extracting the red edge position from reflectance spectra for estimating leaf and canopy chlorophyll contents of cereal crops. *ISPRS J. Photogramm. Remote Sens.* **2017**, *129*, 103–117. [CrossRef]
69. Yao, Z.; Lei, Y.; He, D. Early visual detection of wheat stripe rust using visible/near-infrared hyperspectral imaging. *Sensors* **2019**, *19*, 952. [CrossRef]
70. Xiao, D.; Pan, Y.; Feng, J.; Yin, J.; Liu, Y.; He, L. Remote sensing detection algorithm for apple fire blight based on UAV multispectral image. *Comput. Electron. Agric.* **2022**, *199*, 107137. [CrossRef]

71. Scholten, J.; Klein, M.; Steemers, A.; de Bruin, G. Hyperspectral imaging-A Novel non-destructive analytical tool in paper and writing durability research. In Proceedings of the Art '05–8th International Conference on Non-Destructive Investigations and Microanalysis for the Diagnostics and Conservation of the Cultural and Environmental Heritage, Lecce, Italy, 15–19 May 2005.
72. Timper, P.; Minton, N.; Johnson, A.; Brenneman, T.; Culbreath, A.; Burton, G.; Baker, S.; Gascho, G. Influence of cropping systems on stem rot (Sclerotium rolfsii), Meloidogyne arenaria, and the nematode antagonist Pasteuria penetrans in peanut. *Plant Dis.* **2001**, *85*, 767–772. [CrossRef]

 *agriculture*

Article

Rapid and Non-Destructive Methodology for Measuring Canopy Coverage at an Early Stage and Its Correlation with Physiological and Morphological Traits and Yield in Sugarcane

Raja Arun Kumar [1,*], Srinivasavedantham Vasantha [1], Raju Gomathi [1], Govindakurup Hemaprabha [2], Srinivasan Alarmelu [2], Venkatarayappa Srinivasa [2], Krishnapriya Vengavasi [1], Muthalagu Alagupalamuthirsolai [1], Kuppusamy Hari [1], Chinappagounder Palaniswami [1], Krishnasamy Mohanraj [2], Chinnaswamy Appunu [2], Ponnaiyan Geetha [1], Arjun Shaligram Tayade [1,3], Shareef Anusha [1], Vazhakkannadi Vinu [2], Ramanathan Valarmathi [2], Pooja Dhansu [4] and Mintu Ram Meena [4]

[1] Division of Crop Production, ICAR-Sugarcane Breeding Institute, Coimbatore 641007, India; vasanthavedantham@yahoo.com (S.V.); r.gomathi@icar.gov.in (R.G.); k.vengavasi@icar.gov.in (K.V.); alagu.m@icar.gov.in (M.A.); k.hari@icar.gov.in (K.H.); palaniswami.c@icar.gov.in (C.P.); geetha.p@icar.gov.in (P.G.); arjun.tayade@icar.gov.in (A.S.T.); anusha.s@icar.gov.in (S.A.)

[2] Division of Crop Improvement, ICAR-Sugarcane Breeding Institute, Coimbatore 641007, India; g.hemaprabha@icar.gov.in (G.H.); s.alarmelu@icar.gov.in (S.A.); v.sreenivasa@icar.gov.in (V.S.); k.mohanraj@icar.gov.in (K.M.); c.appunu@icar.gov.in (C.A.); vinu.v@icar.gov.in (V.V.); valarmathi.r@icar.gov.in (R.V.)

[3] ICAR-National Institute of Abiotic Stress Management, Baramati 413115, India

[4] ICAR-Sugarcane Breeding Institute Regional Station, Karnal 132001, India; pooja@icar.gov.in (P.D.); mr.meena@icar.gov.in (M.R.M.)

* Correspondence: r.arun@icar.gov.in or arunkr.plphy@gmail.com

Citation: Kumar, R.A.; Vasantha, S.; Gomathi, R.; Hemaprabha, G.; Alarmelu, S.; Srinivasa, V.; Vengavasi, K.; Alagupalamuthirsolai, M.; Hari, K.; Palaniswami, C.; et al. Rapid and Non-Destructive Methodology for Measuring Canopy Coverage at an Early Stage and Its Correlation with Physiological and Morphological Traits and Yield in Sugarcane. *Agriculture* **2023**, *13*, 1481. https://doi.org/10.3390/agriculture13081481

Academic Editors: Koki Homma, Jibo Yue, Chengquan Zhou, Haikuan Feng, Yanjun Yang and Ning Zhang

Received: 5 June 2023
Revised: 25 June 2023
Accepted: 17 July 2023
Published: 26 July 2023

Abstract: Screening for elite sugarcane genotypes for canopy cover in a rapid and non-destructive way is important to accelerate varietal/clonal selection, and little information is available regarding canopy cover and leaf production, leaf area, biomass production, and cane yield in sugarcane crop. In the present investigation, the digital images of sugarcane crop by using *Canopeo* software was assessed for their correlation with the physiological and morphological parameters and cane yield production. The results revealed that among the studied parameters, canopy coverage has shown a significantly better correlation with the plant height (0.581 **), leaf length (0.853 **), leaf width (0.587 **), and leaf area (0.770 **) in commercial sugarcane clones. Two-way cluster analysis has led to the identification of Co 0238, Co 86249, Co 10026, Co 99004, Co 94008, and Co 95020 with better physiological traits for higher sugarcane yield under changing climate. Additionally, in another field experiment with pre-breeding, germplasm, and interspecific hybrid sugarcane clones, the canopy coverage showed a significantly better correlation with germination, shoot count, leaf weight, leaf area index, and plant height, and finally with biomass ($r = 0.612$ **) and cane yield ($r = 0.458$ **). It has been found that the plant height, total dry matter (TDM), and leaf area index (LAI) had significant correlation with the cane yield, and the canopy cover data from digital images act as a surrogate for these traits, and further it has been observed that CC had better correlation with cane yield compared to the other physiological traits viz., SPAD, total chlorophyll (TC), and canopy temperature (CT) under ambient conditions. Light interception determined using a line quantum sensor had a significant positive correlation ($r = 0.764$ **) with canopy coverage, signifying the importance of determining the latter in a non-destructive way in a rapid manner and low cost.

Keywords: sugarcane clones; canopy cover; light interception; biomass; cane yield

1. Introduction

Sugarcane is one the most important industrial crops in global agriculture, and it has emerged as a multiproduct crop benefiting producers and consumers [1]. Sugarcane is the second most important industrial crop after cotton in India, occupying about 5 million ha of land with a sugar production of 32.38 metric tons [2]. The sugar industry is the second largest agro-industry in India, and it contributes to 1.1% of the national GDP besides providing for 4% of the population residing in rural areas [3]. Due to the burgeoning population and other constraints (abiotic stress), Ref. [3] the cultivated area of sugarcane will mostly remain static; hence, the only option for the increasing production is to go the vertical way/enhance crop productivity. Sugarcane is a C_4 crop that produces four carbon compounds as the primary product in the carbon assimilation cycle, and it is commonly grown from latitude 36.7° N to 31° S and from sea level to 1000 m of altitude, and generally sugarcane grows slowly during the early part of its growing period compared to other tropical gramineous crops, taking up to 4 months to produce a complete leaf canopy which intercepts nearly all the incoming radiation [4–7], while maize (*Zea mays*) and pearl millet (*Pennisetum glaucum*) normally produce a complete leaf canopy within 2 months of sowing [8–10]. Owing to the slow production of a complete leaf canopy, dry biomass production is slow in sugarcane during the early part of the growth period. A comparison of sugarcane and maize made in Zimbabwe [11] showed that the growth in dry mass was faster at 4 months after sowing. Sorghum (*Sorghum bicolor*) and maize grow at similar rates [12], suggesting that sorghum grows faster than sugarcane. On the other hand, Bull and Glasziou [4] showed that early growth in dry mass in sugarcane is slow in most of the regions, and high yields produced by sugarcane are mainly due to an extended growth period rather than superior photosynthetic efficiency.

Canopy cover is a useful trait for monitoring crop productivity [13], and canopy photosynthesis is greatest when the crop reaches its maximum canopy cover to intercept nearly most of the incident light and absorb the required photosynthetic radiation for photo-biochemical processes and yield formation [14,15]. The most common method for measuring canopy cover is by determining the light interception with a line quantum sensor [16,17]. Shepherd et al. [13] reviewed the notion that this system would be time-consuming and costly, as the measurements should be collected near solar noon [16,18].

Another method involves using drone-based digital image capturing and processing to predict the canopy coverage. However, such a facility may not be equally accessible to all in the scientific community.

In this context, a recently developed method of Oklahoma State University for measuring canopy coverage, called *Canopeo*, which rapidly determines the canopy coverage (%) using digital images, employs an application for iOS (Apple) and Android (Google) devices and Matlab (Mathworks) [19]. *Canopeo* (Oklahoma State University App Center, Stillwater, OK, USA) is an automatic colour threshold (ACT) image analysis tool that analyses pixels based on the red-to-green (R/G) and blue-to-green (B/G) colour ratios and an excess green index [13]. *Canopeo* was accurate and faster at computing canopy cover than other software and is widely being used in many crops such as alfalfa, cover crops, soybean, sorghum, wheat, potatoes, and turf grass (https://canopeoapp.com/, accessed on 10 July 2022).

Canopy cover (CC), leaf area, and biomass production are reported to be the most important physiological components resulting in better cane yield in sugarcane; hence, quantification of canopy cover, which is the primary factor for biomass production, is highly essential, and the later requires a lot of labour, resources, and time through a leaf area measurement by destructive sampling or by light interception method using line quantum sensors or by the drone-based image capturing. Several reports are available on various crops regarding canopy cover by *Canopeo*, and little information is available regarding canopy cover, leaf production, leaf area, and biomass production in sugarcane crop. The robustness of the *Canopeo* tool needs to be validated and compared with data generated from line quantum sensor and leaf area measurements for light interception measurements in sugarcane crop. The ICAR-Sugarcane Breeding Institute, Coimbatore, India, a century-old historical institute known for the "Nobilization of cane", has evolved more than 3500 sugarcane clones, and to sustain sugarcane production the canopy coverage (CC) trait is highly essential for screening climate-resilient sugarcane clones. Therefore, the present investigation was carried out to (i) evaluate sugarcane canopy cover measured with *Canopeo* and with the light interception method using a line quantum sensor to find an association between two different methods and (ii) to analyse the canopy cover in sugarcane including commercial, interspecific hybrids and germplasm clones and to establish its correlation with physiological and morphological parameters, biomass, and cane yield traits in field conditions.

2. Materials and Methods

2.1. Plant Material and Crop Management of Commercial Hybrids of Sugarcane Clones

Sugarcane clones of commercial hybrids types viz., CoM 0265, Co 86249, Co 99004, Co 10026, Co 86010, CoC 671, Co 1148, Co 95020, Co 2001-13, Co 86032, Co 7717, Co740, Co 62175, Co 8371, Co 0218, CoLK 8102, BO 91, Co 775, Co 0212, Co 91010, ISH 100, Co 94008, Co 0238, Co 86011, Co 8338, Co 85019, Co 8208, Co 419, CoV 92102, Co 13006, and Co 8021 (Table 1) were grown at the ICAR-Sugarcane Breeding Institute, Coimbatore ($11°0'34''$ N, $76°55'2''$ E, 430 m above mean sea level), Tamil Nadu, India. Two budded sets, thirty-eight per row of 6.0 m, were planted, and a full dose of phosphorous (P_2O_5) was applied in the furrows before planting as basal fertilization, while nitrogen (N) and potassium (K_2O) were applied in two equal measures at 45 days after planting (DAP) and at full earthing-up (90 DAP). Detrashing of dried leaves was done at 5, 7, and 10 months after planting for proper sunlight penetration. The crop stand was free from significant disease or insect damage. The morpho-physiological data, viz., germination percent, leaf length, leaf width, leaf number, leaf area, shoot thickness, and plant height, were determined by following standard procedure.

Table 1. Sugarcane clones (source: Hemaprabha et al., 2018) [20].

No.	Clone [a]	Maturity	Colour	Sucrose (%)
1	BO 91	Mid late	Yellow purple	16.40
2	Co 10026 *	Early	Pinkish yellow orange	19.42
3	Co 13006	Mid late	Yellow orange	19.15
4	Co 0212 *	Mid late	Purple	19.67
5	Co 0218	Mid late	Yellow purple	20.12
6	Co 0238 *	Early	Golden purple	19.25
7	Co 1148	Mid late	Light purple	15.18
8	Co 2001-13 *	Mid late	Purple	19.03
9	Co 419	Mid late	Dark purple	17.09
10	Co 62175 *	Mid late	Greenish purple	17.35
11	Co 740	Mid late	Yellowish green	17.96
12	Co 7717	Early	Purple	17.90
13	Co 775	Early	Light purple	18.32
14	Co 8021	Mid late	Purple	17.86
15	Co 8208	Early	Purplish pink	17.86
16	Co 8338	Early	Dark purple	18.82
17	Co 8371	Mid late	Green yellow	18.18
18	Co 85019 *	Mid late	Purple	16.39
19	Co 86010 *	Mid late	Yellowish green	18.45
20	Co 86011	Early	Purple	19.98
21	Co 86032 *	Mid late	Reddish pink	19.45
22	Co 86249 *	Mid late	Green yellow with purple tinge	18.82
23	Co 91010 *	Mid late	Yellow green with purple tinge	19.89
24	Co 94008	Early	Purple	18.71

Table 1. *Cont.*

No.	Clone [a]	Maturity	Colour	Sucrose (%)
25	Co 95020	Mid late	Yellowish green	18.79
26	Co 99004	Mid late	Yellowish green	20.00
27	CoC 671	Early	Light purple to purple yellow	21.00
28	CoLk 8102	Mid late	Yellowish purple	18.00
29	CoM 0265	Mid late	Green	19.33
30	CoV 92102 *	Mid late	Purple	19.80
31	ISH 100	Mid late	Light purple green	18.20

[a] Asterisks (*) indicate clones suitable for drought conditions [1].

2.1.1. Germination%, Plant Height, and Shoot Thickness

The number of germinants/row was recorded at 30 DAP, and germination % was derived. The plant height was measured from the base to the top most visible transverse mark on the 60 DAP using a measuring tape and the shoot thickness with a digital vernier calliper (Mitutoyo, Kawasaki, Japan) [21].

2.1.2. Leaf Traits

Leaf area (*LA*) was determined in a non-destructive manner by linear measurement method as mentioned by Montgomery (1911):

$$LA = LBK \left(cm^2 \right) \tag{1}$$

where L = maximum length of length, B = maximum breadth, and K = constant (0.75 based on regression analysis).

2.1.3. Biomass

During the formative stage, the biomass samples were collected in a one-meter square area, and all the samples were oven-dried ($60 \pm 5\,°C$) until a constant weight was reached.

2.1.4. Determination of Canopy Cover in Commercial Hybrids of Sugarcane Crop at Early Formative Phase

The non-destructive method of canopy coverage was recorded at 60 DAP using *Canopeo* software installed in Android mobile phone. *Canopeo* is an application for iOS (Apple, Cupertino, CA, USA) and Android (Google, Mountain View, CA, USA) mobile devices and Matlab (Mathworks, Natick, MA, USA) that can rapidly analyse canopy cover (Figure 1a) from pictures [19]. The accuracy of the CC recorded through *Canopeo* software is 91% (correctly classify pixels as green/true positive), and specificity is 89% (non-green/true negative) as mentioned by the original author of the software from Oklahoma State University [19]. The distance between the mobile and sugarcane plant while recording the measurements is 80 cm. In order to facilitate easy CC recording, a 35-inch length selfie stick was also used for a few taller clones.

The captured image was processed rapidly through the *Canopeo* software immediately after image acquisition on an Android mobile device, and the derived canopy coverage (%) was saved as a separate folder for further analysis. The canopy coverage image was captured by keeping the mobile parallel to the soil [19]. However, in our study, another method (keeping mobile perpendicular to the soil) was also followed along with the standard method (keeping mobile parallel to the soil), and finally, both methods were compared by correlation to identify the best method/position for capturing the image in the sugarcane crop.

(a)

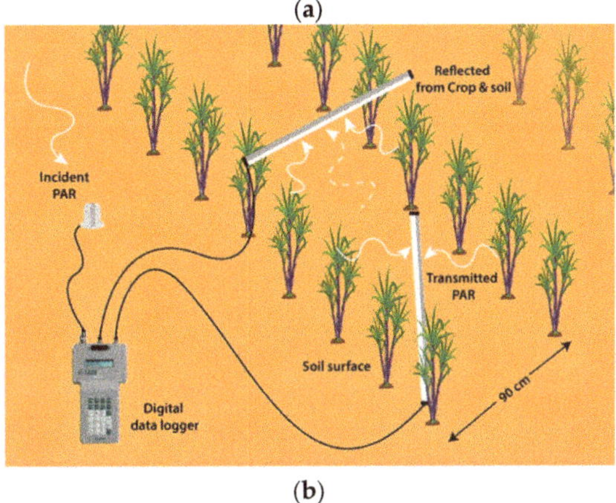

(b)

Figure 1. (**a**) Representative figure of measuring canopy coverage in a sugarcane field. (**b**) Representative figure of measuring light interception in a sugarcane field.

The samples are recorded for the canopy coverage from 9.00 to 11.00 AM, and data are recorded just opposite to the sunlight direction in order to avoid the shade of the observer. The represented values are the average of four observations per replication, i.e., a total of 8 observations per treatment.

2.1.5. Determination of Light Interception in Commercial Hybrids of Sugarcane Crop at Early Formative Phase

The light interception (LI%) was determined using line quantum sensor LI-191SA (LICOR Inc., Lincoln, NE, USA) connected with the LI-1400 a multipurpose datalogger that functions both as a data logging device and a multichannel, auto-ranging meter between 11.00 to 12.00 IST (Figure 1b). The intercepted photosynthetically active radiation (IPAR) for a particular day was computed as the difference between incident PAR at the top and the transmitted PAR received at the bottom of the canopy (the radiation reflected from the crop and soil was also taken into account for deriving the LI%), and the correlation between the CC% and LI % was conducted. Also, the radiation reflected by the soil surface was also determined and finally incorporated for the LI% calculation.

2.2. Plant Material and Crop Management of the Breeding Population, Interspecific Hybrids, and Basic Species Clones of Sugarcane Clones

In order to determine the correlation between the canopy coverage with biomass and cane yield, 38 sugarcane clones including improved breeding population clones (004-73, 04-423, 14-154, 07-520, 04-595, 04-472, 12-127, 97-77, 01-807, 20-158, 20-614, 20-335, 99-45, 98-290, WL 10-40, 14-161, 81 GUK 192, 81 GUK 527, 92 GUK 220, 97 GUK 111, 98 GUK 116, GUK 00-910, GUK 02-91, GUK 06-402, 88 GUK 072, 97 GUK 9, 97 GUK 74, 987 GUK 124, 20-191, 07-776, 99-19, 99-291, 06-013, and 01-803), interspecific hybrids (ISH 107 and ISH 111), and germplasm clones (Kheli and Pathri) were planted in the randomized block design in two replications at the experimental farm of ICAR-Sugarcane Breeding Institute, Coimbatore, India (11°0′34″ N, 76°55′2″ E, 430 m above mean sea level) during the years 2021–2022. The canopy coverage was recorded at 60 DAP using *Canopeo* software and analysed as mentioned in Experiment 1. The germination, shoot thickness, and plant height were determined as mentioned in Experiment 1.

During the formative stage, the fresh biomass samples were collected in a one-meter square area, and all the samples were separated into leaf, sheath, and stem parts and were oven-dried (60 ± 5 °C) for determination of constant weight. The constant dry weight was used for computing the overall dry matter production.

2.2.1. SPAD

Non-destructive chlorophyll estimation was recorded using a SPAD meter (Soil Plant Analysis Development) (atLeaf, Wilmington, Delaware, NC, USA) that computes the chlorophyll content of a leaf by recording the transmission of red light and infrared light at 660 nm and 940 nm, respectively, and converts the reading into a digital signal [1].

2.2.2. Canopy Temperature

The canopy temperature was measured with a thermal imaging infrared camera (FLIR E6) between 11:30 a.m. and 12:00 noon on cloudless days. The image captured was processed through FLIR software (FLIR Tools version 5.1.15036.1001), and the final data were used. The thermal imaging camera was held to view the crop at a 30° angle from horizontal at a 90° angle to the row, with the minimum exposure to the soil, and the emissivity factor of 0.95 was used for the green canopy. Each canopy temperature measurement was the average of three readings at different locations in each clone. Images were registered in the Thermal MSX® mode (FLIR Systems, Wilsonville, OR, USA), and files were saved in standard 14-bit JPG format.

2.2.3. Chlorophyll Fluorescence

Chlorophyll fluorescence (F_v/F_m) was measured in intact sugarcane leaves using a chlorophyll fluorometer (model OS30p, Opti-Sciences, Hudson, NH, USA). The leaves were dark-adapted for 15 min using leaf clips (Opti Sciences), and the (F_v/F_m) readings were recorded by passing a saturating light:

$$\frac{F_v}{F_m} = \frac{F_m - F_o}{F_m} \tag{2}$$

where F_v/F_m = ratio of variable fluorescence to maximal fluorescence, F_m = maximal fluorescence, F_o = minimal fluorescence, and F_v = variable fluorescence of photosystem II [1].

2.2.4. Sucrose, Cane Yield, and CCS

Sugarcane juice was extracted in a crusher with 65% extraction capacity, and the juice quality was analysed as total soluble sugars (TSS) (Brix) and sucrose content (Pol%) according to the standard method [22]. Cane yield was estimated at the 12th month of the crop stage, and the middle 4 rows of canes were harvested and weighed for the plot yield, and the yield per hectare was calculated and expressed as t ha^{-1}. Commercial cane sugar (CCS) was determined and expressed in percentage and t ha^{-1} according to Equations (3) and (4), respectively [21].

$$CCS\% = \frac{\text{Sucrose content} \times 1.022}{\text{TSS} \times 0.292} \tag{3}$$

$$CCS\ (t/ha) = \frac{CCS\% \times \text{Yield}}{100} \tag{4}$$

2.3. Statistical Analysis

Analysis of variance (ANOVA) was performed on the data following the method of Gomez and Gomez (1984) [23], and the least significant difference (LSD) values were calculated at the 5% probability level. Duncan multiple range test (DMRT) was performed to separate significant genotypes, and alphabets were superscripted for easy view. The Pearson-product-moment correlation coefficients (r) between leaf length, leaf width, leaf number, leaf area, shoot thickness, plant height, germination percent, and canopy cover were computed using SAS 9.3 (SAS Institute, Cary, NC, USA) [24]. The scatterplot matrix showing the correlation and frequency counts among the studied parameter, i.e., canopy coverage % (CC), cane yield (CY), dry biomass at formative phase (DWFP), shoot diameter (SD), leaf area (3rd top visible dewlap), leaf area (cm^2), leaf width (cm), leaf length (cm), leaf number (L.No), and plant height (PH), was created using JMP genomics software version 6.1. Two-way cluster analysis with the wards method showing the grouping of sugarcane clones with the studied parameter was also conducted using JMP genomics software. Regression analysis was carried out between the CC and biomass, cane yield, and their corresponding slope (β), and significance was determined following "F" test at 0.05 probability. A correlation diagram displaying the correlation between the studied parameters along with the p-value was conducted through R software version 4.1.3.

3. Results

3.1. Association between Sugarcane Canopy Cover Measured with Canopeo at Different Positions

The canopy cover (CC) data was determined at the early formative phase (60–150 DAP) through *Canopeo*, and both digital images acquired parallel to the ground and perpendicular to the ground were analysed for their association and relevance in sugarcane crop. Among the four phases of the sugarcane crop, the formative phase which starts at 60 DAP is reported to have high relevance to the cane yield; hence, the CC data were recorded at 60 DAP. The correlation between canopy coverage (%) from an image acquired parallel to the ground and perpendicular to the ground is shown in Figure 2a. A significantly better correlation of r = 0.870 ** was observed between the canopy coverage (%) data through images acquired parallel to the ground and perpendicular to the ground of the sugarcane crop. Canopy cover images were taken in properly weeded/weed-free fields to reduce the data error; i.e., the background images of weeds mimic the crop, and this results in an overestimation of CC data. The data revealed a significant linear relationship at 1% probability level between the data captured through two positions.

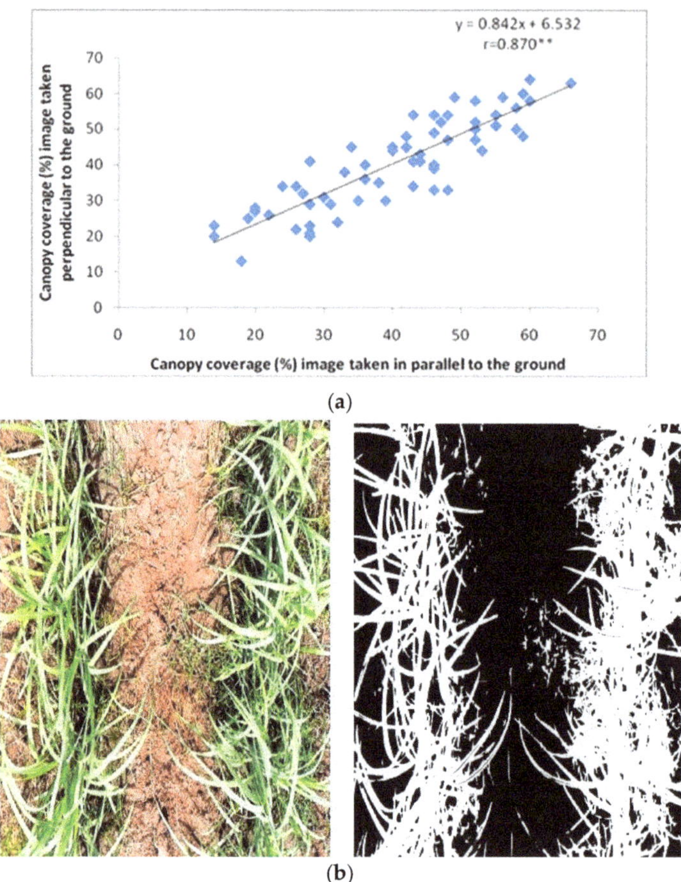

Figure 2. (**a**) Correlation between canopy coverage (%) image acquired parallel to the ground and perpendicular to the ground. ** denotes significant at 1%. (**b**) Canopy coverage (%) images, i.e., original image (left) and classified image (right) of sugarcane clone.

3.2. Association between Sugarcane Canopy Cover Measured with Canopeo and with the Light Interception by PAR (Photosynthetically Active Radiation) Line Quantum Sensor

The light interception (LI) data were recorded simultaneously while capturing the canopy cover images through *Canopeo* using Android mobile. Light interception data were acquired through multi-channelled PAR quantum sensors; i.e., one line quantum sensor was placed diagonally between the rows of sugarcane crop, and another line quantum sensor between and above the crop for measuring the transmitted PAR and reflected PAR simultaneously. Incident PAR measurement was achieved through a point sensor for ease of work.

Further, a significant correlation (r = 0.764 **) was observed between canopy coverage (%) from images acquired in parallel to the ground, and light interception by a line quantum sensor (Figure 3) confirms the accuracy of the CC data of *Canopeo*. A positive coefficient indicates that as the value of the independent variable (canopy cover) increases, the mean of the dependent variable (light interception) also tends to increase. The slope coefficient or β value of the regression was 0.695, and the coefficient represents the mean increase of LI% for every additional increment of CC. In the present regression equation, (Figure 3) for every 0.695 increment in CC, a correspondingly one unit increment in LI was observed, and the model was found statistically significant at 1% probability through the "F" test.

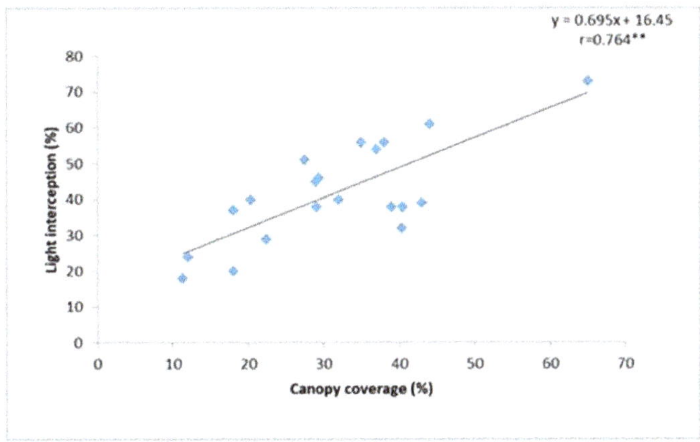

Figure 3. Correlation between canopy coverage (%) image acquired in parallel to the ground and light interception. ** denotes significance at 1%.

3.3. Canopy Cover in Sugarcane Crop, and Its Correlation with Morphological Parameters, Biomass, and Cane Yield Traits in Field Conditions

The results for CC%, germination %, leaf area (cm^2), leaf length (cm), leaf width (cm), leaf number, plant height, and shoot thickness are shown along with LSD at 5% in Table 2.

Table 2. Variation in canopy coverage (CC), germination % (G), and leaf and shoot morphology in sugarcane clones under field condition.

Genotypes	CC%	G%	Leaf Area (cm^2)	LL (cm)	L.No	LW (cm)	PH (cm)	SD (mm)
BO 91	18.86 DEF	50.50 A	584.12 G	76.08 EF	6.00	1.70 F	20.50 EFG	10.33
Co 0212	31.71 ABC	44.95 ABCDE	878.93 BCDEFG	93.25 ABCD	5.83	2.15 E	23.00 BCDEF	11.00
Co 0238	27.80 BCDE	41.48 BCDE	1055.78 ABCD	92.00 ABCD	5.58	2.59 CDEB	21.33 DEFG	12.33
Co 10026	27.95 BCDE	45.00 ABCDE	1088.68 ABCD	93.91 ABC	5.50	2.80 AB	26.08 AB	12.33
Co 1148	18.56 DEF	41.71 ABCDE	818.78 CDEFG	77.83 DEF	6.00	2.30 CDE	19.08 FG	11.00
Co 13006	15.96 F	37.77 DE	772.62 CDEFG	83.16 BCDEF	5.33	2.19 E	21.41 DEFG	12.00
Co 2001-13	24.70 BCDEF	45.32 ABCDE	1031.05 ABCD	88.41 ABCDE	6.08	2.54 CDEB	20.91 DEFG	12.50
Co 62175	26.43 BCDEF	42.82 ABCDE	1000.48 ABCDE	89.33 ABCDE	5.91	2.42 CDEB	25.00 ABCD	12.50
Co 740	22.75 BCDEF	47.03 ABC	865.65 BCDEFG	79.33 CDEF	6.08	2.24 DE	21.25 DEFG	12.66
Co 8021	25.83 BCDEF	46.75 ABC	966.83 ABCDEF	89.25 ABCDE	5.75	2.46 CDEB	24.91 ABCD	12.16
Co 85019	26.03 BCDEF	44.30 ABCDE	1057.50 ABCD	87.08 ABCDE	5.83	2.72 BCD	23.08 BCDEF	13.50
Co 86010	28.71 BCD	41.43 BCDE	946.71 ABCDEFG	80.08 BCDEF	5.91	2.56 CDEB	23.33 BCDE	12.16
Co 86032	16.85 EF	46.52 ABCD	646.62 EFG	71.50 F	5.16	2.17 E	20.50 EFG	11.66
Co 86249	32.80 AB	50.50 A	1124.91 ABC	92.66 ABCD	5.83	2.74 ABC	24.16 ABCDE	12.00
Co 94008	31.25 ABC	49.30 AB	1266.71 A	92.75 ABCD	5.75	3.16 A	23.00 BCDEF	11.66
Co 95020	39.50 A	45.87 ABCDE	1215.46 AB	100.25 A	5.83	2.74 ABC	27.33 A	12.66
Co 99004	26.18 BCDEF	45.83 ABCDE	1144.82 ABC	91.41 ABCDE	5.75	2.70 BCD	25.66 ABC	11.66
CoLk 8102	22.16 BCDEF	38.19 CDE	627.051 FG	84.33 BCDEF	5.75	1.71 F	18.83 G	11.00
CoM 0265	30.88 ABC	37.50 E	912.43 ABCDEFG	95.41 AB	5.41	2.30 CDE	18.50 G	12.00
CoV 92102	20.45 CDEF	41.25 BCDE	720.83 DEFG	80.58 BCDEF	4.66	2.43 CDEB	21.66 CDEFG	11.00
Mean	25.77	44.21	936.3	86.9	5.7	2.4	22.5	11.9
LSD@5%	9.4	7.3	310.4	12.8	NS	0.4	3.4	NS

CC%: Canopy coverage %, G%: Germination %, LL: Leaf length, L.No: Leaf number, LW: leaf width, PH: Plant height, SD: Shoot thickness. NS: Non-significant. n = 3 Values carrying the same letters as superscripts in each column are not significantly different from each other treatment.

3.3.1. Canopy Coverage

The mean canopy coverage (CC%) of the sugarcane crop was 25.7%, and the minimum and maximum CC% were 15.9 and 32.8, respectively (Table 2). Among the studied clones, Co 0212, Co 0238, Co 10026, Co 62175, Co 85019, Co 86010, Co 86249, Co 94008, Co 95020, Co 99004, and CoM 0265 were recorded with better canopy coverage of more than 25%, while Co 13006, BO 91, and Co 1148 indicated a poor CC% of less than 17%.

3.3.2. Plant Height

The mean plant height of the sugarcane crop was 22.5 cm, and the minimum and maximum plant height were 18.5 and 27.33, respectively (Table 2). Among the studied clones, Co 0212, Co 0238, Co 10026, Co 1148, Co 13006, Co 2001-13, Co 62175, Co 740, Co 8021, Co 85019, Co 86010, Co 86032, Co 86249, Co 94008, Co 95020, and Co 99004 were recorded with better plant height of more than the mean plant height (22.5 cm). The clones, viz., Co 95020, Co 10026, Co 62175, and Co 99004, were observed with significantly better plant height compared to other studied clones.

3.3.3. Germination Percentage

The mean data of germination % were 41.25, and the clones, viz., Co 2001-13, Co 86249, Co 94008, Co 10026, CoV 92102, and BO 91, recorded significantly better germination %, while the clones Co 13006, CoLk 8102, and CoM 0265 showed relatively less germination % (<40%) (Table 2).

3.3.4. Leaf Area, Leaf Number, Leaf Length, and Leaf Width

The mean leaf length (3rd top visible dewlap) of the sugarcane clones was 80.58 cm, and the clones, viz., Co 95020, Co 0212, Co 0238, Co 10026, Co 94008, Co 86249, Co 85019, Co 8021, and Co 62175, were recorded with significantly better leaf length (>85 cm) than other clones, while Co 86032, BO 91, Co 740, and Co 1148 observed with poor leaf length (Table 2). The mean leaf no. per shoot was 5.7, and non-significant differences were observed among the studied clones. The mean leaf width (3rd top visible dewlap) of the sugarcane clones was 2.4 cm, and the clones, viz., Co 94008, Co 95020, Co 86249, Co 86010, Co 85019, Co 10026, Co 99004, Co 2001-13, and Co 0238, exhibited better leaf width (>2.4 cm), while BO 91 and CoLk 8102 recorded poor leaf width compared to other clones.

3.3.5. Shoot Thickness

The mean shoot diameter of the sugarcane clones was 11.9 mm, and the clones, viz., Co 85019, Co 740, Co 10026, Co 0238, Co 62175, and Co 95020, were observed with significantly better shoot diameter, while BO 91, Co 0212, Co 1148, and CoLk 8102 recorded less shoot diameter.

3.3.6. Scatterplot Matrix

The scatterplot matrix showing the correlation and frequency counts among the studied parameter, i.e., canopy coverage % (CC), cane yield (CY), dry biomass at formative phase (DWFP), shoot diameter (SD), leaf area (3rd top visible dewlap), leaf area (cm^2), leaf width (cm), leaf length (cm), leaf number (L.No), and plant height (PH), is shown in Figure 4. The canopy coverage % data acquired through the image have shown a significantly better correlation with plant height (0.581 **), leaf length (0.853 **), leaf width (0.587 **), and leaf area (0.770 **).

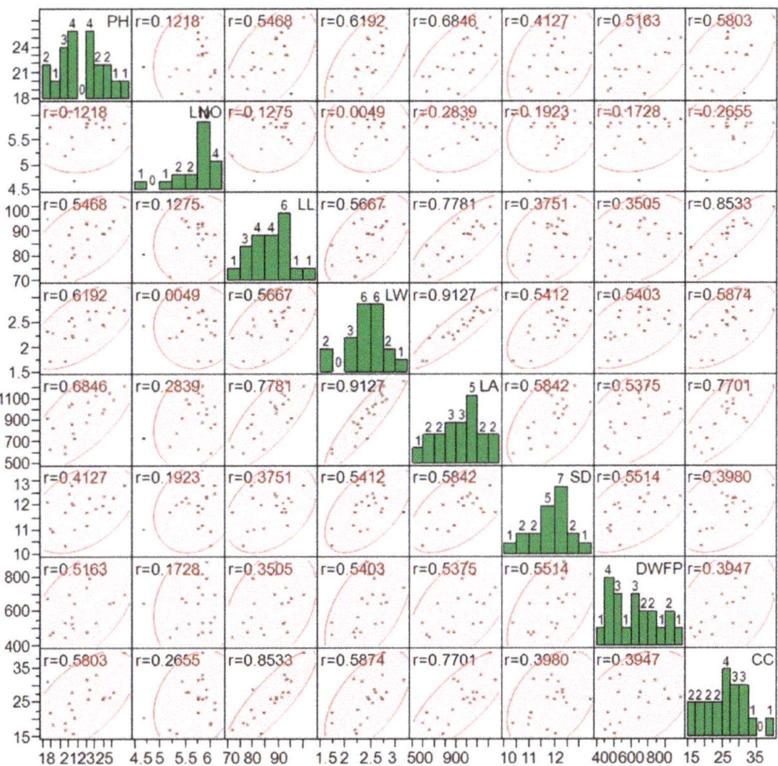

Figure 4. Scatterplot matrix showing the correlation, frequency counts among the studied parameter, i.e., canopy coverage % (CC), dry biomass at formative phase (DWFP), shoot diameter (SD), leaf area (3rd top visible dewlap) (LAI), leaf area (cm^2) (LA2), leaf width (cm) (LW), leaf length (cm) (LL), leaf number (L.No), and plant height (PH).

3.3.7. Two-Way Cluster Analysis

Two-way cluster analysis showing the grouping of sugarcane clones with the studied parameter is shown in Figure 5. The results revealed three distinct clusters: **Cluster I**: BO 91, CoLk 8102, Co 1148, Co 13006, Co 86032, and CoV 92102; **Cluster II**: Co 0212, CoM 0265, Co 0238, Co 86249, Co 10026, Co 99004, Co 94008, and Co 95020; and **Cluster III**: Co 2001-13, Co 85019, Co 62175, Co 86010, Co 8021, and Co 740. Among the three clusters, Cluster I was recorded as relatively lesser in plant height (20.33 cm), leaf number (5.49), leaf length (78.9 cm), leaf width (2.1 cm), total leaf area (695 cm^2), shoot diameter (11.2 mm), dry biomass (510 g dry-weight m^{-2}), and canopy coverage (18%), while Cluster II was observed with better plant height (23.64 cm), leaf length (94 cm), leaf width (2.65 cm), total leaf area (1085 cm^2), and canopy coverage (31%). Cluster III was recorded with better leaf number, dry biomass, and shoot thickness among the studied parameters.

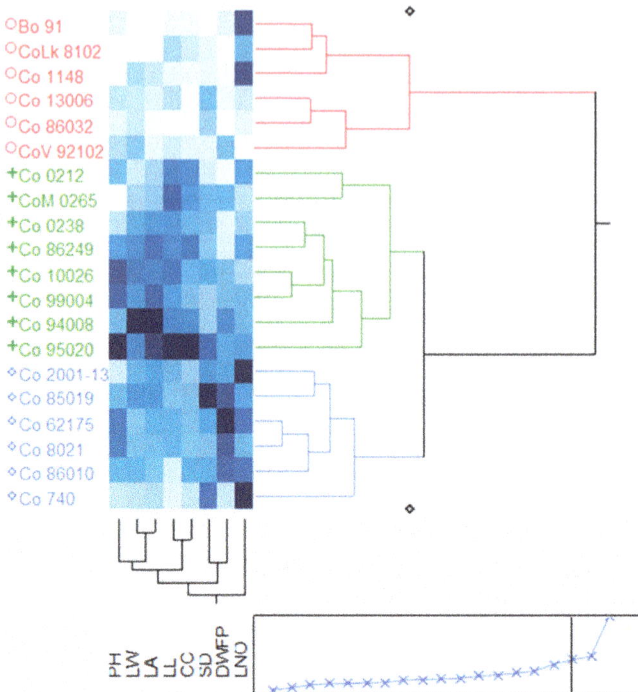

Figure 5. Two-way cluster analysis displaying the ward method grouping of sugarcane clones based on the studied parameter.

3.4. Canopy Cover in Sugarcane Crop (Improved Breeding Population, Interspecific Hybrid, and Basic Germplasm Clone) and Its Correlation with Physiological, Morphological, and Cane Yield Traits

Canopy coverage:

The mean canopy coverage (CC%) of the sugarcane crop was 32.5%, and the minimum and maximum CC% were 17.2 and 49.0, respectively (Table 2). Among the studied clones, the 004-73, 04-423, 14-161, GUK 06-402, and 01-803 were recorded with better canopy coverage of more than 40% (Figure 2b), while 04-595, 97 GUK 111, and 98 GUK 116 indicated a poor CC% of less than 20% (Figure 6).

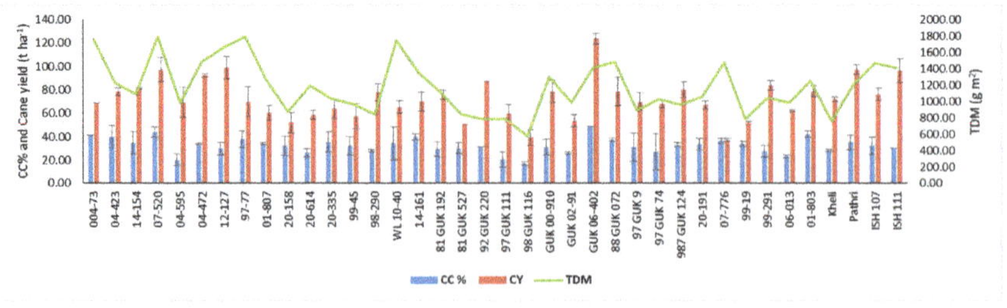

Figure 6. The mean cane canopy (CC%) coverage, cane yield (t/ha), and total dry matter (TDM) of various pre-breeding, germplasm, and interspecific hybrid sugarcane clones.

Cane yield:

The mean, minimum, and maximum cane yield in sugarcane clones (improved breeding population, interspecific hybrid, and basic germplasm) were 72.4, 37.3, and 123.8 (t/ha). Among the studied clones, 07-520, 12-127, GUK 06-402, 987GUK 124, 99-291, Pathri, and ISH 111 recorded better cane yield compared to other clones (Figure 6).

Distribution of the dry matter partitioning and physiological and morphological traits:

The distribution of the dry matter partitioning (LWT: leaf weight, SHWT: sheath weight, STWT: stem weight, and TDM: total dry matter (g .dwt.m^{-2}) and S.Hgt (cm) in the studied sugarcane clones are shown in Figure 7a. The mean LWT, S.Hgt, SHWT, STWT, and TDM were 349, 191, 251,579, and 1179 g .dwt.m^{-2}. The distribution of the SUC%: Juice sucrose, CC: Canopy cover, CCSY: commercial cane sucrose, CT: Canopy temperature (°C) (Figure 7b), CY: cane yield, GC: germination count, LAI: leaf area index, NOC: number of canes, NOL: Number of leaves, SHC: shoot count, and SPAD: Soil Plant Analysis Development ratios are shown (Figure 7b). The mean SUC%, CC, CCSY, CT, CY, GC, LAI, NOC, NOL, SHC, and SPAD were 16, 32, 8.2, 33, 72.4, 12, 2.09, 8.2,77, 61, 23, and 26, respectively. The distribution of the chlorophyll fluorescence (F_v/F_m) and total chlorophyll (mg.cm^{-2}) are shown in Figure 7c. The mean chlorophyll fluorescence (CFL) and total chlorophyll (TC) were 0.609, and 0.0204, respectively.

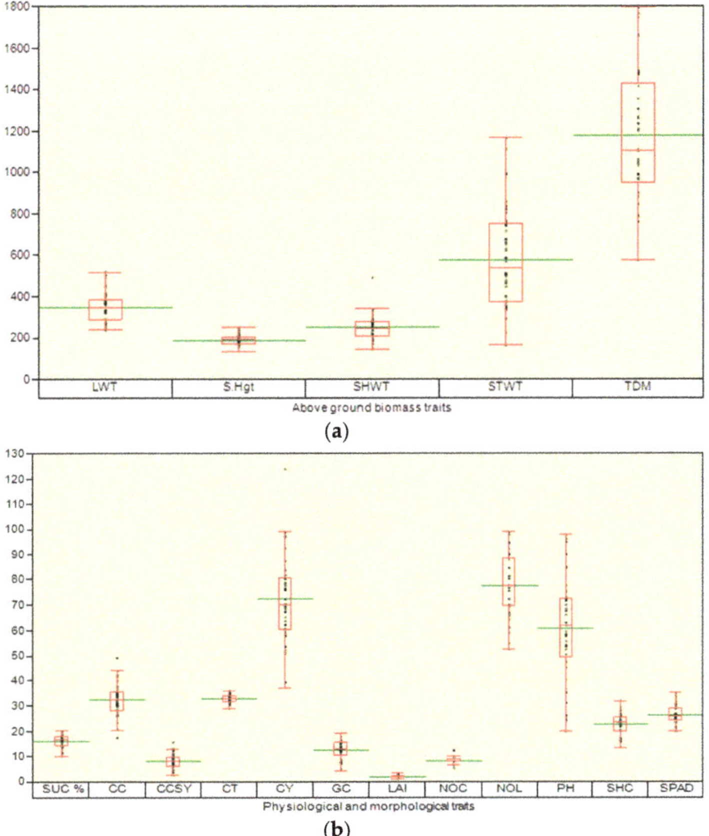

Figure 7. *Cont.*

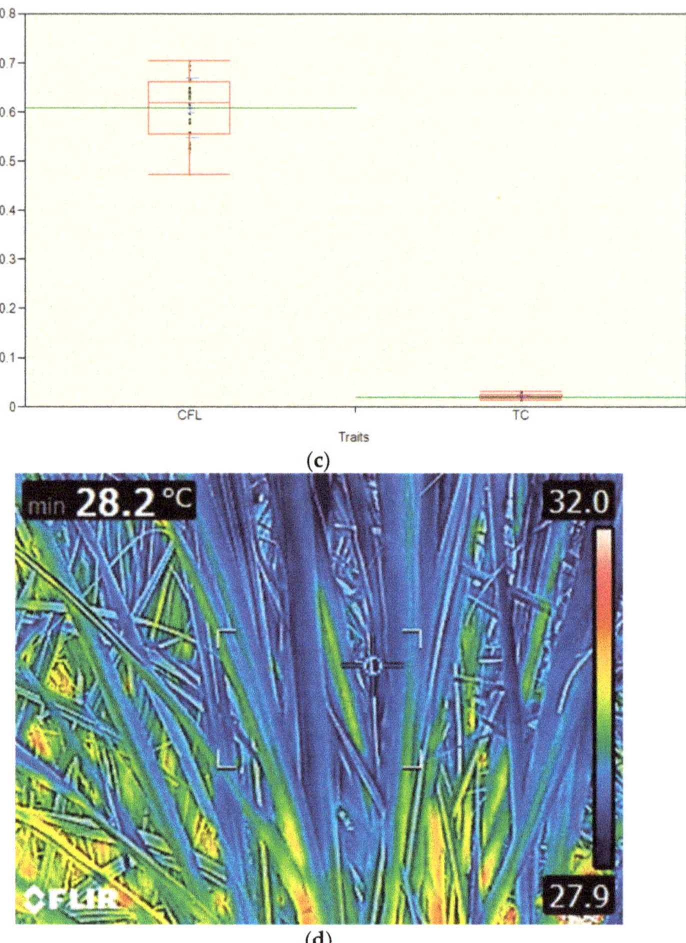

Figure 7. (**a**) Box plot displaying (upper quartile, lower quartile, median, upper extreme, lower extreme, whisker, outlier, and mean (horizontal green line) and the distribution of the dry matter partitioning LWT: leaf weight, SHWT: stem weight: STWT, and TDM: total dry matter (g .dwt.m^{-2}) ratios) and S.Hgt: shoot height (cm). (**b**) Box plot displaying the distribution of the SUC%: Juice sucrose, CC: Canopy cover (%), CCSY: commercial cane sucrose (ton/ha), CT: Canopy temperature (°C) CY: cane yield (ton/ha), GC: germination count/row, LAI: leaf area index, NOC: number of canes/row, NOL: Number of leaves, SHC: shoot count/row, and SPAD: Soil Plant Analysis Development ratios. (**c**) Box plot displaying the distribution of the physiological (CFL: Chlorophyll fluorescence and TC: total chlorophyll content mg·cm^{-2} ratios). (**d**) Thermal image displaying the canopy temperature (°C) sugarcane crop.

3.5. Correlation between Physiological and Morphological with Canopy Coverage

The correlation between physiological and morphological traits and cane yield is shown in Figure 8. Cane yield, SUC%, CCSY, and TDM had significant correlations with canopy coverage (r = 0.46 **, 0.42 **, 0.51 **, 0.62 **, respectively). The germination count, shoot count, initial plant height, and final plant height also had significant correlation with CC (r = 0.46 **, 0.56 **, 0.60 **, and 0.35 *, respectively), while chlorophyll fluorescence (F_v/F_m), canopy temperature (CT), SPAD, and total chlorophyll (TC) showed a non-significant association with CC (r = −0.19, 0.00, 0.00, and −0.01, respectively). The leaf weight and stem weight also revealed a positive correlation (0.36 * and 0.65 **) with CC. Also, the correlation between physiological traits, viz., chlorophyll fluorescence, SPAD, total chlorophyll (TC), and canopy temperature (CT), with cane yield (CY) (r = −0.10 ns, −0.23 ns, −0.23 ns, 0.01 ns, respectively). The leaf area index (LAI), plant height (PH), and total dry matter (TDM) had significant correlations with CC (r = 0.44 **, 0.60 **, and 0.62 **, respectively).

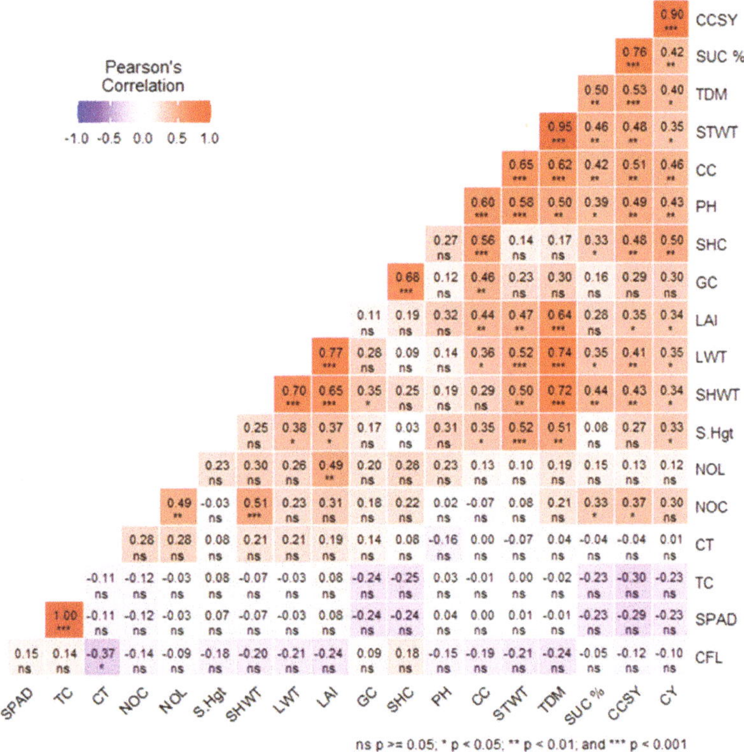

Figure 8. Correlation between various physiological and morphological traits with yield, viz., SPAD, TC:total chlorophyll (mg/cm^2), CT: Canopy temperature (°C), NOC: number of canes/row, NOL: number of leaves, S.Hgt: shoot height (cm) at final stage, SHWT: Sheath weight (g.m^{-2}), LWT: leaf dry weight (g.m^{-2}), LAI: Leaf area index, GC: germination count, SHC: shoot count-early stage, PH: plant height at early stage (cm), CC: canopy coverage (%), STWT: stalk weight (g.m^{-2}), TDM: total above-ground dry matter (g.m^{-2}), SUC%: Juice sucrose%, CCSY: Commercial cane sugar (t ha^{-1}), and CY: Cane yield (t ha^{-1}), *** denotes $p < 0.001$, ** denotes $p < 0.01$; * denotes $p < 0.05$; and ns denotes non-significant $p \geq 0.05$. The intensity of the colour indicates the strength of the correlation.

Based on the canopy coverage data, the simple regression analysis has revealed the prediction of cane yield and total dry matter (TDM) as per Equations (5) and (6) mentioned below:

$$\text{Cane yield} = (33.00 + 1.213 \, \text{CC\%}) \tag{5}$$

$$\text{Total dry matter} = (198.18 + 30.204 \, \text{CC\%}) \tag{6}$$

These simple Equations (5) and (6) suggest the usefulness of the canopy coverage trait in forecasting cane and total dry matter in sugarcane in a rapid and accurate way.

4. Discussion

4.1. Position of the Camera and Comparison of Light Interception (LI) with Canopy Cover (CC)

This paper describes the methodology of canopy cover (CC) determination in sugarcane and its association with dry matter and cane yield. A significant correlation (r = 0.870 **) was observed between canopy coverage (%) from images acquired parallel to the ground and perpendicular to the ground (Figure 2). Also, the canopy coverage (%) from images acquired parallel to the ground had a significantly better correlation with the leaf area thus revealing that the parallel position of the camera for capturing the image for CC% is suited for the sugarcane crop. A significant positive correlation coefficient (0.764 **) between light interception (by line quantum sensor) and canopy cover coverage (by *Canopeo*) indicates (Figure 3) the similarity between both data. There is a strong correlation between the canopy coverage (%) image acquired in parallel to the ground and light interception (Figure 3) by a line quantum sensor which we have observed in sugarcane crops also in the present investigation [13]. Similar to our findings, others have also reported that the ground coverage values estimated from digital images taken above the canopy have been correlated to light interception measurements which are limited by the time of measurements and the presence of clouds [25]. The limitation of this light interception method is that the measurements should be taken in unobstructed sunlight and close to solar noon [26]. The canopy cover methodology for estimating light interception in soybeans has been reviewed to have advantages over the above limitations [27]. In this technique, ground area coverage was determined by digital images taken above the canopy. The canopy coverage values were similar throughout the day and were correlated in a one-to-one relationship with light interception measurements made with a line quantum sensor at solar noon. Shepherd et al. (2018) [13] have also reported a linear relationship between canopy cover measured with pictures (R^2 = 0.94) and videos (R^2 = 0.92) in *Canopeo* and light interception.

4.2. Germination

Better germination of sugarcane sets in the field is often reported to be linked with the early vigour. In our study, the mean germination % was 44.21, and the clones, viz., BO 91, Co 10026, Co 740, Co 8021, Co 86032, Co 86249, Co 94008, CoV 92102, Co 95020, and Co 99004, recorded a significantly better germination (>44) percentage. Several reports [28,29] suggest that, due to the genetic nature and environment, there exists high variability in sett germination percent in sugarcane varieties, and these reports corroborate our findings.

4.3. Leaf Length, Width, Leaf Number, and Leaf Area

The rate of leaf appearance is cultivar-dependent and determined mainly by temperature [30], but it can also be altered by water stress that decelerates expansive growth [31]. Our experiment also confirms the previous report [30] having greater variability in leaf number which suggests that the variation is mainly due to clonal dependence at ambient conditions. Differential thermal requirements for nine sugarcane cultivars to produce the first leaves and the association of the rate of leaf appearance which has the potential for increasing yield [30] are determined by the extent of genetic variation apart from environmental influence.

Leaf arrangement was associated with higher sugar/metric ton, and selection by breeders for higher leaf area indices and for optimum leaf arrangement is suggested [32]. A significant positive correlation between leaf area index and ground cover in potatoes (*Solanum tuberosum*) under different management conditions has been reported [33], and this shows that the canopy coverage (%) image acquired non-destructively through *Canopeo* software using simple android mobile will be useful in determining the leaf area of the sugarcane crop at an early stage rapidly compared to the conventional destructive methods which consume a lot of labour and other resources. *Canopeo* is faster at calculating a canopy cover percentage and can be easily done while in the field. It took less than 1 min to take three pictures or one video per plot, and with the line quantum sensor, data collection time per plot was variable due to cloud cover.

4.4. Dry Matter Production or Biomass

Most of the better-performing sugarcane clones (Co 86010, Co 85019, and Co 10026) identified in this study had a drought-tolerant parent [1], and, in addition to that, Co 62175, Co 85019, and Co 10026 were high-biomass clones. The poor performance of the clones, viz., BO 91, Co 1148, and CoLK 8102, might be plausibly due to their best suitability to subtropical Indian areas rather than a tropical condition in India, while the clones Co 10026, Co 86249, Co 99004, Co 94008, and Co 95020 are of high biomass type with better leaf area production resulting in better canopy coverage.

4.5. Tiller Number and Plant Height

The variability (high tillering and shy tillering) in sugarcane tillering and its relation to sugarcane productivity [34,35] have been widely discussed [36]. It was reported that the number of tillers and plant height at six months after planting are highly correlated with canopy cover ($rg = 0.72$) and canopy height ($rg = 0.69$), respectively [37]. Our results are in line with the previous study of [38] which reported that early biomass had a high genetic correlation with unmanned aerial vehicle (UAV)-derived canopy height (0.810) and canopy cover (0.710). Capturing spectral reflectance by means of UAV at the whole canopy level rather than at the individual leaf level has been an important contributing factor for the high trait-yield correlation compared to individual leaf spot measurements which do not represent whole-canopy dynamics [38].

Canopy cover is a useful trait related to crop growth, water use, and stalk number, and cane yield is considered an important parameter in crop monitoring [37].

4.6. Canopy Temperature and Cane Yield

Canopy temperature, a surrogate trait for canopy conductance, has been previously monitored in sugarcane, and it showed a significant genotypic variation and a strong negative genetic correlation with biomass [39,40]. Our study (Figures 7b,d and 8) observed similar findings ($r = 0.04^{ns}$, $r = 0.01^{ns}$ between CT and TDM, CY, respectively) and also corroborates the report [41] where canopy temperature has been reported as highly negatively correlated with stalk productivity ($r = -0.53^{**}$) under drought stress, while there is a non-significant correlation ($r = -0.18^{ns}$)). Under ambient conditions, the canopy temperature is generally observed with less variability (poor r value with cane yield) among the sugarcane clones, and the better expression of canopy temperature is observed only under abiotic stress conditions where the deeper roots function in tapping of water at deeper zones and support transpiration with subsequent higher canopy conductance, canopy cooling, and better correlation with crop yield.

4.7. Chlorophyll Fluorescence vs. TDM and Cane Yield

Chlorophyll fluorescence is being reported to be one of the best traits for screening the healthy crop under abiotic stress and in the present investigation (Figure 3) where the crop responses under ambient conditions did not translate in the form of TDM ($r = -0.24^{ns}$) and cane yield ($r = -0.10^{ns}$). The chlorophyll fluorescence exhibits a non-significant correlation ($r = 0.02^{ns}$) with cane yield under ambient conditions, while a positive correlation of *Fv/Fm* with stalk productivity ($r = 0.56^{**}$) under drought stress [41].

4.8. SPAD Index vs. TDM and Cane Yield

The SPAD index is a widely discussed trait for the rapid determination of chlorophyll content, and it is also reported to have a significant correlation with crop yield. Chlorophyll is the basic molecule that helps in the absorption of solar radiation and aids in the synthesis of carbohydrates through photosynthesis and finally crop yield. In our experiment, a non-significant correlation of $r = -0.01^{ns}$ and $r = -0.23^{ns}$ was observed between the TDM, cane yield, and SPAD (Figure 3). These findings corroborate the findings of conclusions of Silva (2007) where the SPAD index has been reported to have a non-significant correlation with stalk productivity (r = 0.19 ^{ns}) under ambient condition, while there is a significant correlation (r = 0.36 **). Thus, it reveals that the SPAD index is a useful trait preferably under abiotic conditions, where the stress leads to loss of chlorophyll and declined photosynthesis and reduced synthesis of carbohydrates and finally crop yield.

4.9. Canopy Coverage vs. TDM and Cane Yield

Canopy cover is a valuable trait for monitoring crop productivity [13], and canopy photosynthesis is greatest when the crop reaches its maximum canopy cover to intercept virtually most of the incident light and absorbs the required photosynthetic radiation for photo-biochemical processes and yield formation [14,15]. From the present study, it is clear that the GUK clones had significantly better CC and cane yield compared to other clones. The GUK clones have the parental genes of *Erianthus sps* which is fast growing, with more leaf area, CC, biomass, and cane yield. It has been reported that *Erianthus sps* exhibits vigorous growth, high biomass production, and high tillering ability and is suitable for abiotic stress conditions [42]. Our experiment results (Figure 8) also confirm the previous reports by displaying significant correlations of r = 0.46 **, 0.42 **, 0.51 **, and 0.62 **, respectively, of CY, SUC%, CCSY, and TDM with canopy coverage. The germination count, shoot count, initial plant height, and final plant height also had a significant correlation with CC (r = 0.46 **, 0.56 **, 0.60 **, and 0.35 *, respectively). The leaf weight and stem weight also revealed a positive correlation (0.36 * and 0.65 **) with CC (Figure 8). From the overall discussion, it has been found that the plant height, total dry matter (TDM), and leaf area index (LAI) had significant correlation with the cane yield, and the canopy cover data from digital images act as a surrogate for these traits, and further it has been observed that CC had better correlation (Figure 8) with cane yield compared to the other physiological traits, viz., SPAD, total chlorophyll (TC), and canopy temperature (CT).

4.10. Summary of Key Findings, Advantages, and Limitations

4.10.1. Key Findings

In the present investigation, the canopy covering digital images of sugarcane crop by using *Canopeo* software was evaluated for its correlation with the physiological and morphological parameters and cane yield production. The results show that among the studied parameters, canopy coverage had a significantly better correlation with the plant height (0.581 **), leaf length (0.853 **), leaf width (0.587 **), and leaf area (0.770 **) in commercial-type sugarcane clones (Figure 4).

Canopy cover data of sugarcane clones (improved breeding population, interspecific hybrid, and basic germplasm) also revealed a significant correlation of r = 0.46 **, 0.42 **, 0.51 **, and 0.62 **, respectively, of cane yield (CY), juice sucrose (SUC%), commercial cane sugar yield (CCSY), and total dry matter (TDM) with canopy coverage (CC). The germination count, shoot count, initial plant height, and final plant height also had a significant correlation with CC (r = 0.46 **, 0.56 **, 0.60 **, and 0.35 *), respectively, while the chlorophyll fluorescence, canopy temperature, and SPAD index revealed a poor correlation with TDM and cane yield. The leaf weight and stem weight also revealed a positive correlation (0.36 * and 0.65 **) with CC (Figure 8). From the overall discussion, it has been found that the plant height, total dry matter (TDM), and leaf area index (LAI) had a significant correlation with the cane yield.

4.10.2. Advantages

The traditional light interception method for determining canopy coverage using a line quantum sensor also had a significant positive correlation (r = 0.764 **) with canopy coverage captured through *Canopeo*; thus, our results signify the importance of canopy coverage determination by *Canopeo* in a rapid, non-destructive way and low-cost way.

4.10.3. Limitations

The presence of weeds in the crop field background poses difficulty to classifying or differentiating the crop and weed, and for measuring the canopy coverage, the crop should be in a completely weed-free as well as also detrashed field (removal of senescence leaf) which is more suitable to avoid overestimation of the canopy coverage. If the camera lens were nearer to the crop, then the canopy few crop portions may be excluded in the analysis, and on the other hand, extra sugarcane rows would have been included in the image if the camera lens were positioned at a greater height above the top of the canopy. The vegetation taller than about 2.5 m requires the use of aerial images or special equipment [19].

4.11. Future Research Direction

The canopy coverage data measurement through the drone/unmanned aerial vehicle-based image and the utilization of pix4d software version 4.8.4 and other software are an emerging trend for the determination of canopy coverage which is valuable for yield forecasting in sugarcane and other crops.

5. Conclusions

The present investigation revealed that in commercial sugarcane clones the mean data of canopy cover were 25.77%, and the clones, viz., Co 95020, Co 0212, CoM 0265, and Co 86249, showed significantly better canopy cover % (>30%) compared to other clones, while the clones Co 13006, BO 91, and Co 1148 were observed with poor canopy coverage (<20%). Also, among the observed traits, canopy coverage % data acquired through image have shown a significantly better correlation with the plant height (0.581 **), leaf length (0.853 **), leaf width (0.587 **), and leaf area (0.770 **). Further, there is a significant correlation (r = 0.585 **) between the canopy coverage (%) image acquired in parallel to the ground and the light interception through line quantum sensors which consume more labour and costly instruments/sensors. The canopy coverage (%) image acquired non-destructively through using simple Android mobile will be useful in determining the leaf area of the sugarcane crop at an early stage rapidly compared to the conventional destructive methods which consume a lot of labour and other resources. Two-way cluster analysis revealed that Cluster II comprising Co 0212, CoM 0265, Co 0238, Co 86249, Co 10026, Co 99004, Co 94008, Co 95020 Co 0238, Co 86249, Co 10026, Co 99004, Co 94008, and Co 95020 was observed with better plant height (23.64 cm), leaf length (94 cm), leaf width (2.65 cm), total leaf area (1085 cm^2), and canopy coverage (31%). In a second field experiment with diverse sugarcane clones (improved breeding population, interspecific hybrid, and basic germplasm), the canopy coverage showed a significantly better correlation with biomass (r = 0.612 **) and cane yield (r = 0.458 **), while the chlorophyll fluorescence, canopy temperature, and SPAD index revealed a poor correlation with TDM and cane yield. Light interception determined using a line quantum sensor had a significant positive correlation with canopy coverage signifying the importance of canopy coverage determination in a non-destructive way.

Author Contributions: Project formulation and execution of the experiment: R.A.K., S.V., G.H., K.V., S.A. (Shareef Anusha), S.A. (Srinivasan Alarmelu), K.M. and V.S.; Canopy cover data analysis: R.A.K., P.G. and C.P.; Data analysis: R.A.K. and M.A.; Physiological analysis: R.A.K., K.H., V.V. and R.V., drafting of the manuscript: R.A.K., V.S., K.V., C.A., A.S.T., R.G., P.D. and M.R.M. All authors have read and agreed to the published version of the manuscript.

Funding: This research received no external funding.

Institutional Review Board Statement: Not applicable.

Data Availability Statement: The data presented in this study are available on request from the corresponding author.

Acknowledgments: The authors are thankful to the Director, ICAR-Sugarcane Breeding Institute, Coimbatore, for the constant encouragement and support for carrying out the research work.

Conflicts of Interest: The authors declare no conflict of interest.

References

1. Arun kumar, R.; Vasantha, S.; Tayade, A.S.; Anusha, S.; Geetha, P.; Hemaprabha, G. Physiological Efficiency of Sugarcane Clones Under Water-Limited Conditions. *Trans. ASABE* **2020**, *63*, 133–140. [CrossRef]
2. Anonymous. Sugarcane and Molasses Production at a Glance. *Coop. Sugar* **2019**, *50*, 44–45.
3. Bakshi, R.; Karuppaiyan, R. Status Paper Challenge and Research Priorities for Sugarcane Breeding Programme in India. In Proceedings of the Conference: Annual Convention (2017) and Technical Expo, Lucknow, India, 12–13 May 2017.
4. Bull, T.A.; Glasziou, K.T. Sugarcane. In *Crop Physiology: Some Case Histories*; Evans, L.T., Ed.; Cambridge University Press: Cambridge, UK, 1975; pp. 51–72.
5. Thompson, G.D. *The Growth of Sugarcane Variety N14 at Pongola*; Mount Edgecombe Research Report; South African Sugar Association: Mount Edgecombe, South Africa, 1991; Volume 7.
6. Robertson, M.J.; Wood, A.W.; Muchow, R.C. Growth of Sugarcane Under High Input Conditions in Tropical Australia: I. Radiation Use, Biomass Accumulation and Partitioning. *Field Crop. Res.* **1996**, *48*, 11–25. [CrossRef]
7. Allison, J.C.S.; Pammenter, N.W.; Haslam, R.J. Why Does Sugarcane (*Saccharum* sp. Hybrid) Grow Slowly? *S. Afr. J. Bot.* **2007**, *73*, 546–551. [CrossRef]
8. Allison, J.C.S. Effect of Plant Population on the Production and Distribution of Dry Matter in Maize. *Ann. Appl. Biol.* **1969**, *63*, 135–144. [CrossRef]
9. Wilson, J.H.; Clowes, M.S.J.; Allison, J.C.S. Growth and Yield of Maize at Different Altitudes in Rhodesia. *Ann. Appl. Biol.* **1973**, *73*, 77–84. [CrossRef]
10. Begg, J.E. The Growth and Development of a Crop. of Bulrush Millet (Pennisetum Typhoides S. & H.). *J. Agric. Sci.* **1965**, *65*, 341–349.
11. Allison, J.C.S.; Haslam, R.J. Comparison of Growth of Sugarcane and Maize. *Zimbabwe J. Agric. Res.* **1982**, *20*, 119–127.
12. Muchow, R.C.; Davis, R. Effect of Nitrogen Supply on the Comparative Productivity of Maize and Sorghum in a Semi-arid Tropical Environment: II. Radiation Interception and Biomass Accumulation. *Field Crop. Res.* **1988**, *18*, 17–30. [CrossRef]
13. Shepherd, M.J.; Lindsey, L.E.; Lindsey, A.J. Soybean Canopy Cover Measured with Canopeo Compared with Light Interception. *Agric. Environ. Lett.* **2018**, *3*, 180031. [CrossRef]
14. Lee, C.D. Reducing Row Widths to Increase Yield: Why It Does Not Always Work. *Crop. Manag.* **2006**, *5*, 18. [CrossRef]
15. Wells, R. Soybean Growth Response to Plant Density: Relationships Among Canopy Photosynthesis, Leaf Area, and Light Interception. *Crop. Sci.* **1991**, *31*, 755–761. [CrossRef]
16. De Bruin, J.L.; Pedersen, P. New and Old Soybean Cultivar Responses to Plant Density and Intercepted Light. *Crop. Sci.* **2009**, *49*, 2225–2232. [CrossRef]
17. Hankinson, M.W.; Lindsey, L.E.; Culman, S.W. Effect of Planting Date and Starter Fertilizer on Soybean Grain Yield. *Crop. Forage Turfgrass Mgmt.* **2015**, *1*, 1–6. [CrossRef]
18. Adams, J.E.; Arkin, G.F. A Light Interception Method for Measuring Row Crop. Ground Cover. *Soil Sci. Soc. Am. J.* **1977**, *41*, 789–792. [CrossRef]
19. Patrignani, A.; Ochsner, T.E. Canopeo: A Powerful New Tool for Measuring Fractional Green Canopy Cover. *Agron. J.* **2015**, *107*, 2312–2320. [CrossRef]
20. Hemaprabha, G. Sugarcane Varieties for Abiotic Stress Tolerance. In *Training manual on "Sugarcane Cultivation in Biotic and Abiotic Stress Conditions"*; Nair, V., Gopalasundaram, P., Shanty, T.R., Prathap, P.D., Eds.; Icar-Sugarcane Breeding Institute—641 007: Tamil Nadu, India, 2008.
21. Vasantha, S.; Raja, A.K.; Vengavasi, K.; Tayade, A.S.; Shareef, A.; Govindakurup, H. Sugarcane stalk traits for high throughput phenotyping in restricted irrigation regimes. *Sugar Tech.* **2023**, *25*, 788–796. [CrossRef]
22. Meade, G.P.; Chen, J.C.P. *Cane Sugar Hand Book*, 10th ed.; John Wiley & Sons: New York, NY, USA, 1977.
23. Gomez, K.A.; Gomez, A.A. *Statistical Procedures for Agricultural Research*; Wiley: Hoboken, NJ, USA, 1984.
24. SAS Institute. *The SAS System for Windows, Release 9.3*; SAS Institute: Independence, KS, USA, 2011.
25. Gonias, E.D.; Oosterhuis, D.M.; Bibi, A.C.; Purcell, L.C. Estimating Light Interception by 519 Cotton Using a Digital Imaging Technique. *Am. J. Exp. Agric.* **2012**, *2*, 1–8.
26. Board, J.E.; Kamal, M.; Harville, B.G. Temporal Importance of Greater Light Interception to Increase Yield in Narrow-Row Soybean. *Agron. J.* **1992**, *84*, 575–579. [CrossRef]
27. Purcell, L.C. Soybean Canopy Coverage and Light Interception Measurements Using Digital Imagery. *Crop. Sci.* **2000**, *40*, 834–837. [CrossRef]
28. Humbert, R.P. *The Growing of Sugar Cane*; Elsevier: Amsterdam, The Netherlands, 1968.
29. Donaldson, R.A. Season Effects on the Potential Biomass and Sucrose Accumulation of Some Commercial Cultivars of Sugarcane. Ph.D. Thesis, University of KwaZulu-Natal, KwaZulu-Natal, South Africa, 2009; p. 200.
30. Bonnett, G.D. Rate of Leaf Appearance in Sugarcane, Including a Comparison of a Range of Varieties. *Funct. Plant Biol.* **1998**, *25*, 829–834. [CrossRef]
31. Singels, A.; Inman-Bamber, N.G. Modelling Genetic and Environmental Control of Biomass Partitioning at Plant and Phytomer Level of Sugarcane Grown in Controlled Environments. *Crop. Pasture Sci.* **2011**, *62*, 66–81. [CrossRef]
32. Irvine, J.E. Relations of Photosynthetic Rates and Leaf Canopy Characters to Sugarcane Yield. *Crop. Sci.* **1975**, *15*, 671–676.

33. Boyd, N.S.; Gordon, R.; Martin, R.C. Relationship Between Leaf Area Index and Ground Cover in Potato Under Different Management Conditions. *Potato Res.* **2002**, *45*, 117–129. [CrossRef]
34. Nair, N.V.; Sreenivasan, T.V. Studies on Some Early Attributes of Sugarcane (*Saccharum officinarum.*) in Relation to Yield and Yield Components. *Indian J. Genet. Plant Breed.* **1990**, *50*, 354–358.
35. Kumar, N.; Sow, S.; Rana, L.; Singh, A.K.; Kumar, A.; Kumar, A.; Singh, S.N. Physio-Agronomic Performance of Sugarcane (*Saccharum* spp. Hybrid Complex) Genotypes Under Various Planting Geometry. *Ann. Agric. Res. New S.* **2023**, *44*, 93–98.
36. Kapur, R.; Duttamajumder, S.K.; Rao, K.K. A Breeder's Perspective on the Tiller Dynamics in Sugarcane. *Curr. Sci.* **2011**, *100*, 183–189.
37. Cholula, U.; da Silva, U.J.A.; Marconi, T.; Thomasson, J.A.; Solorzano, J.; Enciso, J. Forecasting yield and lignocellulosic composition of energy cane using unmanned aerial systems. *Agronomy* **2020**, *10*, 718. [CrossRef]
38. Natarajan, S.; Basnayake, J.; Wei, X.; Lakshmanan, P. High-Throughput Phenotyping of Indirect Traits for Early-Stage Selection in Sugarcane Breeding. *Remote Sens.* **2019**, *11*, 2952. [CrossRef]
39. Basnayake, J.; Lakshmanan, P.; Jackson, P.; Chapman, S.; Natarajan, S. Canopy Temperature: A Predictor of Sugarcane Yield for Irrigated and Rainfed Conditions. In Proceedings of the International Society of Sugar Cane Technologists, Chien Mai, Thailand, 5–8 December 2016; pp. 50–57.
40. Chapman, S.; Merz, T.; Chan, A.; Jackway, P.; Hrabar, S.; Dreccer, M.; Holland, E.; Zheng, B.; Ling, T.; Jimenez-Berni, J. Pheno-Copter: A Low-Altitude, Autonomous Remote-Sensing Robotic Helicopter for High-Throughput Field-Based Phenotyping. *Agronomy* **2014**, *4*, 279–301.
41. Silva, M.D.A.; Jifon, J.L.; Da Silva, J.A.; Sharma, V. Use of Physiological Parameters as Fast Tools to Screen for Drought Tolerance in Sugarcane. *Braz. J. Plant Physiol.* **2007**, *19*, 193–201. [CrossRef]
42. Valarmathi, R.; Mahadevaswamy, H.K.; Ulaganathan, V.; Appunu, C.; Karthigeyan, S.; Pazhany, A.S. Assessing the Genetic Diversity and Population Structure of World Germplasm Collection of *Erianthus arundinaceus* (Retz.) Jeswiet Using Sequence-Related Amplified Polymorphic Markers. *Sugar Tech.* **2022**, *24*, 438–447. [CrossRef]

agriculture

Article

Technology of Automatic Evaluation of Dairy Herd Fatness

Sergey S. Yurochka [1], Igor M. Dovlatov [1,*], Dmitriy Y. Pavkin [1], Vladimir A. Panchenko [2], Aleksandr A. Smirnov [1], Yuri A. Proshkin [1] and Igor Yudaev [3]

[1] Federal State Budgetary Scientific Institution, Federal Scientific Agroengineering Center VIM (FSAC VIM), 1st Institutsky Proezd 5, 109428 Moscow, Russia; yurochkasr@gmail.com (S.S.Y.); dimqaqa@mail.ru (D.Y.P.); alexander8484@inbox.ru (A.A.S.); ledsk@bk.ru (Y.A.P.)
[2] Department of Theoretical and Applied Mechanics, Russian University of Transport, 127994 Moscow, Russia; pancheska@mail.ru
[3] Energy Department, Kuban State Agrarian University, 350044 Krasnodar, Russia; etsh1965@mail.ru
* Correspondence: dovlatovim@mail.ru

Abstract: The global recent development trend in dairy farming emphasizes the automation and robotization of milk production. The rapid development rate of dairy farming requires new technologies to increase the economic efficiency and improve production. The research goal was to increase the milk production efficiency by introducing the technology to automatically assess the fatness of a dairy herd in 0.25-point step on a 5-point scale. Experimental data were collected on the 3D ToF camera O3D 303 installed in a walk-through machine on robotic free-stall farms in the period from August 2020 to November 2022. The authors collected data on 182 animals and processed 546 images. All animals were between 450 and 700 kg in weight. Based on the regression analysis, they developed software to find and identify the main five regions of interest: the spinous processes of the lumbar spine and back; the transverse processes of the lumbar spine and the gluteal fossa area; the malar and sciatic tuberosities; the tail base; and the vulva and anus region. The adequacy of the proposed technology was verified by means of a parallel expert survey. The developed technology was tested on 3 farms with a total of 1810 cows and is helpful for the non-contact evaluation of the fatness of a dairy herd within the herd's life cycle. The developed method can be used to evaluate the tail base area with 100% accuracy. The hungry hole can be determined with a 98.9% probability; the vulva and anus area—with a 95.10% probability. Protruding vertebrae—namely, spinous processes and transverse processes—were evaluated with a 52.20% and 51.10% probability. The system's overall accuracy was assessed as 93.4%, which was a positive result. Animals in the condition of 2.5 to 3.5 at 5–6 months were considered healthy. The developed system makes it possible to divide the animals into three groups, confirming their physiological status: normal range body condition, exhaustion, and obesity. By means of a correlation dependence equal to R = 0.849 (Pearson method), the authors revealed that animals of the same breed and in the same lactation range have a linear dependence of weight-to-fatness score. They have developed an algorithm for automated assessment of the fatness of animals with further staging of their physiological state. The economic effect of implementing the proposed system has been demonstrated. The effect of increasing the production efficiency of a dairy farm by introducing the technology of automatic evaluation of the fatness of a dairy herd with a 0.25-point step on a 5-point scale had been achieved. The overall accuracy of the system was estimated at 93.4%.

Keywords: dairy cows; body condition score; 3D TOF sensor; non-contact evaluation; recognize area of interest

Citation: Yurochka, S.S.; Dovlatov, I.M.; Pavkin, D.Y.; Panchenko, V.A.; Smirnov, A.A.; Proshkin, Y.A.; Yudaev, I. Technology of Automatic Evaluation of Dairy Herd Fatness. *Agriculture* **2023**, *13*, 1363. https://doi.org/10.3390/agriculture13071363

Academic Editors: Jibo Yue, Chengquan Zhou, Haikuan Feng, Yanjun Yang and Ning Zhang

Received: 31 May 2023
Revised: 3 July 2023
Accepted: 4 July 2023
Published: 8 July 2023

1. Introduction

Over the last few decades, the global trend in dairy farming has been to automatize and robotize milking processes on commercial farms [1,2]. The common average production period of dairy animals is 3.5 lactations [3]. Due to the rapid development of dairy farming, new technologies are increasingly required to achieve a higher economic efficiency and achieve an improved production [4–6]. On the one hand, intensive production results in an increased milk yield of a cow; on the other hand, intensive production leads to the rapid deterioration of dairy cows—i.e., a reduction in the number of lactations [7]. The reduction of the production life of dairy animals also depends on the premature culling of animals that have a high or low body condition score. Lack of a normal body condition score during lactation is primarily due to dietary deficiencies [7]. Another negative consequence is culling of animals due to poor body condition because an increased body condition score reduces fertility and thereby extends the service period.

A body condition score (BCS) evaluation is important in technological milk production. First and foremost, the BCS score is used to place animals within productivity groups and determine their status. In Russian dairy farm conditions, veterinarians and livestock breeding technicians rotate animals into production groups once a month, provided the milk production technology is well established. The body condition score helps make a decision individually for each cow, based on her current physiological condition, rather than simply on accepted technological norms. In intensive milk production, dairy cows are divided into 5 main groups: group 1, the step-ladder milk yield increasing group, includes new cows from 6 to 100 days after calving, and also cows with a daily milk yield of more than 24 kg per head per day. The total productivity of this group of animals should not be lower than 6000 kg per head per year. The main objectives of the group are: quality feeding with full-fat mixes and good care to achieve the peak milk production by day 40–50; elimination of post-calving complications to inseminate the animals on day 65. During this period, the animals give up to 65–70% of their milk volume during the lactation period. High-yielding cows are transferred to group 1 and should be in group 2, but they need increased nutrition according to milk yield and body condition score. The typical fatness score for group 1 is 3.5 to 3.25 from day 6 to day 30, and 3 to 2.75 from day 31 to day 100. The normal decrease of the body condition score of cows in group 1 is due to an intensive milk production, which requires a large amount of energy. The energy expended cannot be fully compensated by the energy gained from feeding. Therefore, it results in a natural decrease of the body condition score. Maria Ledinek et al., in a study [8], showed that during the calving period, the body condition score decreases, and body fat reserves provide for an increased milk production. By 40–65 days after lactation, animals should be milked as often as possible, and the body condition score should not be reduced by more than 0.5. During this period, the cow consumes up to 12 kg of high energy feed. Insemination takes place when the animal is at peak production and consumes the highest amount of feed and the fatness score is within 3 points. At the same time, the animal's body condition score may not deviate by more than ±0.25 points.

Group 2—milking cows from 101 to 305 days after calving with 24 to 16 kg of milk per head per day. The main objective for the animals in this group is to ensure that the milk yield does not fall by more than 9% per month, and to increase the body condition to 3–3.5 fatness points.

Group 3—milking cows from 101 to 305 days after calving with a milk yield below 16 kg/head/day. The main task for this group is to prevent diseases, correct body weight to a fatness of 3.5–3.75 points and prepare the animals for drying off.

When animals are in the second and third physiological group, from 101 to 305 days after calving, it is necessary to monitor their condition. A cow should have 3.5–3.75 by the start of the dry period. If it is under-conditioned, it should be kept in the first or second group and its milk production should be ignored. Otherwise, under-conditioning can lead to complications during parturition or at the beginning of the next lactation [9]. Overconditioning of a cow above 3.75 will result in an increase in fetal weight. As the cow's weight increases, so does the calf's weight. An increased calf weight at calving causes birth complications and injuries that are equally detrimental to the cow and calf.

Group 4—the first 45 days from day 306 after drying off. During this period, no adjustment is made to the animal's body condition score. It is assumed that the animal already has a body condition score of 3.5–3.75.

Group 5 is the maternity group. The animals are kept 15 days before calving and 5 days after calving. During this period, no adjustment is made to the body condition score of the animals.

The fatness assessment of dairy cows is not only a valuable indicator for evaluating the quality of feeding and the response of the animal to feed, but is also an indirect indicator of its reproductive function. Dairy performance correlates with feed intake. An increase in milk production is associated with a decrease in fertility. During peak lactation, cows require 3.5 times more protein and energy for milk synthesis than protein and energy for life support, as lactation and calf feeding have a higher biological priority than body weight gain and fecundity. At peak milk production, quit estrus and overcalving are the most significant problems. A negative energy balance, which is also affected by decreased body condition dynamics, results in a delayed onset of first heat and ovulation after calving in underfed cows, reduced probability of conception after first insemination, negative effects on follicle growth, corpus luteum function, oocyte quality, impaired intrauterine development, and embryo survival and growth [10].

Cows with a body condition at day 60 of 3.25–2.75 have a 67% chance of conception, and those with a body condition below 2.75 have a 44% chance [11].

Mohamed A.B. Mandour, in a study [12], found that high fatness in first-year heifers increases the risk of ketosis to 3.71%, which is twice as high as in adult cows. The study mentions that cows with a high body condition score consume less feed than cows with a normal body condition score and have a high negative energy balance due to a higher concentration of fatty acids in the plasma, which is associated with an increased risk of ketosis.

Thinawanga Joseph Mugwabana et al., in a study [13], found no relationship between the fatness of cows and the calving rate.

Wynnton C. Meteer [14] found in their study that animals given 70% of the required feed energy had more embryos at the next insemination and a higher probability of insemination than animals that received an energy excess of 130% of the norm. Changing the level of feeding in animals in groups 4 and 5 (middle and late stage before calving) did not significantly affect the amount of pregnancy hormones excreted in the blood.

These studies confirm the above information that the main management of feeding, control, and changing the body condition score of cows should be done during lactation, in animals in groups 1 and 2, to increase the probability of reproductive success in the next insemination of animals.

Poczta W. et al., in their study [15], established a relationship between cow fatness and the likelihood of subclinical ketosis, where cows with a fatness score ≥ 3.25 were more susceptible to the disease than lean cows with a fatness score ≤ 3.

Vanholder T. et al., in a study [16], found a relationship between the body condition score of dairy cows and weight loss within 30–40 days after calving. Of the 47 cows studied, 37 cows lost ≥ 0.75 BCS points at 14 days post calving, and 10 cows lost ≤ 0.75 BCS points. Weight loss is associated with a negative energy balance in the cow after calving and subsequent mobilization of body reserves for recovery.

In [17], the authors found a correlation between the propensity for metritis and the BCS of cows ≤ 3. In [18], the authors evaluated the relationship between BCS points during the transition period and the development of disease and changes in milk yield. A total of 232 cows were assessed and the fatness was scored from 1 to 5 in a 0.25 step. After a blood test, a conclusion on the health of the animals was made. Changes in the body condition of dairy cows using the BCS scale were measured at 21 days before calving and 21 days after calving. The percentage of cows that increased BSC (fatness) during this period was 28%, lost BCS—22%, and retained BCS—50%. Additionally, 18% of the cows that lost BCS during this period had health problems compared to the cows that retained the BCS points.

Furthermore, 28% of the cows that had an increased BCS were less likely to have subclinical ketosis.

The results confirmed that developing ketosis can be detected in an automatic, non-contact method. An alternative way of detecting ketosis is presented in studies [19–21], where blood tests were required to detect disease. On large farms with more than 200 milking herds, the continuous active assessment of animal health by blood testing is not possible, due to the lack of specialists, the time-consuming process, and the need for laboratory equipment. The BCS can be evaluated both manually and automatically.

Studies [18–22] describe the manual method of BCS evaluation. Study [23] gives a detailed review of automatic systems for automatic BCS evaluation. Study [24] describes the development of an automatic BCS evaluation system using a deep learning neural network algorithm using a convolutional neural network. The researchers achieved a recognition accuracy of 94% at a step of 0.5, and 78% at a step of 0.25. In [25], the authors used a convolutional neural network (CNN) to evaluate BCS. The accuracy of the system was assessed using the Kappa index and was within a moderate range (values between 0.41 and 0.60). In [26], the authors also used a convolutional neural network (CNN) to evaluate BCS. The accuracy of the results obtained in the study was 78%, indicating a successful real-time classification. In [27], the authors used the point cloud method to evaluate BCS. Experiments show that the proposed BCS evaluation model achieved an accuracy of 49, 80, and 96% within a deviation of 0, 0.25, and 0.50 points, respectively.

In [28], a dynamic background model (Gaussian Mixture Model, GMM) was used to distinguish the cow from the background. Subsequent Image Processing Algorithms have made it possible to automatically obtain reliable images, to find areas of interest, and to extract image elements without any manual intervention. With 5-fold cross checking, the model has achieved an average accuracy of 56% with a 0.125-point variance, 76% with a 0.25-point variance, and 94% with a 0.5-point variance.

Having studied the modern experience of the world community on the automation of BCS evaluation, our team had set a goal and fully fulfilled the tasks on the development of technology of an automatic system of BCS evaluation. The aim of the research was to improve the production efficiency of dairy farms by implementing the technology of an automatic BCS evaluation of a dairy herd with a 0.25-point step on a 5-point scale.

The main approach we used in developing the technology was to minimize the use of neural network algorithms to find areas of interest. This decision was based on the experience of the team [29,30]. Training neural networks was a labor-intensive and time-consuming process. For example, a trained neural network for standardized breeds of EU countries—Holstein, Brown Latvian, Swiss, etc.—will give a big error during a BCS evaluation of Black-Motley Holstein, Kalmyk breeds, etc. To minimize the error, it was necessary to retrain the neural network. The method we proposed is based on the study of standardized breeds of EU countries to adjust the model and to carry out a further BCS evaluation, avoiding the training of the neural network algorithm on each farm.

Thus, the research resulted in the development of a universal automatic system capable of estimating the BCS of an animal with high accuracy (more than 90%) at a step of 0.25. The developed system was intended for the implementation in automated and robotic free-stall farms. The system was designed to evaluate the BCS (fatness) of animals and to provide recommendations for a wider range of functions to be carried out by a specialist.

2. Materials and Methods

2.1. Farm, Field Data Collection

In earlier studies, we had already achieved a result of algorithmic evaluation, where the system evaluated a fatness score between 2 and 4 with a 10% error, while scores 1 and 5 were evaluated with a 25% error (the results of the automatic evaluation were checked against the results of an expert panel) [29,30]. In the study we conducted, the unsatisfactory result that required further research was on the evaluation of the boundary body condition scores 1 and 5. The difficulty lies in the fact that for the algorithm, cows with a body condition score of 1 and 2 and a score of 4 and 5 are similar. Therefore, in this study, we focused our field data collection on animals with a body condition score of 1 to 3 and 4 to 5.

We selected 3 commercial farms with a total of 182 animals. On the selected farms, all cows have similar traits, the animals are emaciated and of poor performance, and part of the herd features are overweight. Data on animals were collected in the Moscow and Yaroslavl regions.

Field data were collected between August 2020 and November 2022:

- The first farm has a herd of 570 forage cows, which is located in the Yaroslavl region (Central Russia). On this farm, 118 animals were selected at 5–6 months of lactation (from 151 to 180 days of lactation), with a body condition score of 1–4 points. The average annual milk yield per cow per day is 16.8, with an average fat percentage of 3.7%. The animals are of the black-motley breed. The average body condition score of the experimental animals was 2.75. The average weight of the tested animals is 467 kg.
- The second farm has a forage-fed herd of 50 heads, located in the Moscow region (Central Russia). On this farm, we selected 18 animals of 5–6 months of lactation (from 151 to 180 days of lactation) with a body condition score from 2 to 5. The average body condition score of the sample animals was 3.75. The average annual milk yield per cow per day is 15 kg/milk. The average fat content in the milk is 5.2%. The animal breed is the Holsteinized black-motley breed. The high fat content of milk of the black-motley breed can be explained by the fact that additional local selection work was done on this farm to increase the fat content of milk. The milk obtained from the animals is used for cheese production. The average weight of the tested animals is 583 kg.
- the third farm has a herd of 1200 forage-fed animals and is located in the Moscow region (Central Russia); 46 animals of 5–6 months of lactation (from 151 to 180 days of lactation) and with a body condition score of 3–5 were selected on this farm. Experimental animals had an average body condition score of 3.5. The breed of animals were Holsteinized black and mixed breeds. A total of 36 animals were selected, 5–6 months of lactation, which predominantly had a borderline body condition score. The average annual milk yield per cow per day was 28 kg/milk, and the average fat content of the milk was 3.7%. The average weight of the studied animals was 571 kg.

All animals had two milkings per day. Expert panels were formed to assess the body condition score by data collection site. The panel consisted of at least two independent veterinarians and two trained specialists. The average body condition score obtained from all experts was the benchmark value.

In addition, a weighing platform was used as an implement to further increase the accuracy of the body condition score by comparing the values obtained.

The live weight of animals was collected, in particular, by the Klüver–Strauch method [31], and the remaining animals were weighed on the platform. A disadvantage of the Klüver–Strauch method is an error in measuring the live weight of up to 10%. In the experiment, a discount of 1% of the actual weight was taken into account for bulk (mud adhered to the animal), and a discount of 3% for the contents of the gastrointestinal tract, when animals were weighed on the platform. Calculation of the live weight, including the discounts made, was automatic.

Based on the Pearson's correlation coefficient, we obtained proof of the representativeness of the animal samples and the relationship between the body condition score and live weight of the animals.

2.2. Equipment and System

Images of the cows' backs were collected on a 3D commercial camera, the O3D 303 3D ToF camera. The system is powered by 220 V, the power supply converts to 24 V to ensure the 3D sensor is operational. The sensor is mounted at a height of +2.2 m above the floor level (Figure 1).

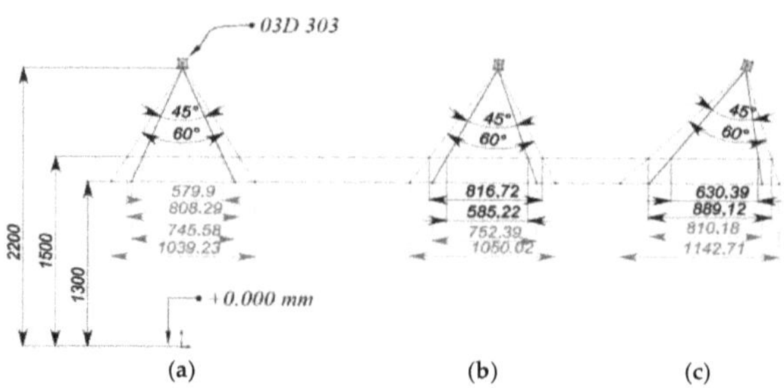

Figure 1. Demonstration of how the distance of the scanned surface varies with the inclination angle of the sensor. (**a**) Position 1 of the optical module 03D "looks" vertically down relative to the back of the cow; (**b**) position 2 of the optical module 03D "looks" at an angle of 5° from the vertical axis; (**c**) position 3 of the optical module 03D "looks" at a 10° angle from the vertical axis.

The height and inclination angle of the three-dimensional sensor are based on four parameters: cow height, cow length, minimum working distance between the camera and the object, and the camera's allowable error. The height of the animals under study ranged between 1300 mm and 1500 mm, the minimum working distance of the camera between the surface and the object under study was 300 mm. The signal from the identification antenna of the animal's RFID tag triggered the three-dimensional image production.

The inclination angle of the sensor taking into account the given 4 parameters is chosen to be 5 degrees, as it can cover a sufficient area of the animal's back under analysis, while keeping the pixel spacing to 0.006 m as the point of interest moves away from the 3D camera. The distance of 0.006 m between pixels is the set distance on which the least squares method is based when forming clusters of points related to areas of interest.

For the correct calculation of the required parameters between the camera and the object under study (coordinates of the received Z-axis pixels), we performed angle normalization (because the tilt angle of the 3D camera relative to the cow's back was introduced), presented in the expression using the R matrix:

where X, Y, Z—the areas of interest point coordinates, and J—the required distance between the interest points areas.

The total dataset contained 546 images from 182 animals with body condition scores from 1 to 5 with a step of 0.25 points (17 classes) (Table 1). Based on the earlier studies, it is sufficient for this camera to take three pictures of each cow, then the images are combined and the system starts determining the fatness.

Table 1. Number of images and proportion of cows for each body condition score.

№	5	4.75	4.5	4.25	4	3.75	3.5	3.25	3	2.75	2.5	2.25	2	1.75	1.5	1.25	1
							Body Score Condition										
*	5	14	4	12	7	5	8	14	18	25	14	11	11	6	12	9	7
**	15	12	12	36	21	15	24	42	52	75	42	33	33	18	36	27	21

*—the number of animals; **—the number of images.

From the data collected, we can see that the predominant body condition scores are 4.75; 4.25; 3.25; 3; 2.5; 2.25; 2; 1.5. The distribution of animals by body condition score was made by the expert panel, whose opinion is considered to be the benchmark (Figure 2).

The 3D camera is able to calculate and output Point Cloud as a multidimensional array I × J × K, where I and J are camera resolution, e.g., 352 × 264, K is X, Y, Z coordinates. Output of received data is in "dat" and ".h5" formats. The recording speed of the video images is 5 fps. Due to this feature, we obtained 3 to 5 images of each cow in the initial image. The images were collected according to the scheme shown in Figure 1. The camera error stated in the manufacturer's specifications is ±0.01 m for each meter between the lens and the object. Therefore, assuming that the working distance between the cow's rump (1.5 m) and the 3D camera lens (2.2 m) is 0.4 m, the error amounted to ≤0.01 m. The optical module was installed at an angle of 5°.

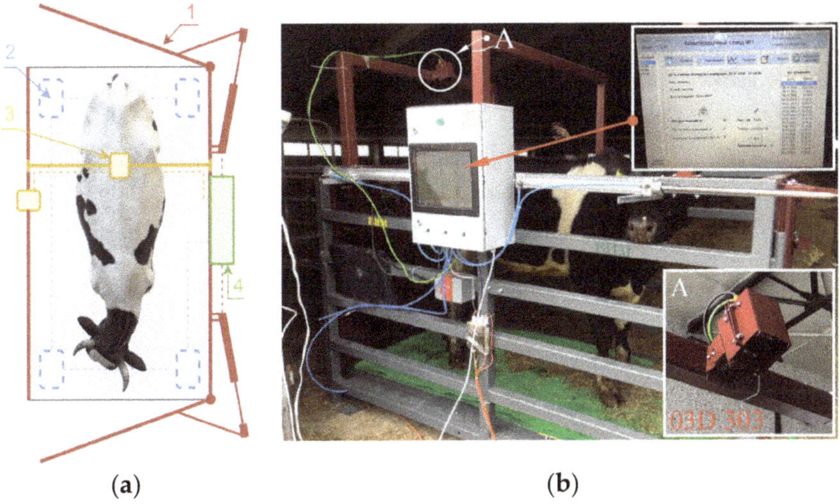

(a) (b)

Figure 2. Developed installation used for field data collecting. (a) Scheme of the developed installation to determine the body condition score, height, and weight of dairy cows up to 1200 kg: 1—automatic gates; 2—weighing module; 3—03D 303 three-dimensional camera; 4—a single control unit; (b) three-dimensional camera for the body condition score evaluation—03D 303 and software.

2.3. Assessment of the Body Condition Score and Analysis of the Results

We used our previously developed software [32] to process the obtained three-dimensional maps and determine the body condition score and standard tools; excel for primary data processing and formatting was used to process the study results.

The results were obtained automatically and those of an expert evaluation were compared manually. The expert evaluation of the body condition score was a benchmark value.

In terms of searching and determining the main areas of interest, the developed software was based on the application of the least squares method (regression analysis) to find the areas of interest.

As the camera was mounted on top of the animal and the points of the cow's back were presented to the data analysis, the points of greatest interest were those near the contour and describing its perimeter. Using the spine of a cow as an example, we can consider the basic expressions to identify it. The algorithm developed is based on the ordinary method of least squares (LS).

The entire surface of a cow's back is represented by an array of points without regard to depth, after which the regression tool is applied. We represent the whole surface as a set of points:

$$(y_1, x_1), (y_2, x_2), \ldots (y_n, x_n) \tag{1}$$

We can apply the method of least squares to minimize the sum of squares of RSS RRS deviations:

$$RSS = \sum_i (y_i - (a + bx_i))^2 \tag{2}$$

To find fixed points for RSS, the following expressions are used:

$$\begin{cases} \dfrac{\partial RSS}{\partial a} = \sum_i 2(y_i - a - bx_i) = 0 \\[2mm] \dfrac{\partial RSS}{\partial b} = \sum_i 2(y_i - a - bx_i) = 0 \end{cases} ; \tag{3}$$

$$\begin{cases} \sum_i y_i - na - b\sum_i x_i = 0 \\[2mm] \sum_i x_i y_i - a\sum_i x_i - b\sum_i x_i^2 = 0 \end{cases} \tag{4}$$

$$\begin{cases} \overline{y} - a - b\overline{x} = 0 \\[2mm] \overline{xy} - a\overline{x} - b\overline{x^2} = 0 \end{cases} \tag{5}$$

$$\begin{cases} a = \overline{y} - b\overline{x} \\[2mm] \overline{xy} - (\overline{y} - b\overline{x})\overline{x} - b\overline{x^2} = 0 \end{cases} \tag{6}$$

$$\begin{cases} a = \overline{y} - b\overline{x} \\[2mm] \overline{xy} - \overline{x}\,\overline{y} + b\left[(\overline{x})^2 - \overline{x^2}\right] = 0 \end{cases} \tag{7}$$

Thus, the regression and refinement of the ridge line to the point cloud produces the result shown in Figure 3.

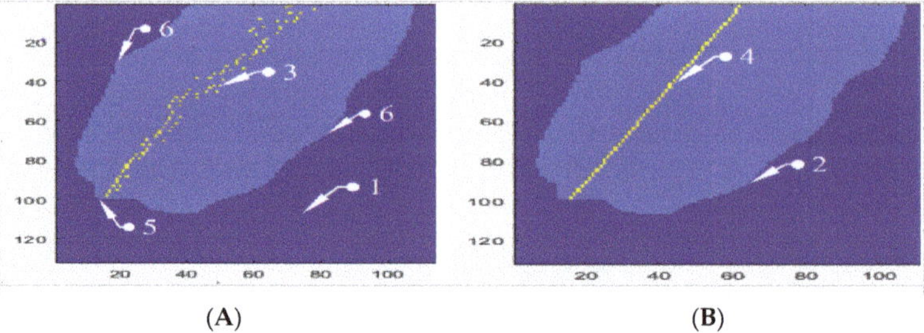

(A)   (B)

Figure 3. Defining the spinal column axis with extracting the area of interest. 1—Filtered area; 2—cow's contour; 3—unspecified animal's spinal column; 4—specified animal's spinal column; 5—tail head; 6—hips. (**A**) initial ridge line plotting by linear regression. (**B**) the ridge construction as a set of points on each longitudinal axis.

Figure 3 showed the results of the regression method. Figure 3B showed the ridge construction as a set of points on each longitudinal axis constructed. The lighter silhouette shows the silhouette of the cow, represented as a cloud of points, disregarding the Z-axis. Figure 3A is an initial ridge line plotting by linear regression.

To determine the cow's height, it is necessary to estimate the coordinates (xyz) of each point along the ridge line and find the extremum along the Z-axis. The point that is the extremum is the withers from which the cow's height is determined.

Table 2 shows the two approaches to BCS evaluation, the upper part was used for BCS evaluation by the expert panel, the lower part of the table was used for automatic evaluation. The numerical values were determined manually by analyzing the resulting field database of animals. The numerical characteristics are the average values for each body condition score and are relevant for the black-motley and the Holstein black-motley breeds raised in Central Russia. For other breeds, the numerical characteristics described in Table 2 may differ [32–34].

Table 2. Methods for assessing areas of interest in determining the body condition score of dairy cows by the expert panel and using automated methods.

Body Condition Rating Table for Experts					
Current body score condition	Obesity	Above average	Medium	Below average	Exhaustion
Score	5	4	3	2	1
Spinous processes of the lumbar and back	The back is rounded, hidden in adipose tissue	The back is straight, do not protrude	Raised back, slightly protruding	Visibly protrude, each process is visible	Customized, prong top
Transverse process of lumbar and hunger hollow	Hidden in adipose tissue, the area of the fossa is rounded, filled	Smooth rounded edge, the area of the fossa is filled, not sore	Viewed separately, viewed pit	They stand out noticeably, you can count them. The hole is clearly visible	They protrude strongly, the vertebral bodies are visible, the emaciated state, the fossa is deep
Hips and pin bone	Hidden in adipose tissue, not visible, ridge is rounded, filled with adipose tissue	Rounded but slightly prominent. Smooth surface	Protrude not sharply, moderately filled	Visibly protruding, thin layer of soft tissue	Protrude strongly
Head of tail	Hidden in adipose tissue	Rounded, moderately in adipose tissue	Smooth, covered with soft tissues, adipose tissue is fragmented	Tail vertebrae protrude	Protrude strongly
Vulva and anus area	Filled and forms a fat fold	filled	In the form of a small cavity	The cavity is deep, rounded	Protrude strongly, deep depression

Body condition score criteria table for automatic scoring																	
Current body score condition	Obesity				Above average				Medium				Below average				Exhaustion
Score	5	4.75	4.5	4.25	4	3.75	3.5	3.25	3	2.75	2.5	2.25	2	1.75	1.5	1.25	1
h1.Spinous processes of the lumbar and back	not allocated				not allocated				<0.01 m				0.01 m	0.013 m	0.016 m	0.02 m	0.02 m<
Transverse process of lumbar	not allocated				not allocated				<0.018 m				0.018 m	0.021 m	0.023 m	0.025 m	<0.025 m
Hips and pin bone	not allocated				145°	143°	140°	138°	135°	132°	129°	127°	125°				125°>
h2. Hunger hollow	0.06 m				0.07 m		0.08 m		0.09 m	0.1 m	0.11 m						0.12 m≥
h3. Head of tail	0.05 m	0.06 m	0.07 m	0.08 m	0.09 m	0.1 m	0.11 m	0.12 m	0.13 m	0.15 m	0.17 m	0.19 m					0.2 m≤
Vulva and anus area	minimum convex				low convex				average convex				maximum convex			

When the system evaluates the spinous processes of the lumbar and back, and the transverse processes of the lumbar and dorsum, the system first draws straight lines along the ridge and parallel to the ridge lines in the area of the transverse processes of the lumbar then measures the pixel height along the lines (Table 2, side view, node B, parameter h1).

In the hip's area, the system assesses the angle: two lines are drawn along the protruding parts of the back and then the angle is assessed (Figure 4, fatness score 3). The angle at 136° is an indication of a body condition score of 3, and the angle 125°> is a fatness score of 1.75 to 1.

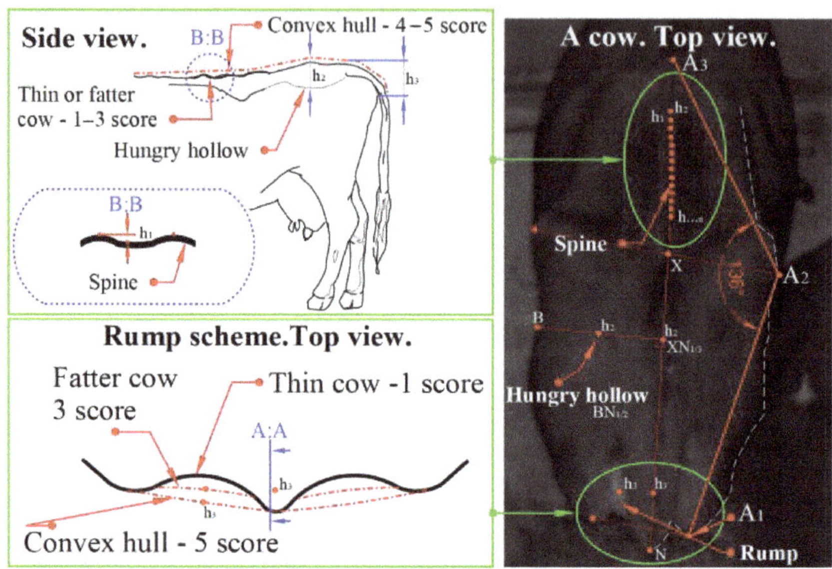

Figure 4. Desired areas of interest.

The points for estimating the angle are plotted on the boundary of the protruding parts of the body: the rump bone is A_1, the hip protrusion is A_2, and the first point at the junction of the transverse processes is A_3. To find the point A_3, we applied neural network tools with the preliminary training on 80 animals in the 5–6 months of lactation with a body condition score of 1–3 points.

To determine the depth of the "hunger hallow", the following procedure was used: step 1—point XN1/3, which is 1/3 of the length of the segment XN; step 2—at 1/2 the length of the segment BXN1/3, set point h2. Then, we compare the difference in height between points h2. The depth of the "hunger hallow " for a 5-point animal is 0.06 m and for a 1-point animal the depth of the hunger hole is 0.12 m.

The h3 points are determined by the lowest point in the tail base and the highest tail base.

As the last step before determining the fatness, the system checks all criteria and determines the body condition score on a 5-point scale in 0.25-point steps.

Figure 5 shows three-dimensional images converted into the black-and-white format. The pictures show animals with a body condition score from 1 to 5 on a 5-point scale and an explanation of which area of the animal's back is manipulated by the algorithm.

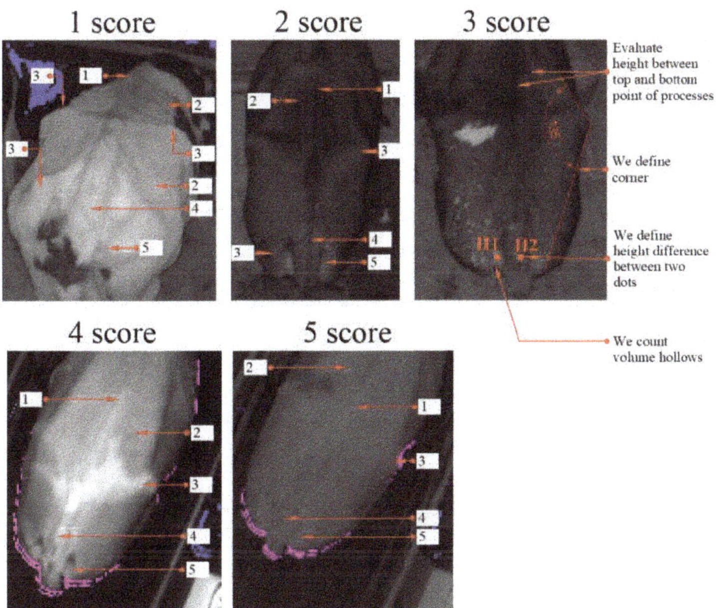

Figure 5. Animals' body condition scores and areas of interest. 1—Spinous processes of lumbar and back/dorsum; 2—transverse processes of lumbar and hunger hollow area; 3—hips and pin bone; 4—head of tail; 5—vulva and anus area.

3. Results and Discussion

3.1. Results

To understand the developed system's overall effectiveness, it was necessary to analyze and evaluate the effectiveness of each area of interest. All the resulting field data were evaluated using the developed method. The results were compared with the experts' evaluation. To understand the overall effectiveness of the developed system, it was necessary to analyze the evaluation effectiveness of each area of interest. To this end, a graph was plotted (Figure 6). The graph shows in the percentage terms the areas of interest and their detection probability, where 0% was not detected in all animals and 100% was detected in all animals.

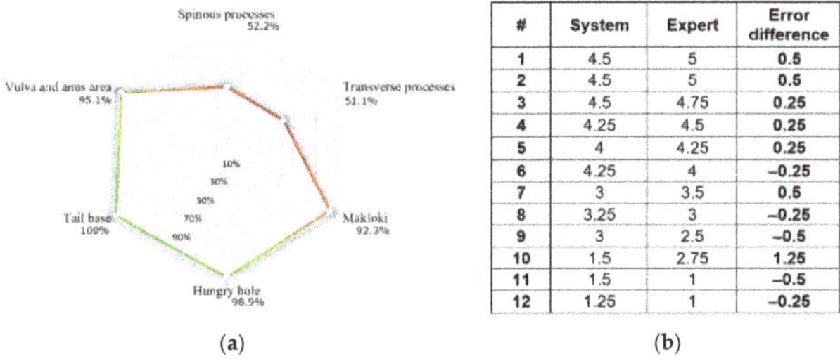

#	System	Expert	Error difference
1	4.5	5	0.5
2	4.5	5	0.5
3	4.5	4.75	0.25
4	4.25	4.5	0.25
5	4	4.25	0.25
6	4.25	4	−0.25
7	3	3.5	0.5
8	3.25	3	−0.25
9	3	2.5	−0.5
10	1.5	2.75	1.25
11	1.5	1	−0.5
12	1.25	1	−0.25

(a)

(b)

Figure 6. Efficiency of the system when detecting the areas of interest in the studied animals. (**a**) measurement efficiency of detecting the areas of interest; (**b**) difference in the BCS evaluation between the automatic measurement of the developed system and the evaluation made by the expert group and the difference between the obtained values.

The graph analysis results of Figure 7 show that the developed method can estimate the tail base area with the 100% accuracy. The hunger hollow is determined with a 98.9%

accuracy and the vulva and anus area with a 95.10% probability. Protruding vertebrae—namely, spinous processes and transverse processes—are evaluated with a 52.20% and a 51.10% accuracy. The accuracy of 50% was explained by the fact that according to Figure 2, these areas were not determined or were determined incorrectly in animals with a body condition score ranging from 3.25 to 5. The overall accuracy of the system was estimated by the experts at 93.4%, which was a positive result.

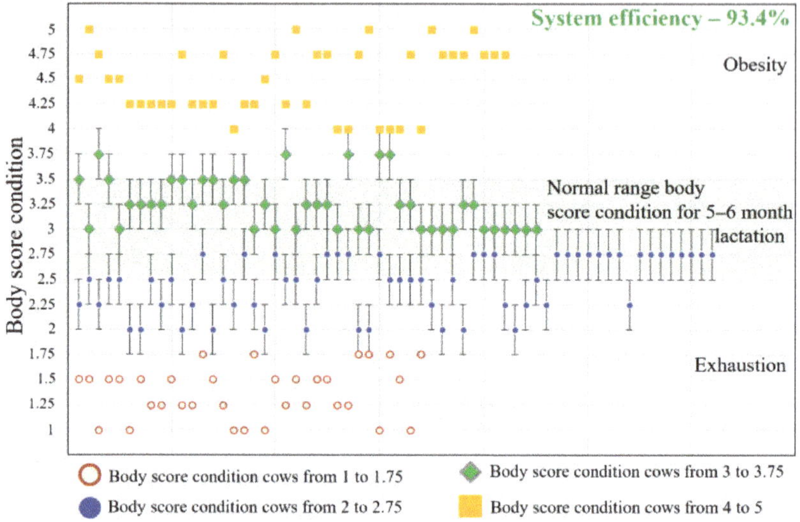

Figure 7. Distribution graph of the fatness of 5–6 month old animals obtained in an automatic evaluation.

Additionally, Figure 6 showed that the evaluation of the system and that of the experts have more discrepancies when the body condition score is 4–5, with the largest error of 1.25 and the smallest error of 0.25.

Figure 7 shows that animals with a condition score of 2.5 to 3.5 at 5–6 months are healthy. The developed system gives reasons to divide the animals into three groups, confirming their physiological status: normal range body condition, exhaustion, and obesity. In this case, it is worth bearing in mind that the system has an accuracy of 93.4%. Then, in this study, 4 animals with a body condition score of 3.75, and 15 animals with a body condition score of 2.5 had a 6.6% probability of belonging to another physiological status group. This is due to the fact that the system was wrong by 0.25. Errors caused by other nutritional scores are not critical, as technologically, an animal is either healthy and does not require any manipulation even though the system gave a nutritional score of 3 ± 0.25, or it has exhaustion/obesity, which requires manipulation of the animal to improve its physiological status.

In our observations, most of the animals with a fatness score of 3.75–5 were on the second farm (percentage of the total herd when ranked by score) with an average annual milk yield per cow per day of 15 kg/milk and a fat content of 5.1–6%. This farm was financially sound, and the main activity was getting milk from the animals for cheese production. When analyzing the cause of overweight animals, it was found that the farm staff were disrupting the feeding ration and the animals were receiving more energy than they needed; the animals were kept in loose housing.

On the second farm, a correlation was established between live weight and body condition score for 32 animals (Figure 8). The correlation determined by Pearson's method is R = 0.849.

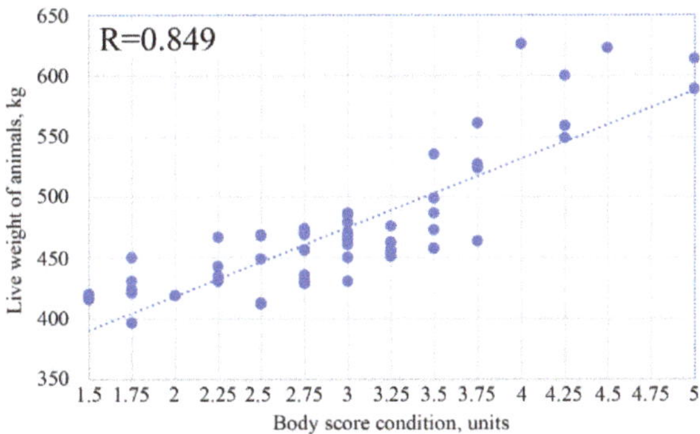

Figure 8. Distribution graph of the fatness of 5–6-month-old animals obtained during an automatic evaluation in 32 animals.

The Pearson correlation tool was chosen to determine if there was a relationship between the live weight and body condition score of the cows under study, as the data obtained have a normal distribution. Discussion of the data shows that part of the values on the scale from 4 BCS points to 5 BCS points and a live weight above 525 kg have a chaotic distribution. The Pearson correlation is R = 0.849, which does not guarantee 100% correlation. This is explained by the following: the live weight of animals consists of basic parameters—the amount of dirt accumulated on the animal, the amount of feed eaten, and the month of pregnancy. Additionally, live weight was obtained using the Klüver–Strauch method [33], where the method itself has a margin of error. This correlation did not allow the evaluation of live weight by the fatness score, but is only an additional signaling indicator that draws attention to live weight gain. Therefore, by analyzing the data by the Pearson correlation, our main aim was to understand if there is a relationship between obesity and weight gain. This was important for the purpose of additional animal monitoring, where the developed software will signal if an animal is overweight, which in turn negatively affects the probability of successful insemination. If it was detected that an animal is gaining excessive live weight, it was therefore necessary to move the animal to another housing group to change the feeding ration.

Studies [33] found that an optimal range of body weight for an increased performance does exist due to the non-linear relationship between milk yield and body weight. Dairy breeds respond more strongly to bodyweight range than dual-purpose breeds. Cows with an average weight are the most productive in the population. Heavy cows (>750 kg) produce much less milk. Special attention should, therefore, be paid to the daily ration, and further increases in body weight of dairy cows should be avoided. Animals with a body condition score of 1 to 2.5, in most cases, were found on a farm with an average annual milk yield per cow per day of 16.8 kg/milk, and a fat content of 3.6 to 3.8%. After examining the keeping conditions of the animals, several criteria influencing the emaciation of the animals were observed. The main criterion was the feed ration. The animals under study received mainly legume–grass hay with the addition of micro and macro nutrients in their diets. The animals were continuously fed a complete daily ration consisting of 4 kg of legume–grass hay, 15 kg of mixed grass silage, 6 kg of root crops, 5 kg of high energy mixed fodder, 1 kg of barley powder, and 50 g of table salt. In addition, it was recorded that animals were kept in concrete buildings, typical for buildings constructed in the 1980s, with a disturbed microclimate and tethered housing without regular walks.

Based on the research results, the algorithm for the automatic evaluation of the animal body condition (fatness), followed by the staging of their physiological condition, was supplemented and modified. The algorithm was divided into two parts and is shown in Figures 9 and 10. The second part of the algorithm is an integral part of the first one. The algorithm was included in the software code of the developed software.

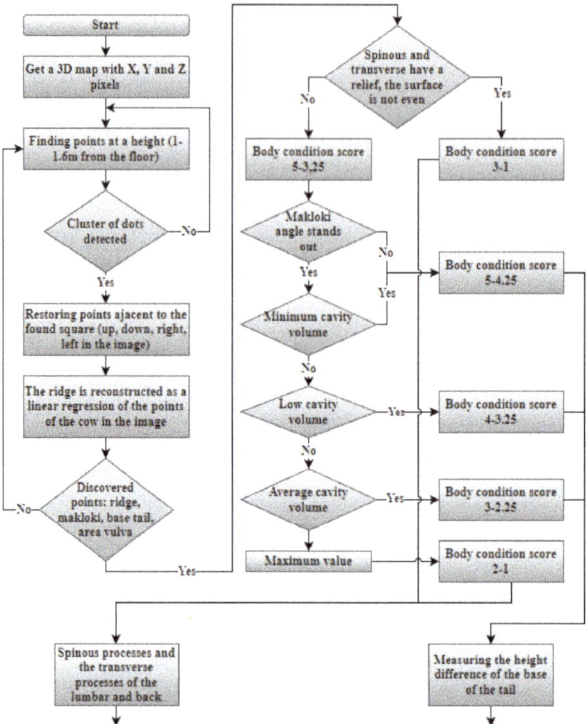

Figure 9. First part of the BCS Algorithm.

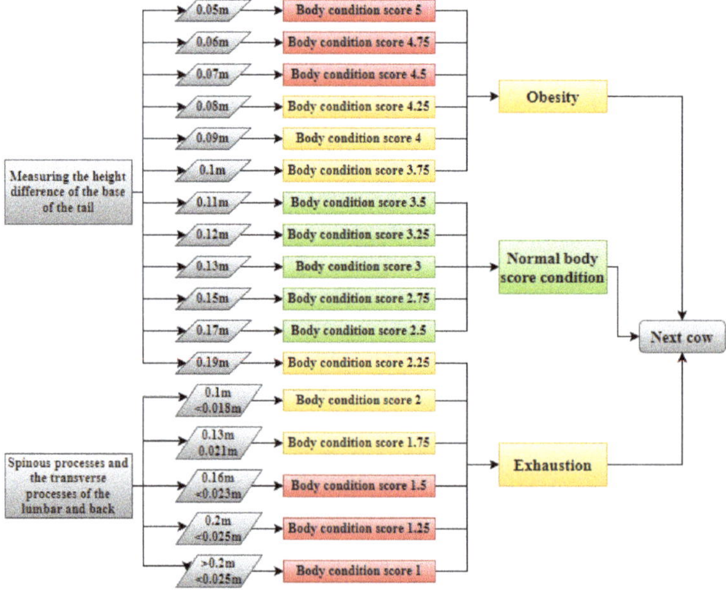

Figure 10. Second part of the BCS algorithm.

The explanation of Figure 9 starts with a three-dimensional map containing X, Y and Z coordinates of each pixel, and then searches for points at height (in the range of 1–1.6 m from the floor). When a cluster of points is detected, the algorithm determines the location of the main features: the spine, the hips, the tail head, and the vulva region. The topography of the spinous and transverse processes relief was then determined and the initial BCS value was determined.

As for Figure 10, for the initial BCS of 2.25–5 with no scalloping of the spinous and transverse processors and a low depression, the difference in tail head height was measured. For the primary BCS values in the range 1–3 with spinous and transverse processes in relief and a large hollow volume, the spinous and transverse processes of the lumbar and dorsum were measured. Depending on the results, the algorithm outputs had three evaluation options—'normal BCS', 'obesity', 'exhaustion'.

The developed software gave information about the animal by scanning RFID tags: date and time of the last measurement, sex, status, date of birth, actual weight, weight excluding animal contamination (bulk), weight excluding GIT contents, animal height, and the BCS value. The software developed for each animal provided more detailed information, traced the dynamics of changes in physical parameters, kept the herd log, and had service settings.

3.2. Research limitation

We refer to several factors as limitations of the research.

Factor 1 is the capability of the machinery and equipment and the environmental conditions in which they were operated. Dairy cows were evaluated both indoors and in the open field. Based on our experience with the equipment, we have found that 3D TOF cameras with a 840 nm wavelength, when shooting animals outdoors in bright sunlight, had noise that prevented effective fatness scoring. As such, 3D TOF cameras at 940 nm may be considered for further research. According to the manufacturers (the study does not specify a specific manufacturer), the 940 nm 3D cameras solve the problem of not being able to produce 3D maps in bright sunlight. In this study, three-dimensional cameras based on 940 nm were not tested.

Factor 2—while evaluating the body condition score of an animal, we could not estimate the animal's weight to an accuracy of 1 kg. We considered it possible to install an additional 3D camera to measure the torso depth of the animal—automatically using the Klüver–Strauch method [32] based on the digital data obtained for the torso depth, height and body condition score. However, this method would also not give accurate information about the animal's live weight, as there was no information about the cow's pregnancy, degree of contamination, and gastrointestinal contents. Additional discounts and coefficients relative to live weight would have to be introduced, but this may result in a high margin of error.

Factor 3 is the use of artificial intelligence to find the areas of interest. The matter is that if all fields of interest are calculated by means of artificial intelligence, then the exploitation of the system with each new breed or farm will demand resources for additional training of the system, which is not practical. Using the proposed research method for code development is a more labor-intensive process than collecting a data array and training artificial intellect. However, this method would definitely prove to be more practical, because it covered dairy cattle breeds, which are bred in Russia.

Factor 4 is the use of specific equipment. For these studies, it was not the specific manufacturer of the 3D cameras that was important, but their characteristics. It was important to choose a 3D camera that has a wavelength of 840 nm and a resolution of 352×264, and the factory error rate of used cameras is not higher than 1 cm per 1 m distance at a distance from the object in question.

3.3. Economic Efficiency

The proposed technology will improve production efficiency on large dairy farms by reducing animal stress, controlling animal nutrition when necessary, and early detecting physical deviations (Table 3).

Table 3. Cost of implementing the technology using the example of the farms under study.

Criteria	Used Solutions			Proposed Technology		
	1st Farm	2nd Farm	3rd Farm	1st Farm	2nd Farm	3rd Farm
The number of cows, heads	560	50	1200	560	50	1200
Milk yield, kg	16.8	15	28	40–50	40–50	40–50
Culling, %	7	7.5	6	4	4	4
Die, %	5.5	6	4	1.2	1.2	1.2
Feed consumption, t/day	16.8	3.3	54	28.2	2.5	60.4
Veterinary care costs, rub/month	157,300	78,650	235,950	18,000	12,000	25,000
Veterinary care costs, rub/year	1,887,600	943,800	2,831,400	The system installation's price		
				2,515,968	2,515,968	2,515,968
Number of calves, head	333	29	714	448	40	960
Calves for sale (80%), heads	266	23	571	358	32	768
Calves for sale (1 month, 60 kg), profit, rub	2,397,600	208,800	5,140,800	3,225,600	288,000	6,912,000
Calves for sale (6 months, 140 kg), profit, rub	3,916,080	341,040	8,396,640	5,268,480	470,400	11,289,600
Calves for sale (12 months, 350 kg), profit, rub	4,195,800	365,400	8,996,400	5,644,800	504,000	12,096,000
Total profit, rub	8,358,180	−116,710	19,238,490	11,565,712	−1,269,068	27,672,632

The percentage of culling and mortality was planned to be reduced by adjusting the ration and improving the general maintenance condition of the animals on the farms. We also proposed to increase the actual milk yield per day.

Often, farms have in-house veterinarians, but with the introduction of the biometric system, costs can be reduced, and external specialists can be called in only when necessary. Feed costs would also be reduced, as feed rations for the animals can be monitored and adjusted.

The main profit increase was expected to come from the improved life quality of the cows, and as a consequence, the birth rate of calves will also increase.

Sales are planned by age groups. The distribution will be as follows: 80% of all calves born on the farm during the year will be sold. Of these, 50% will be sold at the age of 1 month, 35%—6 months, and 15%—12 months.

As far as in the first year, Farm 1 and Farm 3 would make 38.4% and 43.8% more profit, respectively, but this technology did not look profitable on the second farm. We recommend that this biometric system should only be installed on large farms with 560 heads or more.

3.4. Technology Applicability

Having confirmed the cost-effectiveness of the developed BCS estimation technology, we can now describe how we implement automatic BCS evaluation for milk production.

The automatic livestock monitoring system operated in two ways. The first way was stationary, and the second way was mobile.

The stationary method consisted in the fact that on the farm, in the places of the daily pass of the animals, for example, the system of BCS evaluation was mounted behind the milking parlor in the "gallery". The system consisted of a three-dimensional camera and data collection and processing unit, as well as an identification antenna, which read the ID number of the cow. The data were sent to a server.

The mobile method implied bringing the system once a month to the box where a group of animals was kept to scan the BCS score (Figure 11).

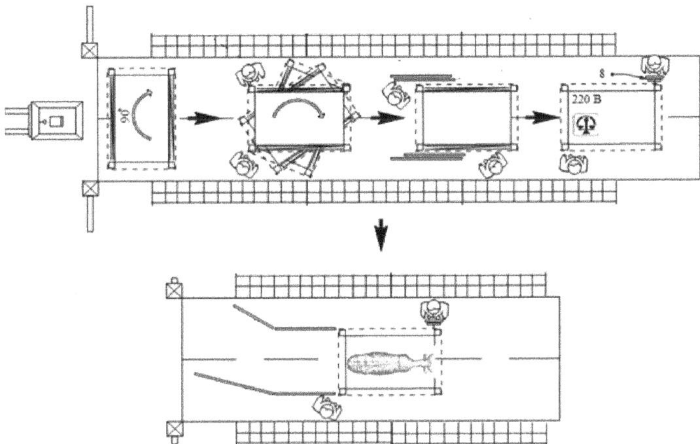

Figure 11. Schematic diagram of the installation of a mobile BCS evaluation system in a cubicle housing a group of animals.

The system was brought in by a forklift or an ATV to the cubicle where the animals were kept in a loose housing. Two staff members then turn the system around; they set the fence in the desired position, pointing to one side or the other. Then, one employee drove the animals in and out of the system, a second employee took care of reading the cow number, assessed the body condition, and recorded the data. When the group was finished, the system was assembled in the transport position and transported to the next group of animals. The data were transferred on a flash drive to a server. This procedure is done on a monthly basis. The advantage of the mobile system is that the fatness estimation can be done while the animals are grazing in the fields.

On the basis of the data obtained, the developed software plots a graph—a diagram of the change in the BCS score—and compares it with the set-required values for the current physiological status of each cow. There were several applications of the technology. The first situation was when we had an animal with an increased body condition score. The system recorded that the BSC score was increased, then queried the following data from the herd management software: current physiological status, which group the animal is in, current milk production, day of lactation, insemination status, and specific ration. For example, an animal was on day 75 of lactation, no conception had occurred, the BCS score was 3.75, milk yield was 17 kg/day, and fatness was 3.7%. Then, an automatic decision was made that the ration should be adjusted by reducing the amount of energy the animal receives without changing the animal's maintenance group, as the animal was at the peak of lactation and its milking requirements should be met. When moving to the next group, a gradual decrease in milking should be observed, accordingly. At the same time, we have to monitor the animal's condition so that by the end of lactation, the animal has a corrected condition. For example, an animal on day 190 of lactation and the conception on day 110, the BCS condition score was 2.5, milk yield 14 kg/day, and fat content 3.5%. In this case, the animal should be moved from group 3 to group 1 or 2 in order to adjust the feeding level to ensure an energy surplus.

The BCS evaluation system was needed as an additional tool to monitor feeding and assist in decision making for each cow when moving them to different housing groups.

The development of an automatic fatness estimation system will make it possible to collect data sets and statistics for each animal. This will make it possible, when collecting data on feeding, animal genetics, breeding material, and diseases, to form animal groups on farms more effectively, revealing their genetic potential in terms of productivity.

4. Conclusions

The effect of increasing the production of dairy farms had been achieved by implementing the technology of an automatic evaluation of the fatness of dairy herds (BCS) in a 0.25 step on a 5-point scale. The developed technology had been tested on 3 farms, with a total herd of 1810 animals, and provided for a non-contact BCS evaluation of a dairy herd required throughout the life of the herd within the farm. The overall accuracy of the system was estimated at 93.4%. The study has demonstrated the economic effect of implementing the proposed system.

Author Contributions: S.S.Y.—project management, methodology, natural data collection, writing—editing. D.Y.P.—conceptualization, software development. I.M.D.—system development, writing—analysis and editing. V.A.P.—visualization, rude preparation. A.A.S.—software testing. Y.A.P.—natural data collecting, editing. I.Y.—natural data collecting, writing—editing. All authors have read and agreed to the published version of the manuscript.

Institutional Review Board Statement: Not applicable.

Informed Consent Statement: Not applicable.

Conflicts of Interest: The authors declare no conflict of interest. The funders had no role in the design of the study; in the collection, analyses, or interpretation of data; in the writing of the manuscript, or in the decision to publish the results.

References

1. Brito, L.; Bedere, N.; Douhard, F.; Oliveira, H.; Arnal, M.; Peñagaricano, F.; Schinckel, A.; Baes, C.; Miglior, F. Review: Genetic selection of high-yielding dairy cattle toward sustainable farming systems in a rapidly changing world. *Animal* **2021**, *15*, 100292. [CrossRef] [PubMed]
2. Newton, J.E.; Nettle, R.; Pryce, J.E. Farming smarter with big data: Insights from the case of Australia's national dairy herd milk recording scheme. *Agric. Syst.* **2020**, *181*, 102811. [CrossRef]
3. Dallago, G.M.; Wade, K.M.; Cue, R.I.; McClure, J.T.; Lacroix, R.; Pellerin, D.; Vasseur, E. Keeping Dairy Cows for Longer: A Critical Literature Review on Dairy Cow Longevity in High Milk-Producing Countries. *Animals* **2021**, *11*, 808. [CrossRef]
4. Alem, H. The Role of Technical Efficiency Achieving Sustainable Development: A Dynamic Analysis of Norwegian Dairy Farms. *Sustainability* **2021**, *13*, 1841. [CrossRef]
5. Liu, D.; He, D.; Norton, T. Automatic estimation of dairy cattle body condition score from depth image using ensemble model. *Biosyst. Eng.* **2020**, *194*, 16–27. [CrossRef]
6. Ledinek, M.; Gruber, L.; Steininger, F.; Fuerst-Waltl, B.; Zottl, K.; Royer, M.; Krimberger, K.; Mayerhofer, M.; Egger-Danner, C. Analysis of lactating cows on commercial Austrian dairy farms: The influence of genotype and body weight on efficiency parameters. *Arch. Anim. Breed.* **2019**, *62*, 491–500. [CrossRef] [PubMed]
7. Buonaiuto, G.; Lopez-Villalobos, N.; Costa, A.; Niero, G.; Degano, L.; Mammi, L.M.E.; Cavallini, D.; Palmonari, A.; Formigoni, A.; Visentin, G. Stayability in Simmental cattle as affected by muscularity and body condition score between calvings. *Front. Veter. Sci.* **2023**, *10*, 1141286. [CrossRef]
8. Ledinek, M.; Gruber, L.; Steininger, F.; Fuerst-Waltl, B.; Zottl, K.; Royer, M.; Krimberger, K.; Mayerhofer, M.; Egger-Danner, C. Analysis of lactating cows in commercial Austrian dairy farms: Interrelationships between different efficiency and production traits, body condition score and energy balance. *Ital. J. Anim. Sci.* **2019**, *18*, 723–733. [CrossRef]
9. Montagner, P.; Krause, A.R.T.; Schwegler, E.; Weschenfelder, M.M.; Maffi, A.S.; Xavier, E.G.; Schneider, A.; Pereira, R.A.; Jacometo, C.B.; Schmitt, E.; et al. Relationship between pre-partum body condition score changes, acute phase proteins and energy metabolism markers during the peripartum period in dairy cows. *Ital. J. Anim. Sci.* **2017**, *16*, 329–336. [CrossRef]
10. Weik, F.; Archer, J.A.; Morris, S.T.; Garrick, D.J.; Miller, S.P.; Boyd, A.M.; Cullen, N.G.; Hickson, R.E. Live weight and body condition score of mixed-aged beef breeding cows on commercial hill country farms in New Zealand. *N. Z. J. Agric. Res.* **2021**, *65*, 172–187. [CrossRef]
11. Petrov, E.B.; Taratorkin, V.M. The main technological parameters of the modern technology of milk production at livestock complexes (farms). In *Recommendations.-M.: FGUNU "Rosinformagrotech"*; 2007; p. 176, ISBN 978-5-7367-0616-7.
12. Mandour, M.A.; Al-Shami, S.A.; Al-Eknah, M.M. Body condition scores at calving and their association with dairy cow performance and health in semiarid environment under two cooling systems. *Ital. J. Anim. Sci.* **2015**, *14*, 3690. [CrossRef]

13. Mugwabana, T.J.; Nephawe, K.A.; Muchenje, V.; Nedambale, T.L.; Nengovhela, N.B. The effect of assisted reproductive technologies on cow productivity under communal and emerging farming systems of South Africa. *J. Appl. Anim. Res.* **2018**, *46*, 1090–1096. [CrossRef]

14. Meteer, W.C.; Wilson, T.B.; Keisler, D.H.; Cardoso, F.C.; Shike, D.W. Effects of prepartum plane of nutrition during mid- or late gestation on beef cow body weight, body condition score, blood hormone concentrations and preimplantation embryo. *Ital. J. Anim. Sci.* **2016**, *15*, 264–274. [CrossRef]

15. Poczta, W.; Średzińska, J.; Chenczke, M. Economic Situation of Dairy Farms in Identified Clusters of European Union Countries. *Agriculture* **2020**, *10*, 92. [CrossRef]

16. Vanholder, T.; Papen, J.; Bemers, R.; Vertenten, G.; Berge, A.C.B. Risk factors for subclinical and clinical ketosis and association with production parameters in dairy cows in the Netherlands. *J. Dairy Sci.* **2015**, *98*, 880–888.

17. Daros, R.R.; Hötzel, M.J.; Bran, J.A.; LeBlanc, S.J.; von Keyserlingk, M.A. Prevalence and risk factors for transition period diseases in grazing dairy cows in Brazil. *Prev. Vet. Med.* **2017**, *145*, 16–22.

18. Barletta, R.; Filho, M.M.; Carvalho, P.; Del Valle, T.; Netto, A.; Rennó, F.; Mingoti, R.; Gandra, J.; Mourão, G.; Fricke, P.; et al. Association of changes among body condition score during the transition period with NEFA and BHBA concentrations, milk production, fertility, and health of Holstein cows. *Theriogenology* **2017**, *104*, 30–36. [CrossRef]

19. Garzón-Audor, A.; Oliver-Espinosa, O. Incidence and risk factors for ketosis in grazing dairy cattle in the Cundi-Boyacencian Andean plateau, Colombia. *Trop. Anim. Health Prod.* **2019**, *51*, 1481–1487. [CrossRef]

20. Senoh, T.; Oikawa, S.; Nakada, K.; Tagami, T.; Iwasaki, T. Increased serum malondialdehyde concentration in cows with subclinical ketosis. *J. Vet. Med. Sci.* **2019**, *81*, 817–820.

21. Sheehy, M.R.; Fahey, A.G.; Aungier SP, M.; Carter, F.; Crowe, M.A.; Mulligan, F.J. A comparison of serum metabolic and production profiles of dairy cows that maintained or lost body condition 15 days before calving. *J. Dairy Sci.* **2017**, *100*, 536–547.

22. Grainger, C.; Wilhelms, G.D.; McGowan, A.A. Effect of body condition at calving and level of feeding in early lactation on milk production of dairy cows. *Aust. J. Exp. Agric.* **1982**, *22*, 9–17.

23. Enevoldsen, C.; Kristensen, T. Estimation of body weight from body size measurements and body condition scores in dairy cows. *J. Dairy Sci.* **1997**, *80*, 1988–1995. [PubMed]

24. Yukun, S.; Pengju, H.; Yujie, W.; Ziqi, C.; Yang, L.; Baisheng, D.; Yonggen, Z. Automatic monitoring system for individual dairy cows based on a deep learning framework that provides identification via body parts and estimation of body condition score. *J. Dairy Sci.* **2019**, *102*, 10140–10151. [PubMed]

25. Kirsanov, V.V.; Pavkin DYu Dovlatov, I.M.; Yurochka, S.S.; Ruzin, S.S. *Developing an Algorithm for Body Condition Scoring of Dairy Cows*; Agricultural Engineering: Moscow, Russia, 2022; Volume 24, pp. 4–8. (In Russian)

26. Pavkin, D.Y.; Yurochka, S.S.; Dovlatov, I.M.; Shilin, D.V.; Polikanova, A.A. Computer Program: A Program for Weight, Height and Fatness Graph Plotting in an Intelligent Robotic Milking System. Certificate of State Registration of a Computer Program/Certificate Number: RU 2022682171/Date of Publication: 21.11.20. (In Russian). Available online: https://www.fips.ru/iiss/document.xhtml?faces-redirect=true&id=cfba880672bf7c933ea4ce0212ddaa3a (accessed on 7 July 2023).

27. Yurochka, S.S.; Pavkin, D.Y.; Dovlatov, I.M.; Ruzin, S.S.; Dolgalev, A.P.; Shilin, D.V. An Intelligent Way of Areas of Interest Detecting When Fatness and the Fatness Score Evaluation Determination. Certificate of Registration of Computer Program 2022615611, 03/31/2022. (In Russian). Available online: https://www.fips.ru/iiss/document.xhtml?faces-redirect=true&id=8b3b27228dff00ab5c510cae2943c228 (accessed on 7 July 2023).

28. Cevik, K.K. Deep Learning Based Real-Time Body Condition Score Classification System. *IEEE Access* **2020**, *8*, 213950–213957. [CrossRef]

29. Thaysen, J.; Boisen, A.; Hansen, O.; Bouwstra, S. Automatic estimation of dairy cow body condition score based on attention-guided 3D point cloud feature extraction. *Comput. Electron. Agric.* **2023**, *206*, 107666. [CrossRef]

30. Kolesen, V.P.; Yurashchik, S.V.; Deshko, I.A.; Dyuba, M.I. Breeding Basics Agricultural Animals. Educational and Methodological Aid. Available online: https://www.ggau.by/index.php?option=com_attachments&task=download&id=247 (accessed on 16 June 2023).

31. Yurochka, S.S.; Dovlatov, I.M.; Matveev, V.Y. Development of a method for determining the fatness score of dairy cows using three-dimensional images. *Vestn. NGIEI* **2023**, *4*, 18–28. [CrossRef]

32. Berry, D.P.; Macdonald, K.A.; Penno, J.W.; Roche, J.R. Association between body condition score and live weight in pasture-based Holstein-Friesian dairy cows. *J. Dairy Res.* **2006**, *73*, 487–491.

33. Paul, A.; Mondal, S.; Kumar, S.; Kumari, T. Body Condition Scoring in Dairy Cows-A Conceptual and Systematic Review. *Indian J. Anim. Res.* **2020**, *54*, 929–935.

34. Nagy, S.; Kilim, O.; Csabai, I.; Gábor, G.; Solymosi, N. Impact Evaluation of Score Classes and Annotation Regions in Deep Learning-Based Dairy Cow Body Condition Prediction. *Animals* **2023**, *13*, 194. [CrossRef]

Article

Deep Learning Application for Crop Classification via Multi-Temporal Remote Sensing Images

Qianjing Li [1,2], Jia Tian [1,3,*] and Qingjiu Tian [1,2]

[1] International Institute for Earth System Science, Nanjing University, Nanjing 210023, China
[2] Jiangsu Provincial Key Laboratory of Geographic Information Science and Technology, Nanjing University, Nanjing 210023, China
[3] School of Instrumentation and Optoelectronic Engineering, Beihang University, Beijing 100191, China
* Correspondence: tianjia@nju.edu.cn

Abstract: The combination of multi-temporal images and deep learning is an efficient way to obtain accurate crop distributions and so has drawn increasing attention. However, few studies have compared deep learning models with different architectures, so it remains unclear how a deep learning model should be selected for multi-temporal crop classification, and the best possible accuracy is. To address this issue, the present work compares and analyzes a crop classification application based on deep learning models and different time-series data to exploit the possibility of improving crop classification accuracy. Using Multi-temporal Sentinel-2 images as source data, time-series classification datasets are constructed based on vegetation indexes (VIs) and spectral stacking, respectively, following which we compare and evaluate the crop classification application based on time-series datasets and five deep learning architectures: (1) one-dimensional convolutional neural networks (1D-CNNs), (2) long short-term memory (LSTM), (3) two-dimensional-CNNs (2D-CNNs), (4) three-dimensional-CNNs (3D-CNNs), and (5) two-dimensional convolutional LSTM (ConvLSTM2D). The results show that the accuracy of both 1D-CNN (92.5%) and LSTM (93.25%) is higher than that of random forest (~91%) when using a single temporal feature as input. The 2D-CNN model integrates temporal and spatial information and is slightly more accurate (94.76%), but fails to fully utilize its multi-spectral features. The accuracy of 1D-CNN and LSTM models integrated with temporal and multi-spectral features is 96.94% and 96.84%, respectively. However, neither model can extract spatial information. The accuracy of 3D-CNN and ConvLSTM2D models is 97.43% and 97.25%, respectively. The experimental results show limited accuracy for crop classification based on single temporal features, whereas the combination of temporal features with multi-spectral or spatial information significantly improves classification accuracy. The 3D-CNN and ConvLSTM2D models are thus the best deep learning architectures for multi-temporal crop classification. However, the ConvLSTM architecture combining recurrent neural networks and CNNs should be further developed for multi-temporal image crop classification.

Keywords: crop type classification; deep learning; multi-temporal; remote sensing

Citation: Li, Q.; Tian, J.; Tian, Q. Deep Learning Application for Crop Classification via Multi-Temporal Remote Sensing Images. *Agriculture* **2023**, *13*, 906. https://doi.org/10.3390/agriculture13040906

Academic Editor: Roberto Alves Braga Júnior

Received: 1 April 2023
Revised: 17 April 2023
Accepted: 19 April 2023
Published: 20 April 2023

1. Introduction

Detailed and accurate information on crop-type cultivation is essential for developing economically and ecologically sustainable agricultural strategies in a changing climate, and for satisfying human food demands [1]. Multi-temporal remote sensing (RS) images acquired throughout the growing season provide an effective method for acquiring crop cover information over large areas [1,2]. Multi-temporal images can be used to distinguish crop growth states and the phenological characteristics of crops. In addition, they provide enriched features that allow more complex and stable crop classification tasks. They have thus seen wide use in the field of agricultural RS [3,4].

Two main strategies are available for multi-temporal crop classification. The first strategy is to stack multi-temporal images by time sequence and classify them with classifiers such as support vector machine (SVM), random forest and maximum likelihood [5,6]. However, this approach does not model temporal correlations and uses features independently, ignoring possible temporal dependencies [6,7]. Most classifiers such as SVM rely heavily on features that are not designed for time-series data, making it difficult to exploit any inherent time-series variability features. In addition, the stacked images increase redundancy and lead to the dimensionality catastrophe with increasing time-series length, which negatively affects classification performance [6,8]. The second strategy is to obtain new images from reflectance images by using spectral indices, such as the normalized difference vegetation index (NDVI) and the enhanced vegetation index (EVI), and then construct time-series data to reveal the temporal pattern of the different features. With this method, crops and other vegetation are classified with high accuracy. However, the classification results of this method are limited strictly by the number of images in the time-series. If the number is too small, then the temporal pattern has little effect on classification performance [8]. In addition, manual feature engineering based on human experience and prior knowledge is essential with this approach, which increases the complexity of processing and computation [7,9]. Moreover, the construction of VIs based on the specific spectral features ignores other spectral bands, which in turn affects the classification performance.

Current multi-temporal RS images are multi-spectral, multi-temporal and multi-spatial. In multi-temporal images, crops are represented via variations in temporal, spectral, and spatial features. These features can be comprehensively included in four-dimensional (4D: time, height, width, and band) data that require classification models to learn and represent temporal, spectral, and spatial features. Multi-temporal images thus pose new challenges to the models used for data processing, so integrating multi-temporal images and continuously improving crop classification accuracy requires continued attention.

Deep learning is a breakthrough technique in machine learning that outperforms traditional algorithms in terms of feature extraction and representation [5–7], which has led to its application in numerous RS classification tasks [8–10]. Convolutional neural networks (CNNs) produce more accurate results than other models in most RS image classification problems [8,9,11]. The one-dimensional CNN (1D-CNN) model is commonly used to extract spectral features from hyperspectral images or temporal features from time-series images, providing an effective and efficient method for crop identification in time-series RS images [12]. The CNN learning process is computationally efficient and insensitive to data shifts such as image translation, allowing CNN models to recognize image patterns in two dimensions (2D) [13]. Three-dimensional (3D) CNN models use the spatial, temporal, and spectral information in multi-temporal images, and therefore are widely used in multi-temporal crop classification [11,14]. Long short-term memory (LSTM), a variant of recurrent neural networks (RNNs), is a natural candidate to represent temporal dependency over various temporal periods with gated recurrent connections [9,15]. LSTM models have been widely used for multi-temporal crop classification because they can analyze sequential data [9,16,17]. For multi-temporal crop classification, both CNN and RNN provide more accurate results than machine learning and traditional classification [5,9,11]. However, various deep learning architectures produce different results when applied to multi-temporal crop classification, feature learning and representation of crop spectral, spatial, and temporal information.

Convolutional LSTM (ConvLSTM) is a type of RNN with internal matrix multiplication replaced by convolution operations [18]. ConvLSTM, integrating both LSTM and CNN structures, shows unexpected adaptability to multi-temporal images [19–21]. However, due to the prevalence of CNNs and RNNs and the requirement for higher data dimensions, the ConvLSTM model is less commonly used in multi-temporal crop classification. Nevertheless, the potential of the ConvLSTM model deserves further exploration.

To summarize, multi-temporal images pose a new challenge to classification models in terms of data processing and feature extraction, but also open new opportunities for using data-driven deep learning to classify RS images. In this work, we use multi-temporal Sentinel-2 RS images as input data, and analyze the advantages of using such data and the structural advantages of various deep learning models. This research investigates (1) the possibility of using multi-temporal images for more accurately classifying crops; (2) the contribution of spectral, temporal, and spatial information to multi-temporal crop classification; and (3) the potential and requirements of using deep learning for multi-temporal crop classification. We also (4) search for a feasible and suitable deep learning model that provides optimum classification accuracy from multi-temporal images. Although such deep learning models have long been used for RS applications, this work compares and analyzes multi-temporal crop classification based on the deep learning architectures of CNN, LSTM, and ConvLSTM.

2. Materials

2.1. Study Area

The study area, Norman county, is located in northwestern Minnesota (Figure 1), which is a highly productive agricultural state in the United States. Minnesota is in the Great Plains of the central United States, and agricultural land covers the vast majority of the study area. The continental climate of the region is cold in the winter and hot and humid in the summer, with 600 mm/year of precipitation. The highest temperatures occur in July, and the lowest in January, with an average of 197 sunny days per year. The climatic and temperature conditions make single-season crop cultivation the main cropping system. The major crops in this region are corn, soybeans, sugarbeets, and spring wheat, which are planted in about 89% of the study area. Corn begins being planted at the end of April, matures in September, and is harvested through October. Soybeans are planted in May and harvested from mid-September through the end of October. Spring wheat is sown in early April and harvested from mid-July through August. Sugarbeets are planted in mid-April, mature in September, and are harvested by the end of October.

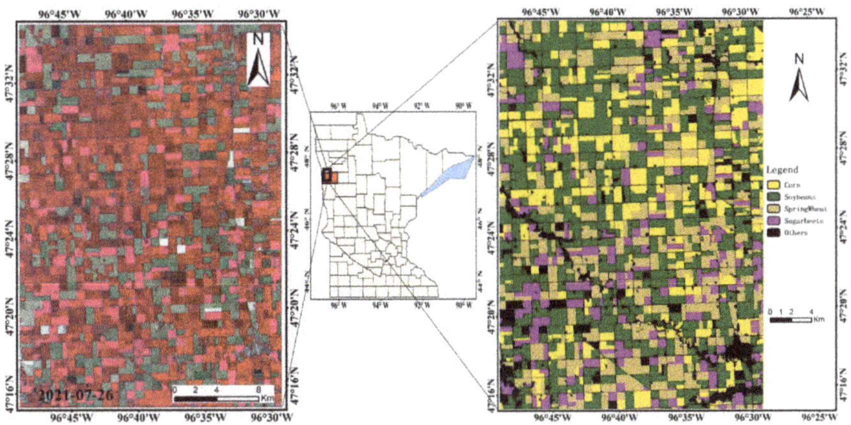

Figure 1. False color image and the Cropland Data Layer (CDL) of study areas.

2.2. Data

2.2.1. Remote Sensing Images

Sentinel-2 images were downloaded from the Sentinel Hub (https://www.sentinel-hub.com/ (accessed on 28 October 2022)). Cloud-free images from April 2021 to October 2021 were selected to encompass the entire crop growing season. A total of 13 Sentinel-2 images (Tables 1 and 2) were selected as the main input data of the experiment. Data preparation involved stacking and resampling the 20 m spectral bands to 10 m and the removal of the coastal band, water vapor, and the cirrus band, accomplished through the Sentinel Application Platform (SNAP).

Table 1. Spectral bands of Sentinel-2 images.

Band Names	Spectral Band	Central Wavelength (nm)	Band Names	Spectral Band	Central Wavelength (nm)
Blue	B2	490	Red-Edge	B7	775
Green	B3	560	NIR	B8	842
Red	B4	665	NIR	B8a	865
Red-Edge	B5	705	SWIR	B11	1610
Red-Edge	B6	740	SWIR	B12	2190

Table 2. Acquisition time of Sentinel-2 images.

Day of Year (DOY)	Acquisition Time	Day of Year (DOY)	Acquisition Time
112	22 April 2021	230	18 August 2021
137	17 May 2021	235	23 August 2021
150	30 May 2021	242	30 August 2021
165	14 June 2021	257	14 September 2021
192	11 July 2021	270	27 September 2021
207	26 July 2021	295	22 October 2021
225	13 August 2021		

2.2.2. Training and Validation Samples

The Cropland Data Layer (CDL) is a crop-type distribution product published by the United States Department of Agriculture and the National Agricultural Statistics Service. The 2021 CDL (Figure 1) for Norman County has a spatial resolution of 30 m, and was obtained from the CropScape website portal (https://nassgeodata.gmu.edu/CropScape/ (accessed on 20 October 2022)). Although the CDL is not the absolute ground truth, it is the most accurate crop-type product available, especially for corn and soybeans, with over 95% accuracy [22]. In Minnesota, the accuracies for several major crop types are close to or above 95% [23]. Therefore, a result of visual interpretation of multi-temporal Sentinel-2 images based on CDL data was used to select the crop samples for training and testing our crop classification model.

Based on the CDL, crop types in the study area were classified as corn, soybeans, sugar beets, spring wheat, and "other." The latter category ("other") includes all surface cover types except for the four major crops. To ensure the representativeness of the samples and the data size requirements of the deep learning model, the samples are selected to ensure that the sample points are distributed throughout the study area, that the central sample pixel type is consistent with the type of surrounding pixels, and that the sample pixel type is the dominant type in the local neighborhood. The sample points were created from a function of randomly created points and labeled by visual interpretation. Table 3 details the samples used for training the classification model and evaluating the accuracy. To train the model, the training and validation samples in Table 3 are randomly divided into training samples and validation samples in a ratio of 7:3.

Table 3. The five categories used in the present study for classification and the number of samples.

Sample Type	Training and Validation Samples	Testing Samples
Corn	1481	4096
Soybeans	1487	4738
Spring Wheat	1445	4674
Sugarbeets	1471	4167
Others	1546	5210

3. Methodology

3.1. Methodological Overview

The overall workflow of this study is shown in Figure 2. Firstly, we selected samples as described in Section 2.2.2. Next, different time-series images were constructed for the subsequent classification experiments (Section 3.2). Multiple deep learning models were constructed (Section 3.5), in which random forest was used as benchmark model. Details of the experiments can be found in Section 3.6. Finally, all classification results were validated, compared and analyzed.

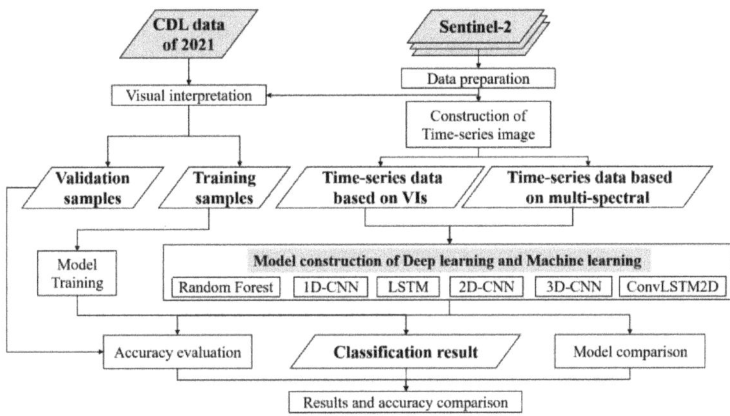

Figure 2. General workflow of this study.

3.2. Temporal Phenological Patterns

Two main strategies are available to represent the temporal patterns of crops for multi-temporal image crop classification: (1) time-series VIs constructed from spectral characteristics, and (2) time-series multi-spectral bands based on spectral stacking [5], which means stacking multi-temporal images by time sequence. Both strategies have been used to construct time-series data to represent the temporal characteristics of crops. Given the sensitivity of the NDVI [24] and EVI [25] to the physiological state of vegetation and their wide application [5,9], these indices have been used to construct time-series data. Their formulas are as follows:

$$\text{NDVI} = (NIR - RED)/(NIR + RED) \tag{1}$$

$$\text{EVI} = G \times (NIR - RED)/(NIR + C_1 \times RED - C_2 \times BLUE + L), \tag{2}$$

where $G = 2.5$, $C_1 = 6.0$, $C_2 = 7.5$, and $L = 1.0$. *NIR*, *RED* and *BLUE* represent the spectral reflectance bands of B8(NIR), B4(Red) and B2(Blue) in Sentinel-2 (Table 1).

3.3. Deep Learning Models

A CNN is a multilayer feed-forward neural network. The advantages of local connectivity and weight sharing not only decrease the number of parameters but also reduce the complexity of the model and make CNNs more suitable for processing numerous images [9,26]. CNNs may be one-dimensional (1D-CNN), two-dimensional (2D-CNN), or three-dimensional (3D-CNN), by having convolution kernels of different dimensions. Sequence data are fed into 1D-CNNs for learning and representing sequence relationships. Patch-based 2D-CNNs can be used for learning and representing spatial and spectral features in images. Cube-based 3D-CNNs correspond to the spectral, spatial, and temporal features in multi-temporal images [12,14]. The LSTM solves the problems of vanishing gradient, exploding gradient, and deficiencies in long-term dependency representation that appear in RNNs. In LSTM, the gate mechanisms, which include the input gate, output gate, and forget gate, enhance or weaken the state of the data in the cell for information protection and control [16,17]. The ConvLSTM model is an improvement and extension of the LSTM model, wherein matrix multiplication in LSTM is replaced by a convolution at each gate [20]. The ConvLSTM model combines the structural advantages of LSTM and CNN, and not only captures the spatial context of the image, but also models the long-term dependencies in the spectral domain. In addition, inter- and intra-layer data transfer enables the ConvLSTM to extract features more efficiently than a CNN or LSTM [18,19].

3.4. Sample Dimensions

Limited by the size and dimensions of samples in multi-temporal RS images, classification samples contain different spectral, temporal, and spatial information. This study uses various deep learning models to learn and represent spectral, temporal, and spatial information from multi-temporal images. The time-series classification data constructed from VI have only temporal characteristics [9], and their samples are one-dimensional vectors (Figure 3a). The time-series data constructed directly using multi-spectral, multi-temporal images are two-dimensional matrices with the shape of (band, time) (Figure 3b). The time-series data constructed from VIs including the spatial neighborhood are three-dimensional matrices (Figure 3c) with the shape of (height, width, time). The multi-spectral features combined with the spatial neighborhood in multi-temporal images produce four-dimensional matrices with the shape of (time, height, width, band) (Figure 3d). The "time" in three- or four-dimensional matrices means the number of temporals in time-series.

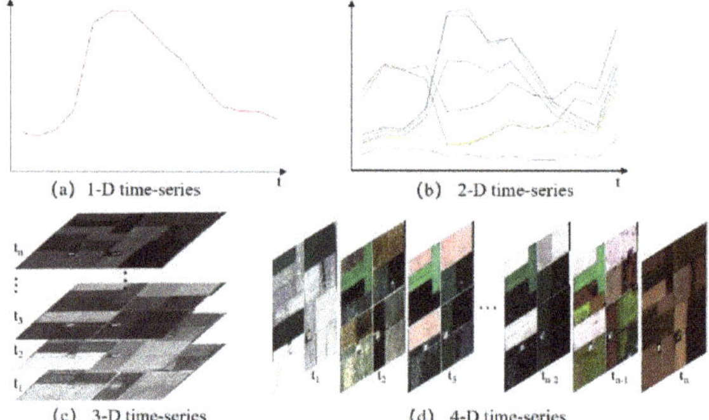

Figure 3. Time-series samples with different dimensions. (**a**) 1-D time-series, (**b**) 2-D time-series, (**c**) 3-D time-series, (**d**) 4-D time-series.

3.5. Deep Learning Architectures

The main deep learning classification models used in the study are 1D-CNN, LSTM, 2D-CNN, 3D-CNN, and ConvLSTM2D. The temporal, spectral, and spatial information of multi-temporal images can be learned and represented by different deep learning models corresponding to different types of samples. Both 1D-CNN and LSTM models can represent temporal features, and the model input corresponds to 1D and 2D samples (Figure 3a,b). 1D-CNN (Conv1D) models acquire the temporal patterns of sequence data through a 1D convolution, and Conv1D layers learn local features by stacking in a shallow network, whereas a deeper network synthesizes more pattern features within a larger receptive field. The representation of sequence patterns by LSTM models at different temporal frequencies is advantageous for analyzing the temporal characteristics within a crop growing season. 3D times-series samples (Figure 3c) are used as 2D-CNN input, and the Conv2D layer captures the crop temporal and spatial variations through convolution of the spatial domain and through time sequences of the multi-temporal images. 3D-CNN convolves multi-temporal images from different dimensions and represents features of shallow and deep temporal, spatial, and spectral information of crops by stacking convolutional layers (Conv3D). Like

LSTM, ConvLSTM2D is sensitive to temporal patterns, and convolutional operations inside the ConvLSTM2D cell efficiently capture spatial information. The structure (ConvLSTM2D) learns and represents temporal, spectral, and spatial information similar to that of the 3D-CNN models. Both 3D-CNN and ConvLSTM2D models use 4D time-series samples (Figure 3d) as model input. Figure 4 shows the different network architectures.

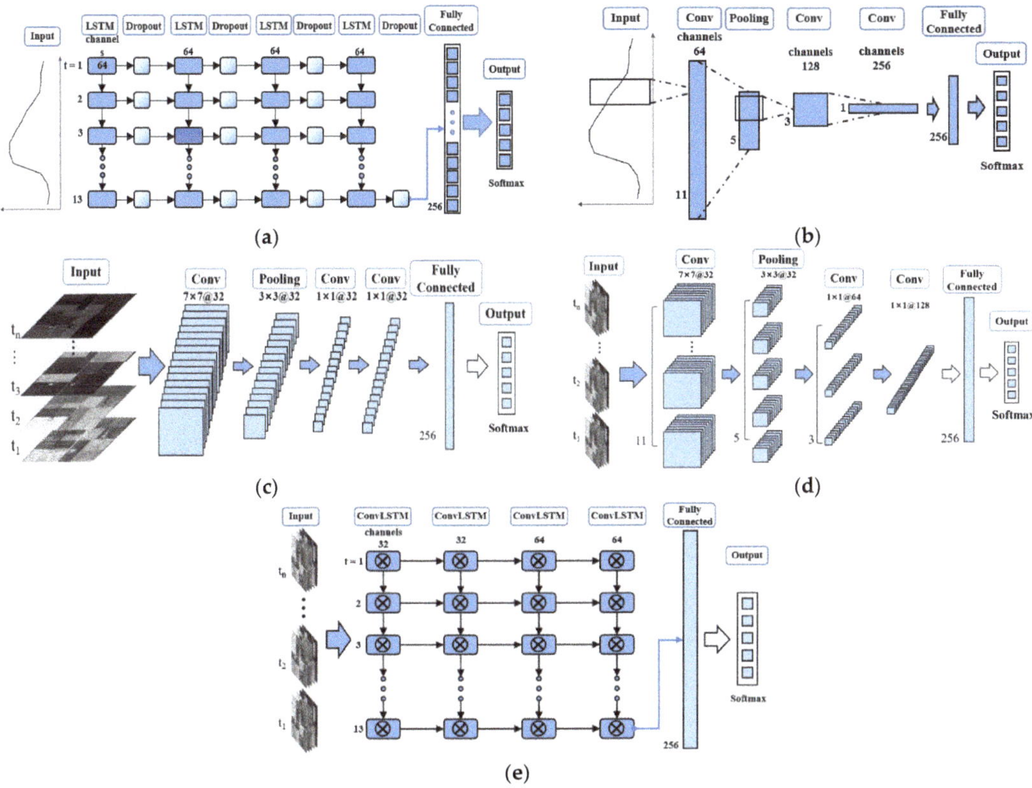

Figure 4. Architectures of (**a**) LSTM, (**b**) 1D-CNN, (**c**) 2D-CNN, (**d**) 3D-CNN, and (**e**) ConvLSTM2D.

Because of the versatility and complexity of deep learning architectures, no standard procedure exists to search for the optimal combination of hyperparameters and the associated layers [18,19]. As a result, an extremely large number of potential network architectures must be considered, making it impossible to try them all. In this paper, the hyperparameter setting and optimization of model are based on strategies from the literature [8,9]. The hyperparameters of the deep learning models include the type and number of hidden layers and the number of neurons in each layer. The layer channels are 16, 32, 64, 128, 256 and the sample sizes are 3, 5, 7, 9. The learning rate is 0.01 or 0.05. The length of the time series is 13. The convolution kernel width is 3 [26,27]. Pooling layers are fixed as max-pooling, with a window size of 2. Dropout with probabilities of 0.3, 0.5, and 0.8 is a regularization technique that randomly drops neurons in a layer during training to prevent the output of the layer from relying on only a few neurons. Each model contains two fully connected layers at the output end. The last layer contains five neurons corresponding to the probability of the five classes.

The hyper-parameters are selected and determined step-by-step, based on numerous training experiments. Each deep learning model (Figure 4) is determined by stepwise optimization and adjustment [9]. A large number of training experiments have shown that the epoch of 400 can meet the training requirements of the model. All deep learning architectures are trained by a backpropagation algorithm, where the stochastic gradient descent is used as the optimizer for model training. The parameters of the stochastic gradient descent are decay = $10 - 5$ and momentum = 0.99. The sample size of the architectures is 9. The learning rate and batch size are 0.01 and 32, respectively. The dropout probability in LSTM is 0.8. Binary cross entropy serves as the loss function. Deep learning models were built using the Keras library and TensorFlow. Finally, the confusion matrix and kappa coefficient from Scikit-learn are metrics for evaluating the accuracy of crop classification. The calculation of VIs and the construction of time-series data are implemented in Python.

3.6. Experiment Design

The multi-temporal images are divided into different experimental groups based on the multiple sample types presented in Section 3.2, and the different deep learning models are used to classify the crops based on multi-temporal images. Additionally, random forest is used as a benchmark model in E1, E2, E3 and E6. See Table 4 for details. The B2348 (Table 4) corresponds to the four spectral bands in Table 1. The same applies to the other features (Table 4).

Table 4. Experiment groups.

Number	Features	Samples Dimensions	Model	
E1	NDVI	1-D time-series	1D-CNN	LSTM
E2	EVI			
E3	B2348	2-D time-series	1D-CNN	LSTM
E4	B2348 + B11 + B12			
E5	B2345678			
E6	All Bands			
E7	NDVI	3-D time-series	2D-CNN	
E8	EVI			
E9	B2348	4-D time-series	3D-CNN	ConvLSTM-2D
E10	All Bands			

E1 and E2 are time-series VI datasets with temporal features constructed from a single VI. E3–E6 are multi-temporal images acquired with different spectral combinations. E3 is a conventional spectral combination of red–green–blue and near-infrared bands. E4 and E5 add shortwave infrared (SWIR) and red-edge spectral bands to E3, respectively. E6 contains the 10 spectral bands of Sentinel-2 images. 1D-CNN and LSTM models are used for crop classification with different spectral combinations and to analyze how multi-spectral and temporal information affect classification accuracy. E7 and E8 are used to classify crops with a 2D-CNN model, and the comparison with E1 and E2 is designed to quantify the contribution of spatial information in multi-temporal crop classification. E9 and E10 are used to classify crops with 3D-CNN and ConvLSTM2D models; E9 uses conventional spectral bands as input and E10 uses the 10 spectral bands of Sentinel-2 images. The comparison and analysis of crop classification with the different experimental groups show how temporal, spectral, and spatial information affect classification accuracy.

4. Results

The accuracy of crop classification via multi-temporal images mainly depends on three factors: time-series data construction, feature extraction, and classification method. Our experiments verify the contribution of time-series data and deep learning models. Various time-series data are constructed based on the strategy presented in Section 3.2 and feed into the deep learning architectures (Figure 4) of Section 3.3 for different experiments. The classification results and accuracies are given in subsequent sections.

4.1. Classification Based on VI Time Series

E1 and E2 in Figure 5 and Table 5 show the results of time-series crop classification based on NDVI and EVI. The classification accuracies produced by the 1D-CNN (Figure 4b) and LSTM (Figure 4a) models for E1 and E2 exceed 92%, and the kappa coefficient is greater than 0.9. The highest overall accuracy (OA) for E2 (LSTM) is close to 94%. Compared with random forest, deep learning models based on 1D-CNN and LSTM have higher accuracy (Table 5) and better performance in local regions (Figure 5). These results show that the 1D-CNN and LSTM models constructed herein are suitable for multi-temporal crop classification based on VI. Compared with E1, the OA for E2 increases by 0.26% and 0.69% for the 1D-CNN and LSTM models, respectively. This reflects the variability of different VIs and the similarity of time-series VI for crop classification. Compared with the 1D-CNN model, the LSTM model is more accurate; the OA improves by 0.75% and 1.18% for E1 and E2, respectively. These results show that both the LSTM and 1D-CNN models can capture temporal features, although the LSTM model is more accurate.

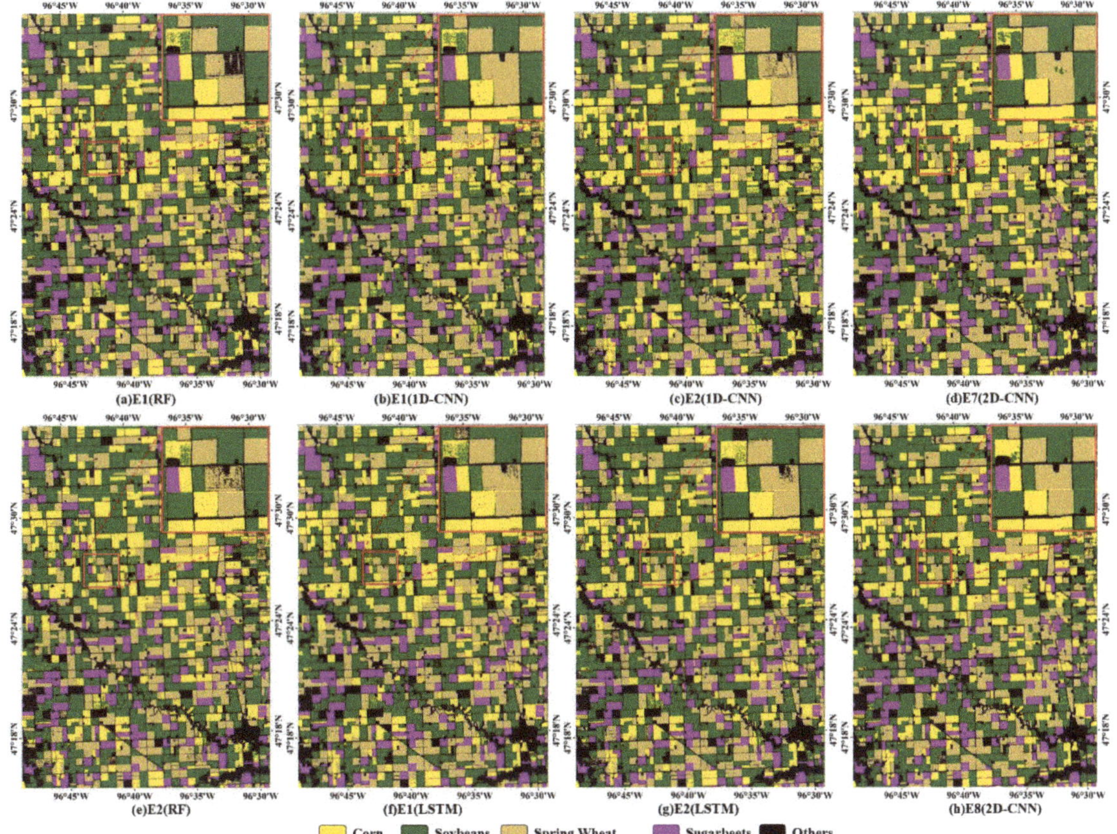

Figure 5. Crop classification results based on VI time-series (see red boxes for more detail).

Table 5. Classification accuracy produced by various models with VI time series.

Number	Model	Accuracy	
		OA	Kappa
	RF	91.02	0.891
E1	1D-CNN	92.50	0.906
	LSTM	93.25	0.915
	RF	91.24	0.893
E2	1D-CNN	92.76	0.909
	LSTM	93.94	0.924
E7	2D-CNN	94.74	0.934
E8	2D-CNN	94.76	0.934

Differences in architecture also affect classification accuracy. Compared with the other results in Figure 5, the RF-based results (Figure 5a,e) are worse locally, while almost no salt-and-pepper noises appear in Figure 5c,h. Compared with E1 and E2, the accuracy of E7 and E8 improved by 0.82% to 2.24%, and the improvement exceeds RF by 3.5%. E7 and E8 classified by the 2D-CNN model (Figure 4c) produce a favorable overall classification accuracy of above 94.7% and a kappa coefficient of 0.934, which is attributed to the effective learning and representation of temporal and spatial information in patch-based time-series VI data by 2D-CNN.

Figure 5 and Table 5 also show that the classification results based on deep learning outperform the random forest. However, the misclassification of crop types in Figure 5 indicates that further optimization is still needed. Based on the same model, there is no significant accuracy difference in E1 and E2. This indicates that improving accuracy solely using time-series data (temporal features) constructed from a single VI is difficult. However, the addition of spatial information not only improves crop classification accuracy but also eliminates salt-and-pepper noise. In addition, the 1D-CNN and LSTM architectures limit the possibility of exploiting spatial information in multi-temporal crop classification, whereas the 2D-CNN model produces more accurate crop classification based on single VI time-series data.

4.2. Classification Based on Multi-Spectral Time Series

Figure 6 and Table 6 show the classification results of E3–E6 based on the time-series data constructed from multi-spectral, multi-temporal images. The crop classification accuracy of the 1D-CNN model is less than that of the LSTM model applied to E3–E6, which is similar to the results of the LSTM model. Therefore, hereinafter, we consider only the crop classification results based on LSTM.

The input data in E3–E6 have both multi-spectral and -temporal features, differing only in the number of multi-spectral bands, as explained in Section 3.4. Table 6 shows that the accuracy of RF-based is lower than deep learning, and Figure 6 also shows that results of deep learning are better in local areas. The OA of E3–E6 is 95.31%, 96.72%, 96.37%, and 96.94%, respectively. Compared with E3, the addition of spectral bands, especially red-edge bands (E5) or SWIR bands (E4), improves the crop classification accuracy, with SWIR bands contributing slightly more than red-edge bands. Using the LSTM model with E6 surprisingly remains the most accurate configuration, with the crop-classification accuracy improving by 1.63% with respect to E3. This indicates that the advantage of the number of spectral bands in multi-spectral images cannot be neglected. With the addition of spectral bands, salt-and-pepper noise is eliminated to varying degrees, with the least salt-and-pepper noise coinciding with the most accurate crop classification (Figure 6f), indicating that the salt-and-pepper phenomenon is weakened but hardly eliminated by using multi-spectral bands. Combined with the presentation in Section 4.1, these results further demonstrate how spatial information affects multi-temporal crop classification.

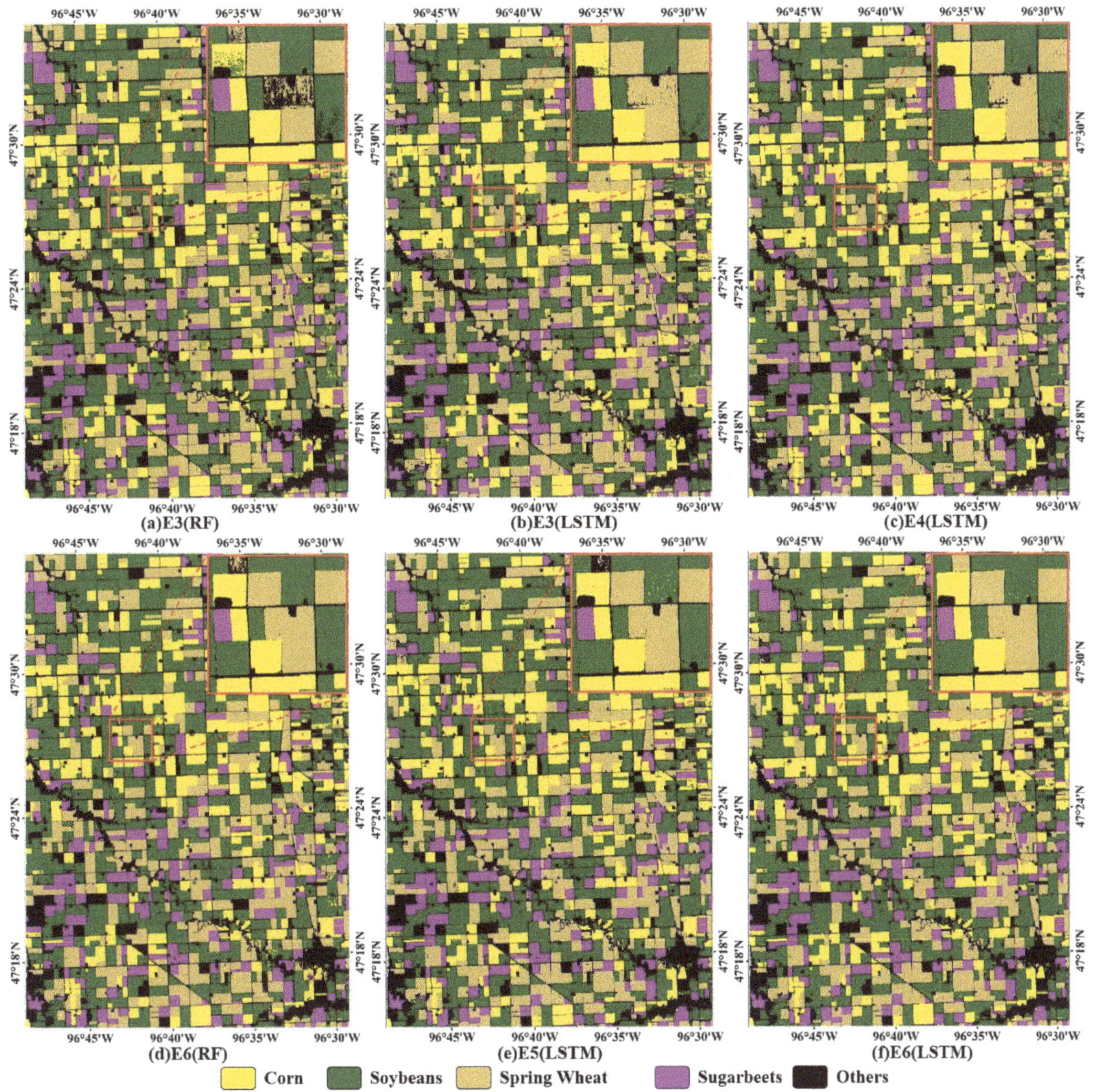

Figure 6. Crop classification results based on multi-spectral time series.

Furthermore, the addition of different spectral bands in E3–E6 increases the diversity of input classification data. In the same experimental group, the accuracy difference between 1D-CNN and LSTM varies from 0.1% to 0.44%, with the minimum difference of 0.1% presented in E6. However, in the different experimental groups, the accuracy difference of the same model varies from 1.06% to 1.95%, with E6 showing an accuracy improvement of nearly 2% compared to E3. In E9 and E3, the spatial information causes differences in the input data. The accuracy difference between different deep learning models with the same input data is small, ranging from 0.21% to 0.42%. In contrast, the accuracy difference between the same models with different input data is larger, ranging from 1.88% to 1.25%.

This indicates that increasing the diversity of input data is more important for improving crop classification accuracy than using different deep learning models.

Table 6. Classification accuracy produced by various models and multi-spectral time-series data.

Number	Model	Accuracy	
		OA	**Kappa**
	RF	93.48	0.918
E3	1D-CNN	94.89	0.936
	LSTM	95.31	0.941
E4	1D-CNN	96.28	0.953
	LSTM	96.72	0.959
E5	1D-CNN	96.02	0.950
	LSTM	96.37	0.955
	RF	95.51	0.944
E6	1D-CNN	96.84	0.960
	LSTM	96.94	0.962
E9	3D-CNN	96.77	0.960
	ConvLSTM2D	96.56	0.957
E10	3D-CNN	**97.43**	**0.968**
	ConvLSTM2D	97.25	0.966

Figure 7 and Table 6 present the classification results of E9 and E10 using the 3D-CNN (Figure 4d) and ConvLSTM2D (Figure 4e) models. The OA of 3D-CNN in E9 and E10 was 96.77% and 96.56%, respectively, with kappa coefficients of 0.960 and 0.957. The OA of ConvLSTM2D in E9 and E10 was 97.43% and 97.25%, respectively, with kappa coefficients of 0.968 and 0.966. The accuracy is slightly greater when using the 3D-CNN model than when using the ConvLSTM2D model. The use of the 3D-CNN model on E10 produces the greatest crop classification accuracy of 97.43%, which translates into an OA improved by 3.69%, 2.67%, 0.49%, and 4.93% with respect to E2 (LSTM), E8 (2D-CNN), E6 (LSTM), and E1 (1D-CNN), respectively. Compared with the E6 (LSTM), the salt-and-pepper noise is eliminated in E9 and E10 (Figure 7b,d), although the improvement in accuracy is not obvious. E10 produces more accurate results than E9 because it contains more spectral bands in the input data.

The classification results of the different experiments verify the feasibility of the model constructed herein (Figure 4) for multi-temporal crop classification. The comparison of the results of the different experiments shows that both the construction of the time-series data and that of the classification model influence the crop classification accuracy. The LSTM model produces more accurate crop classification results than the 1D-CNN model. However, when using time-series data constructed from VIs, the 2D-CNN model produces more accurate results than the 1D-CNN and LSTM models after the elimination of the salt-and-pepper noise. When using time-series data constructed by stacking spectral bands, increasing the number of bands in the input data improves the crop classification accuracy while somewhat reducing the salt-and-pepper noise. Additionally, the LSTM model again produces slightly more accurate crop classifications than the 1D-CNN model, which indicates that the LSTM model is more able to capture temporal features.

E10 treated by the 3D-CNN and ConvLSTM2D models (Figure 4) produces the most accurate crop classification of all experiments. In addition, the architectures of the 3D-CNN and ConvLSTM2D models lead to better learning and representation for multi-temporal crop features, making these models more suitable for crop classification from multi-temporal images.

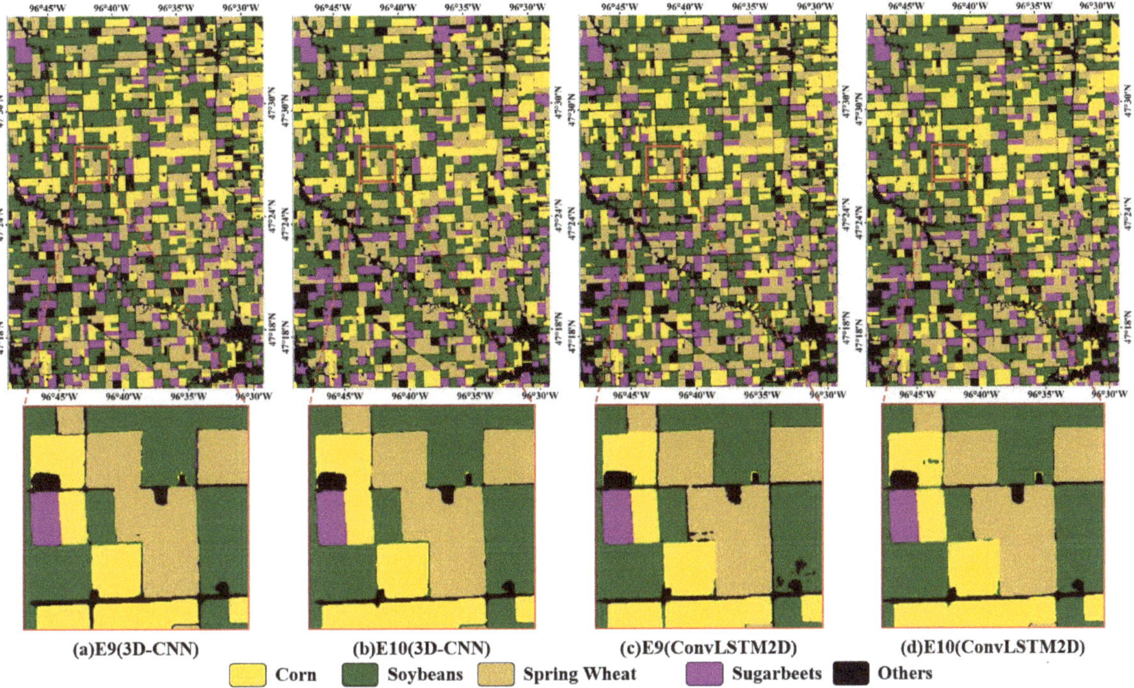

Figure 7. Crop classification results based on temporal, spectral, and spatial information.

Combined with the previous analysis of classification accuracy, VI time-series data using only temporal information only slightly improves the crop classification accuracy. The addition of multi-spectral data based on temporal information improves crop classification accuracy, and the salt-and-pepper noise is more easily alleviated upon increasing the number of spectral bands. As the number of input features increases, the contribution of spatial information in improving classification accuracy decreases. However, the elimination of salt-and-pepper noise through the use of spatial information remains a clear advantage in crop mapping. Therefore, making full use of the temporal, spectral, and spatial information is a more feasible strategy for multi-temporal crop classification. The deep learning architecture fed with 4D data involving multi-temporal images is thus the best model for accurate crop classification based on multi-temporal images.

5. Discussion

5.1. Analysis of Time-Series Profile

Figure 8 shows the temporal profiles of crops produced by VIs and spectra. The buffer areas of crop profiles overlap throughout the growing season, despite the difference in average reflectance or VI values. In the middle of the growing season, the spectral overlap within the crop becomes smaller (~DOY 200–220) than in the early or late growing season. During this period, the temporal curves of crops with one standard deviation are more stable and distinguishable, which indicates that this feature should be useful for differentiating between crops. In addition, the temporal windows always serve for single-temporal crop classification [28]. However, the similarity and overlap of profiles over the whole growing season make it difficult to distinguish crops such as corn and soybeans based solely on single images [1,2].

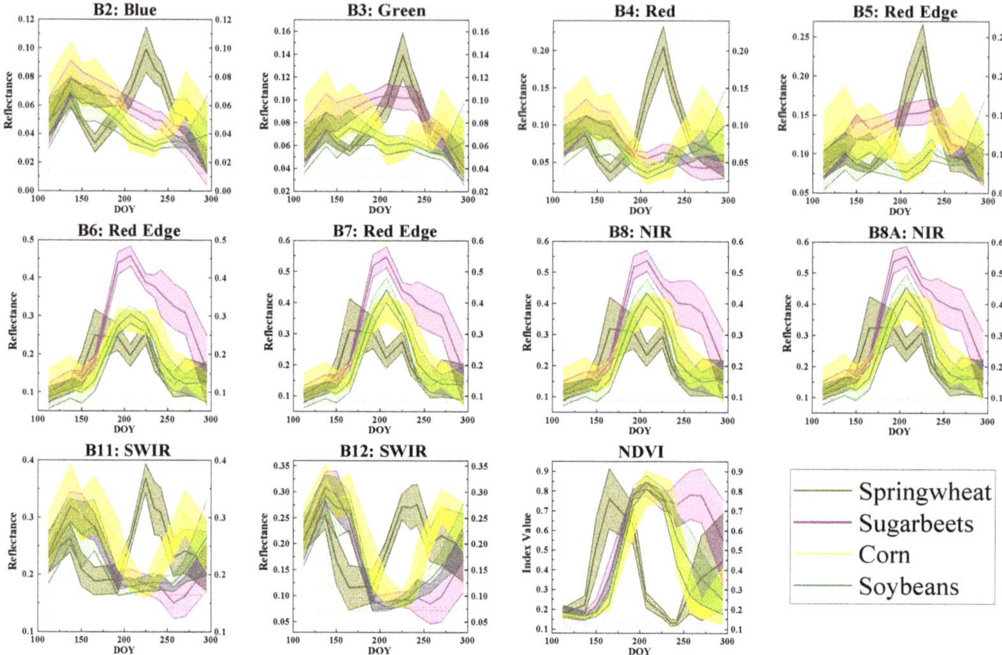

Figure 8. Time-series spectral band and vegetation indices are aggregated for crop fields. The buffers indicate one standard deviation calculated from the fields.

The differences in the time series curves (Figure 8) between crops in different spectral ranges and time periods make it possible to distinguish between crops [29]. For example, the gap in B8 (Figure 8) during the middle growing season ($\approx$DOY 180–220) makes it possible to distinguish between spring wheat and sugar beets. Figure 8 shows that almost no spectral overlap occurs between corn and soybeans in B11 and B12 during the period of time ($\approx$DOY 170–200). The gap observed in the profiles of sugarbeets and other crops in bands B6-B8 and B8A, as shown in Figure 8, occurs during two periods of time, which are around DOY 180–220 and 250–270. Spring wheat can be directly distinguished from profiles in B2–B5 (Figure 8) around DOY 225 and in B11 and B12 in the period DOY 210–240. Corn and soybeans can be differentiated with greater probability in the period DOY 170–200 in B11 and B12. In addition, the overlap in temporal profile based on the NDVI is similar to the other spectra in Figure 8. The profiles of corn and soybeans almost overlap over the entire growing season, which explains the difficulty of distinguishing between these two crops [3,4]. The profiles of sugarbeets and spring wheat clearly differ between DOY 260 and 170.

As previously mentioned, time-series images based on single VI or band are insufficient to accurately distinguish between different crops. However, different crops exhibit spectral differences in the time-series curves of each spectral band (Figure 8), indicating the potential of each spectral band to distinguish between different crops. Better utilization of the advantages of multi-spectral bands has greater potential to improve the accuracy of crop classification [30]. The addition of different types of spectral bands such as red-edge and SWIR has reinforced this conclusion in classification experiments [9].

5.2. Effects of Temporal, Spectral, and Spatial Feature

The effects of temporal, spectral, and spatial information on crop classification are revealed in the different time-series data. The crop classification results due to the different time-series classification data are shown in Figures 5–7 and Tables 5 and 6. Using only temporal features may not be sufficient for accurate crop classification due to salt-and-pepper noise (Figures 5 and 6), which can affect pixel-based classification. Fully exploiting the abundant spectral and spatial information in multi-temporal images can be challenging when using only VI, but it provides more possibilities for improving accuracy. [5,31] pointed out that spatial features such as texture can lead to good classification performance, and a similar result occurs for 2D-CNN classification (Figure 5). In addition, based on the analysis in the previous sections, the contribution to the accuracy of spatial information such as texture [9,32] decreases as the number of input features increases. Moreover, the spatial information contributes significantly to the classification accuracy for a feature input of a single VI. [8] also suggested that more information-dense data are required to improve the crop-classification accuracy based on multi-temporal images. The diversity of information and the differences in time-series data depicted in Figure 8 provide more possibilities for accurate classification and can alleviate the salt-and-pepper phenomenon. Nevertheless, spatial information remains a vital ingredient to eliminate salt-and-pepper noise.

5.3. Comparison of Deep Learning Models

The temporal dependencies in multi-temporal images are long term and complex, and crops have unique temporal, spectral, and spatial features (Figure 8). Sufficient model complexity and automated feature learning and representation satisfy the data-processing needs of models in multi-temporal crop classification [9,12]. Differing from the result that 1D-CNN accuracy is higher than that of LSTM [9], increasing the number of spectral bands in this work causes the accuracy of 1D-CNN to be close to that of LSTM. This indicates that input features and application scenarios (more crop types) may also affect the accuracy of the classification. The architecture of 2D-CNN models is limited by their structure, meaning that they can only accept time-series data constructed by a single VI or spectral band as input. This prevents 2D-CNN models from exploiting multi-spectral information. The analysis in Section 4 also points out that 2D-CNN models are less accurate than 1D-CNN and LSTM models using multiple spectral bands. In contrast with 2D-CNN models, both 3D-CNN and ConvLSTM2D models require 4D data that perfectly fit the temporal, spectral, and spatial features. The classification results (Figures 7 and 9) of 3D-CNN and ConvLSTM2D models are also significantly more accurate and stable than other comparative models. [30] also pointed out that models such as 3D-CNN should be considered for crop classification from multi-temporal images.

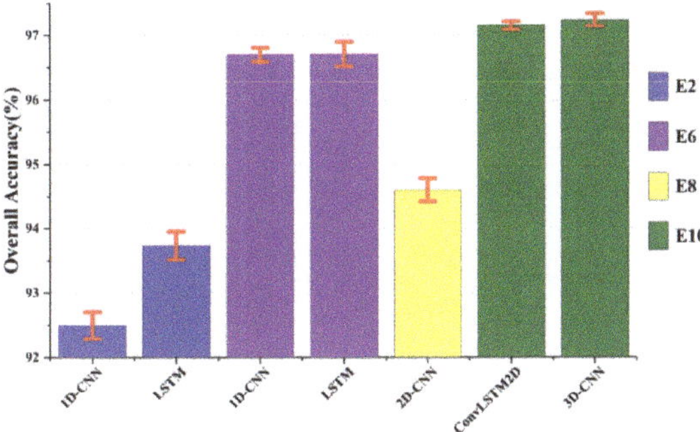

Figure 9. The OA of different deep learning models.

As described in Section 3.5, each model is trained extensively to achieve the best classification results. Therefore, the parameters of deep learning models in this work will likely need to be adjusted to achieve satisfactory accuracy for other classification tasks. Additionally, numerous model training experiments are necessary in this process.

5.4. Potential of 3D-CNN and ConvLSTM2D for Crop Classification from Multi-Temporal Images

Crop classification from multi-temporal RS images often has a time lag due to data acquisition [5,6]. However, time-series data can alleviate this issue, whereby different objects have the same spectrum, and the same objects have different spectra in the background of relatively complex crop cultivations. Previous analyses also revealed that fully exploiting the temporal, spectral, and spatial information in multi-temporal images should be a major avenue to improve classification accuracy. 3D-CNN and ConvLSTM2D models can integrate multi-temporal information and have advantageous structures not found in other models such as 2D-CNN and SVM [11,19]. The best classification accuracies are provided by 3D-CNN and ConvLSTM2D models, and exceed 97% (Table 6). Figure 10 shows the strong correlation between the results obtained herein and the CDL for the area ratio of different crops. It also shows potential applications for crop classification based on multi-temporal images.

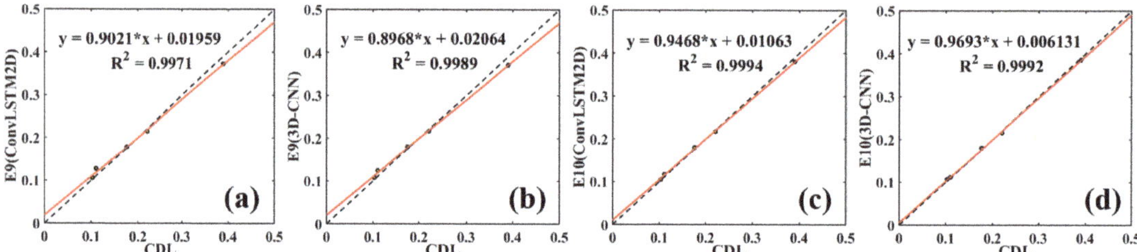

Figure 10. Correlation of crop-area ratio. ((**a**–**d**) correspond to four experiments, as shown in the vertical label. The scatter points mean the fraction of different crop over the study area. The red line reflects the consistency of crop area between the classification results and the CDL.)

Different network structures in deep learning models such as inception [33], dropout [8], and transformer [34] all enhance the feature learning and representation capabilities of the network. Deep learning models (Figure 4) are constructed by simple stacking of modules, so they lack special design for multi-temporal images and cannot treat scale effects [35] in images. In addition, information redundancies (Figure 8) with high inter-band similarity must be considered. Both architectures have inherent advantages for processing multi-temporal images. Although ConvLSTM2D has fewer applications in multi-temporal image crop classification than 3D-CNN [14], the results of this study show that this model approaches the classification capability of 3D-CNN. References [13,36] pointed out that 3D-CNN is not suitable for establishing long-term dependencies of time-series data due to locally computed convolutions, whereas ConvLSTM2D combines the sequence processing capability of LSTM and the structure of CNN, which facilitates the addition of multiple special structures and modules so that it can be exploited to classify crops from multi-temporal images.

6. Conclusions

This paper constructs various time-series datasets based on Sentinel-2 multi-temporal images by VI or spectral stacking, and develops deep learning models with different structures for classifying crops from multi-temporal images. The results lead to the following conclusions:

(1)　Greater data diversity (temporal, spectral and spatial information) is effective in improving crop classification accuracy. The temporal feature only provides limited improvement in the accuracy of crop classification from multi-temporal images. As more spectral information is added, the accuracy can be further improved, and the impact of salt-and-pepper noise can be alleviated. The inclusion of spatial information can eliminate salt-and-pepper noise, and its contribution to accuracy decreases as the number of input features increases.

(2)   Various deep learning models have limitations in crop classification from multi-temporal images. 1D-CNN and LSTM models cannot extract spatial features while integrating temporal and spectral features. Additionally, a 2D-CNN is suitable for crop classification of time-series data given a single feature such as a VI or band because the multi-spectral advantages are hard to consider when combining temporal and spatial information. The 3D-CNN and ConvLSTM2D models are the most accurate for classifying crops and are more suitable for multi-temporal crop classification than other deep learning models.

(3)   The deep learning models based on Conv3D and ConvLSTM2D, which integrate temporal, spectral, and spatial information, are the most accurate models for multi-temporal crop classification. In addition, the advantages of incorporating RNN and CNN and the more flexible structure mean that ConvLSTM should be investigated.

In this paper, smaller areas and simple crop types are used for deep learning multi-temporal crop classification application studies. In future research, crop classification based on deep learning is still needed for large-scale study areas and complex planting systems, such as crop rotation and more crop types. In addition, the impact of clouds on image acquisition is difficult to avoid. While the acquisition of synthetic aperture radar (SAR) is not affected by clouds, which can also increase the diversity of classification data. Therefore, research into crop classification by synergistic SAR and optical images with different acquisition frequencies will be carried out. Additionally, the ConvLSTM model will be used as the classification model to explore its potential in multi-source image crop classification.

Author Contributions: Conceptualization, Q.L. and J.T.; Methodology, Q.L. and J.T.; Software, Q.L.; Validation, Q.L.; Formal analysis, Q.L. and J.T.; Writing—original draft preparation, Q.L.; Writing—review and editing, Q.L., J.T. and Q.T.; Visualization, Q.L.; Project administration, Q.T.; Funding acquisition, J.T. All authors have read and agreed to the published version of the manuscript.

Funding: This research was funded by the National Natural Science Foundation of China (grant number: 42101321 and 41771370) and the Open Fund of State Key Laboratory of Remote Sensing Science (grant number: OFSLRSS202119).

Institutional Review Board Statement: Not applicable.

Data Availability Statement: The data presented in this study are available on request from the author.

Acknowledgments: The authors acknowledge the support provided by their respective institutions.

Conflicts of Interest: The authors declare no conflict of interest.

References

1.   Blickensdörfer, L.; Schwieder, M.; Pflugmacher, D.; Nendel, C.; Erasmi, S.; Hostert, P. Mapping of crop types and crop sequences with combined time series of Sentinel-1, Sentinel-2 and Landsat 8 data for Germany. *Remote Sens. Environ.* **2022**, *269*, 112831. [CrossRef]

2.   Griffiths, P.; Nendel, C.; Hostert, P. Intra-annual reflectance composites from Sentinel-2 and Landsat for national-scale crop and land cover mapping. *Remote Sens. Environ.* **2019**, *220*, 135–151. [CrossRef]

3.   Xu, J.; Zhu, Y.; Zhong, R.; Lin, Z.; Xu, J.; Jiang, H.; Huang, J.; Li, H.; Lin, T. DeepCropMapping: A multi-temporal deep learning approach with improved spatial generalizability for dynamic corn and soybean mapping. *Remote Sens. Environ.* **2020**, *247*, 111946. [CrossRef]

4.   Xu, J.; Yang, J.; Xiong, X.; Li, H.; Huang, J.; Ting, K.C.; Ying, Y.; Lin, T. Towards interpreting multi-temporal deep learning models in crop mapping. *Remote Sens. Environ.* **2021**, *264*, 112599. [CrossRef]

5.   Cai, Y.; Guan, K.; Peng, J.; Wang, S.; Seifert, C.; Wardlow, B.; Li, Z. A high-performance and in-season classification system of field-level crop types using time-series Landsat data and a machine learning approach. *Remote Sens. Environ.* **2018**, *210*, 35–47. [CrossRef]

6.   Belgiu, M.; Csillik, O. Sentinel-2 cropland mapping using pixel-based and object- based time-weighted dynamic time warping analysis. *Remote Sens. Environ.* **2018**, *204*, 509–523. [CrossRef]

7. Pelletier, C.; Webb, G.; Petitjean, F. Temporal convolutional neural network for the classification of satellite image time series. *Remote Sens.* **2019**, *11*, 523. [CrossRef]

8. Dou, P.; Shen, H.; Li, Z.; Guan, X. Time series remote sensing image classification framework using combination of deep learning and multiple classifiers system. *Int. J. Appl. Earth Obs. Geoinf.* **2021**, *103*, 102477. [CrossRef]

9. Zhong, L.; Hu, L.; Zhou, H. Deep learning based multi-temporal crop classification. *Remote Sens. Environ.* **2018**, *221*, 430–443. [CrossRef]

10. Mou, L.; Bruzzone, L.; Zhu, X.X. Learning spectral-spatial-temporal features via a recurrent convolutional neural network for change detection in multispectral imagery. *IEEE Trans. Geosci. Remote Sens.* **2019**, *57*, 924–935. [CrossRef]

11. Ji, S.; Zhang, C.; Xu, A.; Shi, Y.; Duan, Y. 3D convolutional neural networks for crop classification with multi-temporal remote sensing images. *Remote Sens.* **2018**, *10*, 75. [CrossRef]

12. Qu, Y.; Yuan, Z.; Zhao, W.; Chen, X.; Chen, J. Crop classification based on multi-temporal features and convolutional neural network. *Remote Sens. Technol. Appl.* **2021**, *36*, 304–313. [CrossRef]

13. Giannopoulos, M.; Tsagkatakis, G.; Tsakalides, P. 4D U-Nets for Multi-Temporal Remote Sensing Data Classification. *Remote Sens.* **2022**, *14*, 634. [CrossRef]

14. Yang, X.; Ye, Y.; Li, X.; Lau, R.Y.K.; Zhang, X.; Huang, X. Hyperspectral image classification with deep learning models. *IEEE Trans. Geosci. Remote Sens.* **2018**, *56*, 5408–5423. [CrossRef]

15. Hochreiter, S.; Schmidhuber, J. Long short-term memory. *Neural Comput.* **1997**, *9*, 1735–1780. [CrossRef]

16. Sharma, A.; Liu, X.; Yang, X. Land cover classification from multi-temporal, multi-spectral remotely sensed imagery using patch-based recurrent neural networks. *Neural Netw.* **2018**, *105*, 346–355. [CrossRef]

17. Xie, Y.; Zhang, Y.; Xun, L.; Chai, X. Crop classification based on multi-source remote sensing data fusion and LSTM algorithm. *Trans. Chin. Soc. Agric. Eng.* **2019**, *35*, 129–137. [CrossRef]

18. Shi, X.; Chen, Z.; Wang, H.; Yeung, D.Y.; Wong, W.K.; Woo, W.C. Convolutional LSTM Network: A Machine Learning Approach for Precipitation Nowcasting. In Proceedings of the 28th International Conference on Neural Information Processing Systems, Montreal, QC, Canada, 7–12 December 2015.

19. Hu, W.-S.; Li, H.-C.; Pan, L.; Li, W.; Tao, R.; Du, Q. Spatial-Spectral Feature Extraction via Deep ConvLSTM Neural Networks for Hyperspectral Image Classification. *IEEE Trans. Geosci. Remote Sens.* **2020**, *58*, 4237–4250. [CrossRef]

20. Ahmad, R.; Yang, B.; Ettlin, G.; Berger, A.; Rodriguez-Bocca, P. A machine-learning based ConvLSTM architecture for NDVI forecasting. *Int. Trans. Oper. Res.* **2020**, *30*, 2025–2048. [CrossRef]

21. Seydgar, M.; Naeini, A.A.; Zhang, M.; Li, W.; Satari, M. 3-D convolution-recurrent networks for spectral-spatial classification of hyperspectral images. *Remote Sens.* **2019**, *11*, 883. [CrossRef]

22. Boryan, C.; Yang, Z.; Mueller, R.; Craig, M. Monitoring US agriculture: The US Department of Agriculture, National Agricultural Statistics Service, Cropland Data Layer Program. *Geocarto Int.* **2011**, *26*, 341–358. [CrossRef]

23. NASS/USDA. Minnesota Cropland Data Layer. 2021. Available online: https://www.nass.usda.gov/Research_and_Science/Cropland/metadata/meta.php (accessed on 10 March 2023).

24. Tucker, C.J. Red and photographic infrared linear combinations for monitoring vegetation. *Remote Sens. Environ.* **1979**, *8*, 127–150. [CrossRef]

25. Huete, A.; Didan, K.; Miura, T.; Rodriguez, E.; Gao, X.; Ferreira, L. Overview of the radiometric and biophysical performance of the MODIS vegetation indices. *Remote Sens. Environ.* **2002**, *83*, 195–213. [CrossRef]

26. Chen, Y.; Jiang, H.; Li, C.; Jia, X.; Ghamisi, P. Deep feature extraction and classification of hyperspectral images based on convolutional neural networks. *IEEE Trans. Geosci. Remote Sens.* **2016**, *54*, 6232–6251. [CrossRef]

27. Li, Y.; Zhang, H.; Shen, Q. Spectral-spatial classification of Hyper-spectral imagery with 3D convolutional neural network. *Remote Sens.* **2017**, *9*, 67. [CrossRef]

28. Tian, F.; Wu, B.; Zeng, H.; He, Z.; Zhang, M.; Jose, B. Identifying Soybean Cropped Area with Sentinel-2 Data and Multi-Layer Neural Network. *J. Geo-Inf. Sci.* **2019**, *21*, 918–927. [CrossRef]

29. Wang, H.; Zhao, X.; Zhang, X.; Wu, D.; Du, X. Long time series land cover classification in China from 1982 to 2015 based on Bi-LSTM deep learning. *Remote Sens.* **2019**, *11*, 1639. [CrossRef]

30. Yang, S.; Gu, L.; Li, X.; Tao, J. Crop classification method based on optimal feature selection and hybrid CNN-RF networks for multi-temporal remote sensing imagery. *Remote Sens.* **2020**, *12*, 3119. [CrossRef]

31. Sun, Z.; Chen, W.; Guo, B.; Cheng, D. Integration of Time Series Sentinel-1 and Sentinel-2 Imagery for Crop Type Mapping over Oasis Agricultural Areas. *Remote Sens.* **2020**, *12*, 158. [CrossRef]

32. Lu, Y.; Li, H.; Zhang, S. Multi-temporal remote sensing based crop classification using a hybrid 3D-2D CNN model. *Trans. Chin. Soc. Agric. Eng.* **2021**, *37*, 142–151. [CrossRef]

33. Dong, Y.; Zhang, Q. A Combined Deep Learning Model for the Scene Classification of High-Resolution Remote Sensing Image. *IEEE Geosci. Remote Sens. Lett.* **2019**, *16*, 1540–1544. [CrossRef]

34. Zhang, J.; Zhao, H.; Li, J. TRS: Transformers for Remote Sensing Scene Classification. *Remote Sens.* **2021**, *13*, 4143. [CrossRef]

35. Hou, X.; Bai, Y.; Li, Y.; Shang, C.; Shen, Q. High-resolution triplet network with dynamic multiscale feature for change detection on satellite images. *ISPRS J. Photogramm. Remote Sens.* **2021**, *177*, 103–115. [CrossRef]
36. Garnot, V.S.F.; Landrieu, L.; Giordano, S.; Chehata, N. Time-space tradeoff in deep learning models for crop classification on satellite multi-spectral image time series. In Proceedings of the IGARSS 2019—2019 IEEE International Geoscience and Remote Sensing Symposium, Yokohama, Japan, 28 July–2 August 2019.

 *agriculture*

Article

Real-Time Detection of Apple Leaf Diseases in Natural Scenes Based on YOLOv5

Huishan Li, Lei Shi, Siwen Fang and Fei Yin *

College of Information and Management Science, Henan Agricultural University, Zhengzhou 450046, China
* Correspondence: yin.fei@henau.edu.cn

Abstract: Aiming at the problem of accurately locating and identifying multi-scale and differently shaped apple leaf diseases from a complex background in natural scenes, this study proposed an apple leaf disease detection method based on an improved YOLOv5s model. Firstly, the model utilized the bidirectional feature pyramid network (BiFPN) to achieve multi-scale feature fusion efficiently. Then, the transformer and convolutional block attention module (CBAM) attention mechanisms were added to reduce the interference from invalid background information, improving disease characteristics' expression ability and increasing the accuracy and recall of the model. Experimental results showed that the proposed BTC-YOLOv5s model (with a model size of 15.8M) can effectively detect four types of apple leaf diseases in natural scenes, with 84.3% mean average precision (mAP). With an octa-core CPU, the model could process 8.7 leaf images per second on average. Compared with classic detection models of SSD, Faster R-CNN, YOLOv4-tiny, and YOLOx, the mAP of the proposed model was increased by 12.74%, 48.84%, 24.44%, and 4.2%, respectively, and offered higher detection accuracy and faster detection speed. Furthermore, the proposed model demonstrated strong robustness and mAP exceeding 80% under strong noise conditions, such as exposure to bright lights, dim lights, and fuzzy images. In conclusion, the new BTC-YOLOv5s was found to be lightweight, accurate, and efficient, making it suitable for application on mobile devices. The proposed method could provide technical support for early intervention and treatment of apple leaf diseases.

Keywords: smart agriculture; detection of apple leaf diseases; YOLOv5; transformer; CBAM

Citation: Li, H.; Shi, L.; Fang, S.; Yin, F. Real-Time Detection of Apple Leaf Diseases in Natural Scenes Based on YOLOv5. *Agriculture* **2023**, *13*, 878. https://doi.org/10.3390/agriculture13040878

Academic Editor: Filipe Neves Dos Santos

Received: 15 March 2023
Revised: 10 April 2023
Accepted: 14 April 2023
Published: 15 April 2023

1. Introduction

As one of the top four popular fruits in the world, apple is highly nutritious and provides significant medicinal value [1]. In China, apple production has expanded, making it the world's largest apple producer. However, a variety of diseases hamper the healthy growth of apple, seriously affecting the quality and yield of apple and causing significant economic losses. According to statistics, there are approximately 200 types of apple diseases, most of which occur in apple leaf areas. Therefore, to ensure the healthy development of the apple planting industry, accurate and efficient leaf disease identification and control measures are needed [2].

In traditional disease identification techniques, fruit farmers and experts rely on visual examination based on their experience, a method which is inefficient and highly subjective. With the advance of computer and information technology, image recognition technology has been gradually applied in agriculture. Many researchers have applied machine vision algorithms to extract features such as color, shape, and texture from disease images and input them into specific classifiers to accomplish plant disease recognition tasks [3]. Zhang et al. [4] processed apple disease images using HSI, YUV, and gray models; then, the authors extracted features using genetic algorithms and correlation based-feature selection, and ultimately discriminated apple powdery mildew, mosaic, and rust diseases using an SVM classifier with an identification accuracy of more than 90%. However, the complex image background and the feature extraction, dominated by strong experience, make the labor and time costs much higher, as well as makingthe system difficult to promote and popularize.

In recent years, deep learning convolutional neural networks have been widely used in agricultural intelligent detection, with faster detection speeds and higher accuracy compared to traditional machine vision techniques [5]. There are two types of target detection models; the first is the two-stage detection algorithm represented by R-CNN [6] and Faster R-CNN [7]. Xie et al. [8] used an improved Faster R-CNN detection model for real-time detection of grape leaf diseases, introducing three modules (Inception v1, Inception-ResNet-v2, and SE) in the model, and mean average precision (mAP) achieved 81.1%. Deng et al. [9] proposed a method for large-scale detection and localization of pine wilt disease using unmanned remote sensing and artificial intelligence technology, and a series of optimizations to improve detection accuracy to 89.1%. Zhang et al. [10] designed a Faster R-CNN (MF^{3}R-CNN) model with multiple feature fusion for soybean leaf disease detection, achieving an average accuracy of 83.34%. Wang et al. [11] used the RFCN ResNet101 model to detect potato surface defects and achieved an accuracy of 95.6%. This two-stage detection model was capable of identifying crop diseases, but its large network model and slow detection speed made it difficult to apply in real planting industry.

Another type of target detection algorithm is the one-stage algorithm represented by SSD [12] and YOLO [13–16] series. Unlike the two-stage detection algorithm, it does not require the generation of candidate frames. By converting the boundary problem into a regression problem, features extracted from the network are used to predict the location and class of lesions. Due to its high accuracy, fast speed, short training time, and low computational requirement, it is more suitable for agricultural applications. Wang et al. [17] used the SSD-MobileNet V2 model for the detection of scratches and cracks on the surface of litchi, which eventually achieved 91.81% mAP and 102 frame per second (FPS). In the experiments of Chang-Hwan et al. [18], a new attention-enhanced YOLO model was proposed for identifying and detecting plant foliar diseases. Li et al. [19] improved the CSP, feature pyramid networks (FPN), and non-maximum suppression (NMS) modules in YOLOv5 to detect five vegetable diseases and obtained 93.1% mAP, effectively reducing missing and false detections caused by complex background. In complex orchard environments, Jiang et al. [20] proposed an improved YOLOX model to detect sweet cherry fruit ripeness. In improving the model, mAP and recall were both improved by 4.12% and 4.6%, respectively, which effectively solved the interference caused by fruit overlaps and shaded branches and leaves. Li et al. [21] used the improved YOLOv5n model to detect cucumber diseases in natural scenes and achieved higher detection accuracy and speed. While the development of intelligent crop disease detection using one-stage detection algorithms has matured, less research has been carried out for apple leaf disease detection. Small datasets and simple image backgrounds pose problems for most existing studies. Consequently, it is crucial to develop an apple leaf disease detection model with high recognition accuracy and fast detection speed for mobile devices with limited computing power.

Considering the complex planting environment in apple orchards and the various shapes of lesions, this study proposed the use of an improved target detection algorithm based on YOLOv5s. The proposed algorithm aimed to reduce false detections caused by multi-scale lesions, dense lesions, and inconspicuous features in apple leaf disease detection tasks. As a result, the accuracy and efficiency of the model could be enhanced to provide essential technical support for apple leaf disease identification and intelligent orchard management.

2. Materials and Methods

2.1. Materials

2.1.1. Data Acquisition and Annotation

In this study, three datasets were used to train and evaluate the proposed model: the Plant Pathology Challenge 2020 (FGVC7) [22] dataset, the Plant Pathology Challenge 2021 (FGVC8) [23] dataset, and the PlantDoc [24] dataset.

FGVC7 and FGVC8 [22,23] consist of apple leaf disease images used in the Plant Pathology Fine-Grained Visual Categorization competition hosted by Kaggle. The images were captured by Cornell AgriTech using Canon Rebel T5i DSLR and smartphones, with a resolution of 4000 × 2672 pixels for each image. There are four kinds of apple leaf diseases, namely rust, frogeye leaf spot, powdery mildew, and scab. These diseases occur frequently and cause significant losses in the quality and yield of apples. Sample images of the dataset are shown in Figure 1.

| (a) | (b) | (c) | (d) |

Figure 1. FGVC7 and FGVC8 disease images. (**a**) Frogeye leaf spot; (**b**) Rust; (**c**) Scab; (**d**) Powdery mildew.

PlantDoc [24] is a dataset of non-laboratory images constructed by Davinder Singh et al. in 2020 for visual plant disease detection. It contains 2598 images of plant diseases in natural scenes, involving 13 species of plants and as many as 17 diseases. Most of the images in PlantDoc have low resolution, large noise, and an insufficient number of samples, making detection more difficult. In this study, apple rust and scab images were used to enhance and validate the generalization of the proposed model. Examples of disease images are shown in Figure 2.

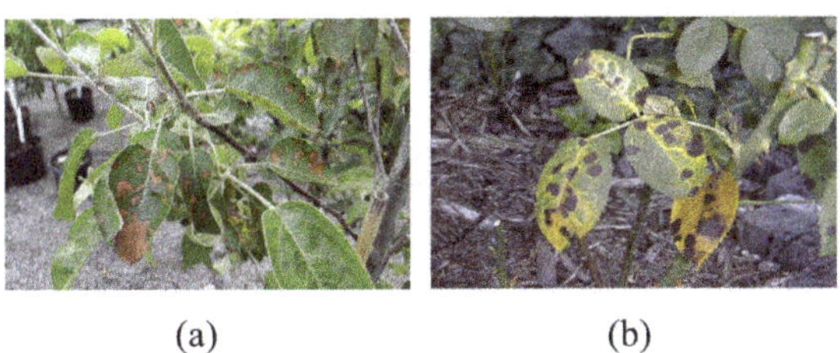

| (a) | (b) |

Figure 2. PlantDoc disease images. (**a**) Rust; (**b**) Scab.

From the collected datasets, we selected (1) images with light intensity varying with the time of day, (2) images capture using different shooting angles, (3) images with different disease intensities, and (4) images from different disease stages to ensure the richness and diversity of the dataset. Finally, a total of 2099 apple leaf disease images were selected. LabelImg software was used to label the images with categories including disease type, center coordinates, width, and height of each disease spots. In total, we annotated 10,727 lesion instances, and annotations are shown in Table 1. The labeled dataset was randomly divided into training and test sets at a ratio of 8:2. This dataset was called ALDD (apple leaf disease data) and was used to train and test the model.

Table 1. Label distribution of ALDD.

Disease Type	Number of Images	Number of Labeled Instances
Scab	498	4722
Frogeye leaf spot	600	3091
Rust	502	2166
Powdery mildew	499	748
Total number	2099	10,727

2.1.2. Data Enhancement

The actual apple orchard in a complex environment contains many disturbances and the currently selected data is far from sufficient. To enrich the image dataset, mosaic image enhancement [16] and online data enhancement were chosen to expand the dataset. Mosaic image enhancement involves a random selection of 4 images from the training set, which are finally combined into one image after rotation, scaling, and hue adjustment. This approach not only enriches the image background and increases the number of instances, but also indirectly boosts the batch size. This accelerates model training and is favorable to improving small target detection performance. Online augmentation is the use of data augmentation in model training, which ensures the invariance of the sample size and the diversity of the overall sample and improves the model's robustness by continuously expanding the sample space. Mainly includes alterations to hue, saturation, brightness transformation, translation, rotation, flip, and other operations. The total number of the dataset is constant; however, the amount of data input to each epoch is changing, and it is more conducive to fast convergence of the model. Examples of enhanced images are shown in Figure 3.

Figure 3. Original and enhanced images. (**a**) Original; (**b**) Flip horizontal; (**c**) Rotation transformation; (**d**) Hue enhancement; (**e**) Saturation enhancement; (**f**) Mosaic enhancement.

2.2. Methods

2.2.1. YOLOv5s Model

Depending on the network depth and feature map width, YOLOv5 can be divided into YOLOv5s, YOLOv5m, YOLOv5l, and YOLOv5x [25]. As the depth and width increase, the number of layers of the network increases as well as the structure becomes more complex. In order to meet the requirements of lightweight deployment and real-time detection, reduce storage space occupied by the model and improve the identification speed, YOLOv5s was selected as the baseline model in this study.

The YOLOv5s was composed of four parts: input, backbone, neck, and prediction. The input section included mosaic data enhancement, adaptive calculation of the anchor box, and adaptive scaling of images. The backbone module performed feature extraction and consisted of four parts: focus, CBS, C3, and spatial pyramid pooling (SPP). There were two types of C3 [26] modules in YOLOv5s for backbone and neck, as shown in Figure 4. The first one used the residual units at the backbone layer, while the second one did not. SPP [27] performed the maximum pooling of feature maps using convolutional kernels of different sizes in order to fuse multiple sense fields and generate semantic information. The neck layer used a combination of (FPN) [28] and path aggregation networks (PANet) [29] to fuse the image features. The prediction included three detection layers, corresponding to 20 × 20, 40 × 40, and 80 × 80 feature maps, respectively, for detecting large, medium, and small targets. Finally, the distance between the predicted boxes and the true boxes was calculated using the complete intersection over union (CIOU) [30] loss function, and the NMS was applied to remove the redundant boxes and retain the detection boxes with the highest confidence. The YOLOv5s network model is shown in Figure 4.

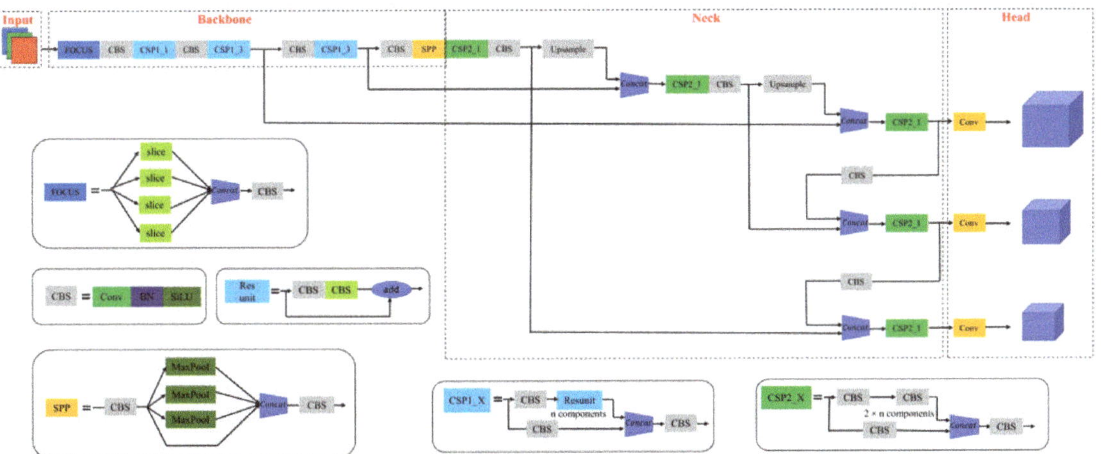

Figure 4. YOLOv5s method architecture diagram.

2.2.2. Bidirectional Feature Pyramid Network

The YOLOv5s combines FPN and PANet for multi-scale feature fusion, with FPN enhancing semantic information in a top-down fashion and PANet enhancing location information from the bottom up. This combination enhances the feature fusion capability of the neck layer. However, when fusing input features at different resolutions, the features are simply summed and their contributions to the fused output features are usually inequitable. To address this problem, Tan et al. [31] developed the BiFPN based on efficient bidirectional cross-scale connections and weighted multiscale feature fusion. The BiFPN introduced learnable weights in order to learn the importance of different input features, while top-down and bottom-up multi-scale feature fusion was applied iteratively. The structure of BiFPN is shown in Figure 5.

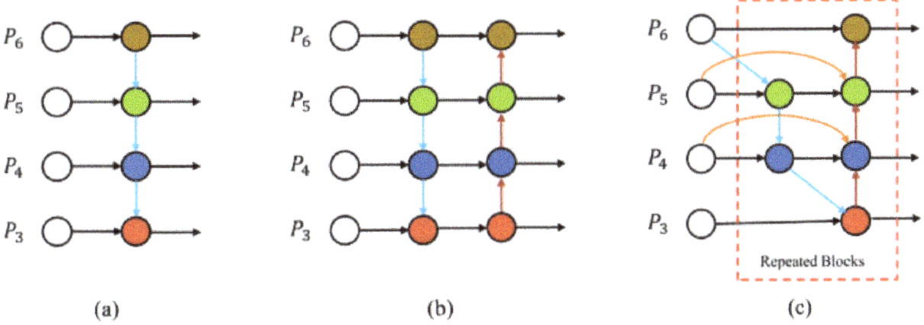

Figure 5. BiFPN network structure diagram, where (**a**) FPN introduces a top-down path to fuse multi-scale features from P3 to P6; (**b**) PANet adds an additional bottom-up path on top of the FPN; (**c**) BiFPN removes redundant nodes and adds additional connections on top of PANet.

The BiFPN removes the node with only one input edge because it does not perform feature fusion. The contribution to the network aim of fusing different features is minimal, and so it is removed and the bidirectional network is simplified. Additionally, an extra edge is added between the input and output nodes that are at the same layer to obtain higher-level fusion features through iterative stacking. The BiFPN introduces a simple and efficient weighted feature fusion mechanism by adding a learnable weight that assigns different degrees of importance to feature maps of different resolutions. The formulas are shown in (1) and (2):

$$P_i^{td} = Conv\left(\frac{w_1 \cdot P_i^{in} + w_2 \cdot Resize\left(P_{i+1}^{in}\right)}{w_1 + w_2 + \epsilon} \right) \tag{1}$$

$$P_i^{out} = Conv\left(\frac{w_1' \cdot P_i^{in} + w_2' \cdot P_i^{td} + w_3' \cdot Resize\left(P_{i-1}^{out}\right)}{w_1' + w_2' + w_3' + \epsilon} \right) \tag{2}$$

where P_i^{in} is the input feature of layer i, P_i^{td} is the intermediate feature on the top-down pathway of layer i, P_i^{out} is the output feature on the bottom-up pathway of layer i, ω is the learnable weight, $\epsilon = 0.0001$ is a small value to avoid numerical instability, Resize is a downsampling or upsampling operation, and $Conv$ is a convolution operation.

The neck layer with BiFPN added a fusion of multi-scale features to provide powerful semantic information to the network. It helped to detect apple leaf diseases of different sizes and alleviated the network's inaccurate identification of overlapping and fuzzy targets.

2.2.3. Transformer Encoder Block

There was a high density of lesions on apple leaves. In order to avoid the problem that the number of lesions and background information increased after mosaic data enhancement, which caused the inability to accurately locate the area where the diseases, the transformer [32] attention mechanism was added to the end of the backbone layer. The transformer module was employed to capture global contextual information and establish long-range dependencies between feature channels and disease targets. The transformer encoder module used a self-attentive mechanism to explore the feature representation capability and an had excellent performance in highly dense scenarios [33]. The self-attention mechanism was designed based on the principles of human vision and allocated resources according to the importance of visual objects. The self-attentive mechanism had a global sensory field, which modeled long-range contextual information, captured rich global semantic information, and assigned different weights to different semantic information to make the network focus more on key information [34]. It was calculated as (3), and contained three basic elements: query, key, and value, denoted by Q, K, and V, respectively.

$$Attention(Q, K, V) = softmax\left(\frac{QK^T}{\sqrt{d_k}} \right) V \tag{3}$$

where d_k is the number of input feature map channel sequences, using normalized data to avoid gradient increment.

Each transformer encoder is composed of a multi-head attention and a feed-forward neural network. The structure of multi-head attention mechanism is shown in Figure 6. It differs from the self-attentive mechanism in that the self-attentive mechanism uses only one set of Q, K, and V values, while it uses multiple sets of Q, K, and V values to compute and stitch multiple matrices together. The different linear transformations feature different vector spaces, which can help the current code to focus on the current pixels and acquire semantic information about the context [35]. The multi-head attention mechanism enhances the ability to extract disease features by capturing long-distance dependent information without increasing the computational complexity and improves the model's detection performance.

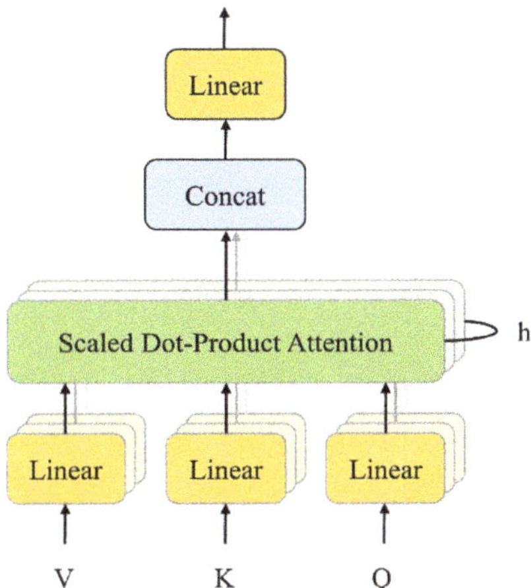

Figure 6. Structure of multi-headed attention mechanism.

2.2.4. Convolutional Block Attention Module

Determining the disease species relies more on local information in the feature map, while the localization of lesions is more concerned with the location information. This model used the CBAM [36] attention mechanism in the improved YOLOv5s to weight the features in space and channels and enhance the model's attention to local and spatial information.

As shown in Figure 7, the CBAM contained two sub-modules: the channel attention module (CAM) and the spatial attention module (SAM), for spatial and channel attention, respectively. The input feature map $F \in R^{C \times H \times W}$ was first passed through the one-dimensional convolution operation $M_c \in R^{C \times 1 \times 1}$ of the CAM, and the convolution result was multiplied with the input features. The output result of CAM was then used as input, the two-dimensional convolution operation $M_s \in R^{1 \times H \times W}$ of the SAM was performed, and then the result was multiplied with the CAM output to obtain the final result. The calculation formulas are as (4) and (5).

$$F' = M_c(F) \otimes F \tag{4}$$

$$F'' = M_s(F') \otimes F' \tag{5}$$

where F denotes the input feature map, M_c denotes the one-dimensional convolution operation of CAM, M_s denotes the two-dimensional convolution operation of SAM, and $\otimes$ denotes element multiplication.

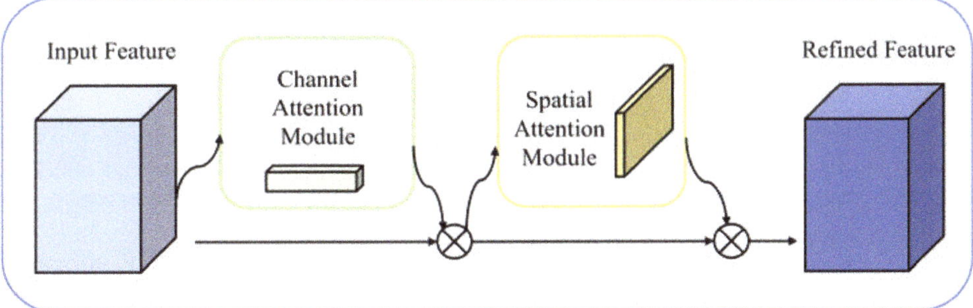

Figure 7. Convolutional block attention module (CBAM).

The CAM in CBAM focused on the weights of different channels and multiplied the channels with the corresponding weights to increase attention to important channels. The feature map F of size H $\times$ W $\times$ C was averaged and maximally pooled to obtain two $1 \times 1 \times$ C channel mappings, respectively, and then a two-layer shared multi-layer perception (*MLP*) operation was performed. The two outputs were summed element by element, and then a sigmoid activation function was applied to output the final result. The calculation process is shown in Equation (6).

$$M_c(F) = \sigma(MLP(AvgPool(F)) + MLP(MaxPool(F))) \tag{6}$$

As shown in Equation (7), the SAM was more concerned with the location information of the lesions. The CAM output was averaged and maximally pooled to obtain two H$'$ $\times$ W$'$ $\times$ 1 channel maps. The final result was obtained by concatenating the two feature maps, followed by a 7×7 convolution operation and a Sigmoid activation function.

$$M_s(F) = \sigma\left(f^{7\times7}([AvgPool(F); MaxPool(F)])\right) \tag{7}$$

2.2.5. BTC-YOLOv5s Detection Model

Based on the original advantages of the YOLOv5s model, this study proposed using an improved BTC-YOLOv5s algorithm for detecting apple leaf diseases. While ensuring the speed of the procedure, it improved the accuracy of identifying apple leaf diseases in a complex environment. The proposed algorithm was improved mainly in three parts: the BiFPN, transformer, and CBAM attention mechanism. Firstly, the CBAM module was added in front of the SPP in the YOLOv5s backbone layer to highlight useful information and suppress useless information in the disease detection task, thereby improving the model's detection accuracy. Secondly, the C3 was replaced with the C3TR module with transformer and improved the ability to extract apple leaf disease features. Thirdly, we replaced the concat layer with the BiFPN layer, and a path from the 6th layer was added to the 20th layer. The features generated by the backbone at the same layer were bidirectionally connected with the features generated by the FPN and the PANet to provide stronger information representation capability. Figure 8 shows the overall framework of the BTC-YOLOv5s model for this study.

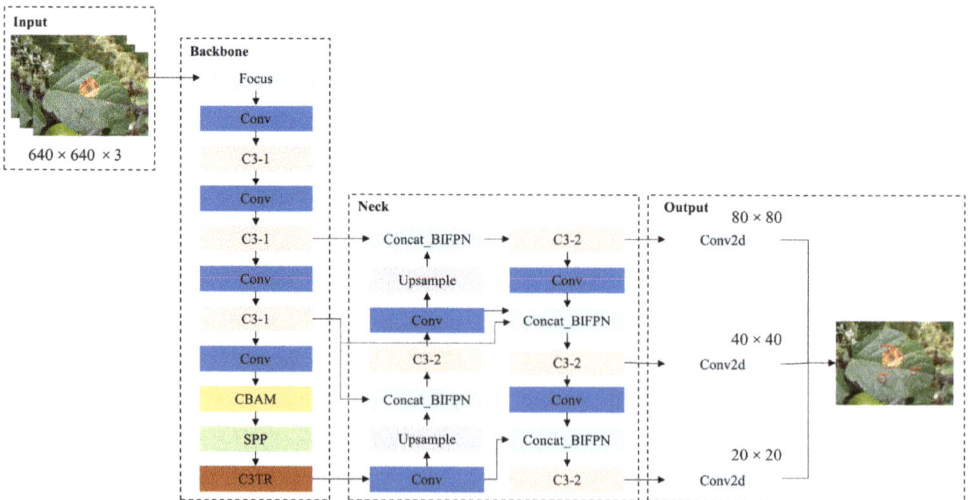

Figure 8. BTC-YOLOv5s model structure diagram.

2.3. Experimental Equipment and Parameter Settings

The model was trained and tested on a Linux system running under the PyTorch 1.10.0 deep learning framework, using the following device specifications: Intel(R) Xeon(R) E5-2686 v4 @ 2.30 GHz processor, 64 GB of memory, and NVIDIA GeForce RTX3090 graphics card with 24 GB of video memory. The software was executed on cuda 11.3, cudnn 8.2.1, and python 3.8.

During training, the initial learning rate was set to 0.01, and the cosine annealing strategy was employed to decrease the learning rate. Additionally, the neural network parameters were optimized using the stochastic gradient descent (SGD) method, with a momentum value of 0.937 and a weight decay index score of 0.0005. The training epoch was 150, the image batch size was set to 32, and the input image resolution was uniformly adjusted to 640 × 640. Table 2 shows the tuned training parameters.

Table 2. Model training parameters.

Parameters	Values
Input size	640 × 640
Batch size	32
Epoch	150
Initial learning rate	0.01
Optimizer	SGD
Momentum	0.937
Weight decay	0.0005

2.4. Model Evaluation Metrics

The evaluation metrics are divided into two aspects: performance assessment and complexity assessment. The model performance evaluation metrics include precision, recall, mAP, and F1 score. The model complexity evaluation metrics include model size, floating point operations (FLOPs), and FPS, which evaluate the computational efficiency and image processing speed of the model.

Precision is the ratio of the correctly predicted positive samples to the total number of samples predicted as positive and is used to measure the classification ability of a model, while the recall measures the ratio of the correctly predicted positive samples to the total number of positive samples. The AP is the integral of precision and recall, and the mAP is the average of AP, which reflects the overall performance of the model for target detection and classification. F1 score is the harmonic mean of precision and recall, and it uses both precision and recall to evaluate the performance of the model. The calculation formulas are shown in Equations (8)–(12).

$$Precision = \frac{TP}{TP + FP} \tag{8}$$

$$Recall = \frac{TP}{TP + FN} \tag{9}$$

where TP is the number of positive samples with correct detection, FP is the number of positive samples with incorrect detection, and FN is the number of negative samples with incorrect detection.

$$AP = \int_0^1 P(R)dR \tag{10}$$

$$mAP = \frac{\sum_{i=1}^{n} AP_i}{n} \tag{11}$$

where n is the number of disease species.

$$F1 = \frac{2 \times Precision \times Recall}{Precision + Recall} \tag{12}$$

The model size refers to the amount of memory required for storing the model. FLOPs is used to measure the complexity of the model, which is the total number of multiplication and addition operations performed by the model. The lower the FLOPs value, the less computation is required for model inference, and the faster model computation will be. The formula for FLOPs is shown in Equations (13) and (14). The FPS indicates the number of pictures processed per second by the model, which can assess the processing speed and is crucial for real-time disease detection. Considering that the model can be implemented on mobile devices with low computational cost, an octa-core CPU without a graphics card was selected to run the test.

$$FLOPs(Conv) = \left(2 \times C_{in} \times K^2 - 1\right) \times W_{out} \times H_{out} \times C_{out} \tag{13}$$

$$FLOPs(Liner) = (2 \times C_{in} - 1) \times C_{out} \tag{14}$$

where C_{in} represents the input channel, C_{out} represents the output channel, K represents the convolution kernel size, and W_{out} and H_{out} represent the width and height of the output feature map, respectively.

3. Results

3.1. Performance Evaluation

The proposed BTC-YOLOv5s model was validated using the constructed ALDD test set. Additionally, the same optimized parameters were used to compare results with YOLOv5s baseline model. As shown in Table 3, the improved model achieved similar AP scores for frogeye leaf spots as the original model, while significantly improving the detection performance for the other three diseases. Notably, scab disease, with its irregular lesion shape, was the most issue to detect, and the improved model achieved a 3.3% increase in AP, which was the largest improvement. These results indicated that the proposed model effectively detected all four diseases with improved accuracy.

Table 3. Comparison of detection results of YOLOv5s and BTC-YOLOv5s.

Models	AP(%)				mAP@0.5(%)	
	Frog	Scab	Powdery	Rust	Spare	Dense
YOLOv5s	93	60.3	88.8	88.7	85.6	80.7
BTC-YOLOv5s	92.9	63.6	90.2	90.3	87.3	81.4

Figure 9 shows evaluation results of precision, recall, mAP@0.5, and mAP@0.5:0.95 for the baseline model YOLOv5s and the improved model BTC-YOLOv5s trained with 150 epochs.

In Figure 9, it is displayed that the precision and recall curves fluctuated within a narrow range after 50 epochs, but that the BTC-YOLOv5s curve remained consistently above the baseline model curve. From the mAP@0.5 curve, it can be seen that the mAP@0.5 curve of the improved model intersected with the baseline model at around 60 epochs. Although the mAP@0.5 of the baseline model increased rapidly in the early stage, the BTC-YOLOv5s model improved steadily in the later stage and showed better results. The mAP@0.5:0.95 curve also demonstrated a similar behavior.

As apple leaf diseases were small and densely distributed, for further verification of the BTC-YOLOv5s model's accuracy, the test sets were divided into two groups based on lesion density, namely sparse distribution and dense distribution of lesions. We compared the detection results of the baseline model and the improved model. The mAP@0.5 of BTC-YOLOv5s model for sparse and dense lesions images was 87.3% and 81.4%, respectively, which was 1.7% and 0.7% higher than that of the baseline model.

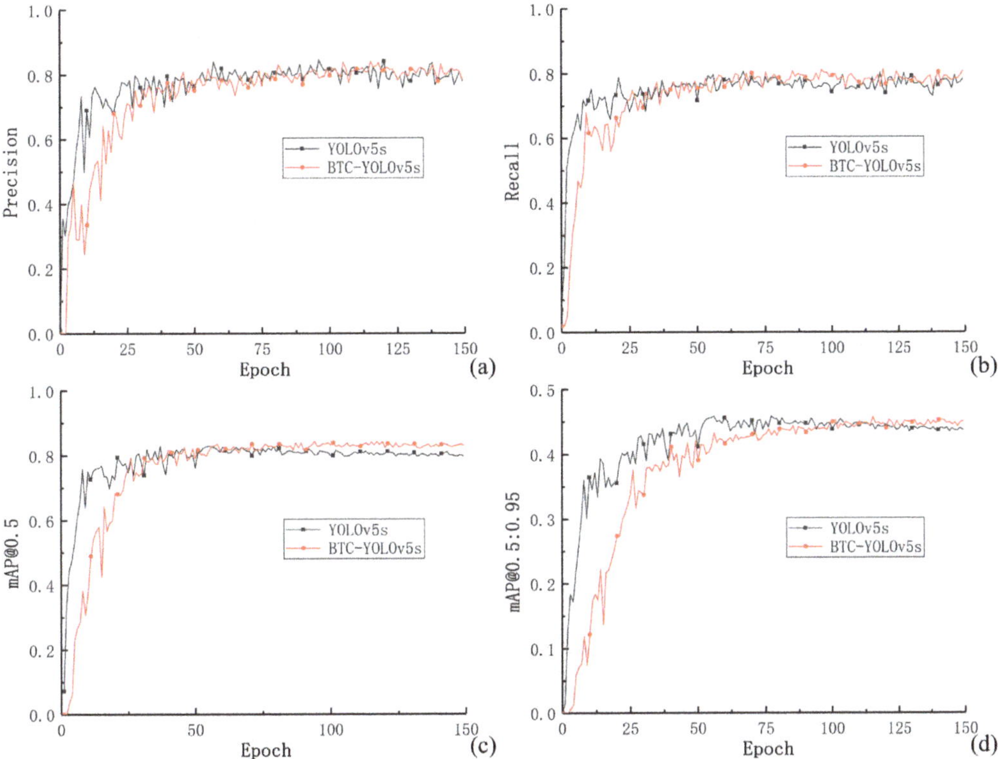

Figure 9. Evaluation metrics of different models, where (**a**) is a comparison of precision curves before and after model improvement; (**b**) comparison of recall curves before and after model improvement; (**c**) comparison of mAP@0.5 curves before and after model improvement; (**d**) comparison of mAP@0.5:0.95 curves before and after model improvement.

As shown in Figure 10, yellow circles represent missed detections and red circles represent false detections. It can be seen that, irrespective of whether the disease is sparse or dense, the baseline model YOLOv5s missed small or blurred lesions (the first row of images in Figure 10a,b). However, the improved model resolved this issue and detected small lesions or diseases on the leaves that were not in the focus range (the second row of images in Figure 10a,b). Additionally, the BTC-YOLOv5s model had higher confidence levels. The baseline model also mistakenly detected the non-diseased parts such as apples, background, and other irrelevant objects (Figure 10(a3,b1)), and there was a false detection whereby the scab was mistakenly detected as rust (Figure 10(b5)). The improved model could concentrate more on diseases and extract the gap characteristics between different diseases at a deeper level to avoid the above errors. Furthermore, the lesions of frogeye leaf spot, scab, and rust were small, dense, and distributed in different parts of the leaves, while powdery mildew typically affected the whole leaf. This led to the scale of the model detection box changing from large to small, and the proposed model was able to adapt well to the scale changes of different diseases.

Therefore, the BTC-YOLOv5s model could not only adapt to the detection of different disease distributions but could also adapt to the changes in apple leaf diseases with different scales and characteristics, showing excellent detection results.

(a)

(b)

Figure 10. Comparison of detection effect of lesion (sparse and dense) before and after model improvement. (**a**) Sparse distribution; (**b**) Dense distribution. Where yellow circles represent missed detections and red circles represent false detections. Lines 1 and 3 are the YOLOv5s baseline model, and lines 2 and 4 are the improved BTC-YOLOv5s model. Numbers 1 and 2 are frogeye leaf spot, numbers 3 and 4 are rust, numbers 5 and 6 are scab, and numbers 7 and 8 are powdery mildew.

3.2. Results of Ablation Experiments

This study verified the effectiveness of different optimization modules via ablation experiments. We constructed several improved models by adding the BiFPN module (BF), transformer module (TR), and CBAM attention module sequentially to the baseline model YOLOv5s and compared the results on the same test data. The experimental results are shown in Table 4.

In Table 4, the precision and mAP@0.5 of the baseline model YOLOv5s were 78.4% and 82.7%. By adding three optimization modules, namely the BiFPN module, transformer module, and CBAM attention module, both precision and mAP@0.5 were improved compared to the baseline model. Specifically, the precision increased by 3.3%, 3.3%, and 1.1%, respectively, and the mAP@0.5 increased by 0.5%, 1%, and 0.2%, respectively. The final combination of all three optimization modules achieved the best results, with precision, mAP@0.5 and mAP@0.5:0.95 all reaching the highest values, which were 5.7%, 1.6%, and 0.1% higher than those of the baseline model, respectively. By fusing cross-channel information with spatial information, the CBAM attention mechanism focused on important features while suppressing irrelevant ones. Additionally, the transformer module used the self-attention mechanism to establish a long-range feature channel with the disease features. The BiFPN module fused the above features across scales to improve the identification of overlapping and fuzzy targets. As a result of the combination of three modules, the BTC-YOLOv5s model achieved the best performance.

Table 4. Results of ablation experiments.

Models	Precision (%)	Recall (%)	mAP@0.5 (%)	mAP@0.5:0.95 (%)
YOLOv5s	78.4	79.7	82.7	45.8
YOLOv5s + BF	81.7	78.4	83.2	45.3
YOLOv5s + CBAM	81.7	79.7	83.7	45.7
YOLOv5s + TR	79.5	78.9	82.9	45.6
YOLOv5s + BF + CBAM	81	81	84.3	44.9
YOLOv5s + BF + TR	83.5	77.6	83	45.1
YOLOv5s + BF + TR + CBAM (proposed)	84.1	77.3	84.3	45.9

Where BF and TR represent the BiFPN module and transformer module, respectively.

3.3. Analysis of Attention Mechanisms

In order to assess the effectiveness of the CBAM attention mechanism module, other structures of the BTC-YOLOv5s model were retained as experimental parameter settings, and only the CBAM module was replaced with other mainstream attention mechanism modules, such as SE [37], CA [38], and ECA [39] modules, for comparison purposes.

Table 5 shows that the attention mechanism could significantly improve the accuracy of the model. The mAP@0.5 of SE, CA, ECA, and CBAM models reached 83.4%, 83.6%, 83.6%, and 84.3%, respectively, which was 0.4%, 0.6%, 0.6%, and 1.3% higher than that of YOLOv5s + BF + TR model. Each attention mechanism improved the mAP@0.5 to varying degrees, with the CBAM model performing the best and reaching 84.3%, which was 0.9%, 0.7%, and 0.7% higher than that of SE, CA, and ECA models, respectively, and the mAP @ 0.5: 0.95 was also the highest among the four attention mechanisms. The SE and ECA attention mechanisms only took into account the channel information in the feature map, while the CA attentional mechanism encoded the channel relations using the location information. In contrast, the CBAM attention mechanism combined spatial and channel attention, emphasizing the information on disease features in the feature map, which was more conducive to disease identification and localization.

Table 5. Performance comparison of different attention mechanisms.

Attention Mechanisms	mAP@0.5 (%)	mAP@0.5:0.95 (%)	Model Size (MB)	FLOPs (G)
SE	83.4	45.3	15.7	17.5
CA	83.6	45.1	15.8	17.5
ECA	83.6	44.8	15.7	17.5
CBAM	84.3	45.9	15.8	17.5

Moreover, the attention module did not increase the model size or FLOPs, indicating that it was a lightweight module. The BTC-YOLOv5s model with the CBAM module achieved improved recognition accuracy while maintaining the same model size and computational cost.

3.4. Comparison of State-of-the-Art Models

The current mainstream two-stage detection model Faster R-CNN and the one-stage detection models SSD, YOLOv4-tiny, and YOLOx-s were selected for comparison experiments. The ALDD dataset was used for training and testing, with the same experimental parameters across all models. The experimental results are shown in Table 6.

Table 6. Performance comparison of mainstream detection models.

Models	mAP@0.5 (%)	F1 (%)	Model Size (MB)	FLOPs (G)	FPS
SSD	71.56	60.77	92.1	274.70	1.15
Faster R-CNN	35.46	35.83	108	401.76	0.16
YOLOv4-tiny	59.86	55.79	22.4	16.19	8.21
YOLOx-s	80.10	77.36	34.3	26.64	4.08
YOLOv5s	82.70	79.04	13.7	16.40	9.80
BTC-YOLOv5s	84.30	80.56	15.8	17.50	8.70

Among all models, the mAP@0.5 and F1 score of Faster R-CNN were lower than 50%, with a large model size and computational effort, resulting in only 0.16 FPS, making it unsuitable for real-time detection of apple leaf diseases. The one-stage detection model SSD had an mAP@0.5 value of 71.56% and a model size of 92.1 MB, which did not meet the detection requirements in terms of model accuracy and complexity. In the YOLO model series, YOLOv4-tiny had an mAP@0.5 of only 59.86%, and the accuracy was too low. The YOLOx-s achieved 80.1% mAP@0.5, but the FLOPs were 26.64 G, and there were only 4.08 pictures per second. Neither of them was not conducive to mobile deployment. The proposed BTC-YOLOv5s model had the highest mAP@0.5 and F1 score among all models, exceeding SSD, Faster R-CNN, YOLOv4-tiny, YOLOx-s, and YOLOv5s by 12.74%, 48.84%, 24.44%, 4.2%, and 1.6%, respectively. The model size and FLOPs were similar to the baseline model, and FPS reached 8.7 frames per second to meet real-time detection of apple leaf diseases in real scenarios.

As seen in Figure 11, the BTC-YOLOv5s model outperformed the other five models in terms of detection accuracy. Additionally, the BTC-YOLOv5s model exhibited comparable model size, computational effort, and detection speed to the other lightweight models. In summary, the overall performance of the BTC-YOLOv5s model was excellent and could accomplish accurate and efficient apple leaf disease detection tasks in real-world scenarios.

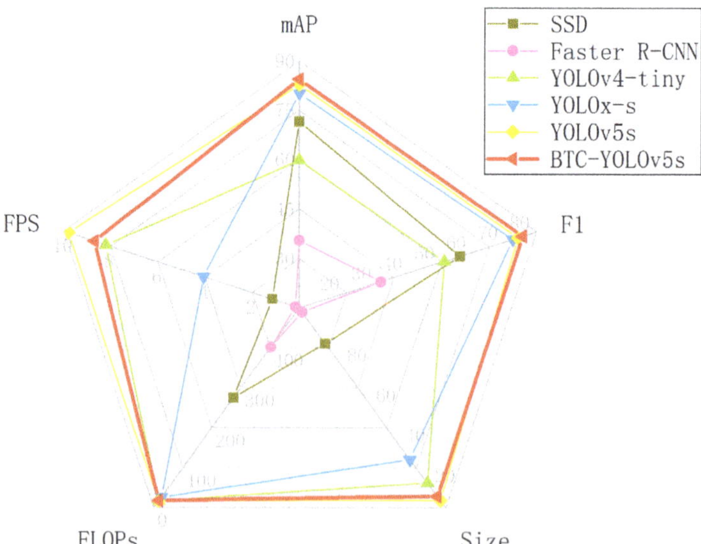

Figure 11. Performance comparison of different detection algorithms.

3.5. Robustness Testing

In the actual production, the detection of apple leaf diseases may be interfered with by various objective environmental factors such as overexposure, dim light, and low-resolution images. In this study, the test set images were simulated by enhancing brightness, reducing brightness, and adding Gaussian noise, resulting in a total of 1191 images (397 images per case). We evaluated the robustness of the optimized BTC-YOLOv5s model under a variety of interference environments to determine its detection effectiveness. Additionally, we tested the model's ability to detect concurrent diseases by adding 50 images containing multiple diseases. Experimental results are shown in Figure 12.

(a)   (b)   (c)   (d)

Figure 12. Robustness test results under three extreme conditions. (**a**) Original; (**b**) Bright light; (**c**) Dim light; (**d**) Blurry. Where first to fifth rows show results for apple frogeye leaf spot, rust, scab, powdery mildew, and multiple diseases, respectively.

From the detection results, the model could accurately detect frogeye leaf spot, rust, and powdery mildew images under all three noise conditions (bright light, dim light, and blurry), with few missing detections. The scab disease was also correctly identified, but a certain degree of missing detections occurred in dim light and blurry conditions. This is mainly because the scab lesions appeared to be black, the overall background of the image has similar color to the lesions under dim light conditions. As shown in the fifth row of Figure 12, the model also demonstrated detection capabilities for images with concurrent onset, although a few missing detections occurred in the blurry condition. The experimental results achieved more than 80% of mAP. Overall, the BTC-YOLOv5s model still exhibited strong robustness under extreme conditions, such as blurred images and insufficient light.

4. Discussion

4.1. Multi-Scale Detection

Multi-scale detection is a challenging task in apple leaf disease detection due to the varying sizes of the lesions. In this study, frogeye leaf spot, scab, and rust lesions are typically small and dense, while powdery mildew is a whole lesion distributed over the leaf. The size of the spots that need to be detected relative to the proportion of the whole image can vary widely between images or even within the same image. To address this issue, this study introduced the BiFPN into YOLOv5s based on the idea of multi-scale feature fusion to improve the model's ability. The BiFPN stacks the entire feature pyramid framework multiple times, providing the network with strong feature representation capabilities. It also performs weighted feature fusion, allowing the network to learn the significance of different input features. In the field of agricultural detection, multi-scale detection has been a popular research topic. For example, Li et al. [21] accomplished multi-scale cucumber disease detection by adding a set of anchors matching small instances. Cui et al. [40] used a squeeze-and-excitation feature pyramid network to fuse multi-scale information, retaining only the 26×26 detection head for pinecone detection. However, the current study still faces the challenge of significantly degraded detection accuracy for very large- or very small-scale targets. Future studies will focus on exploring how models can be applied to different scales of disease spots.

4.2. Attentional Mechanisms

The attention mechanism assigns weight to the image features extracted by the model, enabling the network to focus on target regions with important information, while suppressing other irrelevant information and reducing interference caused by irrelevant backgrounds on detection results. The introduction of the attention mechanism can effectively enhance the detection model's feature learning ability, and many researchers have incorporated it to improve model performance. For example, Liu et al. [41] added the SE attention module to YOLOX to enhance the extraction of the cotton boll feature details. Bao et al. [42] added a dual-dimensional mixed attention (DDMA) to the detection model Neck, which parallelizes coordinate attention with channel and spatial attention to reduce missed and false detections caused by dense blade distribution. This study used the CBAM attention mechanism to enhance the BTC-YOLOv5s model's feature extraction ability. CBAM comprised two modules, SAM and CAM, and using the two submodules alone yielded an accuracy of 83.2% and 83.1%, respectively, inferior to the performance of the model using CBAM. As SAM and CAM are only spatial and channel attention modules alone, whereas CBAM combines both, it considers useful information from both feature channels and spatial dimensions, making it more beneficial for the model to locate and identify lesions.

4.3. Outlook

Although the proposed model can accurately identify apple leaf diseases, there are still some issues that deserve attention and further study. Firstly, the dataset used in this study only contains images of four disease types, whereas there are approximately 200 apple diseases in total. Therefore, future research will include images of more species and different disease stages. Secondly, the accuracy of model is not good in case of dense disease and decreases significantly compared to the performance in the sparse case. The detection results showed that scab had the highest error rate, mainly due to its irregular lesion shape and non-obvious border which interfered with the model detection. In the future, scab disease will be considered as a separate research topic to improve the model's detection accuracy.

5. Conclusions

This study proposed an improved detection model BTC-YOLOv5s based on YOLOv5s aimed at addressing the issues of missing and false detection caused by different shapes of diseased spots, multi-scale, and dense distribution of apple leaf lesions. To enhance the overall detection performance of the original YOLOv5s model, the study introduced the BiFPN module, which increases the fusion of multi-scale features and provides more semantic information. Additionally, the transformer and CBAM attention modules were added to improve the ability to extract disease features. Results indicated that the BTC-YOLOv5s model achieved an mAP@0.5 of 84.3% on the ALDD test set, with a model size of 15.8 M and detection speed of 8.7 FPS on an octa-core CPU device. Additionally, it still maintained good performance and robustness under extreme conditions. The improved model has high detection accuracy, fast detection speed and low computational requirements, making it suitable for deployment on mobile devices for real-time monitoring and the intelligent control of apple diseases.

Author Contributions: Conceptualization, H.L. and F.Y.; methodology, H.L.; software, H.L. and S.F.; validation, H.L., L.S. and S.F.; formal analysis, L.S.; investigation, H.L.; resources, H.L.; data curation, H.L. and S.F.; writing—original draft preparation, H.L.; writing—review and editing, H.L. and F.Y.; visualization, S.F.; supervision, L.S.; project administration, L.S.; funding acquisition, F.Y. All authors have read and agreed to the published version of the manuscript.

Funding: This research was funded by Natural Science Foundation of Henan Province (No.222300420463); Henan Provincial Science and Technology Research and Development Plan Joint Fund (No.222301420113); the Collaborative Innovation Center of Henan Grain Crops, Zhengzhou and by the National Key Research and Development Program of China (No. 2017YFD0301105).

Institutional Review Board Statement: Not applicable.

Informed Consent Statement: Not applicable.

Data Availability Statement: The data presented in this study are available on request from the corresponding author.

Conflicts of Interest: The authors declare no conflict of interest.

References

1. Zhong, Y.; Zhao, M. Research on deep learning in apple leaf disease recognition. *Comput. Electron. Agric.* **2020**, *168*, 105146. [CrossRef]
2. Bi, C.; Wang, J.; Duan, Y.; Fu, B.; Kang, J.-R.; Shi, Y. MobileNet Based Apple Leaf Diseases Identification. *Mob. Netw. Appl.* **2022**, *27*, 172–180. [CrossRef]
3. Abbaspour-Gilandeh, Y.; Aghabara, A.; Davari, M.; Maja, J.M. Feasibility of Using Computer Vision and Artificial Intelligence Techniques in Detection of Some Apple Pests and Diseases. *Appl. Sci.* **2022**, *12*, 906. [CrossRef]
4. Zhang, C.; Zhang, S.; Yang, J.; Shi, Y.; Chen, J. Apple leaf disease identification using genetic algorithm and correlation based feature selection method. *Int. J. Agric. Biol. Eng.* **2017**, *10*, 74–83. [CrossRef]
5. Liu, Y.; Lv, Z.; Hu, Y.; Dai, F.; Zhang, H. Improved Cotton Seed Breakage Detection Based on YOLOv5s. *Agriculture* **2022**, *12*, 1630. [CrossRef]
6. Girshick, R.; Donahue, J.; Darrell, T.; Malik, J. Rich feature hierarchies for accurate object detection and semantic segmentation. In Proceedings of the 2014 IEEE Conference on Computer Vision and Pattern Recognition, Columbus, OH, USA, 23–28 June 2014; pp. 580–587.
7. Ren, S.; He, K.; Girshick, R.; Sun, J. Faster R-CNN: Towards real-time object detection with region proposal networks. *IEEE Trans. Pattern Anal. Mach. Intell.* **2017**, *39*, 1137–1149. [CrossRef] [PubMed]
8. Xie, X.; Ma, Y.; Liu, B.; He, J.; Li, S.; Wang, H. A Deep-Learning-Based Real-Time Detector for Grape Leaf Diseases Using Improved Convolutional Neural Networks. *Front. Plant Sci.* **2020**, *11*, 751. [CrossRef]
9. Deng, X.; Tong, Z.; Lan, Y.; Huang, Z. Detection and Location of Dead Trees with Pine Wilt Disease Based on Deep Learning and UAV Remote Sensing. *AgriEngineering* **2020**, *2*, 294–307. [CrossRef]
10. Zhang, K.; Wu, Q.; Chen, Y. Detecting soybean leaf disease from synthetic image using multi-feature fusion faster R-CNN. *Comput. Electron. Agric.* **2021**, *183*, 106064. [CrossRef]
11. Wang, C.; Xiao, Z. Potato Surface Defect Detection Based on Deep Transfer Learning. *Agriculture* **2021**, *11*, 863. [CrossRef]
12. Liu, W.; Anguelov, D.; Erhan, D.; Szegedy, C.; Reed, S.; Fu, C.-Y.; Berg, A.C. SSD: Single shot MultiBox detector. In Proceedings of the Computer Vision—ECCV 2016, Amsterdam, The Netherlands, 11–14 October 2016; Springer: Cham, Switzerland; pp. 21–37.

13. Redmon, J.; Divvala, S.; Girshick, R.; Farhadi, A. You only look once: Unified, real-time object detection. In Proceedings of the 2016 IEEE Conference on Computer Vision and Pattern Recognition (CVPR), Las Vegas, NV, USA, 27–30 June 2016; pp. 779–788.
14. Redmon, J.; Farhadi, A. YOLO9000: Better, Faster, Stronger. In Proceedings of the 2017 IEEE Conference on Computer Vision and Pattern Recognition (CVPR), Honolulu, HI, USA, 21–26 July 2017; pp. 6517–6525.
15. Redmon, J.; Farhadi, A. Yolov3: An incremental improvement. *arXiv* **2018**, arXiv:1804.02767.
16. Bochkovskiy, A.; Wang, C.-Y.; Liao, H.-Y.M. Yolov4: Optimal speed and accuracy of object detection. *arXiv* **2020**, arXiv:2004.10934.
17. Wang, C.; Xiao, Z. Lychee Surface Defect Detection Based on Deep Convolutional Neural Networks with GAN-Based Data Augmentation. *Agronomy* **2021**, *11*, 1500. [CrossRef]
18. Son, C.-H. Leaf Spot Attention Networks Based on Spot Feature Encoding for Leaf Disease Identification and Detection. *Appl. Sci.* **2021**, *11*, 7960. [CrossRef]
19. Li, J.; Qiao, Y.; Liu, S.; Zhang, J.; Yang, Z.; Wang, M. An improved YOLOv5-based vegetable disease detection method. *Comput. Electron. Agric.* **2022**, *202*, 107345. [CrossRef]
20. Li, Z.; Jiang, X.; Shuai, L.; Zhang, B.; Yang, Y.; Mu, J. A Real-Time Detection Algorithm for Sweet Cherry Fruit Maturity Based on YOLOX in the Natural Environment. *Agronomy* **2022**, *12*, 2482. [CrossRef]
21. Li, S.; Li, K.; Qiao, Y.; Zhang, L. A multi-scale cucumber disease detection method in natural scenes based on YOLOv5. *Comput. Electron. Agric.* **2022**, *202*, 107363. [CrossRef]
22. Thapa, R.; Zhang, K.; Snavely, N.; Belongie, S.; Khan, A. The Plant Pathology Challenge 2020 data set to classify foliar disease of apples. *Appl. Plant Sci.* **2020**, *8*, e11390. [CrossRef]
23. Plant Pathology 2021-FGVC8. Available online: https://www.kaggle.com/competitions/plant-pathology-2021-fgvc8 (accessed on 14 March 2023).
24. Singh, D.; Jain, N.; Jain, P.; Kayal, P.; Kumawat, S.; Batra, N. PlantDoc: A dataset for visual plant disease detection. In Proceedings of the 7th ACM IKDD CoDS and 25th COMAD, Hyderabad, India, 5–7 January 2020; pp. 249–253.
25. Dong, X.; Yan, S.; Duan, C. A lightweight vehicles detection network model based on YOLOv5. *Eng. Appl. Artif. Intell.* **2022**, *113*, 104914. [CrossRef]
26. Park, H.; Yoo, Y.; Seo, G.; Han, D.; Yun, S.; Kwak, N. C3: Concentrated-comprehensive convolution and its application to semantic segmentation. *arXiv* **2018**, arXiv:1812.04920.
27. He, K.; Zhang, X.; Ren, S.; Sun, J. Spatial pyramid pooling in deep convolutional networks for visual recognition. *IEEE Trans. Pattern Anal. Mach. Intell.* **2015**, *37*, 1904–1916. [CrossRef] [PubMed]
28. Lin, T.-Y.; Dollár, P.; Girshick, R.; He, K.; Hariharan, B.; Belongie, S. Feature pyramid networks for object detection. In Proceedings of the 2017 IEEE Conference on Computer Vision and Pattern Recognition (CVPR), Honolulu, HI, USA, 21–26 July 2017; pp. 2117–2125.
29. Liu, S.; Qi, L.; Qin, H.; Shi, J.; Jia, J. Path aggregation network for instance segmentation. In Proceedings of the 2018 IEEE/CVF Conference on Computer Vision and Pattern Recognition, Salt Lake City, UT, USA, 18–23 June 2018; pp. 8759–8768.
30. Zheng, Z.; Wang, P.; Liu, W.; Li, J.; Ye, R.; Ren, D. Distance-iou loss: Faster and better learning for bounding box regression. In Proceedings of the Thirty-Fourth AAAI Conference on Artificial Intelligence (AAAI-20), New York, NY, USA, 7–12 February 2020; Volume 34, pp. 12993–13000.
31. Tan, M.; Pang, R.; Le, Q.V. Efficientdet: Scalable and efficient object detection. In Proceedings of the 2020 IEEE/CVF Conference on Computer Vision and Pattern Recognition (CVPR), Seattle, WA, USA, 13–19 June 2020; pp. 10781–10790.
32. Dosovitskiy, A.; Beyer, L.; Kolesnikov, A.; Weissenborn, D.; Zhai, X.; Unterthiner, T.; Dehghani, M.; Minderer, M.; Heigold, G.; Gelly, S. An image is worth 16×16 words: Transformers for image recognition at scale. *arXiv* **2020**, arXiv:2010.11929.
33. Zhu, X.; Lyu, S.; Wang, X.; Zhao, Q. TPH-YOLOv5: Improved YOLOv5 Based on Transformer Prediction Head for Object Detection on Drone-captured Scenarios. In Proceedings of the IEEE/CVF International Conference on Computer Vision, Montreal, BC, Canada, 11–17 October 2021; pp. 2778–2788.
34. Carion, N.; Massa, F.; Synnaeve, G.; Usunier, N.; Kirillov, A.; Zagoruyko, S. End-to-end object detection with transformers. In Proceedings of the European Conference on Computer Vision, Glasgow, UK, 23–28 August 2020; pp. 213–229.
35. Nediyanchath, A.; Paramasivam, P.; Yenigalla, P. Multi-head attention for speech emotion recognition with auxiliary learning of gender recognition. In Proceedings of the ICASSP 2020—2020 IEEE International Conference on Acoustics, Speech and Signal Processing (ICASSP), Barcelona, Spain, 4–8 May 2020; pp. 7179–7183.
36. Woo, S.; Park, J.; Lee, J.Y.; Kweon, I.S. Cbam: Convolutional block attention module. In Proceedings of the European Conference on Computer Vision (ECCV), Munich, Germany, 8–14 September 2018; pp. 3–19.
37. Hu, J.; Shen, L.; Sun, G. Squeeze-and-excitation networks. In Proceedings of the IEEE Conference on Computer Vision and Pattern Recognition, Salt Lake City, UT, USA, 18–23 June 2018; pp. 7132–7141.
38. Hou, Q.; Zhou, D.; Feng, J. Coordinate attention for efficient mobile network design. In Proceedings of the 2021 IEEE/CVF Conference on Computer Vision and Pattern Recognition (CVPR), Nashville, TN, USA, 20–25 June 2021; pp. 13713–13722.
39. Wang, Q.; Wu, B.; Zhu, P.; Li, P.; Zuo, W.; Hu, Q. ECA-Net: Efficient Channel Attention for Deep Convolutional Neural Networks. In Proceedings of the 2020 IEEE/CVF Conference on Computer Vision and Pattern Recognition (CVPR), Seattle, WA, USA, 13–19 June 2020; pp. 11531–11539.
40. Cui, M.; Lou, Y.; Ge, Y.; Wang, K. LES-YOLO: A lightweight pinecone detection algorithm based on improved YOLOv4-Tiny network. *Comput. Electron. Agric.* **2023**, *205*, 107613. [CrossRef]

41. Liu, Q.; Zhang, Y.; Yang, G. Small unopened cotton boll counting by detection with MRF-YOLO in the wild. *Comput. Electron. Agric.* **2023**, *204*, 107576. [CrossRef]
42. Bao, W.; Zhu, Z.; Hu, G.; Zhou, X.; Zhang, D.; Yang, X. UAV remote sensing detection of tea leaf blight based on DDMA-YOLO. *Comput. Electron. Agric.* **2023**, *205*, 107637. [CrossRef]

Article

UAV-Based Remote Sensing for Soybean FVC, LCC, and Maturity Monitoring

Jingyu Hu [1], Jibo Yue [1], Xin Xu [1], Shaoyu Han [1,2], Tong Sun [1], Yang Liu [2,3], Haikuan Feng [2,4] and Hongbo Qiao [1,*]

[1] College of Information and Management Science, Henan Agricultural University, Zhengzhou 450002, China
[2] Key Laboratory of Quantitative Remote Sensing in Agriculture of Ministry of Agriculture, Beijing Research Center for Information Technology in Agriculture, Beijing 100097, China
[3] Key Lab of Smart Agriculture System, Ministry of Education, China Agricultural University, Beijing 100083, China
[4] College of Agriculture, Nanjing Agricultural University, Nanjing 210095, China
* Correspondence: qiaohb@henau.edu.cn

Abstract: Timely and accurate monitoring of fractional vegetation cover (FVC), leaf chlorophyll content (LCC), and maturity of breeding material are essential for breeding companies. This study aimed to estimate LCC and FVC on the basis of remote sensing and to monitor maturity on the basis of LCC and FVC distribution. We collected UAV-RGB images at key growth stages of soybean, namely, the podding (P1), early bulge (P2), peak bulge (P3), and maturity (P4) stages. Firstly, based on the above multi-period data, four regression techniques, namely, partial least squares regression (PLSR), multiple stepwise regression (MSR), random forest regression (RF), and Gaussian process regression (GPR), were used to estimate the LCC and FVC, respectively, and plot the images in combination with vegetation index (VI). Secondly, the LCC images of P3 (non-maturity) were used to detect LCC and FVC anomalies in soybean materials. The method was used to obtain the threshold values for soybean maturity monitoring. Additionally, the mature and immature regions of soybean were monitored at P4 (mature stage) by using the thresholds of P3-LCC. The LCC and FVC anomaly detection method for soybean material presents the image pixels as a histogram and gradually removes the anomalous values from the tails until the distribution approaches a normal distribution. Finally, the P4 mature region (obtained from the previous step) is extracted, and soybean harvest monitoring is carried out in this region using the LCC and FVC anomaly detection method for soybean material based on the P4-FVC image. Among the four regression models, GPR performed best at estimating LCC (R^2: 0.84, RMSE: 3.99) and FVC (R^2: 0.96, RMSE: 0.08). This process provides a reference for the FVC and LCC estimation of soybean at multiple growth stages; the P3-LCC images in combination with the LCC and FVC anomaly detection methods for soybean material were able to effectively monitor soybean maturation regions (overall accuracy of 0.988, mature accuracy of 0.951, immature accuracy of 0.987). In addition, the LCC thresholds obtained by P3 were also applied to P4 for soybean maturity monitoring (overall accuracy of 0.984, mature accuracy of 0.995, immature accuracy of 0.955); the LCC and FVC anomaly detection method for soybean material enabled accurate monitoring of soybean harvesting areas (overall accuracy of 0.981, mature accuracy of 0.987, harvested accuracy of 0.972). This study provides a new approach and technique for monitoring soybean maturity in breeding fields.

Keywords: UAV; chlorophyll; fractional vegetation cover; maturity monitoring; anomaly detection

Citation: Hu, J.; Yue, J.; Xu, X.; Han, S.; Sun, T.; Liu, Y.; Feng, H.; Qiao, H. UAV-Based Remote Sensing for Soybean FVC, LCC, and Maturity Monitoring. *Agriculture* **2023**, *13*, 692. https://doi.org/10.3390/agriculture13030692

Academic Editor: David Oscar Yawson

Received: 23 February 2023
Revised: 13 March 2023
Accepted: 15 March 2023
Published: 16 March 2023

1. Introduction

Soybean, the world's most important source of plant protein, plays a vital role in global food security [1]. Physiological parameters of soybean such as leaf chlorophyll content (LCC), vegetative cover (FVC), and yield are closely linked [2,3]. In addition, soybean maturity is a crucial indicator for harvesting, and harvesting too early or too late can also impact yield [4]. Therefore, it is essential to quickly and accurately estimate soybean FVC and LCC information and monitor soybean maturity.

Crop maturity is a significant factor affecting crop seed yield and is an essential indicator for agricultural decision makers in judging suitable varieties [5]. LCC is a crucial driver of photosynthesis in green plants. Its content is closely related to the photosynthetic capacity, growth and development, and nutrient status of vegetation [3,6–9]. FVC is the percentage of vegetation in the study area, and can visually reflect the growth status of surface vegetation [10,11]. As the crop matures, crop LCC gradually decreases due to degradation, and leaves turn yellow and begin to fall off (FVC decline). The change in crop LCC and FVC can be used to characterize the degree of maturity. Therefore, the crop's maturity can be quantified using LCC and FVC.

Traditional manual methods of measuring LCC and FVC are inefficient, costly, destructive [12,13], and challenging with which to achieve accurate estimation of LCC and FVC over large areas. On the other hand, traditional manual discrimination of crop maturity relies on its color and hardness. This process is time consuming and subject to human bias [14]. Previous studies have shown that crop LCC, FVC, and maturity can be estimated and monitored using remote sensing technology [15–21]. Remote sensing technology provides methods for crop monitoring on a large scale, especially satellite remote sensing [22]. However, satellite remote sensing images' low resolution and long revisit time make them unsuitable for accurately monitoring crops [23,24]. UAVs have received increasing attention due to their ability to cover a large area in a short time while simultaneously performing tasks at high frequency [25,26]. Additionally, UAVs are able to minimize measurement errors caused by environmental factors [27,28].

In recent decades, many studies have been performed to estimate soybean LCC and FVC on the basis of remote sensing techniques such as UAVs. The methods for estimating LCC and FVC are as follows: (1) Physical modeling. This is based on the physical principles of radiative transfer to establish a physical model such as PROSAIL [29,30]. However, the various parameters in the physical model are usually not easily accessible, which limits the practical application of the estimated crop parameters [31]. (2) Empirical methods. These use parameters based on spectral reflectance, and the vegetation index (VI) acts on the regression model. The emergence of machine learning (ML) provides superior regression models such as GPR [32], RF [33], and ANN [34] to perform fast and accurate estimation of crop parameters. (3) Hybrid methods. Hybrid models combine the first two techniques. For example, Xu et al. [35] coupled the PROSAIL model and the Bayesian network model to infer rice LCC. Although the hybrid method is able to improve the estimation accuracy, the instability of the hybrid method is a critical problem, which requires the balance of the interface between the inversion algorithm and the physical model to be addressed. However, this can substantially increase the complexity of the work. Research work on crop maturity monitoring and identification has also continued. These methods include colorimetric methods [36], fluorescence labeling methods [37], nuclear magnetic resonance imaging [38], electronic nose [39], and spectral device imaging [40]. Early spectral imaging is widely used for crop maturity identification. For example, Khodabakhshian et al. [41] created a maturity monitoring model for pears based on a hyperspectral imaging system in the chamber. However, early spectral imaging devices are similar to colorimetry, among other things, and are limited in their use, mainly being restricted to the laboratory. The subsequent advent of UAVs has made it possible to accurately monitor crop maturity at the regional scale in the field. The methods of crop maturity monitoring by UAV include (1) those based on spectral indices and machine learning models, (2) those based on image processing with deep learning (DL), and (3) those based on transfer learning. The former mainly select the spectral indices related to maturity combined with machine learning models to achieve maturity monitoring. Volpato et al. [42] input the greenness index (GLI) into a nonparametric local polynomial model (LOESS) and a segmented linear model (SEG) to monitor soybean maturity on the basis of RGB images acquired by UAV. In addition, Makanza et al. [43] found a senescence index for maturity identification. Other machine learning models used in this context include the partial least squares regression model (PLSR) [4] and the generalized summation model (GAM) [44]. Although these spectral indices and machine learning models are simple and fast, they are unstable, and each crop's characteristic spectral indices differ. With respect to deep learning, Zhou et al. [45] used YOLOv3 for strawberry maturity recognition, achieving a maximum classification average accuracy of 0.93 for fully mature strawberries. In addition, deep learning for maturity monitoring also includes BPNN [14] and VGG16 [46]. Compared with machine learning, deep-learning-based monitoring of crop maturity can obtain higher accuracy. However, its superior performance requires a large amount of sample image data to support it, which increases the challenge of field data collection. Migration learning makes it possible to use the original pre-trained model in other related studies. For example, Mahmood et al. [47] performed migration learning using two deep learning pre-training processes, which was eventually able to classify dates into three maturity levels (immature, mature, and overripe). However, the migration of the pre-trained model is based on the premise that the target domain needs to be highly relevant to it, which places a higher demand on the generality of the data used to train the model. Although deep learning and transfer learning bring a new aspect to crop maturity monitoring. However, their features originate from the original images and ignore the potential of images of crop physiological parameters (e.g., LCC, FVC). During crop maturation, LCC and FVC change accordingly. Especially in breeding fields, early maturing lines lead to significant variations in overall FVC and LCC. These early maturing lines are out of the population distribution and become outliers. Therefore, crop maturity can be monitored based on the changes in the pixel distribution of crop FVC and LCC images.

The objective of this study was to perform soybean maturity monitoring using FVC and LCC. FVC and LCC were quickly estimated using spectral indices combined with regression models. Soybean field data were obtained from RGB images of four periods taken by UAV. The following objectives were identified, to be achieved using these data: (1) to estimate FVC and LCC from image data and multiple regression models, and to map them using the best model, and (2) to propose a new method for soybean maturity monitoring by detecting soybean LCC and FVC anomalies.

2. Study Area and Data

2.1. Study Area

The study site is located in Jiaxiang County, Jining City, Shandong Province, China (Figure 1a,b). Jiaxiang County is located at longitude $116°22'10''$–$116°22'20''$ E and latitude $35°25'50''$–$35°26'10''$ N. It has a continental climate in the warm-temperate monsoon region, with an average annual temperature of 13.9 °C, an average daily minimum temperature of -4 °C, and an average daily maximum temperature of 32 °C. The average elevation is 35–39 m, and the annual rainfall is about 701.8 mm. The trial site is a soybean breeding field (Figure 1c), planted at a density of 195,000 plants ha^{-1} with 15 cm row spacing, and 532 soybean lines were grown.

2.2. Ground Data Acquisition

Ground data collection included four stages: pod set on 13 August 2015 (P1), early bulge on 31 August 2015 (P2), peak bulge on 17 September 2015 (P3), and maturity on 28 September 2015 (P4). Forty-two sets of data were measured for each period. Twenty-three data points were collected at P4, the harvest period when some of the early maturing soybeans had already been harvested. During the data processing, we removed four abnormal data. In addition, to reduce soil background's influence on LCC and FVC mapping, we added eight soil points. Finally, a total of 153 sampling points were retained for this experiment.

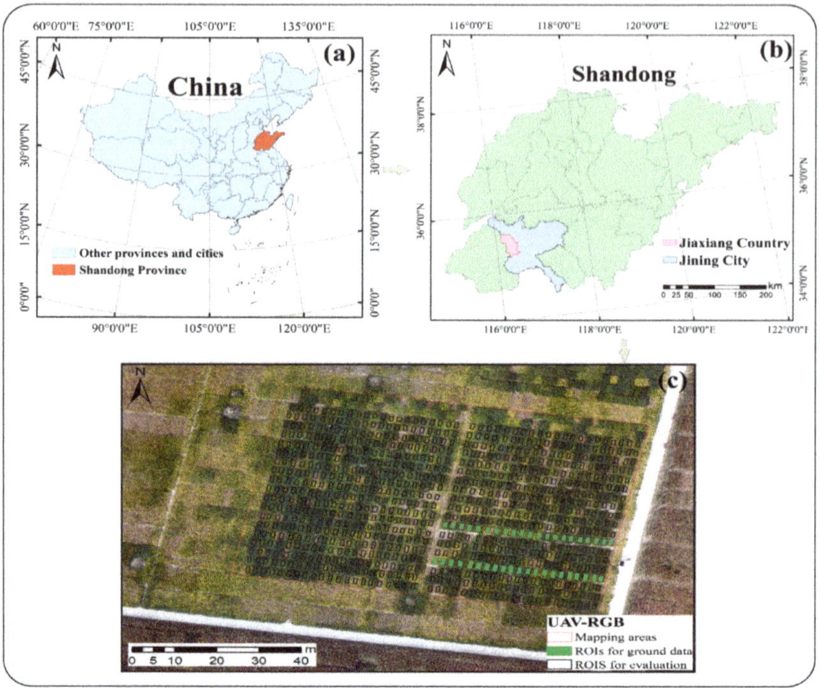

Figure 1. Study area and experimental sites: (**a**) location of the study area in China; (**b**) map of Jiaxiang County, Jining City, Shandong Province; (**c**) UAV RGB images and actual data collection sites. Note: The green ROI in (**c**) is the ground data collection area, and the 800 black boxes ROIs are used for monitoring and evaluation (including the area where the ground data collection ROIs are located).

2.2.1. Soybean LCC Data

Soybean canopy chlorophyll content (LCC) was obtained using measurements from portable Dualex scientific sensors (Dualex 4; Force-A; Orsay, France). The operation was repeated five times in the center of each soybean plot, and the mean value was taken. The results of the analysis of the soybean canopy chlorophyll data set are presented in Table 1.

Table 1. Results of soybean LCC field measurements (Dualex units).

Data (2015)	Stage	n	Min	Max
8.13	P1	41	20.99	28.92
8.31	P2	42	29.27	42.37
9.17	P3	41	6.52	38.28
9.28	P4	21	8.81	36.05
-	P1–P4	149	6.52	42.37

n, min, and max represent the number of soybean plots measured and the minimum and maximum values of LCC, respectively. Note: In this study, LCC was obtained with Dualex 4 device measurements, and by convention, we replaced Dualex units.

2.2.2. Soybean FVC Data

In this experiment, soybean LAI was measured with the LAI-2200C Plant Canopy Analyzer (Li-Cor Biosciences, Lincoln, NE, USA). Finally, the LAI was converted to FVC using PROSAIL [48,49]. The conversion equation is shown as Equation (1). Table 2 shows the results of the analysis of the FVC dataset in soybean fields. *G* is the leaf-projection factor for a spherical orientation of the foliage, Ω is the clumping index, LAI is the leaf area index, and θ is the viewing zenith angle (in this experiment, $G = 0.5$, $\theta = 0$, $\Omega = 1$).

$$FVC = 1 - e^{-G \times \Omega \times \frac{LAI}{\cos(\theta)}}$$

(1)

Table 2. Results of soybean FVC field measurements.

Data (2015)	Stage	n	Min	Max
8.13	P1	41	0.76	0.99
8.31	P2	42	0.68	0.99
9.17	P3	41	0.41	0.96
9.28	P4	21	0.64	0.96
-	P1–P4	145	0.41	0.99

2.2.3. Soybean Maturity Survey

This work used visual interpretation of RGB images from drones to obtain soybean maturity information, and the specific criteria are shown in Table 3.

Table 3. Criteria for determining the maturity of soybean plots.

Category	Description
Harvested	The soybean planting area has been harvested (Figure 2b).
Mature	More than half of the upper tree crown and leaves are yellow (Figure 2c).
Immature	More than half of the upper tree crown and leaves are green (Figure 2d).

Figure 2. Maturity information: (**a**) P3-RGB; (**b**) mature region; (**c**) immature region; (**d**) harvested region. Note: Plot1, Plot2, and Plot3 in (**a**) correspond to (**b,c**), and (**d**), respectively.

2.3. UAV RGB Image Acquisition and Processing

In this work, the sensor platform used was an eight-rotor aerial photography vehicle, DJI 000 UAV (Shenzhen DJI Technology Co., Ltd., Guangdong, China), equipped with a Sony DSC-QX100 [50] high-definition digital camera for RGB image acquisition functions. In addition, a Trimble GeoXT6000 GPS receiver was used to determine the test field ground control point (GCP).

The soybean field UAV RGB images were acquired from 11:00 a.m. to 2:00 p.m. The UAV required three radiation calibrations and flight parameter settings before takeoff. The altitude was set to approximately 50 m (calculating a spatial resolution of approximately

1.17 cm on the ground). The RGB images were obtained and stitched together using AgiSoft Photoscan (AgiSoft LLC, St. Petersburg, Russia) to produce RGB digital orthophoto maps (DOMS). ArcGIS and ENVI handled DOMS.

3. Method

3.1. Soy-Based Material LCC and FVC Anomaly Detection

The grayscale histograms of LCC and FVC ground measurements during P2, P3, and P4 are shown in Figure 3. Because both the LCC and FVC values of soybean crops are significantly lower at maturity. Additionally, both deviated from the original normal distribution. Therefore, it is possible to evaluate mature and immature soybean samples by analyzing the distribution of soybean LCC and FVC gray histogram.

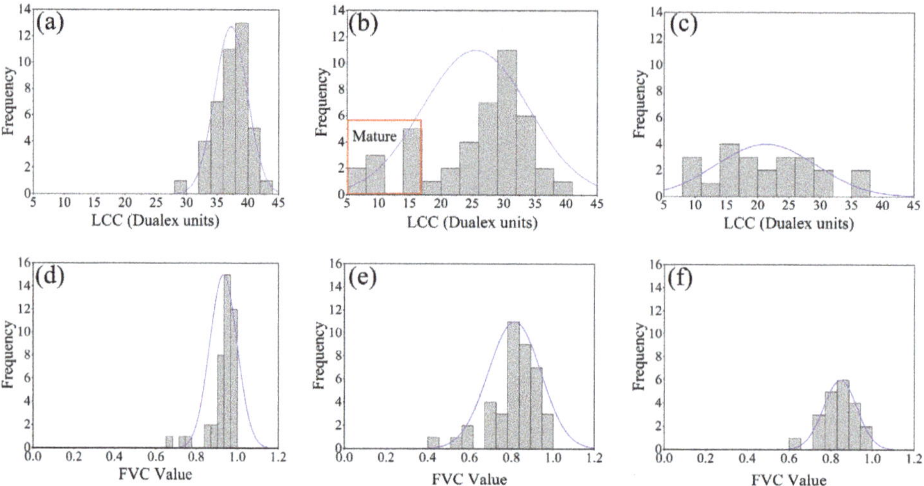

Figure 3. Histograms of the measured LCC and FVC statistics for P2 and P3: (**a**) P2-LCC; (**b**) P3-LCC; (**c**) P4-LCC; (**d**) P2-FVC; (**e**) P3-FVC; (**f**) P4-FVC. Note: The P4 partially mature soybeans had been harvested, so the amount of ground data was different from that in the cases of P2 and P3.

This was undertaken as follows: (1) The soybean LCC image pixels were read and presented as a frequency histogram (Figure 4a). (2) Groups that deviate from the normal distribution are distributed in the tails of the histogram. Removing the tail values normalizes the histogram. Next, the absolute values of the kurtosis and the histogram skewness are combined. The combination is used as a criterion to assess normality. Repeated iterations remove the tails. When the combination reaches a minimum value, the histogram is considered to have reached the most normal distribution (Figure 4b). (3) The expected target soybean region is obtained by extracting the threshold value corresponding to the most normal distribution. The region below the threshold value in the P3-LCC mapping is the mature soybean region (Figure 4c). In practice, since different soybean maturity categories exist at different times, the soybean category corresponding to the histogram threshold needs to be determined on a case-by-case basis. The above procedure was implemented in the Python 3.8 environment.

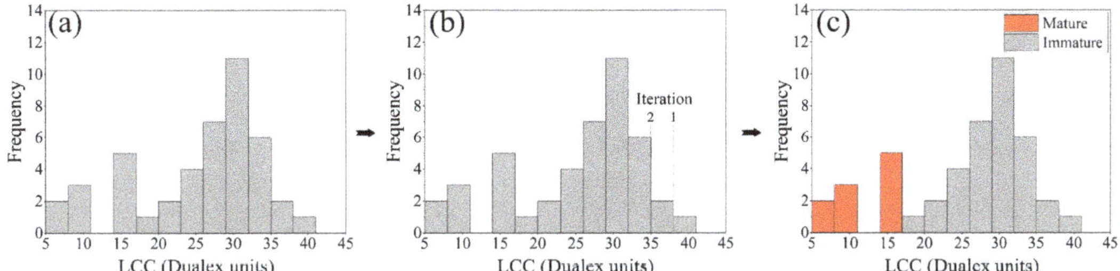

Figure 4. Soybean LCC and FVC anomaly detection methods: (**a**) P3-LCC image element histogram; (**b**) iterative process; (**c**) final results.

3.2. FVC and LCC Remote Sensing Estimation

3.2.1. Color Index

Vegetation indices (VIs) provide a simple and effective measurement of crop growth. They are widely used to estimate FVC and LCC. On the basis of previous relevant studies, we selected 20 color-based VIs, the details of which are presented in Table 4.

Table 4. Vegetation index details.

Vegetation Index	Formula	Reference
DN value of Red Channel (R)	R	[51]
DN value of Green Channel (G)	G	[51]
DN value of Blue Channel (B)	B	[51]
Normalized Redness Intensity (r)	R/(R + G + B)	[52]
Normalized Greenness Intensity (g)	G/(R + G + B)	[52]
Normalized Blueness Intensity (b)	B/(R + G + B)	[52]
Red–Blue Ratio Index (RBRI)	R/B	[53]
Green–Blue Ratio Index (GBRI)	G/B	[53]
Green–Red Ratio Index (GRRI)	G/R	[54]
Blue–Red Ratio Index (BRRI)	B/R	[54]
Blue–Green Ratio Index (BGRI)	B/G	[54]
Normalized Red–Blue Difference Index (NRBDI)	(R − B)/(R + B)	[55]
Normalized Green–Red Difference Index (NGRDI)	(G − R)/(G + R)	[55]
Normalized Green–Blue Difference Index (NGBDI)	(G − B)/(G + B)	[55]
Excess Red Index (EXR)	1.4R − G	[56]
Excess Green Index minus Excess Red Index (EXG-EXR)	2G − R − B − (1.4R − G)	[57]
Visible Atmospherically Resistant Index Normalized blueness (VARI)	(G − R)/(G + R − B)	[58]
R + G	R + G	[59]
(G + B − R)/2B	(G + B − R)/2B	[54]
(R − G)/(R + G + B)	(R − G)/(R + G + B)	[54]

Note: R, G, and B in the formula in the table represent their corresponding DN values, respectively.

3.2.2. Regression Model

Partial least squares (PLS) is able to provide a more stable estimate than least squares, and the standard deviation of the regression coefficients is smaller than that estimated by least squares [60]. For example, suppose there are two matrices, X (VIs) and Y (LCC or FVC). Usually, X and Y are normalized to find the projection of VI on the principal components and maximize the covariance of $p1$ and $q1$, see Equations (2) and (3), and solve the objective function to establish the regression equation. $p1$ is the first principal component of X, and $q1$ is the first principal component of Y.

$$u1 = Xp1, v1 = Xp2 \tag{2}$$

$$Cov(u1, v1) = \sqrt{Var(u1)Var(v1)}\mathrm{Corr}(u1, v1) \rightarrow Max \tag{3}$$

Soybean FVC and LCC are often associated with multiple VIs. This also means that a dependent variable Y, corresponding to multiple independent variables X, often accompanies such studies. The principle of stepwise multivariate analysis is to analyze, in a stepwise manner, the contribution of all independent variables X to the dependent variable [61]. If the contribution is significant, this variable is considered essential and is retained, or, conversely, it is removed if the contribution is insignificant. Finally, a regression model is built based on the analysis. Equation (4) is the regression equation of MSR, e is the error term, and βn is the constant term regression coefficient corresponding to the nth VI.

$$Y = \beta_0 + \beta_1 X_1 + \beta_2 X_2 + \ldots + \beta_n X_n + e \tag{4}$$

Random Forest is an extended variant of bagging. It builds bagging integration with decision trees as learners and further introduces "random attribute selection" into the training process to give it better generalization performance [62,63]. The final prediction result of the random forest is the mean of the prediction results of all CART regression trees. In addition, RF is able to calculate the out-of-bag (OOB) data prediction error rate and replace other VIs in order to calculate the variable importance (VIM) during training to build decision trees. The specific results are shown in Section 3.1. VI importance is calculated using Equation (5), where j is some VI, and i is the ith tree.

$$VIM_j^{(OOB)} = \frac{\sum_{i=1}^{n} VIM_{ij}^{(OOB)}}{n} \tag{5}$$

GPR is usually used for regression problems with low and small samples, and is better able to handle nonlinear problems [64]. GPR assumes that the learning data are sampled using a Gaussian process (GP), and that the prediction results are closely related to the kernel function (covariance function) [65]. The standard Gaussian kernel functions are the radial basis function kernel, the rational quadratic kernel, the sine square kernel, and the dot product kernel. In GPR, the kernel function can find a corresponding mapping, making the data linearly separable in high-dimensional space. The probability density function of GPR is given in Equation (6).

$$p(x_1, x_2 \ldots, x_n) = \frac{1}{2\pi^{\frac{n}{2}} \sigma_1 \sigma_2 \ldots \sigma_n} exp\left(-\frac{1}{2}\left[\frac{(x_1 - \mu_1)^2}{\sigma_1^2} + \frac{(x_2 - \mu_2)^2}{\sigma_2^2} \ldots + \frac{(x_n - \mu_n)^2}{\sigma_n^2}\right]\right) \tag{6}$$

3.3. Technical Route and Accuracy Evaluation

3.3.1. Technical Route

This study focuses on estimating soybean LCC and FVC and producing a mapping of two physiological parameters using four machine-learning techniques. Finally, soybeans were monitored for early maturity and harvesting. The technical route is shown in Figure 5, and the details of the study are as follows.

(1) Soybean FVC and LCC estimation and mapping. The FVC and LCC of soybean were estimated using PLSR, MSR, RF, and GPR, and the best regression model was found and used for FVC and LCC mapping.

(2) Soybean maturation monitoring. The soybean material LCC and FVC anomaly detection method was used to determine the LCC of P3. A threshold value was obtained for the mature region for the monitoring of the LCC of the mature region. This threshold was also used for soybean maturation monitoring at P4 (i.e., during the maturity stage).

(3) Soybean harvesting monitoring. LCC and FVC anomaly detection of soybean material was carried out for P4 mature plots. Complete the identification of the soybean harvesting area where the mature plots of P4 were obtained from (2).

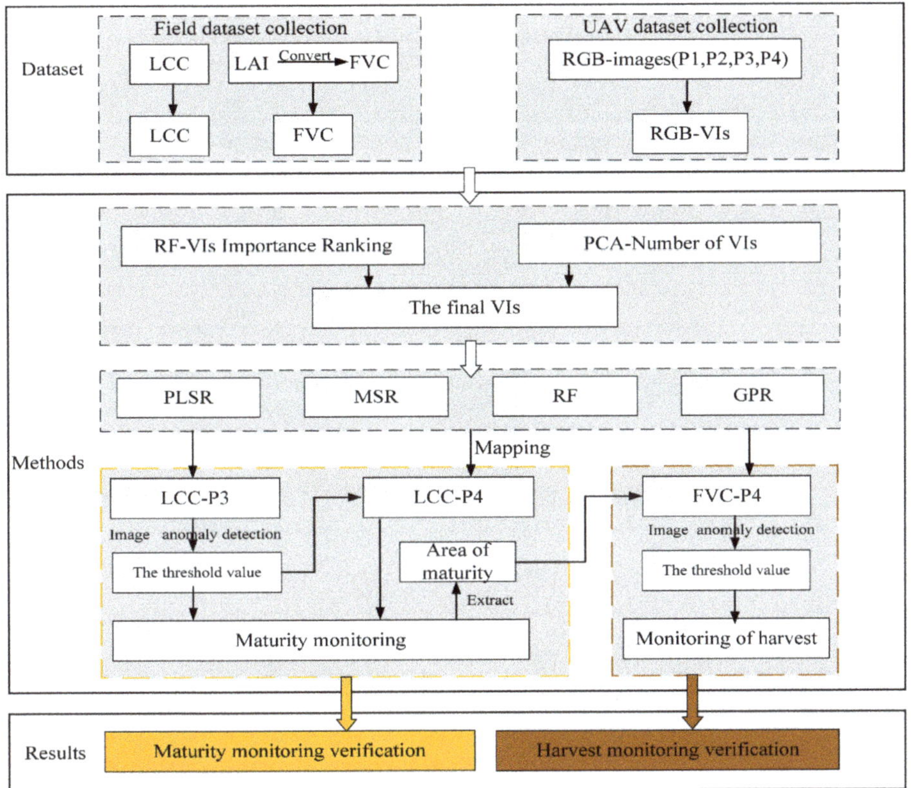

Figure 5. Technical route.

3.3.2. Precision Evaluation

To ensure that the final model has a high generalization capability, the 153 data points generated in this work were randomly divided into two groups (in a ratio of 7:3). We evaluated the ability of PLSR, MSR, RF, and GPR to predict LCC and FVC by means of the coefficient of determination (R^2), and root mean square error (RMSE), with R^2 values in the range [0–1], whereby higher R^2 values correspond to smaller RMSE. Smaller values of RMSE represent higher accuracy in the values of LCC and FVC predicted by the models. The calculation procedure is shown in Equations (7) and (8):

$$R^2 = 1 - \frac{\sum_{i=1}^{n}(y_i - \hat{y}_i)^2}{\sum_{i=1}^{n}(y_i - \overline{y})^2} \tag{7}$$

$$RMSE = \sqrt{\frac{\sum_{i=1}^{n}(y_i - \hat{y}_i)^2}{n}} \tag{8}$$

where n is the number of samples input into the model, y_i represents the measured values of LCC and FVC in the soybean field, $\overline{y}$ is the mean value of measured values, and $\hat{y}_i$ the predicted value.

The experimental method was evaluated on the basis of the confusion matrix. The Accuracy and the Precision were calculated. The higher of the two values corresponds to the higher accuracy. Accuracy and Precision were calculated using Equations (9) and (10).

$$\text{Accuracy} = \frac{TP + TN}{TP + TN + FP + FN} \tag{9}$$

$$\text{Precision} = \frac{TP}{TP + FP} \tag{10}$$

4. Results

4.1. Vegetation Index Correlation and Importance Analysis

We initially selected 20 VIs. Then, calculate their Pearson coefficient and importance. The different colors and sizes of the circles in Figure 6 represent different correlation coefficients. The results show that the correlation performance of VI with LCC and FVC differed slightly. Among the 20 VI correlation studies with LCC, R had the highest correlation coefficient with LCC (-0.72), followed by R + G (-0.68), EXR (-0.62), (R − G)/(R + G + B) (-0.58), NGRDI (0.57), etc. Among the selected VIs correlation studies with FVC, GRRI had the best performance (r = 0.81) and NGRDI (0.80). NGRDI (0.80), (R − G)/(R + G + B) (-0.80), VARI (0.77), and EXR (-0.77) also showed high correlation coefficients.

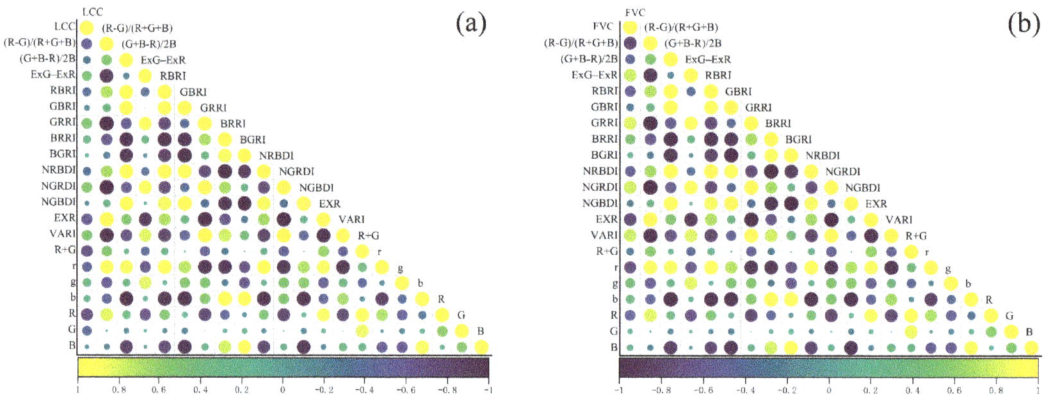

Figure 6. Pearson correlation coefficients between soybean LCC, FVC, and VIs: (**a**) LCC-Vis; (**b**) FVC-VIs.

To determine the feature inputs, we determined the importance of each of the Vis using random forest (Figure 7). Using RF, it was possible to see that R contributed the most to LCC, and VARI contributed the most to FVC, followed by (R − G)/(R + G + B). Finally, in combination with the principal component analysis, it was decided to use R, R + G, EXG-EXR as the characteristic inputs for estimating LCC and VARI, (R − G)/(R + G + B), (G + B − R)/2B, EXG-EXR, EXR as the independent variables for the FVC estimation model.

4.2. Soybean FVC and LCC Estimation and Mapping

The results of the prediction of LCC and FVC using PLSR, MSR, RF, and GPR are shown in Table 5. The best results for the prediction of LCC (R^2: 0.88; RMSE: 3.36 Dualex units) during the modeling of estimated LCC were obtained using GPR. During the validation phase of LCC estimation, R^2, RMSE varied between 0.54 and 0.84, and 3.15 Dualex units and 6.07 Dualex units, respectively, where GPR still maintains the highest estimation accuracy (R^2: 0.84; RMSE: 3.99 Dualex units). With respect to FVC predictive modeling, GPR showed promising results (R^2: 0.94; RMSE: 0.08). In the validation phase, R^2 varied between 0.83 and 0.96. The GPR and RF techniques predicted the results most accurately.

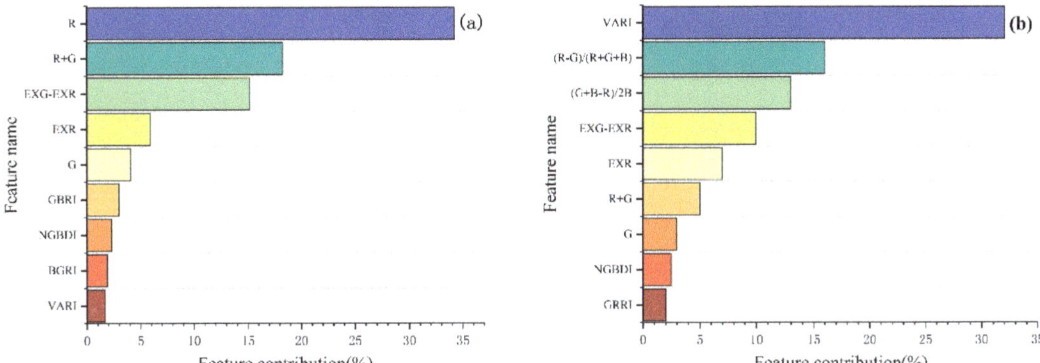

Figure 7. Importance ranking of VIs: (**a**) LCC-VIs; (**b**) FVC-VIs. Note: Only the top nine VIs in the importance ranking are shown here.

Table 5. LCC and FVC estimation results.

Dataset	Methods	LCC		FVC	
		R^2	RMSE	R^2	RMSE
Calibration	PLSR	0.53	6.91	0.80	0.11
	MSR	0.52	6.99	0.80	0.11
	RF	0.86	3.72	0.92	0.09
	GPR	0.88	3.36	0.94	0.09
Validation	PLSR	0.55	6.84	0.83	0.11
	MSR	0.54	6.86	0.83	0.11
	RF	0.82	4.32	0.96	0.08
	GPR	0.84	3.99	0.96	0.08

Figure 8 shows the relationship between the predicted values and ground measurements of LCC and FVC for soybean. Most points in Figure 8d are near the 1:1 line, and the underestimation is more prominent in the soil point data (LCC minimum near 0.85 Dualex units). The results in Figure 8g,h indicate that GPR works best at predicting FVC for soybean and soil data, so GPR was used as the regression model for the estimation of LCC and FVC.

The spatial distribution of LCC and FVC is plotted in Figure 9. Most of the soybeans were at peak growth during P1, and P2, so Figure 9b,e show a balanced distribution of LCC. However, the images presented in Figure 9g indicate the beginning of differentiation in soybean maturity, a change caused by the maturation of early maturing soybeans, and also explain why the P3-P4 LCC mapping (Figure 9h,k) showed significant heterogeneity in the same plots.

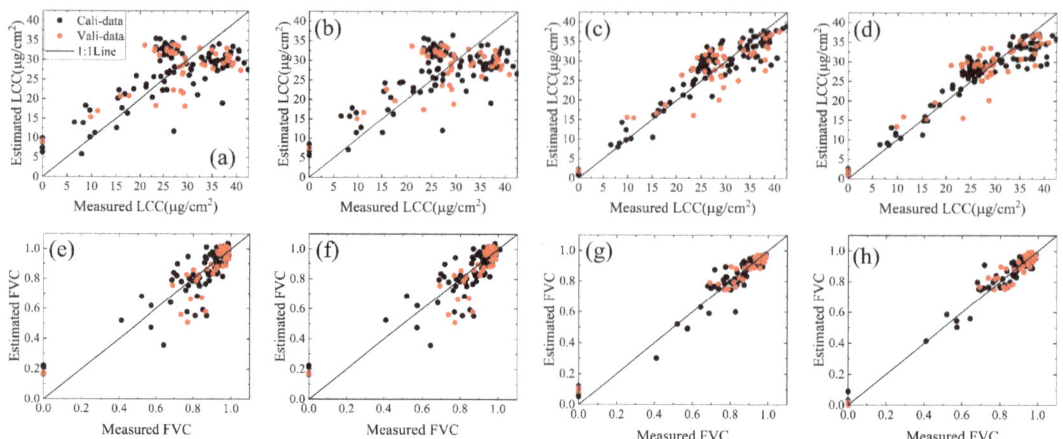

Figure 8. Relationship between predicted and measured soybean LCC and FVC: (**a**) PLSR-LCC; (**b**) MSR-LCC; (**c**) RF-LCC; (**d**) MSR-LCC; (**e**) PLSR-FVC; (**f**) MSR-FVC; (**g**) RF-FVC; (**h**) GS-FVC.

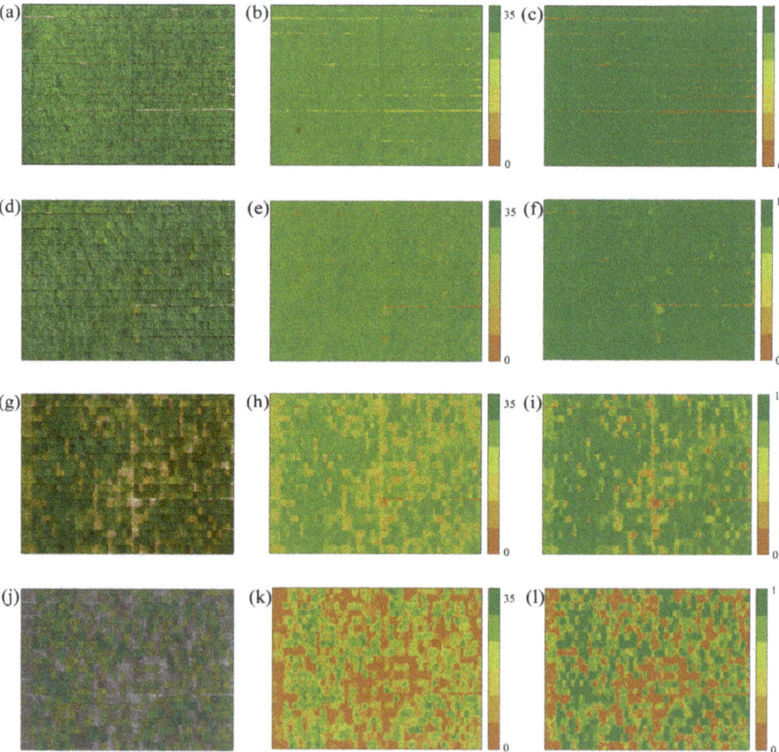

Figure 9. Full-period UAV RGB images with LCC and FVC spatial distribution maps: (**a,d,g,j**) RGB images of the P1–P4 periods, respectively. (**b,e,h,k**) LCC spatial distribution maps of P1–P4, respectively. (**c,f,i,l**) FVC spatial distribution maps of P1–P4, respectively.

4.3. Soybean Maturity and Harvest Monitoring and Mapping

4.3.1. Soybean Population Canopy LCC Histogram Analysis and Maturity Monitoring

Figure 10a shows the grayscale histogram obtained from the detection of the P3-LCC distribution using soybean LCC and FVC anomaly detection methods. The results correspond to the measured data analyzed in Section 3.1 and confirm our hypothesis. The red area of the histogram is the low threshold region that escaped the normal distribution, i.e., the region of maturity caused by early maturing soybean strains, with a threshold value of 18.89 Dualex units. Finally, the actual maturity of the soybean plots was compared with the monitored maturity using a confusion matrix to calculate the results. The results of this monitoring (Figure 10b) showed that LCC using P3 (non-maturity) could be used to accurately monitor soybean, with a total accuracy of 0.988, an accuracy in the mature area of 0.951, and an accuracy in the common area of 0.987. The results of the P3-LCC monitoring visualization are shown in Figure 10f.

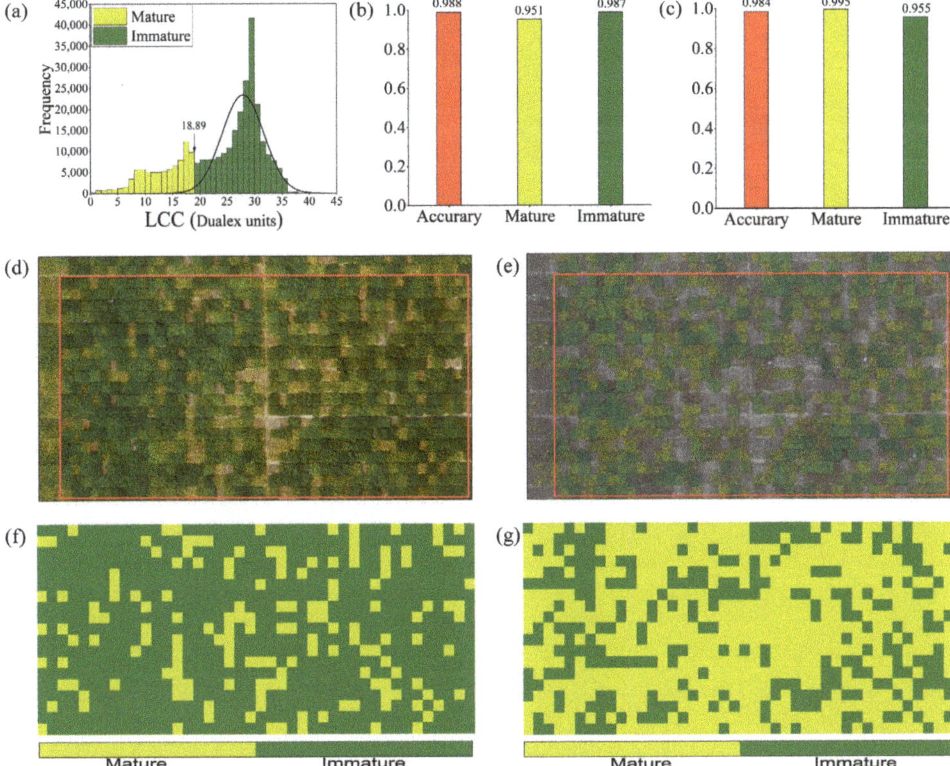

Figure 10. P3 soybean early maturity monitoring: (**a**) Histogram of P3-LCC anomaly distribution; (**b**) P3 maturity monitoring accuracy; (**c**) P4 maturity monitoring accuracy; (**d**) P3-RGB; (**e**) P4-RGB; (**f**) P3-LCC monitoring visualization results; (**g**) P4-LCC monitoring visualization results. Note: The red boxed areas in (**d**,**e**) are the areas used for accuracy evaluation.

To further validate the applicability of the LCC threshold (18.89 Dualex units) for the mature region extracted at P3 (immature stage), we applied this threshold to the LCC at P4 (mature stage) to perform soybean maturity monitoring. The overall monitoring precision was 0.984, the precision in the mature region was 0.995, and the precision in the immature region was 0.955 (see Figure 10c). The results of the monitoring visualization are shown in

Figure 10g. The results indicate that the LCC maturity soybean thresholds obtained from P3 are feasible for use in P4 soybean maturity monitoring.

4.3.2. Soybean Population Canopy FVC Histogram Analysis and Harvest Monitoring

In this section, the mature soybean regions monitored were extracted, as shown in Figure 10d, and the harvest monitoring of soybean in these regions during the P4 period was performed using FVC. The grayscale histogram of the FVC anomaly distribution is shown in Figure 11a, and the red grayscale histogram on the left denotes the harvest monitoring region, with a threshold of 0.609. The evaluation of the confusion matrix visualized under this harvest threshold revealed the results presented in Figure 11c, with a total accuracy of 0.981, a harvest accuracy of 0.972, and a maturity accuracy of 0.987. This indicates the relative sensitivity of the harvest region when monitoring soybeans using FVC. After validating the results, it was found that this error was caused by different soybean managers leaving stubble on the lower part of the soybean stalk when performing harvesting.

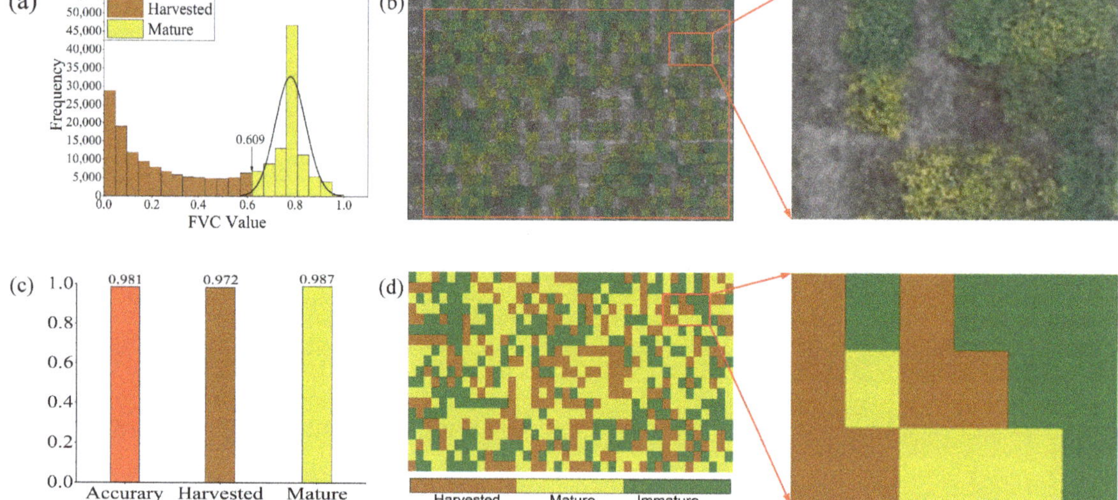

Figure 11. Harvest monitoring results for P4 (mature) soybean: (**a**) histogram of FVC anomaly distribution for P4; (**b**) P4-RGB; (**c**) P4 harvest monitoring accuracy; (**d**) P4-FVC threshold visualization results. Note: This section discusses harvest monitoring in the mature soybean region only (i.e., containing the harvest and maturity regions), rather than the entire region, during the P4 period.

5. Discussion

5.1. Multi-Period LCC and FVC Estimation

In this study, four regression models were selected in order to predict soybean LCC and FVC. GPR had the best stability and accuracy when predicting LCC and FVC in soybean fields (see Table 5), and was superior to the three machine learning models, PLSR, MSR, and RF. In previous studies using PLSR to predict crop parameters [66], PLSR showed excellent prediction ability. However, Figure 8 shows that the LCC predicted using PLSR deviated from the field survey data (RMSE: 6.80). This may be because the VI and LCC used in our study were not purely linear. PLSR, as a linear regression method, cannot effectively determine the nonlinear relationship between VI and LCC. Including the prediction results of MSR for LCC in this experiment can explain this phenomenon more reasonably. The results of FVC estimation showed that all four selected regression models showed good predictive ability, and GPR still constituted the optimal regression model. In a related study, Atzberger et al. [67] used hyperspectral data and regression techniques such as PLSR to predict LCC. However, hyperspectral data are more expensive than RGB images and are not generalizable. In two other studies [68,69], the LCC and FVC were estimated using the PROSALL physical model. Although good results were obtained, this method is usually susceptible to the initial model parameters and requires a priori information. Liang et al. [70] used a hybrid approach (i.e., the PROSALL model in combination with RF) to predict LCC, obtaining a high accuracy. However, the stable coupling required to employ this method remains a challenge. In contrast, the method for estimating FVC and LCC reported in this paper is simple and accurate, and the results are within a reasonable range.

Our ultimate goal was to investigate the effectiveness of four regression techniques for the estimation of soybean FVC and LCC and for monitoring soybean maturity using soybean LCC and FVC anomaly detection methods. This requires us to consider the effects caused by the harvesting area of soybean fields. That is, while exploring a high-precision prediction model for vegetated areas, attention should also be paid to the estimation of bare and near-bare areas (i.e., areas with small amounts of mature soybean stubble or areas influenced by lateral branches of surrounding soybean plants). Although we have tried to make the model converge as much as possible when constructing the GPR, there is still an error of 0.85 Dualex units (LCC) and 0.10 (FVC) in the non-vegetated areas. Of course, this does not exclude the shadows produced at noon on both sides of the soil, which cause the pixel channels to affect the adjacent image elements. Nevertheless, the final results show that GPR is an excellent prediction model for significant coefficients of variation in LCC and FVC.

5.2. Soybean Maturity Monitoring Study Analysis

There is a vast difference in the soybean growth cycle in breeding fields. This causes anomalies in the distribution of LCC and FVC in soybeans before and after the growth period. Capturing such anomalies enables soybean maturity monitoring. Castillo-Villamor et al. [71] used optical vegetation indices as input, then monitored crop growth by anomaly detection and combined it with yield analysis. Although this method has also been used in agriculture, its potential for crop maturity monitoring has been overlooked. Hence, in this work, we detected soybean LCC and FVC distribution. As a result, soybean maturity monitoring was achieved. In a previous study on crop maturity monitoring, Yu et al. [72] achieved 93% accuracy using a novel random forest model to monitor mature regions. However, such methods using spectral indices combined with ML often provide erratic monitoring. Moeinizade et al. [1] achieved 95% accuracy in monitoring soybean maturity using a CNN-LSTM model. Ashtiani et al. [73] used transfer learning based on CNN to monitor mulberry maturity, achieving an overall accuracy of 98.03%. Although DL and transfer learning-based crop maturity monitoring perform better, these methods require a large amount of sample image data for support, necessitating the challenge of collecting data in the field. Moreover, the model automatically extracts the original image features, ignoring the potential of crop FVC.LCC images for soybean maturity monitoring. In contrast, our present work considered the maturity information brought by the change in the distribution of soybean FVC and LCC images. The three monitoring accuracies obtained in this study ranged from 98.1% to 98.8%, further demonstrating the potential of the method for soybean maturity monitoring in breeding fields.

In this study, although we achieved high accuracy in monitoring soybean maturity. However, there are still some limitations. For example, in the P3 period, even though most of the early maturing soybeans were mature. However, there were still some unripe early maturing strains of soybean. This is one of the reasons for the reduced monitoring efficiency. The overall LCC of soybean gradually shifted to the left with time from P2 to P3 until the early maturing region moved away from the normal distribution during P2. This process is dynamic, and the optimal threshold does not necessarily arise at P2 but perhaps 2–3 days before and after P2 (the same is true for the soybean harvesting area monitored by FVC in this study). We monitored whether the soybeans had been harvested in the mature area of P4 using the FVC and LCC anomaly detection methods. Although we have achieved better identification results (Figure 11c), some things still need to be corrected. The different harvesting criteria of different soybean managers are the leading cause of these errors. In our study, LCC and FVC of immature soybean showed a normal distribution. Whether this is the case for all breeding field crops is worth exploring. In addition, the environment of our experiment unfolded in a soybean breeding site with high heterogeneity among soybean fields. Hence, the applicability of this method to monitor crop maturity in specific fields needs to be further explored.

5.3. Future Work

Both parts of the work conducted in this study showed promising results. However, these results are still influenced and limited by some uncertainty factors, including the following: (1) Uncertainty of image acquisition: The study is centered on images. Therefore, even though UAV images have high resolution and solid temporal reconstruction capability, the effects of light changes and camera positions in the same space–time cannot be avoided during the image acquisition process. (2) Uncertainty in the ground data environment: From the P3 images, the presence of leaf stagnation following pod senescence in harvest stubble areas is evident, leading to a reduction in the accuracy of monitoring using the soybean LCC and FVC anomaly detection methods, and increasing the error in the harvest and maturity areas. These uncertainties affect the study, such that these can be added to the study to be performed as part of our follow-up work.

6. Conclusions

In this study, we completed a two-part experiment based on four regression techniques for the rapid and accurate estimation of soybean FVC and LCC, and the monitoring of maturity information based on methods for detecting LCC and FVC abnormalities in soybean material. The experiment was conducted in a multi-strain soybean breeding field covering four soybean growth stages (P1–P4). The results were as follows.

(1) The combination of low-altitude drone technology and machine learning regression models can be used to furnish high-performance soybean FVC and LCC estimation results. Soybean FVC and LCC were estimated using PLSR, MSR, RF, and GPR, respectively, and GPR exhibited the best performance. The LCC prediction results were as follows: R^2: 0.84; RMSE: 3.36 Dualex units. The FVC prediction results were R^2: 0.96; RMSE: 0.08.

(2) The analysis of LCC and FVC anomalies detected in soybean material detection can provide highly accurate monitoring results regarding the maturity of soybean material. The total monitoring accuracies of P3 and P4 mature and immature soybeans were 0.988 and 0.984, respectively. The monitoring accuracy for the P4 mature and harvested area was 0.981.

(3) On the basis of the results of this research process, the frequency of image acquisition between P3 and P4 will be increased with the aim of investigating the relationship between the time interval of image acquisition and the maturity monitoring effect.

Author Contributions: Conceptualization, J.H. and J.Y.; Data curation, J.Y.; Investigation, X.X., S.H., T.S. and H.F.; Methodology, J.H. and J.Y.; Resources, S.H.; Software, J.H., J.Y., X.X., S.H., T.S. and Y.L.; Validation, X.X. and Y.L.; Visualization, H.F.; Writing—original draft, J.H.; Writing—review and editing, H.Q. All authors have read and agreed to the published version of the manuscript.

Funding: This study was supported by the Henan Province Science and Technology Research Project (232102111123), the National Natural Science Foundation of China (grant number 42101362, 41801225), the Joint Fund of Science and Technology Research Development program (Application Research) of Henan Province, China (222103810024).

Institutional Review Board Statement: Not applicable.

Data Availability Statement: The authors do not have permission to share data.

Conflicts of Interest: The authors declare no conflict of interest.

References

1. Moeinizade, S.; Pham, H.; Han, Y.; Dobbels, A.; Hu, G. An applied deep learning approach for estimating soybean relative maturity from UAV imagery to aid plant breeding decisions. *Mach. Learn. Appl.* **2022**, *7*, 100233. [CrossRef]
2. Brantley, S.T.; Zinnert, J.C.; Young, D.R. Application of hyperspectral vegetation indices to detect variations in high leaf area index temperate shrub thicket canopies. *Remote Sens. Environ.* **2011**, *115*, 514–523. [CrossRef]
3. Zhang, Y.; Hui, J.; Qin, Q.; Sun, Y.; Zhang, T.; Sun, H.; Li, M. Transfer-learning-based approach for leaf chlorophyll content estimation of winter wheat from hyperspectral data. *Remote Sens. Environ.* **2021**, *267*, 112724. [CrossRef]
4. Zhou, J.; Yungbluth, D.; Vong, C.N.; Scaboo, A.; Zhou, J. Estimation of the maturity date of soybean breeding lines using UAV-based multispectral imagery. *Remote Sens.* **2019**, *11*, 2075. [CrossRef]
5. Melo, L.C.; Pereira, H.S.; Faria, L.C.d.; Aguiar, M.S.; Costa, J.G.C.d.; Wendland, A.; Díaz, J.L.C.; Carvalho, H.W.L.d.; Costa, A.F.d.; Almeida, V.M.d. BRS FC104-Super-early carioca seeded common bean cultivar with high yield potential. *Crop Breed. Appl. Biotechnol.* **2019**, *19*, 471–475. [CrossRef]
6. Wang, M.; Niu, X.; Chen, S.; Guo, P.; Yang, Q.; Wang, Z. Inversion of chlorophyll contents by use of hyperspectral CHRIS data based on radiative transfer model. In Proceedings of the IOP Conference Series: Earth and Environmental Science, Montreal, Canada, 22–26 September 2014; p. 012073.
7. Zhang, Y.; Ta, N.; Guo, S.; Chen, Q.; Zhao, L.; Li, F.; Chang, Q. Combining Spectral and Textural Information from UAV RGB Images for Leaf Area Index Monitoring in Kiwifruit Orchard. *Remote Sens.* **2022**, *14*, 1063. [CrossRef]
8. Juarez, R.I.N.; da Rocha, H.R.; e Figueira, A.M.S.; Goulden, M.L.; Miller, S.D. An improved estimate of leaf area index based on the histogram analysis of hemispherical photographs. *Agric. Forest Meteorol.* **2009**, *149*, 920–928. [CrossRef]
9. Yue, J.; Feng, H.; Tian, Q.; Zhou, C. A robust spectral angle index for remotely assessing soybean canopy chlorophyll content in different growing stages. *Plant Methods* **2020**, *16*, 104. [CrossRef]
10. Amin, E.; Verrelst, J.; Rivera-Caicedo, J.P.; Pipia, L.; Ruiz-Verdú, A.; Moreno, J. Prototyping Sentinel-2 green LAI and brown LAI products for cropland monitoring. *Remote Sens. Environ.* **2021**, *255*, 112168. [CrossRef]
11. Li, X.; Lu, H.; Yu, L.; Yang, K. Comparison of the spatial characteristics of four remotely sensed leaf area index products over China: Direct validation and relative uncertainties. *Remote Sens.* **2018**, *10*, 148. [CrossRef]
12. Atzberger, C.; Darvishzadeh, R.; Immitzer, M.; Schlerf, M.; Skidmore, A.; Le Maire, G. Comparative analysis of different retrieval methods for mapping grassland leaf area index using airborne imaging spectroscopy. *Int. J. Appl. Earth Observat. Geoinf.* **2015**, *43*, 19–31. [CrossRef]
13. Yue, J.; Feng, H.; Jin, X.; Yuan, H.; Li, Z.; Zhou, C.; Yang, G.; Tian, Q. A comparison of crop parameters estimation using images from UAV-mounted snapshot hyperspectral sensor and high-definition digital camera. *Remote Sens.* **2018**, *10*, 1138. [CrossRef]
14. Mutha, S.A.; Shah, A.M.; Ahmed, M.Z. Maturity Detection of Tomatoes Using Deep Learning. *SN Comput. Sci.* **2021**, *2*, 441. [CrossRef]
15. Yue, J.; Guo, W.; Yang, G.; Zhou, C.; Feng, H.; Qiao, H. Method for accurate multi-growth-stage estimation of fractional vegetation cover using unmanned aerial vehicle remote sensing. *Plant Methods* **2021**, *17*, 51. [CrossRef]
16. Zhou, C.; Ye, H.; Xu, Z.; Hu, J.; Shi, X.; Hua, S.; Yue, J.; Yang, G. Estimating maize-leaf coverage in field conditions by applying a machine learning algorithm to UAV remote sensing images. *Appl. Sci.* **2019**, *9*, 2389. [CrossRef]
17. Yu, K.; Leufen, G.; Hunsche, M.; Noga, G.; Chen, X.; Bareth, G. Investigation of leaf diseases and estimation of chlorophyll concentration in seven barley varieties using fluorescence and hyperspectral indices. *Remote Sens.* **2014**, *6*, 64–86. [CrossRef]
18. Trevisan, R.; Pérez, O.; Schmitz, N.; Diers, B.; Martin, N. High-throughput phenotyping of soybean maturity using time series UAV imagery and convolutional neural networks. *Remote Sens.* **2020**, *12*, 3617. [CrossRef]
19. Shen, L.; Gao, M.; Yan, J.; Wang, Q.; Shen, H. Winter Wheat SPAD Value Inversion Based on Multiple Pretreatment Methods. *Remote Sens.* **2022**, *14*, 4660. [CrossRef]
20. Gevaert, C.M.; Suomalainen, J.; Tang, J.; Kooistra, L. Generation of spectral–temporal response surfaces by combining multispectral satellite and hyperspectral UAV imagery for precision agriculture applications. *IEEE J. Select. Top. Appl. Earth Observ. Remote Sens.* **2015**, *8*, 3140–3146. [CrossRef]
21. Broge, N.H.; Leblanc, E. Comparing prediction power and stability of broadband and hyperspectral vegetation indices for estimation of green leaf area index and canopy chlorophyll density. *Remote Sens. Environ.* **2001**, *76*, 156–172. [CrossRef]
22. Tao, H.; Feng, H.; Xu, L.; Miao, M.; Long, H.; Yue, J.; Li, Z.; Yang, G.; Yang, X.; Fan, L. Estimation of crop growth parameters using UAV-based hyperspectral remote sensing data. *Sensors* **2020**, *20*, 1296. [CrossRef]
23. Maimaitijiang, M.; Sagan, V.; Sidike, P.; Daloye, A.M.; Erkbol, H.; Fritschi, F.B. Crop monitoring using satellite/UAV data fusion and machine learning. *Remote Sens.* **2020**, *12*, 1357. [CrossRef]
24. Liu, Y.; Feng, H.; Yue, J.; Li, Z.; Yang, G.; Song, X.; Yang, X.; Zhao, Y. Remote-sensing estimation of potato above-ground biomass based on spectral and spatial features extracted from high-definition digital camera images. *Comput. Electron. Agric.* **2022**, *198*, 107089. [CrossRef]
25. Tayade, R.; Yoon, J.; Lay, L.; Khan, A.L.; Yoon, Y.; Kim, Y. Utilization of spectral indices for high-throughput phenotyping. *Plants* **2022**, *11*, 1712. [CrossRef]
26. Yue, J.; Yang, H.; Yang, G.; Fu, Y.; Wang, H.; Zhou, C. Estimating vertically growing crop above-ground biomass based on UAV remote sensing. *Comput. Electron. Agric.* **2023**, *205*, 107627. [CrossRef]

27. Liu, Y.; Hatou, K.; Aihara, T.; Kurose, S.; Akiyama, T.; Kohno, Y.; Lu, S.; Omasa, K. A robust vegetation index based on different UAV RGB images to estimate SPAD values of naked barley leaves. *Remote Sens.* **2021**, *13*, 686. [CrossRef]
28. Kanning, M.; Kühling, I.; Trautz, D.; Jarmer, T. High-resolution UAV-based hyperspectral imagery for LAI and chlorophyll estimations from wheat for yield prediction. *Remote Sens.* **2018**, *10*, 2000. [CrossRef]
29. Jacquemoud, S.; Verhoef, W.; Baret, F.; Bacour, C.; Zarco-Tejada, P.J.; Asner, G.P.; François, C.; Ustin, S.L. PROSPECT+ SAIL models: A review of use for vegetation characterization. *Remote Sens. Environ.* **2009**, *113*, S56–S66. [CrossRef]
30. Berger, K.; Atzberger, C.; Danner, M.; D'Urso, G.; Mauser, W.; Vuolo, F.; Hank, T. Evaluation of the PROSAIL model capabilities for future hyperspectral model environments: A review study. *Remote Sens.* **2018**, *10*, 85. [CrossRef]
31. Yue, J.; Feng, H.; Yang, G.; Li, Z. A Comparison of Regression Techniques for Estimation of Above-Ground Winter Wheat Biomass Using Near-Surface Spectroscopy. *Remote Sens.* **2018**, *10*, 66. [CrossRef]
32. Fu, Y.; Yang, G.; Li, Z.; Song, X.; Li, Z.; Xu, X.; Wang, P.; Zhao, C. Winter wheat nitrogen status estimation using UAV-based RGB imagery and gaussian processes regression. *Remote Sens.* **2020**, *12*, 3778. [CrossRef]
33. Yue, J.; Yang, G.; Tian, Q.; Feng, H.; Xu, K.; Zhou, C. Estimate of winter-wheat above-ground biomass based on UAV ultrahigh-ground-resolution image textures and vegetation indices. *ISPRS J. Photogr. Remote Sens.* **2019**, *150*, 226–244. [CrossRef]
34. Benmouna, B.; Pourdarbani, R.; Sabzi, S.; Fernandez-Beltran, R.; García-Mateos, G.; Molina-Martínez, J.M. Comparison of Classic Classifiers, Metaheuristic Algorithms and Convolutional Neural Networks in Hyperspectral Classification of Nitrogen Treatment in Tomato Leaves. *Remote Sens.* **2022**, *14*, 6366. [CrossRef]
35. Xu, X.; Lu, J.; Zhang, N.; Yang, T.; He, J.; Yao, X.; Cheng, T.; Zhu, Y.; Cao, W.; Tian, Y. Inversion of rice canopy chlorophyll content and leaf area index based on coupling of radiative transfer and Bayesian network models. *ISPRS J. Photogr. Remote Sens.* **2019**, *150*, 185–196. [CrossRef]
36. Baltazar, A.; Aranda, J.I.; González-Aguilar, G. Bayesian classification of ripening stages of tomato fruit using acoustic impact and colorimeter sensor data. *Comput. Electron. Agric.* **2008**, *60*, 113–121. [CrossRef]
37. Cerovic, Z.G.; Goutouly, J.-P.; Hilbert, G.; Destrac-Irvine, A.; Martinon, V.; Moise, N. Mapping winegrape quality attributes using portable fluorescence-based sensors. *Frutic* **2009**, *9*, 301–310.
38. Zhang, L.; McCarthy, M.J. Measurement and evaluation of tomato maturity using magnetic resonance imaging. *Postharvest. Biol. Technol.* **2012**, *67*, 37–43. [CrossRef]
39. Brezmes, J.; Fructuoso, M.L.; Llobet, E.; Vilanova, X.; Recasens, I.; Orts, J.; Saiz, G.; Correig, X. Evaluation of an electronic nose to assess fruit ripeness. *IEEE Sens. J.* **2005**, *5*, 97–108. [CrossRef]
40. Zhao, W.; Yang, Z.; Chen, Z.; Liu, J.; Wang, W.C.; Zheng, W.Y. Hyperspectral surface analysis for ripeness estimation and quick UV-C surface treatments for preservation of bananas. *J. Appl. Spectrosc.* **2016**, *83*, 254–260. [CrossRef]
41. Khodabakhshian, R.; Emadi, B. Application of Vis/SNIR hyperspectral imaging in ripeness classification of pear. *Int. J. Food Prop.* **2017**, *20*, S3149–S3163. [CrossRef]
42. Volpato, L.; Dobbels, A.; Borem, A.; Lorenz, A.J. Optimization of temporal UAS-based imagery analysis to estimate plant maturity date for soybean breeding. *Plant Phenom. J.* **2021**, *4*, e20018. [CrossRef]
43. Makanza, R.; Zaman-Allah, M.; Cairns, J.E.; Magorokosho, C.; Tarekegne, A.; Olsen, M.; Prasanna, B.M. High-throughput phenotyping of canopy cover and senescence in maize field trials using aerial digital canopy imaging. *Remote Sens.* **2018**, *10*, 330. [CrossRef]
44. Marcillo, G.S.; Martin, N.F.; Diers, B.W.; Da Fonseca Santos, M.; Leles, E.P.; Chigeza, G.; Francischini, J.H. Implementation of a generalized additive model (Gam) for soybean maturity prediction in african environments. *Agronomy* **2021**, *11*, 1043. [CrossRef]
45. Zhou, X.; Lee, W.S.; Ampatzidis, Y.; Chen, Y.; Peres, N.; Fraisse, C. Strawberry maturity classification from UAV and near-ground imaging using deep learning. *Smart Agric. Technol.* **2021**, *1*, 100001. [CrossRef]
46. Simonyan, K.; Zisserman, A. Very deep convolutional networks for large-scale image recognition. *arXiv* **2014**, arXiv:1409.1556.
47. Mahmood, A.; Singh, S.K.; Tiwari, A.K. Pre-trained deep learning-based classification of jujube fruits according to their maturity level. *Neural Comput. Appl.* **2022**, *34*, 13925–13935. [CrossRef]
48. Nilson, T. A theoretical analysis of the frequency of gaps in plant stands. *Agric. Meteorol.* **1971**, *8*, 25–38. [CrossRef]
49. Song, W.; Mu, X.; Ruan, G.; Gao, Z.; Li, L.; Yan, G. Estimating fractional vegetation cover and the vegetation index of bare soil and highly dense vegetation with a physically based method. *Int. J. Appl. Earth Observ. Geoinf.* **2017**, *58*, 168–176. [CrossRef]
50. Goulas, Y.; Cerovic, Z.G.; Cartelat, A.; Moya, I. Dualex: A new instrument for field measurements of epidermal ultraviolet absorbance by chlorophyll fluorescence. *Appl. Opt.* **2004**, *43*, 4488–4496. [CrossRef]
51. Kloog, I.; Nordio, F.; Coull, B.A.; Schwartz, J. Predicting spatiotemporal mean air temperature using MODIS satellite surface temperature measurements across the Northeastern USA. *Remote Sens. Environ.* **2014**, *150*, 132–139. [CrossRef]
52. Pearson, R.L.; Miller, L.D.; Tucker, C.J. Hand-held spectral radiometer to estimate gramineous biomass. *Appl. Opt.* **1976**, *15*, 416–418. [CrossRef] [PubMed]
53. Kawashima, S.; Nakatani, M. An algorithm for estimating chlorophyll content in leaves using a video camera. *Ann. Bot.* **1998**, *81*, 49–54. [CrossRef]
54. Sellaro, R.; Crepy, M.; Trupkin, S.A.; Karayekov, E.; Buchovsky, A.S.; Rossi, C.; Casal, J.J. Cryptochrome as a sensor of the blue/green ratio of natural radiation in Arabidopsis. *Plant Physiol.* **2010**, *154*, 401–409. [CrossRef] [PubMed]
55. Verrelst, J.; Schaepman, M.E.; Koetz, B.; Kneubühler, M. Angular sensitivity analysis of vegetation indices derived from CHRIS/PROBA data. *Remote Sens. Environ.* **2008**, *112*, 2341–2353. [CrossRef]

56. Peñuelas, J.; Gamon, J.; Fredeen, A.; Merino, J.; Field, C. Reflectance indices associated with physiological changes in nitrogen-and water-limited sunflower leaves. *Remote Sens. Environ.* **1994**, *48*, 135–146. [CrossRef]

57. Meyer, G.E.; Neto, J.C. Verification of color vegetation indices for automated crop imaging applications. *Comput. Electron. Agric.* **2008**, *63*, 282–293. [CrossRef]

58. Gitelson, A.A.; Kaufman, Y.J.; Stark, R.; Rundquist, D. Novel algorithms for remote estimation of vegetation fraction. *Remote Sens. Environ.* **2002**, *80*, 76–87. [CrossRef]

59. Baret, F.; Guyot, G.; Major, D. TSAVI: A vegetation index which minimizes soil brightness effects on LAI and APAR estimation. In Proceedings of the 12th Canadian Symposium on Remote Sensing and IGARSS'90, Vancouver, Canada, 10–14 July 1989.

60. Hu, H.; Zhang, J.; Sun, X.; Zhang, X. Estimation of leaf chlorophyll content of rice using image color analysis. *Can. J. Remote Sens.* **2013**, *39*, 185–190. [CrossRef]

61. Wang, X.; Xu, L.; Chen, H.; Zou, Z.; Huang, P.; Xin, B. Non-Destructive Detection of pH Value of Kiwifruit Based on Hyperspectral Fluorescence Imaging Technology. *Agriculture* **2022**, *12*, 208. [CrossRef]

62. Ji, S.; Gu, C.; Xi, X.; Zhang, Z.; Hong, Q.; Huo, Z.; Zhao, H.; Zhang, R.; Li, B.; Tan, C. Quantitative Monitoring of Leaf Area Index in Rice Based on Hyperspectral Feature Bands and Ridge Regression Algorithm. *Remote Sens.* **2022**, *14*, 2777. [CrossRef]

63. Han, S.; Zhao, Y.; Cheng, J.; Zhao, F.; Yang, H.; Feng, H.; Li, Z.; Ma, X.; Zhao, C.; Yang, G. Monitoring Key Wheat Growth Variables by Integrating Phenology and UAV Multispectral Imagery Data into Random Forest Model. *Remote Sens.* **2022**, *14*, 3723. [CrossRef]

64. Pasolli, L.; Melgani, F.; Blanzieri, E. Gaussian process regression for estimating chlorophyll concentration in subsurface waters from remote sensing data. *IEEE Geosci. Remote Sens. Lett.* **2010**, *7*, 464–468. [CrossRef]

65. Liang, J.; Liu, D. Automated estimation of daily surface water fraction from MODIS and Landsat images using Gaussian process regression. *Int. J. Remote Sens.* **2021**, *42*, 4261–4283. [CrossRef]

66. Ma, J.; Wang, L.; Chen, P. Comparing Different Methods for Wheat LAI Inversion Based on Hyperspectral Data. *Agriculture* **2022**, *12*, 1353. [CrossRef]

67. Atzberger, C.; Guérif, M.; Baret, F.; Werner, W. Comparative analysis of three chemometric techniques for the spectroradiometric assessment of canopy chlorophyll content in winter wheat. *Comput. Electron. Agric.* **2010**, *73*, 165–173. [CrossRef]

68. Ding, Y.; Zhang, H.; Zhao, K.; Zheng, X. Investigating the accuracy of vegetation index-based models for estimating the fractional vegetation cover and the effects of varying soil backgrounds using in situ measurements and the PROSAIL model. *Int. J. Remote Sens.* **2017**, *38*, 4206–4223. [CrossRef]

69. Jay, S.; Maupas, F.; Bendoula, R.; Gorretta, N. Retrieving LAI, chlorophyll and nitrogen contents in sugar beet crops from multi-angular optical remote sensing: Comparison of vegetation indices and PROSAIL inversion for field phenotyping. *Field Crops Res.* **2017**, *210*, 33–46. [CrossRef]

70. Liang, L.; Qin, Z.; Zhao, S.; Di, L.; Zhang, C.; Deng, M.; Lin, H.; Zhang, L.; Wang, L.; Liu, Z. Estimating crop chlorophyll content with hyperspectral vegetation indices and the hybrid inversion method. *Int. J. Remote Sens.* **2016**, *37*, 2923–2949. [CrossRef]

71. Castillo-Villamor, L.; Hardy, A.; Bunting, P.; Llanos-Peralta, W.; Zamora, M.; Rodriguez, Y.; Gomez-Latorre, D.A. The Earth Observation-based Anomaly Detection (EOAD) system: A simple, scalable approach to mapping in-field and farm-scale anomalies using widely available satellite imagery. *Int. J. Appl. Earth Obs. Geoinf.* **2021**, *104*, 102535. [CrossRef]

72. Yu, N.; Li, L.; Schmitz, N.; Tian, L.F.; Greenberg, J.A.; Diers, B.W. Development of methods to improve soybean yield estimation and predict plant maturity with an unmanned aerial vehicle based platform. *Remote Sens. Environ.* **2016**, *187*, 91–101. [CrossRef]

73. Ashtiani, S.-H.M.; Javanmardi, S.; Jahanbanifard, M.; Martynenko, A.; Verbeek, F.J. Detection of mulberry ripeness stages using deep learning models. *IEEE Access* **2021**, *9*, 100380–100394. [CrossRef]

Article

Extraction of Cropland Spatial Distribution Information Using Multi-Seasonal Fractal Features: A Case Study of Black Soil in Lishu County, China

Qi Wang [1,2,†], Peng Guo [1,†], Shiwei Dong [2,3,*], Yu Liu [2,3], Yuchun Pan [2] and Cunjun Li [2]

1. College of Information Science and Engineering, Shandong Agricultural University, Jinan 271018, China
2. Research Center of Information Technology, Beijing Academy of Agriculture and Forestry Sciences, Beijing 100097, China
3. Key Laboratory of Land Satellite Remote Sensing Application, Ministry of Natural Resources, Nanjing 210013, China

* Correspondence: dongsw@nercita.org.cn
† These authors contributed equally to this work.

Abstract: Accurate extraction of cropland distribution information using remote sensing technology is a key step in the monitoring, protection, and sustainable development of black soil. To obtain precise spatial distribution of cropland, an information extraction method is developed based on a fractal algorithm integrating temporal and spatial features. The method extracts multi-seasonal fractal features from the Landsat 8 OLI remote sensing data. Its efficiency is demonstrated using black soil in Lishu County, Northeast China. First, each pixel's upper and lower fractal signals are calculated using a blanket covering method based on the Landsat 8 OLI remote sensing data in the spring, summer, and autumn seasons. The fractal characteristics of the cropland and other land-cover types are analyzed and compared. Second, the ninth lower fractal scale is selected as the feature scale to extract the spatial distribution of cropland in Lishu County. The cropland vector data, the European Space Agency (ESA) WorldCover data, and the statistical yearbook from the same period are used to assess accuracy. Finally, a comparative analysis of this study and existing products at different scales is carried out, and the point matching degree and area matching degree are evaluated. The results show that the point matching degree and the area matching degree of cropland extraction using the multi-seasonal fractal features are 90.66% and 96.21%, and 95.33% and 83.52%, respectively, which are highly consistent with the statistical data provided by the local government. The extracted accuracy of cropland is much better than that of existing products at different scales due to the contribution of the multi-seasonal fractal features. This method can be used to accurately extract cropland information to monitor changes in black soil, and it can be used to support the conservation and development of black soil in China.

Keywords: cropland; multi-seasonal; fractal feature; feature extraction; accuracy evaluation; black soil

Citation: Wang, Q.; Guo, P.; Dong, S.; Liu, Y.; Pan, Y.; Li, C. Extraction of Cropland Spatial Distribution Information Using Multi-Seasonal Fractal Features: A Case Study of Black Soil in Lishu County, China. *Agriculture* **2023**, *13*, 486. https://doi.org/10.3390/agriculture13020486

Academic Editor: Rui Costa Martins

Received: 11 January 2023
Revised: 14 February 2023
Accepted: 16 February 2023
Published: 18 February 2023

1. Introduction

Black soil, which is marked by black or dark black humus topsoil, is a valuable natural resource and the most fertile soil in the world [1]. Due to the impact of global warming and human activities, black soil has been exposed for a long time in some areas, and its soil structure degrades as wind and water erosions intensify [2,3]. This poses a severe challenge to the sustainable development of agriculture and food security in China. In response to the urgency to protect black soil, the Action Plan for Conservation Tillage in Northeast China (2020–2025) is jointly issued by the Ministry of Agriculture and Rural Affairs and the Ministry of Finance. This action plan is issued to deploy comprehensive promotion and application of conservation tillage in appropriate areas to ensure the sustainable development of black soil. The quality of black soil has changed and degraded markedly due to frequent human activities on the global and regional scales. Accurately determining the amount and spatial distribution of cropland on black soil is beneficial to the national government when implementing special protection measures to reduce the loss of black soil and improve the quality of black land. Information extraction is essential to implement conservation measures for the spatial distribution of cropland in black soil areas.

Remote sensing technology is an efficient way to realize large-scale cropland monitoring. Current research mainly focuses on extracting ground object information accurately. The spectral features, temporal features, and spatial features of remote sensing data are used for classification. Spectral features are the physical properties of natural materials, which generally refer to the absorption, reflection, and transmission of electromagnetic radiation of ground objects. Temporal features are the features that change in different time phases. Spatial features refer to the laws of spatial relationships between ground object pixels in remote sensing images through numerical operations. Classification methods based on spectral features generally analyze the spectral curves of ground objects to classify them. Machine learning-based algorithms, such as decision trees, support vector machines, random forests, and deep learning-based algorithms, have been used broadly [4–9]. These methods require mass training samples and significant time spent controlling the samples' quality and adjusting complex model parameters to obtain the optimal results. Classification methods based on temporal features mainly focus on the analysis of time series to obtain the changes in the features of ground objects to reduce the influence of incomplete information brought by the use of a single temporal phase. However, the frequently used low-resolution MODIS data have limitations for features with a more fragmented distribution [10]. Moreover, it is necessary to reduce the influence of data redundancy of long-time series of remote sensing images. Classification methods based on spatial features are independent of mixed pixels and can directly extract the gray structure features of images. Examples are the gray-level co-occurrence matrix [11], fractal analysis [12], Fourier transform, wavelet transform, gray edge detection, variance function, and so on. However, there are some limitations in applying these methods to regions without directivity and regularity of texture features of images, such as optical images of mining areas [13]. Regarding the selection of data sources, some studies use many high-resolution images to obtain sufficient spatial features of their targets [14–16], such as planet 4, GF-2, and WorldView-2. Compared to a single feature, numerous studies have begun to combine multiple features to achieve higher accuracy. Combining spectral and temporal features can use spectral diversity and improve recognition ability of changes in ground object features [17,18]. Integrating temporal and spatial features can retain the law of ground objects changing with time and reduce the influence of mixed pixels [19,20].

The distribution area of black soil in Northeast China, which comprises complex and heterogeneous environmental conditions and vegetation growth environments, is not suitable for extracting information using traditional methods. However, the main crops in this area have a concentrated growing period that differs significantly from natural vegetation, which is appropriate for extracting cropland information using remote sensing images' multi-seasonal features. Multi-seasonal remote sensing data are widely used for change detection and information extraction. The fusion of multi-seasonal data can compensate for the lack of information in single-temporal data so that the seasonal change information of ground objects can be effectively used for improving accuracy. When combined with various information extraction techniques, the feature information of ground objects is enhanced, and the accuracy of information extraction is improved [21–23]. To reduce the effect of mixed pixels, spatial feature classification methods, such as the fractal method, can be combined with multi-seasonal features to improve extraction accuracy. Fractal is a regional algorithm for the iterative processing of surface textures without selecting training samples. The texture information of natural objects may show a certain degree of statistical self-similarity within a limited range, which demonstrates that the fractal method can be used for iterative processing when extracting information from ground objects to narrow the scope and highlight the features of ground objects. Existing results suggest that the fractal method can reveal important differences in land use and land-cover types [12], improve classification accuracy, and reduce computational time to some extent [24,25]. Various studies have been conducted using fractal algorithms based on large-scale data with low spatial resolution [26]. However, for small-scale studies, the temporal variation of cropland has been neglected [27].

In this study, the multi-seasonal features of the Landsat 8 OLI remote sensing data were introduced into a fractal algorithm to improve classification accuracy, taking both temporal and spatial features into account, and the developed method was developed with existing products. The remainder of this paper is organized as follows: Section 2 introduces the data and the method used. The accuracy of the processing results was evaluated and compared with existing products at different scales, and the results are presented in Section 3. Section 4 discusses the applicability and uncertainty of the method developed in this study, and Section 5 provides a summary of this study.

2. Materials and Methods

2.1. Study Area

Lishu County (Figure 1), which is located in the western part of Jilin province, China, has a temperate humid and semi-humid monsoon climate with a low annual temperature, with plains in the north and hills in the southeast. Many soil types in this area, mainly including black soil, black calcium soil, light black calcium soil, and brown soil, belong to the typical thin black soil area of Northeast China [28]. The main land-cover types include cropland, forest, grassland, impervious surfaces, bare land, and water. This area has a large cropland, accounting for more than 80% of the county. It is a veritable central grain-producing area in this region and plays an irreplaceable role in ensuring China's food security. Maize, rice, and soybean are the main crops, and deciduous trees are the main natural vegetation. The growth period of each crop is concentrated and different from that of natural vegetation, which is suitable for extracting information from croplands using multi-seasonal data. Since 2007, the national government has established a research demonstration area in Lishu County in conjunction with various scientific research institutes and put forward a black soil protection project named the Lishu Model, which is committed to protecting the sustainable development of black soil. The complex spatial heterogeneity of the region and frequent human activities lead to land cover changes in the area.

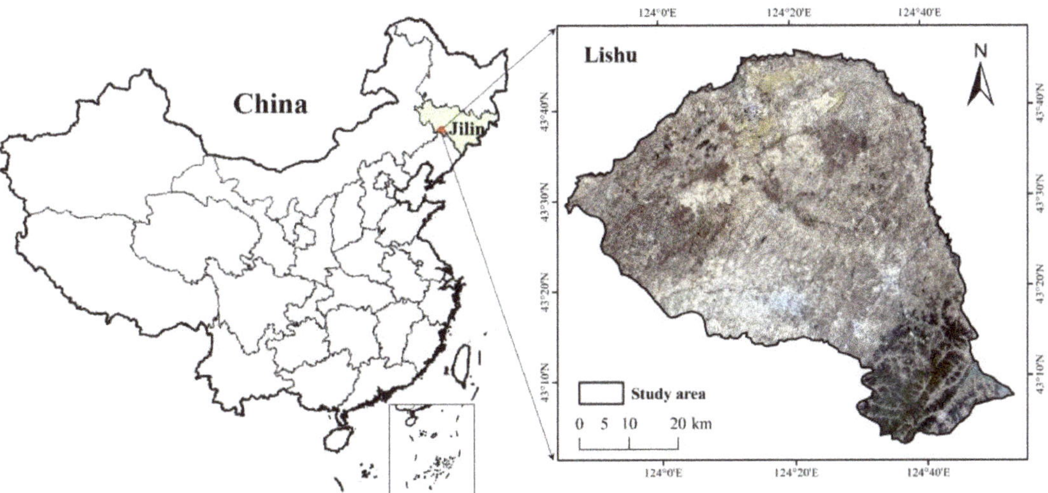

Figure 1. Location of the study area.

2.2. Data and Data Processing

2.2.1. Remote Sensing Data

The Landsat 8 OLI data covering 2020 were freely downloaded from https://www.gscloud.cn (accessed on 7 October 2021), a data cloud computing and product distribution platform provided by the Geospatial Data Cloud site, Computer Network Information Center, Chinese Academy of Sciences. The data used in this study were level 1T standard terrain correction products, which were accurately corrected using ground control sites and digital elevation model data. The principle of data selection was cloudless or partly cloudy (<2%) to ensure monthly coverage as much as possible. According to Figure 2, the main crops are planted from April to May. The peak growth period for the crops is July to August, and the harvest period is September to October. Therefore, the Landsat 8 OLI satellite products on 1 April 2020, 22 July 2020, and 10 October 2020, were selected as the basic data to represent the spring, summer, and autumn seasons, respectively. Seven multispectral bands of the Landsat 8 data for each season were selected, and the detailed information is shown in Table 1. A sequence dataset with 21 bands was obtained in the order of spring, summer, and autumn, which was used for fractal processing.

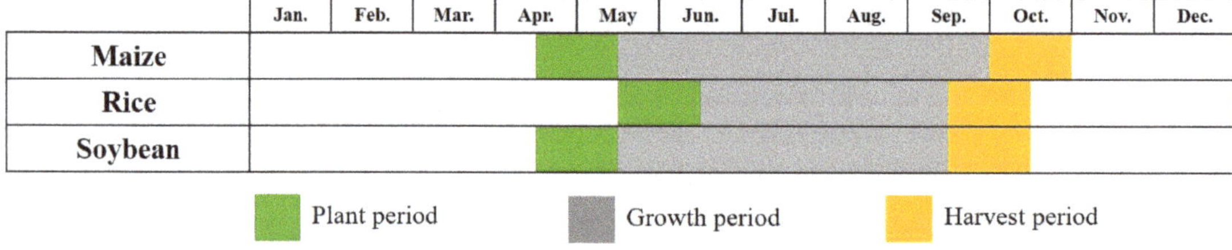

Figure 2. The main crop periods in the study area.

Table 1. The remote sensing images selected in this study.

Acquisition Dates	Season	Satellite Sensors	Band Name	Bandwidth (μm)	Resolution (m)
1 April 2020	Spring		Band 1 Coastal	0.43–0.45	
			Band 2 Blue	0.45–0.51	
			Band 3 Green	0.53–0.59	
22 July 2020	Summer	Landsat 8 OLI	Band 4 Red	0.64–0.67	30
			Band 5 NIR	0.85–0.88	
			Band 6 SWIR 1	1.57–1.65	
10 October 2020	Autumn		Band 7 SWIR 2	2.11–2.29	

2.2.2. Reference Data

The reference data included a statistical yearbook, the vector data, the European Space Agency (ESA) WorldCover data, and three land-cover products from the same period. The statistical yearbook, the vector data, and the ESA WorldCover data were used for accuracy evaluation, and the other three products were used for the comparative analysis in this study. Detailed information of the selected data is shown in Table 2.

Table 2. Detailed information of the selected data for Lishu County in 2020.

Data Set	Data Type	Resolution/Scale	Sensor
Statistical yearbook	Text	/	/
Vector data	Vector	1:100,000	Landsat
ESA WorldCover data	Raster	10 m	Sentinel-1 and Sentinel-2
Esri land cover dataset	Raster	10 m	Sentinel-2
GlobeLand30 dataset	Raster	30 m	Landsat 8/GF-1/HJ-1
CNLUCC	Raster	1000 m	Landsat 8

(1) Statistical yearbook of Lishu County in 2020. The statistical yearbook was produced by the Government of Lishu County and obtained through questionnaires, field visits, and field measurements; thus, it provides highly suitable data for accuracy evaluation.

(2) Vector data of Lishu (2020). The vector data of cropland in Lishu County in 2020 were produced based on a human–computer interactive interpretation using the Landsat images from the Institute of Geographic Sciences and Natural Resources Research, Chinese Academy of Sciences, with a mapping scale of 1:100,000 and including 6 classes and 25 subclasses [29]. Standard quality control and integration checking for each dataset were implemented using many field survey photographs and records during the same period to ensure high-quality and consistent interpretation. Therefore, the vector data are the most reliable and comparable data available in the area during the same period, and the data had been widely applied to estimate the accuracies of different classification results [30,31]. The vector data (Figure 3) were used as the main reference data for the accuracy evaluation of information extraction.

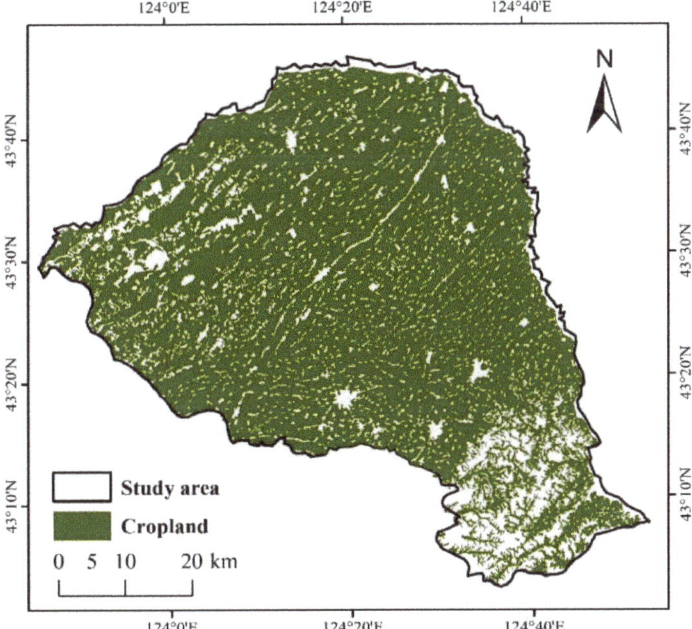

Figure 3. The distribution of cropland in the vector data.

(3) ESA WorldCover data (2020). The ESA WorldCover data provide a global land cover map for 2020 at a 10 m resolution based on the Sentinel-1 and Sentinel-2 data [32], The dataset contains 11 different land-cover classes, including tree cover, shrubland, grassland, cropland, built-up, bare/sparse vegetation, snow and ice, permanent water bodies, herbaceous wetland, mangroves, moss and lichen, and achieves an overall accuracy of 74.4%. Figure 4 is the cropland distribution in the ESA WorldCover data for Lishu County, and the data were used for accuracy evaluation.

(4) Esri land cover dataset (2020). A global land-cover map using the Sentinel-2 images was produced by a deep learning model trained using over 5 billion hand-labeled Sentinel-2 pixels and sampled from over 20,000 sites distributed across all major biomes of the world, with a resolution of 10 m [33]. It provides a 10-class map of the surface, including water, tree, grass, flooded vegetation, crop, built area, bare ground, shrub, snow/ice, and clouds, and it achieves an overall accuracy of 85% across the ten classes. In this study, the distribution of cropland (Figure A1) was used as a reference for the comparative analysis.

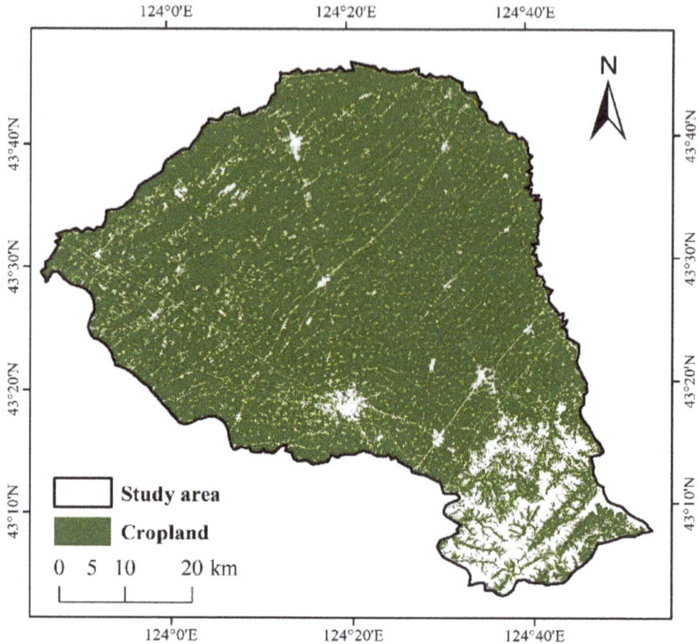

Figure 4. The distribution of cropland in the ESA WorldCover data.

(5) GlobeLand30 dataset (2020). A global land-cover data product with a spatial resolution of 30 m was provided by the National Geographic Information Centre of China [34], which mainly includes ten land-cover types: cropland, forest, grassland, shrubland, wetland, water, tundra, artificial land, bare land, and glacier/permanent snow. The overall accuracy of the GlobeLand30 dataset in 2020 was 85.72%, and the kappa coefficient was 0.82. This product (Figure A2) was also used for the comparative analysis of cropland information extraction.

(6) China Land Use and Land Cover Dataset (CNLUCC) (2020). This dataset was generated by the Resources and Environmental Science and Data Center (RESDC) of the Chinese Academy of Sciences based on Landsat 8 images through manual visual interpretation [35]. The land-cover types include cropland, woodland, grassland, water, residential land, unused land, and 25 secondary classifications, with a spatial resolution of 1000 m. This dataset (Figure A3) was used as a reference for the comparative analysis.

2.2.3. Data Processing

The downloaded remote sensing data and reference data for the study area were first converted by file formatting and re-projected into the UTM Zone 51 N with the WGS84 datum using nearest neighbor resampling. A spatial subset was extracted according to the boundary of Lishu County. Next, all raster data were converted into vector data using the Conversion Tools. Data processing was supported by ENVI 5.3 and ArcGIS 10.6, and the fractal programming operations were performed using IDL 8.5.

2.3. Methods

Figure 5 shows the flowchart of the information extraction method developed in this study. Firstly, data preprocessing was achieved, and fractal processing of the multi-seasonal images was used to select the feature-scale image of cropland. Secondly, cropland information was extracted using the degree of separation between cropland and other land-cover types in the statistical curve of the feature-scale image. An accuracy evaluation of the information extraction results was conducted using overlay analysis. Thirdly, through the comparison with other products, especially the local comparative analysis, the advantages and disadvantages of the method were summarized.

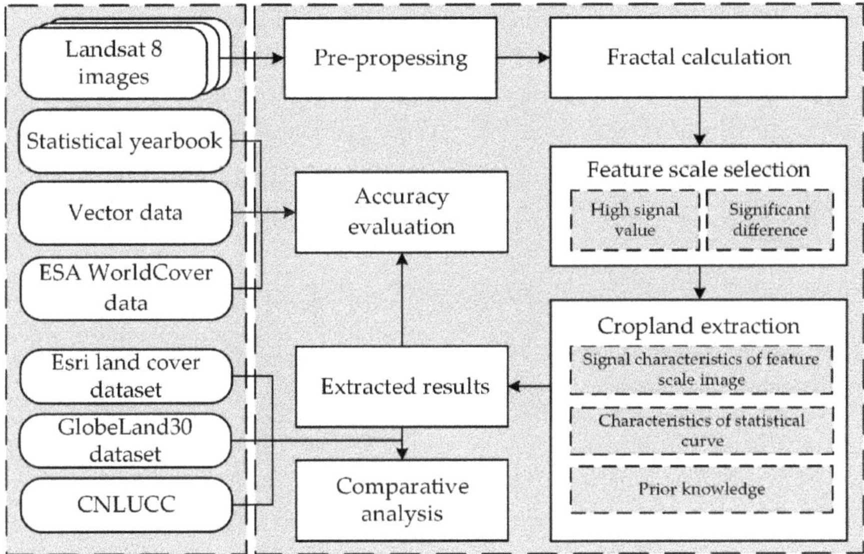

Figure 5. Flowchart of the information extraction method.

This method was divided into four aspects, including the principle of the blanket covering method, the feature-scale selection method, the information extraction method, and the accuracy evaluation metrics.

2.3.1. Blanket Covering Method

The blanket covering method can be used in remote sensing for texture analysis, pattern recognition, and image classification. The purpose of the method is to treat a remote sensing image as a three-dimensional space, with the gray value of each pixel representing the height of the three-dimensional surface, and then sandwich the terrain surface with two blankets, with both the upper and lower blankets at a distance of ε from the terrain surface. The fractal dimension can be calculated from the relationship between the area of the blanket and the volume of the space surrounded by these two blankets [36]. This study used a mathematical transformation iterative analysis from the perspective of signal analysis to select the feature scales of different land-cover types. The fractal dimension of each image's element spectral curve was calculated from the mathematical relationship between the area enclosed by the upper and lower two-dimensional curves and the lengths of these two curves. The specific calculation details are shown in [36,37].

The spectral curve is formulated as a function of $f(m)$ ($m = 1, 2, 3, \ldots, k$, where k is the number of samples in the band series selected), with two curves at a distance ε above and below the curve, which are called the upper fractal curve ($u_\varepsilon(m)$) and the lower fractal curve ($d_\varepsilon(m)$), respectively, and ε is the measurement scale.

$$u_\varepsilon(m) = \max\left\{ u_{\varepsilon-1}(m) + 1, \max_{|n-m \leq 1|} u_{\varepsilon-1}(m) \right\} \tag{1}$$

$$d_\varepsilon(m) = \max\left\{ d_{\varepsilon-1}(m) + 1, \min_{|n-m \leq 1|} d_{\varepsilon-1}(m) \right\} \tag{2}$$

where n is the value of discrete points in close proximity to m.

According to the polygon area surrounded by these two curves and Mandelbrot's definition of curve length, the upper curve length $L_u(\varepsilon)$ and the lower curve length $L_d(\varepsilon)$ can be calculated using the following formula:

$$
\begin{aligned}
L_u(\varepsilon) &= s_\varepsilon^u - s_{\varepsilon-1}^u \\
&= \sum_m (u_\varepsilon(m) - f(m)) - \sum_m (u_{\varepsilon-1}(m) - f(m)) \\
&= \sum_m (u_\varepsilon(m) - u_{\varepsilon-1}(m))
\end{aligned}
\tag{3}
$$

$$
\begin{aligned}
L_d(\varepsilon) &= s_\varepsilon^d - s_{\varepsilon-1}^d \\
&= \sum_m (f(m) - d_\varepsilon(m)) - \sum_m (f(m) - d_{\varepsilon-1}(m)) \\
&= \sum_m (d_{\varepsilon-1}(m) - d_\varepsilon(m))
\end{aligned}
\tag{4}
$$

where s_ε^u is the area of the upper curve enclosed by the proposed curve at measurement scale ε, and $s_{\varepsilon-1}^u$ is the area of the upper curve enclosed by the proposed curve at measurement scale $\varepsilon - 1$. Similarly, s_ε^d is the area of the lower curve enclosed by the proposed curve at measurement scale ε, and $s_{\varepsilon-1}^d$ is the area of the lower curve enclosed by the proposed curve at measurement scale ε.

According to Equations (3) and (4), the measurement scale ε ($\varepsilon = 2, 3, 4, \ldots, n$) and the left and right neighbors are taken from the upper and lower curves, respectively, and three points $(\log(\varepsilon - 1), \log(L(\varepsilon - 1)))$, $(\log(\varepsilon), \log(L(\varepsilon - 1)))$, $(\log(\varepsilon + 1), \log(L(\varepsilon + 1)))$ are obtained. The slope of the line $S(\varepsilon)$ is the fractal signal value of the current scale ε. Finally, the upper and lower fractal signal values for each image are calculated to obtain the upper and lower fractal images. The fractal signal of each pixel in the remote sensing image is calculated using different measurement scales.

2.3.2. Feature Scale Selection Method

The fractal signal images and fractal signal variation curves were obtained at different measurement scales. The signal value of each pixel in the fractal signal image reflects the complexity of the variation of the time series curve comprising 21 bands of three seasons for a certain measurement scale. The time series curve for ground objects with more complex variation has a much higher fractal signal value.

The fractal signal images and the fractal signal variation curves were combined for a comprehensive evaluation to select feature scales of different targets. The scale with a high signal value of the land-cover type and a significant difference from other land-cover types is the feature scale of this land-cover type.

2.3.3. Information Extraction Method

An appropriate threshold range for information extraction determines the accuracy of the final extraction results. The feature-scale images selected can adequately distinguish the target land-cover type from other land-cover types, so the steps of information extraction based on the feature-scale images were carried out in this study. First, the rough distribution interval of the fractal signal value of the target land-cover type was determined based on the fractal signal curve of the sampling statistics and the feature-scale images. Second, all pixels of the feature-scale images were counted to obtain a statistical curve, and the suitable threshold value of image segmentation was selected according to the change characteristics of the curve. Finally, by combining prior knowledge, information extraction was carried out according to the determined threshold.

2.3.4. Accuracy Evaluation Metrics

This study measured the extraction accuracy using a spatial analysis algorithm. Firstly, a spatial location analysis was carried out using the overlay analysis of the data to be evaluated and the reference data. Secondly, the area's similarity was compared, and the two indicators, including the point matching degree and the area matching degree, were combined for a comprehensive evaluation.

The point matching degree refers to the degree to which the extracted results match the reference data space. The extraction results and the reference vector data of cropland were matched at the spatial boundary, and their intersection was obtained. The spatial position attributes were counted and compared to the reference vector. The point matching degree reflects the spatial relationship between the extracted results and the reference vector. The higher the point matching degree, the higher the coincidence degree of the two kinds of data.

The area matching degree refers to the similarity between the extracted results and the reference data. The calculation method used was the ratio of the extracted area of cropland to the cropland area in the reference data, in which the cropland area was calculated using vector geometric statistics, as shown in Equation (5):

$$S_c = (1 - \left| \frac{S_t - S_z}{S_z} \right|) \times 100\% \tag{5}$$

where S_t is the extracted area of cropland; S_z is the cropland area of the reference data; and S_c is the area matching degree, which reflects the relationship between the extracted result and the area value of the reference data. The higher the area matching degree, the closer the area value of the two sets of data.

3. Results

3.1. Fractal Processing and Feature Analysis

According to the survey data and the land-cover classification products, six typical land-cover types, including cropland, grassland, impervious surface, forest, water, and bare land, were selected for the fractal feature analysis. Six pixels of each type were randomly selected from the upper and lower fractal signal images, respectively. Their average was considered as the fractal signal value of the variation curves in Figure 6. As shown in Figure 6, the variation curves of the upper and lower fractal signal values of different land-cover types with different scales were calculated, and the horizontal axis "scale" n denotes the nth iteration based on Equations (1) and (2).

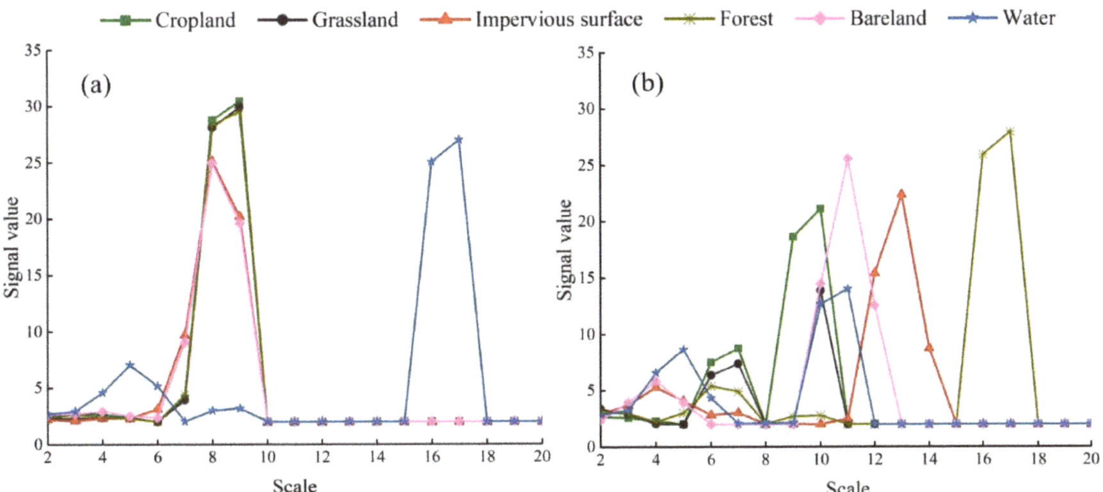

Figure 6. The variation curves of upper (**a**) and lower (**b**) fractal signals of different land-cover types in Lishu County in 2020.

The fractal features of different land-cover types were analyzed based on the variation curves of the upper and lower fractal signals and have different change features.

(1) Both the upper and lower fractal signal values of different land-cover types are different at the same scale, and the fractal signal value of the same land-cover type significantly differs at different scales.

(2) For the variation curve of the upper fractal signal, the variations are concentrated at the third to tenth and fifteenth to eighteenth scales. The variations in the lower fractal signal variation curve are mainly concentrated at the second to eighteenth scales.

(3) The fractals can selectively highlight the features of different land-cover types at specific scales. Taking cropland as an example, cropland is reflected at the eighth and ninth scales of the upper fractal signal curve, and at the ninth and tenth scales of the lower fractal signal curve. According to the method of feature scale selection and significant differences in fractal features for different land-cover types, the ninth scale of the lower fractal was selected as the fractal feature scale of cropland, as depicted in Figure 7.

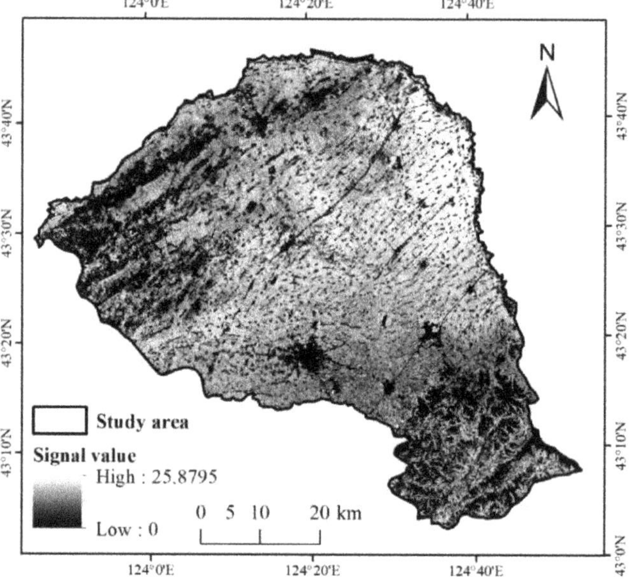

Figure 7. Lower fractal image at the ninth scale.

3.2. Cropland Information Extraction and Accuracy Evaluation

According to Figure 6b and the feature-scale image in Figure 7, the fractal signal values of cropland and other land-cover types differ and have an obvious separation. Figure 6b shows that the signal value of cropland is concentrated around 20, while those of other land-cover types are concentrated around 2. However, the result of Figure 6b was calculated based on the sampling sites and only represents the approximate range of signal values of each land-cover type. Therefore, we plotted a statistical curve of the fractal signal values of all pixels at the feature scale, which reflects the relationship between the fractal signal value and the number of pixels, as shown in Figure 8. As the signal value increases, the number of pixels shows the characteristics of sharp increase, sharp decrease, slow increase, and slight decrease, and finally tends to be smooth. Specifically, the number of pixels reaches the highest value at a signal value of 2 and decreases sharply to a trough at a signal value of 7.30 (blue point in Figure 8). Then, the number of pixels starts to increase slowly with an increase in the signal value and reaches a peak at a signal value of 17.68, which is generally consistent with the result obtained for cropland, as shown in Figure 6b. Finally, the number of pixels begins to decrease slowly. After the signal value of 21.58 (green point in Figure 8), the number of pixels begins to smooth again. Combined with the actual spatial distribution of cropland in the remote sensing image, we determined that the signal segmentation threshold of cropland is from 7.30 to 21.58 in the feature-scale image, and we extracted the spatial distribution of cropland, as shown in Figure 9.

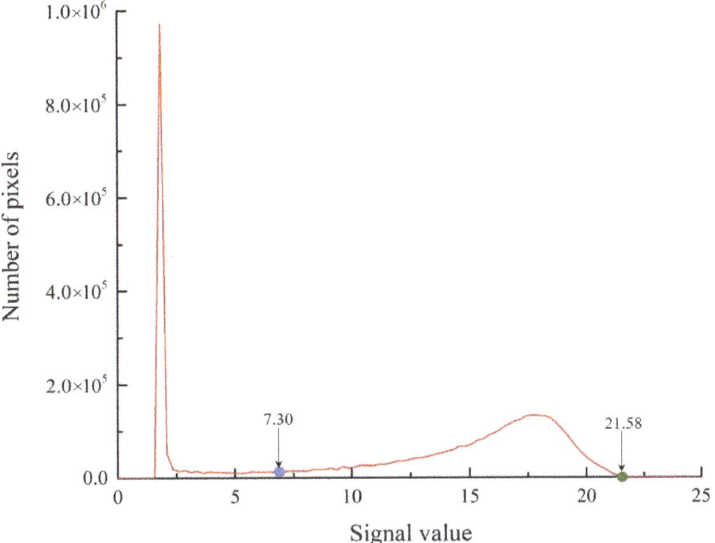

Figure 8. Statistical curve of the feature-scale image.

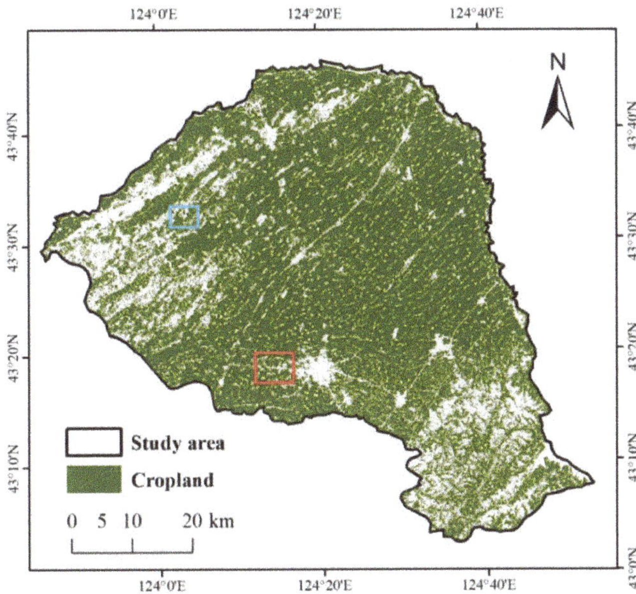

Figure 9. Spatial distribution of cropland based on fractal extraction. The subsets (blue and red boxes) are used for detailed exhibition in Figures 10 and 11.

Figure 10. Comparison of the fractal extraction results (**a**), vector data (**b**), ESA WorldCover data (**c**), Google Earth image (**d**), Esri land cover dataset (**e**), GlobeLand30 dataset (**f**), and CNLUCC (**g**) located in the blue box in Figure 9.

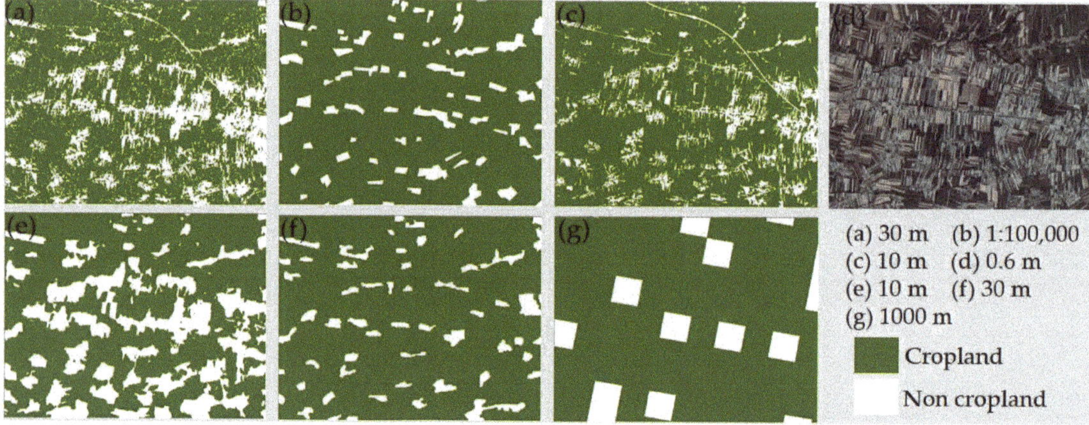

Figure 11. Comparison of the fractal extraction results (**a**), vector data (**b**), ESA WorldCover data (**c**), Google Earth image (**d**), Esri land cover dataset (**e**), GlobeLand30 dataset (**f**), and CNLUCC (**g**) located in the red box in Figure 9.

An accuracy assessment of the results was conducted using the vector data (Figure 3) and the ESA WorldCover data (Figure 4) of cropland in Lishu County in 2020. The extracted area of cropland based on the fractal method is 2759.86 km^2, and the total areas of cropland in the vector data and the ESA WorldCover data are 2659.10 km^2 and 3304.54 km^2, respectively. Compared to the vector data and the ESA WorldCover data, according to Equation (5) and the matching methods developed in this study, the calculated area matching degree of cropland extraction is 96.21% and 83.52%, respectively, and the point matching degree is 90.66% and 95.33%, respectively. The extracted results show that cropland located in the central, eastern, and northern plain areas has a high extraction accuracy, while cropland located in the southeastern mountainous and hilly areas and northwestern plain areas has a low extraction accuracy.

3.3. Comparative Analysis of Fractal Extracted Results with Existing Products

3.3.1. Comparative Analysis for the Extracted Area of Cropland

A comparative analysis of this study and existing products was performed. The existing products were selected, including the Esri land cover dataset, the GlobeLand30 dataset, and the CNLUCC, and the statistical yearbook, vector data, and ESA WorldCover data of Lishu County were employed for evaluating the accuracies. The area matching degree and point matching degree were used to evaluate the accuracy of the comparative analysis, and the comparison results are shown in Table 3.

Table 3. Comparative analysis results.

Reference Data	Data Set	Area/km^2	Area Matching Degree/%	Point Matching Degree/%
Statistical yearbook	Extracted data	2759.86	94.88	/
	Esri land cover dataset	3074.49	82.89	/
	GlobeLand30 dataset	3151.74	79.95	/
	CNLUCC	3021.39	84.91	/
Vector data	Extracted data	2759.86	96.21	90.66
	Esri land cover dataset	3074.49	84.38	98.74
	GlobeLand30 dataset	3151.74	81.45	97.17
	CNLUCC	3021.39	86.38	95.86
ESA WorldCover data	Extracted data	2759.86	83.52	95.33
	Esri land cover dataset	3074.49	93.04	96.49
	GlobeLand30 dataset	3151.74	95.38	94.02
	CNLUCC	3021.39	91.43	89.55

For the statistical yearbook data, the cropland area is 2625.33 km^2, and the cropland areas of the extracted data and the other three products are 2759.86 km^2, 3074.49 km^2, 3151.74 km^2, and 3021.39 km^2, respectively. Compared to the cropland area of the statistical yearbook, the area matching degree of the extracted data is 94.88%, which is much bigger than other values, as shown in Table 3. For the vector data of Lishu County, the area matching degrees of three products are lower than that of the extracted data (96.21%), ranging from 81.45% to 86.38%. The area matching degree of the extracted data increases by 9.83–14.76%. However, the point matching degrees of the three products are higher than that of the extracted data (90.66%), ranging from 95.86% to 98.74%, because the existing three products have excessive extraction results of the cropland, as shown in Table 3. For the ESA WorldCover data, the area matching degrees of three products are higher than that of the extracted data (83.52%), ranging from 91.43% to 95.38%. However, the point matching degree of the extracted data has the second highest accuracy (95.33%) out of the four datasets. Therefore, given both the area matching degree and point matching degree, these three comparative results suggest that the extraction accuracy of cropland in this study is better than that of existing products at different scales because of the contribution of multi-seasonal fractal features.

3.3.2. Comparative Analysis for the Spatial Distribution of Cropland

The area matching degrees of the extracted results in this study are consistent with the results of the statistical yearbook and the vector data. Still, the point matching degree is slightly lower than that of existing products of different scales. The main feature of fractal geometry can describe irregular or fragmented natural features [37]. Because of multi-seasonal fractal features, many field ridges and roads were finely divided into other land-cover types, leading to fragmentation of the spatial distribution of cropland in this study. However, field ridges and roads are all classed as croplands in the vector data, the ESA WorldCover data, and the existing three products, including the Esri land cover dataset, the GlobeLand30 dataset, and the CNLUCC, as depicted in Figures 10 and 11.

Two typical subsets were selected and analyzed. The first subregion in the blue box in Figure 9 is comprised of cropland, forest, and impervious surface in Lishu County, as depicted in Figure 10. According to Figure 10, the fractal algorithm clearly distinguishes cropland, forest, and impervious surface, and field roads are also accurately identified. In the CNLUCC, the delineations of cropland and other land-cover types need to be more accurate due to their low resolution. The spatial distributions of cropland in the ESA WorldCover data, the Esri land cover dataset, and the GlobeLand30 dataset, are relatively consistent, but part of the forest land is misclassified as cropland. The limited accuracy for these land-cover types might be attributed to the need for training samples in this area and seasonal variations of vegetation used for remote sensing classification.

The second subregion, located in the red box in Figure 9, is mainly dominated by greenhouses in Lishu County, as depicted in Figure 11. According to Figure 11, for the spatial distribution of cropland, the fractal extracted results are much better than the results of the CNLUCC, worse than the results of the vector data, the GlobeLand30 dataset, and the Esri land cover dataset, and are relatively consistent with the ESA WorldCover data. However, compared to the other products, the fractal method developed in this study could extract cropland outside the greenhouses with high accuracy and clearly identify the outline of the greenhouses according to the Google Earth image, as depicted in Figure 11a,d.

Therefore, the comparison experiments demonstrated that the fractal extraction method based on multi-seasonal remote sensing data could better distinguish cropland and other land-cover types.

4. Discussion

4.1. Theoretical Assumptions of the Fractal Method Proposed in this Study

The cropland information extracted by the fractal method based on multi-seasonal remote sensing data developed in this study was effective. Firstly, fractal analysis methods are sensitive to regional variations in land-cover types [38]. In fractal calculation, land cover with more complex variation is easier to distinguish. The impact of a curve's complexity is embodied in the fractal calculation process. According to Equations (1) and (2), the upper fractal curve tends toward a gradually narrowing trough and smoothing peak, while the inverse is true for the lower fractal curve [39]. After fractal processing, different land-cover types have different feature scales in the upper fractal or lower fractal. For example, water is reflected at the seventeenth scale of the upper fractal signal curve, whilst cropland is reflected at the ninth scale of the lower fractal signal curve in this study, which has been determined by the change characteristic curves of different ground objects [40].

Secondly, Lishu County is mainly dominated by plains with flat terrain, which are suitable for implementing various planting measures. Under the guidance of the "Lishu mode", the quality of black soil has been improved, which is reflected in the well growth status of crops. The time divisions of sowing, heading, and maturity are obvious, and the spectral curve and texture characteristics differ from those of natural vegetation. Compared to the other land-cover types, cropland has more complex feature curve changes at different time stages, and the corresponding texture information is richer. In the fractal calculation, the boundary of cropland converges faster than the other land-cover types, and the fractal features are much easier to distinguish [26].

Finally, by combining multi-seasonal remote sensing images with a fractal analysis algorithm, the method proposed in this study can accurately obtain the spatial distribution of cropland and reduce the time required to select samples. Moreover, the Chinese governments have attached great importance to the conservation of black soil in Northeast China and clearly stated that effective measures should be taken to protect this precious resource. Therefore, supported by the National Key Research and Development Program of China, this method is being applied to extract cropland information for spatial and temporal analysis to evaluate the effectiveness and sustainability of local government projects.

4.2. Uncertainty Analysis of Fractal Method

Although the proposed method achieved satisfactory results, further improvement can be conducted from three aspects. The first one is the spatial resolution of the remote sensing images used. The Landsat 8 OLI data with a resolution of 30 m might have led to additional uncertainty in this study. High-resolution data, such as GF-2 or WorldView-3, should be integrated for the information extraction of cropland. Secondly, the inconsistency of land use and land cover nomenclature for different remote sensing data products might have affected the accuracy assessment of cropland extraction. Thirdly, although the main reference data were obtained using the vector data with a scale of 1:100,000 and the ESA WorldCover data at a 10 m resolution, systematic validations at different scales and regions should be employed to enhance the suitability of the method developed in this study for future implementation. In addition, the relation of the feature scale with a number of sampling sites results in poor comparability for different remote sensing images that have originated from different phases or sensors.

5. Conclusions

This study proposed an information extraction method of cropland based on multi-seasonal fractal features, and its performance was demonstrated in a case study of Lishu County, China. The results showed that fractals could reveal clear separations of different land-cover types at different scales, and the ninth scale of the lower fractal signal was selected as the fractal feature scale for cropland. Compared to the vector data and the ESA WorldCover data, the point matching degree and the area matching degree of cropland extraction based on multi-seasonal fractal features were 90.66% and 96.21%, and 95.33% and 83.52%, respectively, which were highly consistent with the data derived from the statistical yearbook. The extracted accuracy of cropland in this study was much better than that of existing products at different scales. This method can accurately extract cropland information and provide technical support for change monitoring, conservation, and development of black soil in China.

Author Contributions: Conceptualization, S.D., P.G. and Q.W.; methodology, Q.W.; software, Q.W.; validation, Q.W.; formal analysis, P.G.; investigation, Y.L. and Y.P.; resources, C.L.; data curation, Q.W.; writing—original draft preparation, Q.W.; writing—review and editing, P.G. and S.D.; visualization, S.D.; supervision, P.G.; project administration, S.D.; funding acquisition, S.D. All authors have read and agreed to the published version of the manuscript.

Funding: This research was funded by the National Key Research and Development Program of China (Grant Number 2021YFD1500203), and the Key Laboratory of Land Satellite Remote Sensing Application, Ministry of Natural Resources of the People's Republic of China (Grant Number KLSMNR-G202219).

Institutional Review Board Statement: Not applicable.

Informed Consent Statement: Not applicable.

Data Availability Statement: Not applicable.

Conflicts of Interest: The authors declare no conflict of interest.

Appendix A

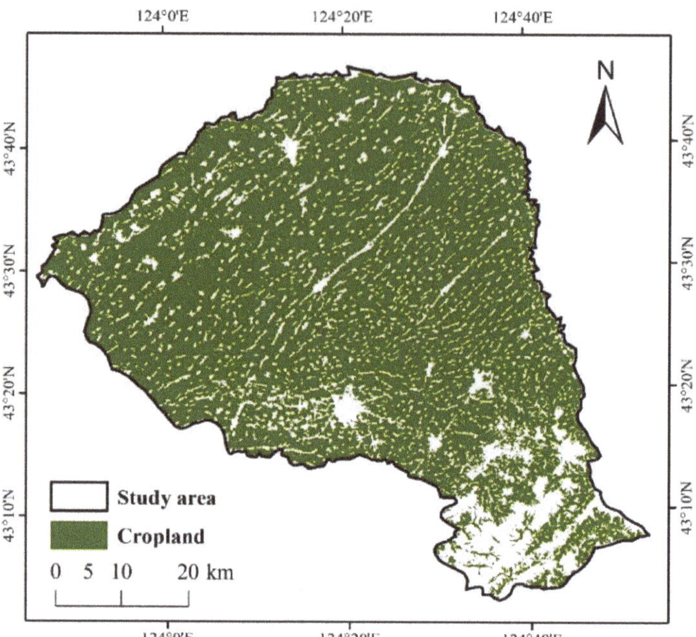

Figure A1. The distribution of cropland in the Esri land cover dataset.

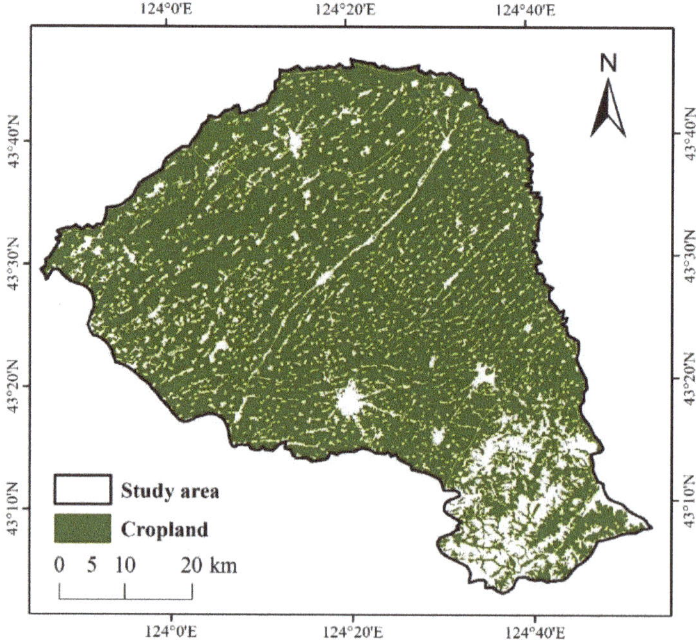

Figure A2. The distribution of cropland in the GlobeLand30 dataset.

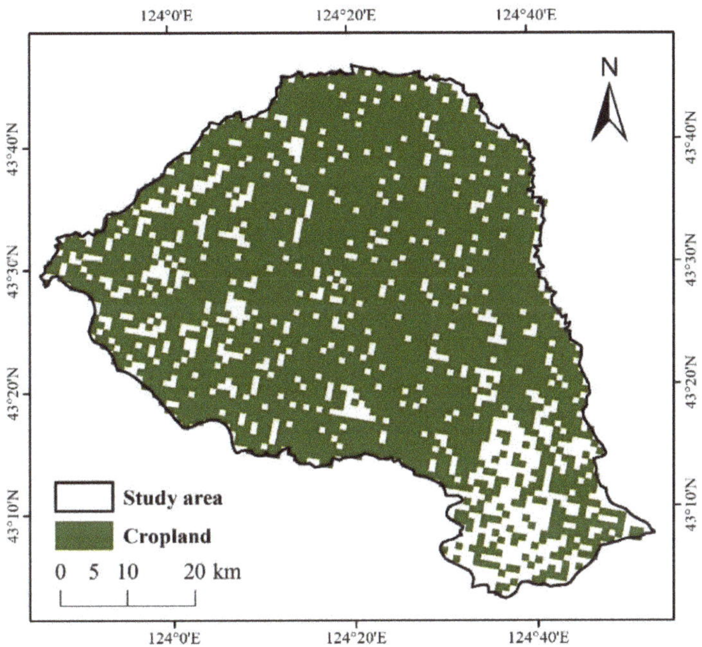

Figure A3. The distribution of cropland in the CNLUCC.

References

1. Yao, Q.; Liu, J.; Yu, Z.; Li, Y.; Jin, J.; Liu, X.; Wang, G. Three years of biochar amendment alters soil physiochemical properties and fungal community composition in a black soil of northeast China. *Soil Biol. Biochem.* **2017**, *110*, 56–67. [CrossRef]
2. Xie, Y.; Lin, H.; Ye, Y.; Ren, X. Changes in soil erosion in cropland in northeastern China over the past 300 years. *Catena* **2019**, *176*, 410–418. [CrossRef]
3. Yang, X.M.; Zhang, X.P.; Deng, W.; Fang, H.J. Black soil degradation by rainfall erosion in Jilin, China. *Land Degrad. Dev.* **2003**, *14*, 409–420. [CrossRef]
4. Khatami, R.; Mountrakis, G.; Stehman, S.V. A meta-analysis of remote sensing research on supervised pixel-based land-cover image classification processes: General guidelines for practitioners and future research. *Remote Sens. Environ.* **2016**, *177*, 89–100. [CrossRef]
5. Liu, Y.; Zhang, B.; Wang, L.; Wang, N. A self-trained semisupervised SVM approach to the remote sensing land cover classification. *Comput. Geosci.* **2013**, *59*, 98–107. [CrossRef]
6. Belgiu, M.; Drăguţ, L. Random forest in remote sensing: A review of applications and future directions. *ISPRS J. Photogramm.* **2016**, *114*, 24–31. [CrossRef]
7. Tan, Q.; Guo, B.; Hu, J.; Dong, X.; Hu, J. Object-oriented remote sensing image information extraction method based on multi-classifier combination and deep learning algorithm. *Pattern Recogn. Lett.* **2021**, *141*, 32–36. [CrossRef]
8. Martins, V.S.; Kaleita, A.L.; Gelder, B.K.; Da Silveira, H.L.; Abe, C.A. Exploring multiscale object-based convolutional neural network (multi-OCNN) for remote sensing image classification at high spatial resolution. *ISPRS J. Photogramm.* **2020**, *168*, 56–73. [CrossRef]
9. Heydari, S.S.; Mountrakis, G. Meta-analysis of deep neural networks in remote sensing: A comparative study of mono-temporal classification to support vector machines. *ISPRS J. Photogramm.* **2019**, *152*, 192–210. [CrossRef]
10. Qu, L.A.; Chen, Z.; Li, M. CART-RF Classification with Multifilter for Monitoring Land Use Changes Based on MODIS Time-Series Data: A Case Study from Jiangsu Province, China. *Sustainability* **2019**, *11*, 5657. [CrossRef]
11. Sulochana, S.; Vidhya, R. Texture based image retrieval using framelet transform-gray level co-occurrence matrix (GLCM). *Int. J. Adv. Res. Artif. Intell.* **2013**, *2*, 68–73. [CrossRef]
12. Dong, P. Fractal signatures for multiscale processing of hyperspectral image data. *Adv. Space Res.* **2008**, *41*, 1733–1743. [CrossRef]
13. Li, H.K. *Study on Remote Sensing Monitoring the Rare Earth Mining and Its Environment Impacts and Evaluation in South China*; China University of Mining and Technology (Beijing): Beijing, China, 2016. (In Chinese)

14. Xu, L.; Ming, D.P.; Zhou, W.; Bao, H.Q.; Chen, Y.Y.; Ling, X. Farmland extraction from high spatial resolution remote sensing images based on stratified scale pre-estimation. *Remote Sens.* **2019**, *11*, 108. [CrossRef]

15. Lei, T.C.; Wan, S.; Wu, Y.C.; Wang, H.P.; Hsieh, C.W. Multi-Temporal Data Fusion in MS and SAR Images Using the Dynamic Time Warping Method for Paddy Rice Classification. *Agriculture* **2022**, *12*, 77. [CrossRef]

16. Li, Z.Q.; Chen, S.B.; Meng, X.Y.; Zhu, R.F.; Lu, J.Y.; Cao, L.S.; Lu, P. Full Convolution Neural Network Combined with Contextual Feature Representation for Cropland Extraction from High-Resolution Remote Sensing Images. *Remote Sens.* **2022**, *14*, 2157. [CrossRef]

17. Peña, M.A.; Liao, R.; Brenning, A. Using spectrotemporal indices to improve the fruit-tree crop classification accuracy. *ISPRS J. Photogramm.* **2017**, *128*, 158–169. [CrossRef]

18. Zheng, B.; Myint, S.W.; Thenkabail, P.S.; Aggarwal, R.M. A support vector machine to identify irrigated crop types using time-series Landsat NDVI data. *Int. J. Appl. Earth Obs.* **2015**, *34*, 103–112. [CrossRef]

19. Duan, M.Q.; Song, X.Y.; Liu, X.W.; Cui, D.J.; Zhang, X.G. Mapping the soil types combining multi-temporal remote sensing data with texture features. *Comput. Electron. Agric.* **2022**, *200*, 107230. [CrossRef]

20. Wang, W.J.; Zhang, X.; Zhao, Y.D.; Wang, S.D. Cotton extraction method of integrated muti-features based on multitemporal Landsat 8 images. *J. Remote Sens.* **2017**, *21*, 115–124. (In Chinese)

21. Chen, T.H.K.; Prishchepov, A.V.; Fensholt, R.; Sabel, C.E. Detecting and monitoring long-term landslides in urbanized areas with nighttime light data and multi-seasonal Landsat imagery across Taiwan from 1998 to 2017. *Remote Sens. Environ.* **2019**, *225*, 317–327. [CrossRef]

22. Senf, C.; Leitão, P.J.; Pflugmacher, D.; van der Linden, S.; Hostert, P. Mapping land cover in complex Mediterranean landscapes using Landsat: Improved classification accuracies from integrating multi-seasonal and synthetic imagery. *Remote Sens. Environ.* **2015**, *156*, 527–536. [CrossRef]

23. Clark, M.L. Comparison of simulated hyperspectral HyspIRI and multispectral Landsat 8 and Sentinel-2 imagery for multi-seasonal, regional land-cover mapping. *Remote Sens. Environ.* **2017**, *200*, 311–325. [CrossRef]

24. Pant, T.; Singh, D.; Srivastava, T. Advanced fractal approach for unsupervised classification of SAR images. *Adv. Space Res.* **2010**, *45*, 1338–1349. [CrossRef]

25. Tzeng, Y.C.; Fan, K.T.; Chen, K.S. A parallel differential box-counting algorithm applied to hyperspectral image classification. *IEEE Geosci. Remote Sens. Lett.* **2012**, *9*, 272–276. [CrossRef]

26. Dong, S.W.; Li, H.; Sun, D.F. Fractal feature analysis and information extraction of woodlands based on MODIS NDVI time series. *Sustainability* **2017**, *9*, 1215. [CrossRef]

27. Dong, S.W.; Li, X.H.; Li, H.; Sun, D.F.; Zhang, W.W.; Zhou, L.D. Extraction of cultivated land using ETM+ image based on multiscale fractal signature. *Trans. CSAE* **2011**, *27*, 213–218. (In Chinese)

28. Nearing, M.A.; Xie, Y.; Liu, B.; Ye, Y. Natural and anthropogenic rates of soil erosion. *Int. Soil Water Conserv.* **2017**, *5*, 77–84. [CrossRef]

29. Liu, J.Y.; Liu, M.L.; Tian, H.Q.; Zhuang, D.F.; Zhang, Z.X.; Zhang, W.; Tang, X.M.; Deng, X.Z. Spatial and temporal patterns of China's cropland during 1990–2000: An analysis based on Landsat TM data. *Remote Sens. Environ.* **2005**, *98*, 442–456. [CrossRef]

30. Zhang, Z.X.; Wang, X.; Zhao, X.L.; Liu, B.; Yi, L.; Zuo, L.J.; Wen, Q.K.; Liu, F.; Xu, J.Y.; Hu, S.G. A 2010 update of National Land Use/Cover Database of China at 1: 100000 scale using medium spatial resolution satellite images. *Remote Sens. Environ.* **2014**, *149*, 142–154. [CrossRef]

31. Du, Y.J.; He, X.L.; Li, X.L.; Li, X.Q.; Gu, X.C.; Yang, G.; Li, W.J.; Wu, Y.G.; Qiu, J. Changes in landscape pattern and ecological service value as land use evolves in the Manas River Basin. *Open Geosci.* **2022**, *14*, 1092–1112. [CrossRef]

32. Zanaga, D.; Van De Kerchove, R.; De Keersmaecker, W.; Souverijns, N.; Brockmann, C.; Quast, R.; Wevers, J.; Grosu, A.; Paccini, A.; Vergnaud, S.; et al. ESA WorldCover 10 m 2020 v100. Zenodo. Available online: https://zenodo.org/record/5571936#.Y_WE5x9ByUk (accessed on 10 January 2023).

33. Karra, K.; Kontgis, C.; Statman-Weil, Z.; Mazzariello, J.C.; Mathis, M.; Brumby, S.P. Global land use/land cover with Sentinel 2 and deep learning. In Proceedings of the 2021 IEEE International Geoscience and Remote Sensing Symposium IGARSS, Brussels, Belgium, 11–16 July 2021; pp. 4704–4707.

34. Chen, J.; Cao, X.; Peng, S.; Ren, H. Analysis and Applications of GlobeLand30: A Review. *ISPRS Int. J. Geo-Inf.* **2017**, *6*, 230. [CrossRef]

35. Xu, X.L.; Liu, J.Y.; Zhang, S.W.; Li, R.D.; Yan, C.Z.; Wu, S.X. China Land Use/Cover Change. Chinese Academy of Sciences Resource and Environmental Science Data Center. Available online: https://www.resdc.cn/DOI/doi.aspx?DOIid=54 (accessed on 7 October 2021).

36. Peleg, S.; Naor, J.; Hartley, R.; Avnir, D. Multiple Resolution Texture Analysis and Classification. *IEEE Trans. Pattern Anal. Mach. Intell.* **1984**, *PAMI-6*, 518–523. [CrossRef] [PubMed]

37. Sun, W.; Xu, G.; Gong, P.; Liang, S. Fractal analysis of remotely sensed images: A review of methods and applications. *Int. J. Remote Sens.* **2006**, *27*, 4963–4990. [CrossRef]

38. Nayak, S.R.; Mishra, J.; Palai, G. Analysing roughness of surface through fractal dimension: A review. *Image Vis. Comput.* **2019**, *89*, 21–34. [CrossRef]

39. Zhou, Z.Y.; Bu, Q. Research on fractal signature feature of Hyperion hyperspectral image. *J. Remote Sens.* **2011**, *15*, 173–182. (In Chinese)
40. Tang, C.; Chen, J.P.; Cui, J.; Wen, B.T. Lithology feature extraction of CASI hyperspectral data based on fractal signal algorithm. *Spectrosc. Spect. Anal.* **2014**, *34*, 1388–1393.

 *agriculture*

Article

Monitoring of Wheat Fusarium Head Blight on Spectral and Textural Analysis of UAV Multispectral Imagery

Chunfeng Gao [1], Xingjie Ji [2,3], Qiang He [1], Zheng Gong [1], Heguang Sun [1], Tiantian Wen [1] and Wei Guo [1,*]

[1] College of Information and Management Science, Henan Agricultural University, Zhengzhou 450046, China
[2] Henan Key Laboratory of Agrometeorological Support and Applied Technique, Zhengzhou 450003, China
[3] Henan Institute of Meteorological Sciences, Zhengzhou 450003, China
* Correspondence: guowei@henau.edu.cn

Abstract: Crop disease identification and monitoring is an important research topic in smart agriculture. In particular, it is a prerequisite for disease detection and the mapping of infected areas. Wheat fusarium head blight (FHB) is a serious threat to the quality and yield of wheat, so the rapid monitoring of wheat FHB is important. This study proposed a method based on unmanned aerial vehicle (UAV) low-altitude remote sensing and multispectral imaging technology combined with spectral and textural analysis to monitor FHB. First, the multispectral imagery of the wheat population was collected by UAV. Second, 10 vegetation indices (VIs)were extracted from multispectral imagery. In addition, three types of textural indices (TIs), including the normalized difference texture index (NDTI), difference texture index (DTI), and ratio texture index (RTI) were extracted for subsequent analysis and modeling. Finally, VIs, TIs, and VIs and TIs integrated as the input features, combined with k-nearest neighbor (KNN), the particle swarm optimization support vector machine (PSO-SVM), and XGBoost were used to construct wheat FHB monitoring models. The results showed that the XGBoost algorithm with the fusion of VIs and TIs as the input features has the highest performance with the accuracy and F1 score of the test set being 93.63% and 92.93%, respectively. This study provides a new approach and technology for the rapid and nondestructive monitoring of wheat FHB.

Keywords: unmanned aerial vehicle; multispectral imagery; fusarium head blight; texture indices; machine learning

Citation: Gao, C.; Ji, X.; He, Q.; Gong, Z.; Sun, H.; Wen, T.; Guo, W. Monitoring of Wheat Fusarium Head Blight on Spectral and Textural Analysis of UAV Multispectral Imagery. *Agriculture* **2023**, *13*, 293. https://doi.org/10.3390/agriculture13020293

Academic Editor: Jose L. Gabriel

Received: 8 January 2023
Revised: 21 January 2023
Accepted: 23 January 2023
Published: 26 January 2023

1. Introduction

Wheat is one of the three major grain crops in the world, and it is also the second-largest grain crop in China [1]. It is also a staple food for about two-thirds of the world's population, which is of a great significance to ensure national food security [2,3]. Fusarium head blight (FHB), also known as scab, is an economically destructive wheat disease mainly caused by Fusarium graminearum, which mainly damages wheat ears [4]. The prevention and control of FHB is extraordinarily important because FHB cannot only cause a serious yield reduction but it can also lead to the deterioration of the wheat's quality [5–7]. More seriously, infected wheat will produce mycotoxins, especially deoxynivalenol (DON) and zearalenone (ZEA), which are detrimental to humans and animals and can lead to acute poisoning symptoms, the destruction of immunity, and even death [8]. Therefore, the effective monitoring of FHB in time and space is particularly important in the investigation of crop health and food security.

The traditional disease assessment and investigation is mainly based on a field visual investigation, which is not only time-consuming and laborious but also has a certain subjectivity and cannot ensure the authenticity and accuracy of the investigation data [9,10]. It is difficult to meet the current requirements for the rapid and accurate detection and real-time monitoring of crop diseases in large-scale planting areas [11]. Remote sensing technology has alleviated this problem to a certain extent, so more and more researchers are attempting to apply remote sensing technology to disease monitoring [12,13]. These studies are based on the theory that disease infection will change the transpiration rate, leaf color, chlorosis, and morphology of crops [13,14]. In particular, UAV remote sensing technology has been widely developed in the field of agricultural monitoring because of its high flexibility, low cost, fast image acquisition, and ability to carry multiple sensors [15].

Some studies have been conducted to use UAV images to retrieve the growth parameters of different crops [16–19]. In recent years, multispectral imagery provides new concepts and methods for crop disease monitoring. Compared with the traditional methods and hyperspectral imagery, multispectral imagery has the advantages of relatively rich spectral information, simple data processing, and a low computing cost in disease detection and it has a certain potential in crop disease monitoring applications. In addition, it has a red-edge (RE) band, which is located between the maximum red absorption and high reflectivity in the near infrared (NIR) region. It is an important spectral feature of vegetation, where the transformation from chlorophyll absorption to cell scattering takes place [3,20]. So far, UAV multispectral images have been used to estimate the chlorophyll content, nitrogen content, biomass, and leaf area index (LAI) [21–24]. In addition, they have been also used by some scholars to monitor the diseases of different crops. Lei et al. [25] achieved the severity monitoring of the yellow leaf disease of areca nut using VIs such as the normalized difference vegetation index (NDVI) and normalized difference red-edge index (NDRE) and using support vector machine (SVM) and decision trees algorithms. Zhao et al. [26] used VIs to monitor rice sheath blight and the results showed that using multispectral imagery was more accurate and sensitive (R^2 = 0.624, RMSE = 0.801), which was better than visible light imagery (R^2 = 0.580, RMSE = 0.847). Rodriguez et al. [27] used five machine learning algorithms, including random forest (RF) and a linear support vector classifier, to monitor potato late blight based on UAV multispectral imagery. Ye et al. [28] used artificial neural network (ANN), RF, and SVM classification algorithms to monitor banana fusarium wilt using UAV multispectral imagery. These studies fully illustrate the potential of using high-resolution UAV multispectral images in the agricultural field. Additionally, the majority of studies on the disease monitoring of crops used the spectral information of UAV images, but the inherent spatial information in the form of texture has not been fully explored. Therefore, it would be promising to take full advantage of the textural feature for the disease monitoring of crops.

Textural analysis is an image processing technique that is widely used for classification tasks [29,30]. The textural feature reflects the visual roughness of ground objects through gray spatial change and its repeatability, which can fully reflect the image characteristics. Different objects generally show different texture types, which can be used to describe and identify ground objects. The overall representation of the same category of characteristics seems similar, but the local detail is different [31]. It has a certain effect on the recognition of crop diseases and the improvement of their accuracy [32]. In recent years, textural analysis has also been used for the estimation of crop biomass and LAI [33–35]. Zheng et al. [30] compared the performance of VIs, raw textural features, the NDTI, and combinations of VIs and the NDTI for estimating the aboveground biomass of rice using UAV multispectral data and found that integrating the NDTI with VIs significantly improved the accuracy compared to using spectral information alone. Li et al. [35] combined color indices and textural features for estimating rice LAI and exhibited the best estimation accuracy when the VIs and textural features were combined as the inputs. Some scholars have also introduced primitive textural features for disease identification and monitoring [13,36]. These studies all showed the potential of combining spectral information with textural information.

However, most of the studies used only raw textural features, and the contribution of textural features did not reach satisfactory results. In addition, few research scholars have focused on the potential of TIs for disease monitoring. So, in this study, we proposed a method that integrated VIs and TIs to monitor wheat FHB. UAV multispectral imagery was used to monitor wheat FHB. The specific work of this study is as follows: (1) 10 commonly used VIs (VARI, CIgreen, CIrededge, DVI, DVIRE, EVI, NDRE, NDVI NPCI, and RVI) were extracted. In addition, three types of TIs, which are NDTI, DTI, and RTI, were constructed to make the most use of the textural information of the imagery. (2) The obtained features were screened to obtain the features sensitive to wheat FHB. (3) Nine wheat FHB monitoring models were constructed with VIs, TIs, and integrated VIs and TIs as the input features to explore the effects of different feature inputs on wheat FHB monitoring. (4) The best FHB monitoring models were applied to map the distribution of wheat diseases in the study area and evaluate the potential of using UAV multispectral imagery to monitor wheat diseases.

2. Materials and Methods

2.1. Study Area

The experiment site was conducted on May 18, 2021 at the experimental farm (34°08′23″ N, 113°47′57″ E) on the Xuchang Campus of Henan Agricultural University, Xuchang City, Henan Province. At this time, the wheat was growing in the wheat field and it was at the grain filling stage. Xuchang is located in the central part of Henan Province. It has a typical temperate and continental monsoon climate. The annual average temperature ranges from 14.3 °C to 14.6 °C and the annual average precipitation is between 671 mm and 736 mm.

Figure 1 demonstrates the study area. The terrain of the experimental farm was relatively flat and the soil belonged to loam. In the previous season, maize was the main grain crop. The study area consisted of 60 experimental plots; they were divided into 3 rows for planting and each row contained 20 experimental plots. The length of each experimental plot was about 1.5 m and the width was about 1 m. During the period from 2019 to 2020, the experimental wheat varieties were sown in autumn. The management measures, such as irrigation and fertilization, in the experimental plots were all the same. At the early stage of wheat flowering in April 2021, professionals randomly selected some wheat plants in each experimental plot and used a micropipette to inject the spore suspension made of fusarium oxysporum into the florets in the middle and upper part of the wheat ear. The inoculated wheat ears were marked with awn cutting and bagged for 1~7 days. The incidence mainly depended on an artificial drip and mutual infection.

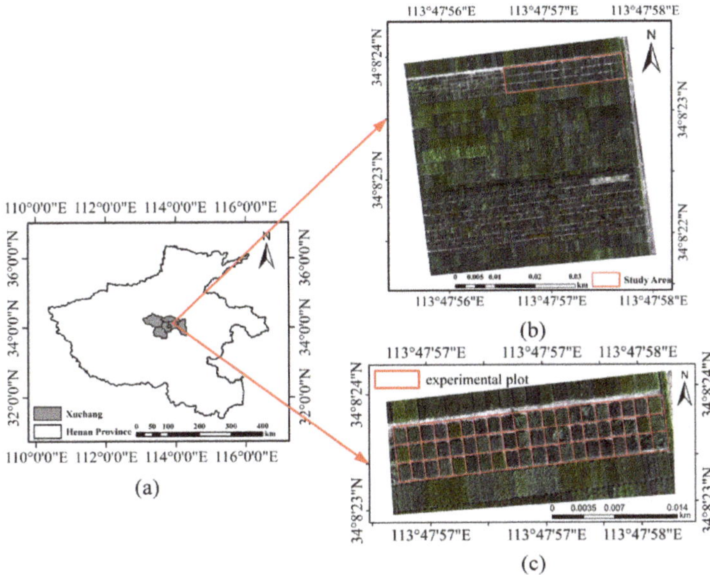

Figure 1. Overview of the study area. (**a**) Location of Xuchang City in Henan Province; (**b**) distribution and location of the study area in the experimental farm; (**c**) distribution and location of the experimental plots.

2.2. Field Data Acquisition and Preprocessing

2.2.1. Remote Sensing Image Acquisition and Preprocessing

In this study, the UAV remote sensing platform used was Phantom 4 Multispectral (P4M). There are several built-in sensors in the P4M which make it a dedicated and customized UAV for the detection and identification of plants or crops. Multispectral cameras were available in the P4M and included six CMOS sensors, one of which was used for RGB visible light imaging, and the other five monochrome sensors were used for multispectral imaging (blue (B), green (G), red (R), RE and NIR). The UAV had a takeoff weight of 1487 g, a maximum ascending speed of 6 m/s, a maximum descending speed of 3 m/s, and a flight time of approximately 27 min. To detect millisecond errors in the camera imaging time, the TimeSync time synchronization system was adopted. The remote sensing images of the study area in five bands were obtained on 18 May 2021. The UAV remote sensing operation was carried out on a sunny day with a low wind speed. The flight time was between 9:00 a.m. and 11:00 a.m., the flight altitude was 20.3 m, the heading overlap and the lateral overlap were 80%, and the ground resolution was 1 cm.

Using Pix4Dmapper, the original images captured by the UAV were spliced together. First, with the flight POS data, the same-named points were found and then the real positions and splicing parameters of the original images were calculated through a space–time measurement to establish the point cloud model. Finally, according to the calibration ground panel used before and after the flight, the pixel values were converted into the surface reflectivity of each spectrum and the imagery is automatically calibrated and generated into orthophoto imagery by optimizing the image content and using the block adjustment technology [37].

2.2.2. Selection of Survey Sampling Points

In this study, canopy images of 60 experimental plots were taken with mobile camera equipment as the auxiliary data for the selection of the sample points. At a height of about 1.2 m vertically above the canopy, images were acquired in bright weather, and each image was taken on a vivo iQOO Neo3 mobile phone, which has 48 million pixels in the rear camera. The images were taken with a fixed shooting direction to ensure that the canopy images of each plot corresponded to the corresponding plot of the multispectral imagery. Some of the typical experimental plots with the corresponding plots of the UAV multispectral imagery are shown in Figure 2. Three categories of sample points were selected: healthy, diseased, and background. In the diseased plots, 470 FHB-infected sampling points were selected, and in the healthy plots, 450 healthy sampling points were selected. In addition, 415 background sampling points were selected. These three types of sample points were used for subsequent model training and verification.

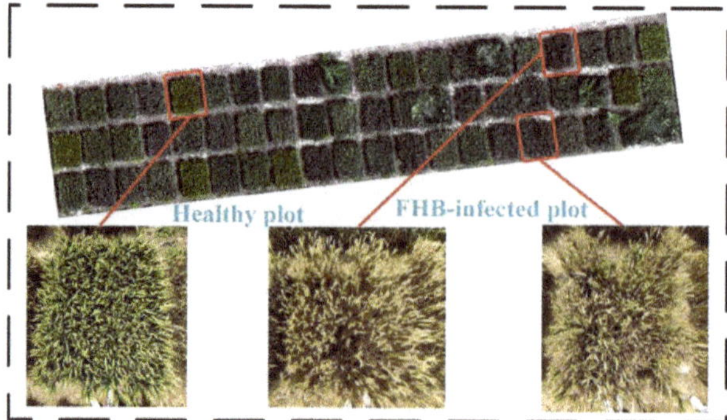

Figure 2. Distribution of images of the canopy experimental plots with the plots corresponding to the UAV multispectral imagery.

2.3. Methods

Our research process was conducted in two sections (Figure 3). The first section was a feature extraction to prepare the input features for the wheat FHB monitoring models and the second section was the construction and validation of the wheat FHB monitoring models; and the best feature combination and rapid wheat FHB monitoring method can be found through the study of these two sections. The two sections are described below.

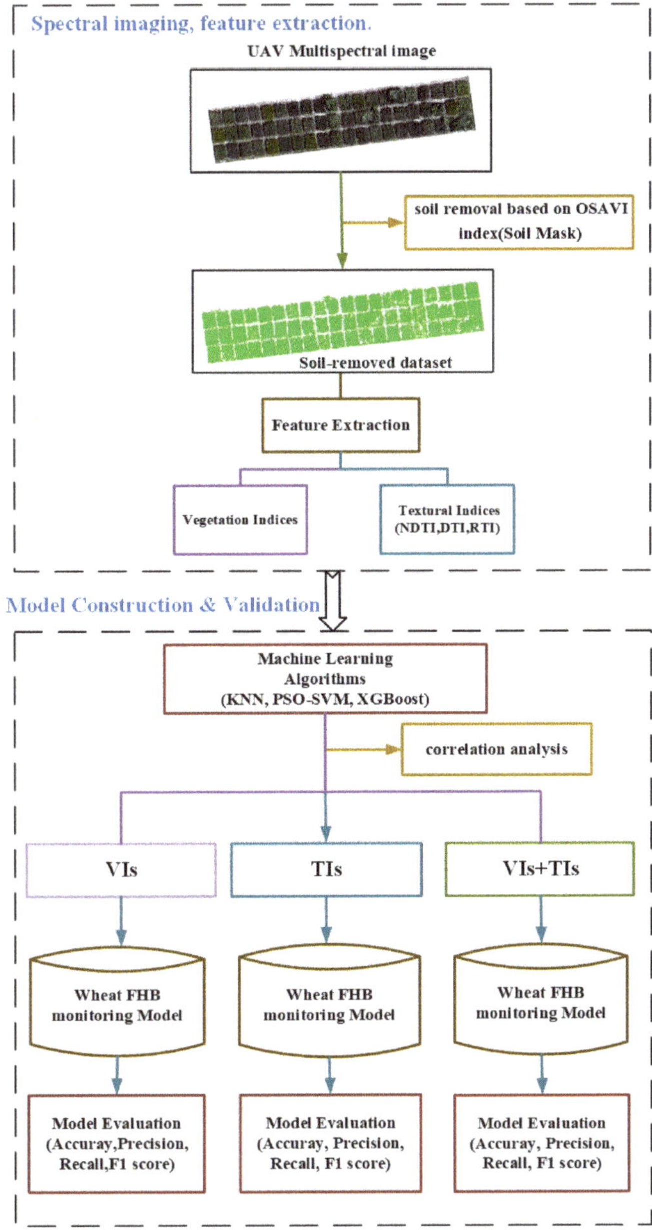

Figure 3. Workflow diagram for feature extraction and model construction and validation.

Considering that the soil may affect the performance of the models, this study conducted a series of studies based on multispectral imagery after removing the soil area. First, the process of removing the soil area from the study area was as follows: the optimized soil-adjusted vegetation index (OSAVI) was used to segment the soil area and wheat area in the multispectral imagery by setting the suitable threshold [38], and the final threshold range was determined through multiple adjustments to construct a binary mask image. This mask was used to remove the soil region. The OSAVI was calculated as follows:

$$(\text{NIR} - \text{R})/(\text{NIR} + \text{R} + 0.16) \tag{1}$$

Subsequently, we calculated 10 commonly used VIs and extracted three TIs (NDTI, DTI, and RTI). The correlation coefficient analysis was used to screen the sensitive classification features to explore the impact of VIs, TIs, and integrated VIs and TIs on the model's accuracy. Then, three machine learning algorithms (KNN, PSO-SVM, and XGBoost) were used for training and classification. The overall recognition effect of each classification algorithm was analyzed and evaluated through the accuracy, precision, recall, and F1 score. Finally, based on the UAV multispectral imagery, the optimal feature combination and classification algorithm for the recognition of FHB in the farmland were obtained. The pixel-level region recognition of FHB based on the best wheat FHB monitoring model was realized. The overall research scheme is shown in Figure 3.

2.4. Feature Extraction

2.4.1. Extraction of VIs

The spectral information from UAVs is mainly used in the form of VIs [39]. VIs represent the mathematical transformation of reflectance of two or more bands to characterize the canopy spectral characteristics of crops [39,40]. To obtain the desired classification accuracy, a group of 10 VIs were calculated based on five spectral bands of UAV imagery (Table 1). These varieties were selected because they may help to distinguish between symptomatic and asymptomatic wheat. The formula and corresponding reference of the selected VIs are given in Table 1. These VIs include the traditional VIs and the red-edge VIs. The traditional VIs (NDVI, RVI, and DVI) are often used to monitor the growth status of crops [41,42]. CIgreen, CIrededge, and NPCI are often used to estimate the chlorophyll content of crops. The red-edge VIs include DVIRE and NDRE, which are similar to DVI and NDVI, but the red band is replaced by the red-edge band. According to the literature review, these VIs have been used to identify crop diseases [41]. In addition, VIs are simple to calculate and their potential for disease monitoring has been discussed by many scholars.

Table 1. Formulas and sources of spectral VIs for monitoring wheat FHB.

VIs Name	Calculation Formula	Reference
Visible atmospherically resistant index (VARI)	$(\text{G} - \text{R})/(\text{G} + \text{R} - \text{B})$	[43]
Green chlorophyll index (CIgreen)	$\text{NIR}/\text{G} - 1$	[44]
Red-edge chlorophyll index (CIrededge)	$\text{NIR}/\text{RE} - 1$	[44]
Difference vegetation index (DVI)	$\text{NIR} - \text{R}$	[45]
Red-edge difference vegetation index (DVIRE)	$\text{NIR} - \text{RE}$	[46]
Enhanced vegetation index (EVI)	$2.5(\text{NIR} - \text{R})/(\text{NIR} + 6\text{R} - 7.5\text{B} + 1)$	[47]
Normalized difference red-edge index (NDRE)	$(\text{NIR} - \text{RE})/(\text{NIR} + \text{RE})$	[48]
Normalized difference vegetation index (NDVI)	$(\text{NIR} - \text{R})/(\text{NIR} + \text{R})$	[49]
Normalized pigment chlorophyll index (NPCI)	$(\text{RE} - \text{B})/(\text{RE} + \text{B})$	[46]
Ratio vegetation index (RVI)	NIR/R	[46]

2.4.2. Extraction of TIs

When wheat is infected with FHB, the ear of the wheat will turn yellow and dry and certain brown spots will appear. With time, the brown spots will gradually expand and eventually spread to the whole ear [3,13]. Wheat canopy infected by FHB and wheat canopy not infected by FHB have different textural characteristics. Therefore, using the textural information reflected by the textural characteristics can effectively solve the problem that characteristics are difficult to distinguish from spectral features and it can also can effectively improve the classification accuracy.

Among several texture algorithms, the commonly used GLCM [50,51] was selected to explore the potential of textural information for wheat FHB monitoring. In this study, 40 textural features of 5 spectral bands were extracted from UAV multispectral imagery. Based on the GLCM, eight textural features of each band, including the mean, variance, homogeneity, contrast, dissimilarity, entropy, second moment, and correlation, were obtained. Since wheat is a row planting crop, usually, the row spacing of wheat planting is about 0.2–0.3 m. Considering the spatial resolution of UAV multispectral imagery was 0.01 m, this study used a 3×3 window size for the extraction of the textural features. The details of the textural features are shown in Table 2.

Table 2. Calculation formulas of textural features.

Textural Features	Calculation Formula
Mean	$\sum_i \sum_j P(i,j)i$
Variance	$\sum_i \sum_j (i - \text{mean})^2 P(i,j)$
Homogeneity	$\sum_i \sum_j P(i,j) \frac{1}{1+(i-j)^2}$
Contrast	$\sum_i \sum_j P(i,j)(i-j)^2$
Dissimilarity	$\sum_i \sum_j P(i,j)|i-j|$
Entropy	$-\sum_i \sum_j P(i,j)\log(P(i,j))$
Second moment	$\sum_i \sum_j P(i,j)^2$
Correlation	$\sum_i \sum_j \frac{(i-\text{mean})(j-\text{mean}) \times P(i,j)^2}{\text{variance}}$

Where $P(i,j)$ represents the image element value of the image at the point (i,j).

To improve the correlation between the textural features and wheat FHB, three TIs (NDTI, DTI, and RTI) were constructed following the thought of NDVI, DVI, and RVI. Combining eight textural features from five spectral bands (40 features in total), all possible combinations of the two textural features were constructed to explore their ability to identify wheat FHB. Finally, 1560 combinations were obtained for each TI and the best combination form was selected to constitute that TI. The three TIs were defined as follows.

$$\text{NDTI} = (T_1 - T_2)/(T_1 + T_2) \tag{2}$$

$$\text{DTI} = T_1 - T_2 \tag{3}$$

$$\text{RTI} = T_1/T_2 \tag{4}$$

where T_1 and T_2 represent the textural feature values in five random bands.

2.5. Training and Evaluation of Machine Learning Models

Based on the three inputs of VIs, TIs, and VIs and TIs integrated, a total of 1335 sampling points were selected, including 450 sampling points in the healthy area, 470 sampling points in the FHB-infected area, and 415 sampling points in the background area (considering the soil removal, the image was still disturbed by other external objects as well as shadows, so the background sampling points were retained). The training and test set were randomly divided according to the ratio of 8:2, and the KNN, PSO-SVM, and XGBoost were used to identify the infected FHB area.

2.5.1. KNN Model

The KNN is a typical supervised learning method and is widely used in classification tasks [52]. The basic principle is to calculate the distance between the sample to be classified as x and all the samples in the training set based on the distance metric, and the k samples with the smallest distance from the sample to be classified are taken as the k nearest neighbor samples of x. Finally, the classification category of x is determined based on the vote. The selection of the k value has a significant impact on the classification result of the KNN algorithm. If the value of k is too small, the phenomenon of overfitting will easily occur and the prediction error will be large, leading to a wrong prediction; if the value of k is too large, the phenomenon of underfitting will occur. So, this study used five-fold cross-validation to select the k value to ensure that a more appropriate k value was chosen.

2.5.2. PSO-SVM Model

Particle swarm optimization (PSO) was first proposed by Eberhart and Kennedy in 1995 [53], which simulated the clustering behavior of insects, birds, and fish for global optimization. SVM is a machine learning algorithm for supervised classification, which has certain advantages in solving small samples, and nonlinear and high-dimensional pattern recognition [54,55]. It first searches for a maximum marginal hyperplane and maps the low-dimensional data to the high-dimensional space through the kernel function [56], so as to turn the linearly inseparable samples into linearly separable samples, and introduces the model penalty factor to improve the generalization of the classification model. However, this method has a large workload and a low efficiency [57,58]. In addition, radial basis function (RBF) was used in this study, in which the kernel function parameter gamma and penalty factor c have a great impact on the accuracy of the model [58]. Therefore, PSO was used to find the appropriate gamma and c to reduce the model's complexity and accelerate the model's convergence.

2.5.3. XGBoost Model

XGBoost [59] is a novel gradient tree boosting method introduced by Chen and Guestrin in 2016. It is an improvement of the gradient boosting algorithm for enhancing the speed and performance of decision trees using gradients [60]. The thought of XGBoost is to adopt a group of classification and regression trees as weak learners and subsequently improve the performance of the trees by creating a cluster of trees that minimizes the regular objective function.

The objective function consists of two parts: training loss and regularization. The representation of the objective function is shown in the following equation.

$$\mathrm{obj}(\theta) = \mathrm{TL}(\theta) + \mathrm{R}(\theta) \tag{5}$$

TL represents the training loss and R represents the regularization term. TL is used to measure the predictive power of the model. Regularization has the advantage of retaining the complexity of the model within the desired range, eliminating problems such as over-stacking or over-fitting of data, and XGBoost can optimize the results by simply adding the predictions from all trees formed from the dataset.

2.5.4. Model Performance Evaluation Metrics

In this study, the accuracy, precision, recall, and F1 score will be used to evaluate the performance of the model. The calculation formulas are as follows.

$$\mathrm{Accuracy} = \frac{\mathrm{TP} + \mathrm{TN}}{\mathrm{TP} + \mathrm{TN} + \mathrm{FP} + \mathrm{FN}} \tag{6}$$

$$Precision = \frac{TP}{TP + FP} \tag{7}$$

$$Recall = \frac{TP}{TP + FN} \tag{8}$$

$$F1 = \frac{2 * Precision * Recall}{Precision + Recall} \tag{9}$$

where true positive (TP) and true negative (TN) represent the number of correctly classified positive samples and the number of correctly classified negative samples; false positive (FP) and false negative (FN) represent the number of misclassified positive samples and the number of misclassified negative samples.

3. Results

3.1. Correlation between Different Modeling Features and Wheat FHB

Correlation analysis is also widely used in studies of pest and disease monitoring [42]. So, in the study, the correlation between the different modeling features and wheat FHB was analyzed. In this study, Spearman correlation was adopted to measure the ability of VIs and TIs to identify wheat FHB. Spearman correlation differs from Pearson correlation in that it allows the variables to be categories and has a stronger robustness [61,62]. From Table 3, it can be seen that the correlation coefficient R between VIs and wheat FHB was between -0.580 and -0.882 and the vegetation index with the highest correlation coefficient was EVI; the correlation coefficient R between TIs and wheat FHB was between -0.866 and -0.893 and the textural index with the highest correlation coefficient was DTI. Compared with VIs, only DTI was higher than EVI, which had the highest correlation coefficient. The P value between the different features and wheat FHB was less than 0.01, indicating that the extraction of VIs and TIs based on UAV multispectral imagery were significantly different from wheat FHB. VIs and TIs can be used as input features for constructing wheat FHB monitoring models.

Table 3. Correlation analysis result between different modeling feature and wheat FHB.

Feature	R	*P* Value
VARI	-0.580	**
CIgreen	-0.757	**
CIrededge	-0.747	**
DVI	-0.879	**
DVIRE	-0.872	**
EVI	-0.882	**
NDRE	-0.757	**
NDVI	-0.861	**
NPCI	-0.805	**
RVI	-0.807	**
NDTI	-0.866	**
DTI	-0.893	**
RTI	-0.869	**

** indicates that the correlation is highly significant at the 0.01 levels.

3.2. Model Analysis and Evaluation

In this study, KNN, PSO-SVM, and XGBoost were used for the modeling. We selected the three VIs (EVI, DVI, and DVIRE) with the highest correlation coefficients as the input features for VIs, combined NDTI, DTI, and RTI as the input features for TIs (these three TIs are made up of their respective best combinations), and integrated these VIs and TIs as the input features to construct wheat FHB monitoring models, respectively. A total of 267 sampling points were used for the test set, including 106 healthy sampling points, 82 sampling points infected with FHB, and 79 background sampling points. In KKN, a five-fold cross-validation was adopted to find the appropriate K value, and in PSO-SVM, the PSO algorithm was utilized to optimize the parameters gamma and c of the model, finding the best gamma and c in each different combination of features. In XGBoost, the parameters of the model were determined through several tuning attempts.

The accuracy, precision, recall, and F1 score were used to evaluate the effect of the monitoring results of the three models and the final parameter setting of the monitoring models are shown in Table 4. From Table 4, the accuracy of the training set and the test set showed that there was no overfitting or underfitting of the models. It can be seen that when VIs were used as the input, the accuracy of the models reached 84.64%–85.02% and the F1 score reached 82.75%–83.09%. When TIs were used as the input, the accuracy of the models reached 91.76%–92.51% and the F1 score reached 90.84%–91.68%. When VIs and TIs were used as the inputs, the accuracy of the models reached 92.13%–93.63% and the F1 score reached 91.29%–92.93%. It can be seen that the models using only VIs as the input performed the worst, lower than the other two forms of feature combinations. This result indicated that TIs outperformed VIs under a single type of feature input, probably because TIs were richer in showing the textural information of FHB-infected wheat, which was different from the healthy wheat canopy. Under both types of feature inputs, the combined use of the spectral and textural information of the imagery enhanced the performance of the models compared with using only VIs or TIs as the inputs, with XGBoost showing the highest performance and outperforming the other two models with an accuracy of 93.63%. It was shown that the performance of wheat FHB monitoring could be improved by taking full advantage of different features and suitable model.

Table 4. Evaluation metrics of wheat FHB monitoring models.

| Features | Models | Training Set | | Test Set | | | |
		Parameters	Accuracy/%	Accuracy/%	Precision/%	Recall/%	F1 Score/%
VIs	KNN	K = 5	81.93	84.64	84.46	83.19	82.75
	PSO-SVM	Gamma = 0.14, c = 9.31	82.11	84.64	84.63	83.20	82.77
	XGBoost	Estimators = 10, max depth = 3	83.05	85.02	85.36	83.63	83.09
TIs	KNN	K = 9	89.79	91.76	91.3	90.81	90.84
	PSO-SVM	Gamma = 0.15, c = 3.70	90.63	92.13	91.80	91.22	91.25
	XGBoost	Estimators = 10, max depth = 3	91.10	92.51	92.00	91.65	91.68
VIs+TIs	KNN	K = 7	90.07	92.51	92.14	91.64	91.68
	PSO-SVM	Gamma = 1.64, c = 7.53	91.85	92.13	91.52	91.25	91.29
	XGBoost	Estimators = 10, max depth = 3	93.16	93.63	93.19	92.90	92.93

3.3. Analysis of Monitoring Effect

Figure 4 shows the confusion matrix of the three models with different inputs. From the confusion matrix, it could be seen that the misclassified sampling points of the models were basically concentrated between the sampling points infected with FHB and the background, and the healthy sampling points were better classified, probably because the healthy sampling points are more different from the sampling points infected with FHB and the background sampling points, while the sampling points of FHB will gradually show the symptoms of whitening and drying on the wheat canopy due to the infection by FHB, thus causing a loss of pigment and being easily confused with the background area.

When only VIs were used as the input, the misclassification between FHB-infected sampling points and the background sampling points was more serious, indicating that the spectral information of the images alone could not monitor wheat FHB well. The misclassification was improved to some extent when only TIs were adopted as the input, probably because the textural information of the canopy of wheat infected with FHB was different from that of the background sample points, and the TIs improved the phenomenon that the spectral features were difficult to distinguish detailed information. The integration of VIs and TIs as the input further improved the misclassification of the samples and enhanced the performance of the models, among which the XGBoost achieved satisfactory results with only 17 misclassified samples, the least misclassified samples, and the model also has the advantage of being fast, so it is well suited for the monitoring of wheat FHB.

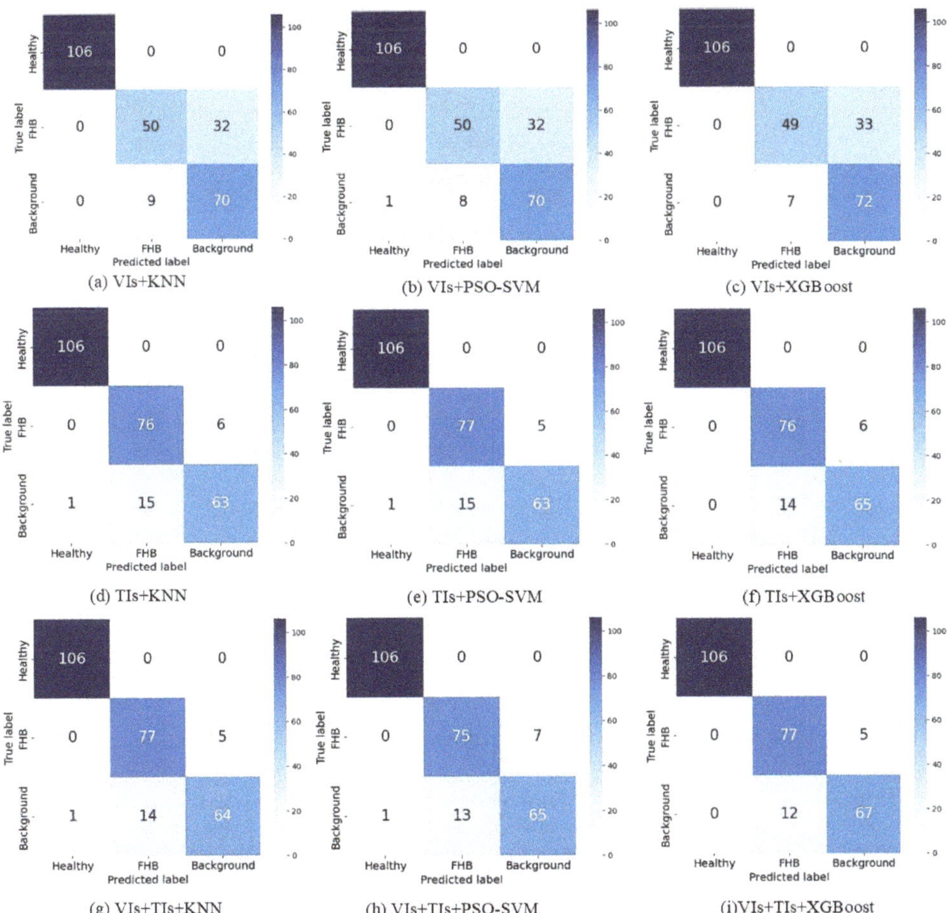

Figure 4. Confusion matrix for the three models with different inputs.

In this study, different feature inputs as well as KNN, PSO-SVM, and XGBoost were used for the monitoring of wheat FHB, and it was clear from the analysis that XGBoost with VIs and TIs as the inputs achieved the best performance, so this model was used for spatial distribution mapping of wheat FHB (Figure 5). The trained XGBoost was used to perform a pixel-level classification of the UAV multispectral imagery. From Figure 5, we can see that the overall FHB incidence in the study area was heavy, probably because wheat FHB is a climatic disease, mainly affected by temperature and humidity, and the images of the study area were acquired during the wheat grain filling stage, which was the peak of the wheat FHB outbreak, making the incidence more serious. In addition, we could see that some background areas and areas infected with FHB were confused with each other, which may be related to the gradual drying of FHB after its incidence. Despite this phenomenon, XGBoost achieved satisfactory results and could be used to achieve the monitoring of wheat FHB. This study provides a new approach for the rapid and nondestructive monitoring of wheat FHB.

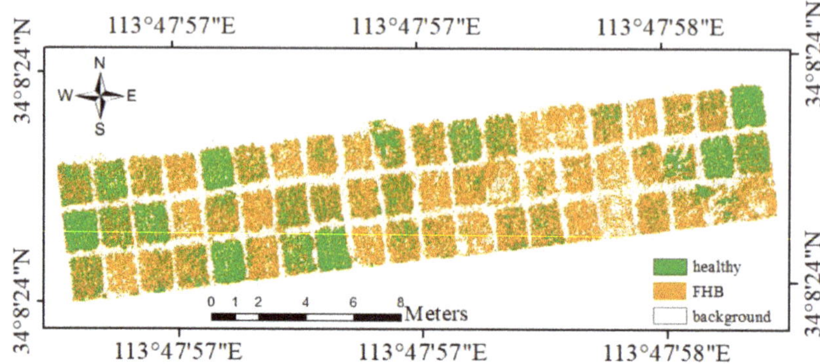

Figure 5. Mapping of the spatial distribution of FHB in wheat.

4. Discussion

Many previous studies, which have used multispectral images from UAVs for plant or crop pests and diseases, have been conducted. Some research scholars have used multispectral images from UAVs to try to monitor citrus huanglongbing (citrus greening) by extracting VIs that were sensitive to the disease, combined with models such as KNN and SVM [63,64]. Other research scholars have used multispectral images for monitoring other diseases such as wheat yellow rust [65], potato late blight [27], and Flavescence dorée [66]. These studies well demonstrated the potential of low-altitude multispectral images for the rapid monitoring of crop diseases. The traditional remote sensing monitoring of pests and diseases, especially based on UAV images [67–69], mostly uses some VIs as the input features. This method only considered the changes in the host conditions and neglects the local detailed textural information of remote sensing images [70,71]. Textural features in UAV remote sensing images can describe the spatial distribution of the brightness of adjacent pixels and unique textural information and are increasingly used in the monitoring of pests and diseases.

Therefore, this study used spectral information and textural information extracted from UAV multispectral imagery to try to monitor wheat FHB. First, to reduce the influence of soil on the monitoring results, OSAVI was used to construct a mask file and set an appropriate threshold to remove the soil areas from the image. Second, we analyzed the correlation of 10 commonly used VIs and three TIs on wheat FHB, and through correlation analysis, we selected three VIs that were significantly correlated with FHB as the input features for the models, which were EVI, DVIRE, and DVI. These VIs were all associated with either NIR or RE, which may be related to the stress state of the crops or plants. After being stressed by pests and diseases, crops will show differential absorption and reflection characteristics in different bands, causing changes in the crops' pigments, water, morphology, and structure [72]. Wheat FHB mainly infects wheat ears, making wheat ears yellow and dry, thus causing the loss of chlorophyll [73], and this symptom can be well reflected by the red-edge band [74,75].

Considering that FHB-infected wheat canopies may present different structures and textures from healthy wheat canopies as well as the background, to further enhance the description of wheat FHB by the textural features, three TIs (NDTI, DTI, and RTI) were constructed instead of the original textural features as the input features for the models. It was found that the TIs were significantly correlated with wheat FHB, probably because the constructed TIs were combinations of different textural features and better-utilized textural information to describe wheat FHB, where the correlation coefficient between DTI and wheat FHB reached -0.893, which was better than VIs, indicating that TIs could also be used as the input features for the wheat FHB monitoring models, and the performance may be better than VIs. Finally, based on the multispectral imagery after the removal of the soil, we used three VIs, three TIs, and integrated VIs and TIs as the input features and selected KNN, PSO-SVM, and XGBoost to construct the wheat FHB monitoring models. Through our analysis, we found that XGBoost, which integrated VIs and TIs as the inputs, could better achieve wheat FHB monitoring with an accuracy of 93.63% and an F1 score as high as 92.93%. The reason may be that XGBoost has the advantage of transforming weak learners into strong learners, and its regularization parameters can ensure the accuracy while avoiding the problem of over fitting. In addition, the model is faster, so the model can be applied to the monitoring of wheat FHB.

In this study, multispectral imaging technology combined with machine learning has achieved great results in wheat FHB monitoring, but there are still some problems that need to be improved. Wheat FHB is one of the most harmful diseases. The infection of wheat FHB will bring irreparable harm to the wheat's quality. Therefore, the early detection of FHB in wheat is particularly important. The research field of this study is relatively single, and further research is needed in more fields to verify the spatial and generalization capabilities of the models used. In addition, we should also consider using the images of multiple stages and key growth periods to further explore the disease characteristics of wheat FHB so as to achieve the goal of an early detection and control. At present, deep learning technology also has a very broad application prospect in plant or crop pest detection, thus the potential of deep learning technology in disease monitoring needs to be further explored.

5. Conclusions

This research proposed a wheat FHB monitoring method combining VIs, TIs, and an XGBoost model. First, based on the multispectral imagery obtained by UAV, OASVI was used to reduce the interference of the soil area. Second, we made full use of the VIs and TIs of UAV multispectral imagery and explored the ability of KNN, PSO-SVM, and XGBoost to monitor wheat FHB under different feature combinations. Lastly, combined with the accuracy evaluation index of the models, the XGBoost model with VIs and TIs as the inputs had the best performance, with an accuracy of 93.63% and an F1 score of 92.93%. The results showed that the fusion of VIs and TIs could improve the accuracy of the model, and XGBoost could quickly and accurately monitor wheat FHB. This research provides technical support and reference for the rapid and nondestructive monitoring of wheat FHB.

Author Contributions: Conceptualization, C.G., X.J. and Z.G.; methodology, C.G. and Z.G.; formal analysis, X.J. and Z.G.; validation, C.G., X.J., Z.G., T.W., Q.H., H.S. and W.G.; investigation, C.G., Z.G., T.W., Q.H. and H.S.; resources, W.G. and C.G.; data curation, W.G.; writing—original draft preparation, C.G. and W.G.; writing—review and editing, W.G. and C.G.; visualization, Z.G. and X.J.; supervision, W.G. and C.G.; funding acquisition, W.G. and X.J. All authors have read and agreed to the published version of the manuscript.

Funding: This research was funded by Henan Province Science and Technology Research Project (212102110028, 22102320035); the National Engineering Research Center for Argo-ecological Big Data Analysis and Application (AE202005); the Science and Technology Innovation Fund of Henan Agricultural University (KJCX2021A16); National Natural Science Foundation of China (32271993).

Institutional Review Board Statement: Not applicable.

Data Availability Statement: Not applicable.

Conflicts of Interest: The authors declare no conflict of interest.

References

1. Palazzini, J.; Fumero, V.; Yerkovich, N.; Barros, G.; Cuniberti, M.; Chulze, S. Correlation between *Fusarium graminearum* and deoxynivalenol during the 2012/13 wheat Fusarium head blight outbreak in Argentina. *Cereal Res. Commun.* **2015**, *43*, 627–637. [CrossRef]
2. Cao, Z.S.; Yao, X.; Liu, H.Y.; Liu, B.; Cheng, T.; Tian, Y.C.; Cao, W.X.; Zhu, Y. Comparison of the abilities of vegetation indices and photosynthetic parameters to detect heat stress in wheat. *Agric. Forest Meteorol.* **2019**, *265*, 121–136. [CrossRef]
3. Liu, L.; Dong, Y.; Huang, W.; Du, X.; Ren, B.; Huang, L.; Zheng, Q.; Ma, H. A disease index for efficiently detecting wheat fusarium head blight using sentinel-2 multispectral imagery. *IEEE Access* **2020**, *8*, 52181–52191. [CrossRef]
4. Salgado, J.D.; Madden, L.V.; Paul, P.A. Quantifying the effects of Fusarium head blight on grain yield and test weight in soft red winter wheat. *Phytopathology* **2015**, *105*, 295–306. [CrossRef] [PubMed]
5. Palacios, S.A.; Erazo, J.G.; Ciasca, B.; Lattanzio, V.M.; Reynoso, M.M.; Farnochi, M.C.; Torres, A.M. Occurrence of deoxynivalenol and deoxynivalenol-3-glucoside in durum wheat from Argentina. *Food Chem.* **2017**, *230*, 728–734. [CrossRef] [PubMed]
6. Bai, G.; Shaner, G. Scab of wheat: Prospects for control. *Plant Dis.* **1994**, *78*, 760–766.
7. Semagn, K.; Iqbal, M.; Jarquin, D.; Crossa, J.; Howard, R.; Ciechanowska, I.; Henriquez, M.A.; Randhawa, H.; Aboukhaddour, R.; McCallum, B.D. Genomic Predictions for Common Bunt, FHB, Stripe Rust, Leaf Rust, and Leaf Spotting Resistance in Spring Wheat. *Genes* **2022**, *13*, 565. [CrossRef]
8. Gilbert, J.; Tekauz, A. Recent developments in research on Fusarium head blight of wheat in Canada. *Can. J. Plant Pathol.* **2000**, *22*, 1–8. [CrossRef]
9. Qiu, R.; Yang, C.; Moghimi, A.; Zhang, M.; Steffenson, B.J.; Hirsch, C.D. Detection of fusarium head blight in wheat using a deep neural network and color imaging. *Remote Sens.* **2019**, *11*, 2658. [CrossRef]
10. Huang, L.; Zhang, H.; Huang, W.; Dong, Y.; Ye, H.; Ma, H.; Zhao, J. Identification of Fusarium head blight in wheat ears using vertical angle-based reflectance spectroscopy. *Arab. J. Geosci.* **2021**, *14*, 423. [CrossRef]
11. Bauriegel, E.; Herppich, W.B. Hyperspectral and chlorophyll fluorescence imaging for early detection of plant diseases, with special reference to Fusarium spec. infections on wheat. *Agriculture* **2014**, *4*, 32–57. [CrossRef]
12. Zhang, J.; Huang, Y.; Pu, R.; Gonzalez-Moreno, P.; Yuan, L.; Wu, K.; Huang, W. Monitoring plant diseases and pests through remote sensing technology: A review. *Comput. Electron. Agric.* **2019**, *165*, 104943. [CrossRef]
13. Liu, L.; Dong, Y.; Huang, W.; Du, X.; Ma, H. Monitoring wheat fusarium head blight using unmanned aerial vehicle hyperspectral imagery. *Remote Sens.* **2020**, *12*, 3811. [CrossRef]
14. Mutanga, O.; Dube, T.; Galal, O. Remote sensing of crop health for food security in Africa: Potentials and constraints. *Remote Sens. Appl. Soc. Environ.* **2017**, *8*, 231–239. [CrossRef]
15. Chu, H.; Zhang, D.; Shao, Y.; Chang, Z.; Guo, Y.; Zhang, N. Using HOG Descriptors and UAV for Crop Pest Monitoring. In Proceedings of the 2018 Chinese Automation Congress (CAC), Xi'an, China, 30 November–2 December 2018; pp. 1516–1519.
16. Guo, Y.H.; Yin, G.D.; Sun, H.Y.; Wang, H.X.; Chen, S.Z.; Senthilnath, J.; Wang, J.Z.; Fu, Y.S. Scaling Effects on Chlorophyll Content Estimations with RGB Camera Mounted on a UAV Platform Using Machine-Learning Methods. *Sensors* **2020**, *20*, 5130. [CrossRef]
17. Zhang, J.; Cheng, T.; Guo, W.; Xu, X.; Qiao, H.; Xie, Y.; Ma, X. Leaf area index estimation model for UAV image hyperspectral data based on wavelength variable selection and machine learning methods. *Plant Methods* **2021**, *17*, s13007–s13021. [CrossRef]
18. Kou, J.; Duan, L.; Yin, C.; Ma, L.; Chen, X.; Gao, P.; Lv, X. Predicting Leaf Nitrogen Content in Cotton with UAV RGB Images. *Sustainability* **2022**, *14*, 9259. [CrossRef]
19. Li, B.; Xu, X.; Zhang, L.; Han, J.; Bian, C.; Li, G.; Liu, J.; Jin, L. Above-ground biomass estimation and yield prediction in potato by using UAV-based RGB and hyperspectral imaging. *ISPRS J. Photogramm. Remote Sens.* **2020**, *162*, 161–172. [CrossRef]
20. Kim, M.S.; Daughtry, C.; Chappelle, E.; McMurtrey, J.; Walthall, C. The Use of High Spectral Resolution Bands for Estimating Absorbed Photosynthetically Active Radiation (A par). In Proceedings of 6th International Symposium on Physical Measurements and Signatures in Remote Sensing, Val D'Isere, France, 17–22 January 1994.
21. Xu, X.; Lu, J.; Zhang, N.; Yang, T.; He, J.; Yao, X.; Cheng, T.; Zhu, Y.; Cao, W.; Tian, Y. Inversion of rice canopy chlorophyll content and leaf area index based on coupling of radiative transfer and Bayesian network models. *ISPRS J. Photogramm. Remote Sens.* **2019**, *150*, 185–196. [CrossRef]
22. Zheng, H.; Li, W.; Jiang, J.; Liu, Y.; Cheng, T.; Tian, Y.; Zhu, Y.; Cao, W.; Zhang, Y.; Yao, X. A comparative assessment of different modeling algorithms for estimating leaf nitrogen content in winter wheat using multispectral images from an unmanned aerial vehicle. *Remote Sens.* **2018**, *10*, 2026. [CrossRef]
23. Han, X.; Wei, Z.; Chen, H.; Zhang, B.; Li, Y.; Du, T. Inversion of winter wheat growth parameters and yield under different water treatments based on UAV multispectral remote sensing. *Front. Plant Sci.* **2021**, *639*. [CrossRef] [PubMed]
24. Han, S.; Zhao, Y.; Cheng, J.; Zhao, F.; Yang, H.; Feng, H.; Li, Z.; Ma, X.; Zhao, C.; Yang, G. Monitoring Key Wheat Growth Variables by Integrating Phenology and UAV Multispectral Imagery Data into Random Forest Model. *Remote Sens.* **2022**, *14*, 3723. [CrossRef]
25. Lei, S.; Luo, J.; Tao, X.; Qiu, Z. Remote Sensing Detecting of Yellow Leaf Disease of Arecanut Based on UAV Multisource Sensors. *Remote Sens.* **2021**, *13*, 4562. [CrossRef]
26. Zhao, X.-y.; Zhang, J.; Zhang, D.-y.; Zhou, X.-g.; Liu, X.-h.; Xie, J. Comparison between the Effects of Visible Light and Multispectral Sensor Based on Low-Altitude Remote Sensing Platform in the Evaluation of Rice Sheath Blight. *Spectrosc. Spectr. Anal.* **2019**, *39*, 1192–1198. [CrossRef]

27. Rodriguez, J.; Lizarazo, I.; Prieto, F.; Angulo-Morales, V. Assessment of potato late blight from UAV-based multispectral imagery. *Comput. Electron. Agric.* **2021**, *184*. [CrossRef]
28. Ye, H.; Huang, W.; Huang, S.; Cui, B.; Dong, Y.; Guo, A.; Ren, Y.; Jin, Y. Identification of banana fusarium wilt using supervised classification algorithms with UAV-based multi-spectral imagery. *Int. J. Agric. Biol. Eng.* **2020**, *13*, 136–142. [CrossRef]
29. Yang, M.D.; Huang, K.S.; Kuo, Y.H.; Tsai, H.P.; Lin, L.M. Spatial and Spectral Hybrid Image Classification for Rice Lodging Assessment through UAV Imagery. *Remote Sens.* **2017**, *9*, 583. [CrossRef]
30. Zheng, H.; Cheng, T.; Zhou, M.; Li, D.; Yao, X.; Tian, Y.; Cao, W.; Zhu, Y. Improved estimation of rice aboveground biomass combining textural and spectral analysis of UAV imagery. *Precis. Agric.* **2019**, *20*, 611–629. [CrossRef]
31. Haralick, R.M.; Shanmugam, K.; Dinstein, I.H. Textural features for image classification. *IEEE Trans. Syst. Man Cybern.* **1973**, 610–621. [CrossRef]
32. Arivazhagan, S.; Shebiah, R.N.; Ananthi, S.; Varthini, S.V. Detection of unhealthy region of plant leaves and classification of plant leaf diseases using texture features. *Agric. Eng. Int. CIGR J.* **2013**, *15*, 211–217.
33. Yue, J.; Yang, G.; Tian, Q.; Feng, H.; Xu, K.; Zhou, C. Estimate of winter-wheat above-ground biomass based on UAV ultrahigh-ground-resolution image textures and vegetation indices. *ISPRS J. Photogramm. Remote Sens.* **2019**, *150*, 226–244. [CrossRef]
34. Turgut, R.; Gunlu, A. Estimating aboveground biomass using Landsat 8 OLI satellite image in pure Crimean pine (*Pinus nigra* JF Arnold subsp. pallasiana (Lamb.) Holmboe) stands: A case from Turkey. *Geocarto Int.* **2022**, *37*, 720–734. [CrossRef]
35. Li, S.; Yuan, F.; Ata-Ui-Karim, S.T.; Zheng, H.; Cheng, T.; Liu, X.; Tian, Y.; Zhu, Y.; Cao, W.; Cao, Q. Combining Color Indices and Textures of UAV-Based Digital Imagery for Rice LAI Estimation. *Remote Sens.* **2019**, *11*, 1763. [CrossRef]
36. Huang, L.; Li, T.; Ding, C.; Zhao, J.; Zhang, D.; Yang, G. Diagnosis of the Severity of Fusarium Head Blight of Wheat Ears on the Basis of Image and Spectral Feature Fusion. *Sensors* **2020**, *20*, 2887. [CrossRef] [PubMed]
37. Berber, M.; Munjy, R.; Lopez, J. Kinematic GNSS positioning results compared against Agisoft Metashape and Pix4dmapper results produced in the San Joaquin experimental range in Fresno County, California. *J. Geod. Sci.* **2021**, *11*, 48–57. [CrossRef]
38. Rondeaux, G.; Steven, M.; Baret, F. Optimization of soil-adjusted vegetation indices. *Remote Sens. Environ.* **1996**, *55*, 95–107. [CrossRef]
39. Liu, J.; Zhu, Y.; Tao, X.; Chen, X.; Li, X. Rapid prediction of winter wheat yield and nitrogen use efficiency using consumer-grade unmanned aerial vehicles multispectral imagery. *Front. Plant Sci.* **2022**, *13*, 1032170. [CrossRef]
40. Qiu, R.; Wei, S.; Zhang, M.; Li, H.; Sun, H.; Liu, G.; Li, M. Sensors for measuring plant phenotyping: A review. *Int. J. Agric. Biol. Eng.* **2018**, *11*, 1–17. [CrossRef]
41. Zheng, Q.; Huang, W.; Cui, X.; Shi, Y.; Liu, L. New spectral index for detecting wheat yellow rust using Sentinel-2 multispectral imagery. *Sensors* **2018**, *18*, 868. [CrossRef]
42. Zhang, D.; Zhou, X.; Zhang, J.; Lan, Y.; Xu, C.; Liang, D. Detection of rice sheath blight using an unmanned aerial system with high-resolution color and multispectral imaging. *PLOS ONE* **2018**, *13*, e0187470. [CrossRef]
43. Gitelson, A.A.; Kaufman, Y.J.; Stark, R.; Rundquist, D. Novel algorithms for remote estimation of vegetation fraction. *Remote Sens. Environ.* **2002**, *80*, 76–87. [CrossRef]
44. Gitelson, A.A.; Viña, A.; Ciganda, V.; Rundquist, D.C.; Arkebauer, T.J. Remote estimation of canopy chlorophyll content in crops. *Geophys. Res. Lett.* **2005**, *32*, L08403. [CrossRef]
45. Tucker, C.J. Red and photographic infrared linear combinations for monitoring vegetation. *Remote Sens. Environ.* **1979**, *8*, 127–150. [CrossRef]
46. Patrick, A.; Pelham, S.; Culbreath, A.; Holbrook, C.C.; De Godoy, I.J.; Li, C. High throughput phenotyping of tomato spot wilt disease in peanuts using unmanned aerial systems and multispectral imaging. *IEEE Instrum. Meas. Mag.* **2017**, *20*, 4–12. [CrossRef]
47. Huete, A.; Didan, K.; Miura, T.; Rodriguez, E.P.; Gao, X.; Ferreira, L.G. Overview of the radiometric and biophysical performance of the MODIS vegetation indices. *Remote Sens. Environ.* **2002**, *83*, 195–213. [CrossRef]
48. Barnes, E.; Clarke, T.; Richards, S.; Colaizzi, P.; Haberland, J.; Kostrzewski, M.; Waller, P.; Choi, C.; Riley, E.; Thompson, T. Coincident Detection of Crop Water Stress, Nitrogen Status and Canopy Density Using Ground Based Multispectral Data. In Proceedings of the Fifth International Conference on Precision Agriculture, Bloomington, MN, USA, 16–19 July 2000; p. 6.
49. Rouse Jr, J.; Haas, R.; Schell, J.; Deering, D. *Monitoring Vegetation Systems in the Great Plains with ERTS. Third Earth Resources Technology Satellite-1 Symposium: Volume 1; Technical Presentations, Section B*; Freden, S., Mercanti, E., Becker, M., Eds.; NASA Special Publication: Washington, DC, USA, 1974; Document ID: 19740022592; pp. 1–9.
50. Liu, L.Y.; Fan, X.J. The Design of System to Texture Feature Analysis Based on Gray Level Co-Occurrence Matrix. In Proceedings of the Applied Mechanics and Materials, The Island of Crete, Greece, 30 May 2015; pp. 904–907.
51. Haralick, R.M. Statistical and structural approaches to texture. *Proc. IEEE* **1979**, *67*, 786–804. [CrossRef]
52. Wolff, J.; Backofen, R.; Grüning, B. Robust and efficient single-cell Hi-C clustering with approximate k-nearest neighbor graphs. *Bioinformatics* **2021**, *37*, 4006–4013. [CrossRef] [PubMed]
53. Venter, G.; Sobieszczanski-Sobieski, J. Particle swarm optimization. *AIAA J.* **2003**, *41*, 1583–1589. [CrossRef]
54. Yang, S.; Hu, L.; Wu, H.; Ren, H.; Qiao, H.; Li, P.; Fan, W. Integration of crop growth model and random forest for winter wheat yield estimation from UAV hyperspectral imagery. *IEEE J. Sel. Top. Appl. Earth Obs. Remote Sens.* **2021**, *14*, 6253–6269. [CrossRef]
55. Guo, A.; Huang, W.; Ye, H.; Dong, Y.; Ma, H.; Ren, Y.; Ruan, C. Identification of Wheat Yellow Rust Using Spectral and Texture Features of Hyperspectral Images. *Remote Sens.* **2020**, *12*, 1419. [CrossRef]

56. Peng, X.; Chen, D.; Zhou, Z.; Zhang, Z.; Xu, C.; Zha, Q.; Wang, F.; Hu, X. Prediction of the Nitrogen, Phosphorus and Potassium Contents in Grape Leaves at Different Growth Stages Based on UAV Multispectral Remote Sensing. *Remote Sens.* **2022**, *14*, 2659. [CrossRef]
57. Han, Z.Y.; Zhu, X.C.; Fang, X.Y.; Wang, Z.Y.; Wang, L.; Zhao, G.X.; Jiang, Y.M. Hyperspectral Estimation of Apple Tree Canopy LAI Based on SVM and RF Regression. *Spectrosc. Spectr. Anal.* **2016**, *36*, 800–805. [CrossRef]
58. Huang, L.; Wu, K.; Huang, W.; Dong, Y.; Ma, H.; Liu, Y.; Liu, L. Detection of Fusarium Head Blight in Wheat Ears Using Continuous Wavelet Analysis and PSO-SVM. *Agriculture* **2021**, *11*, 998. [CrossRef]
59. Chen, T.; Guestrin, C. Xgboost: A Scalable Tree Boosting System. In Proceedings of the 22nd ACM SIGKDD International Conference on Knowledge Discovery and Data Mining, San Francisco, CA, USA, 13–17 August 2016; pp. 785–794.
60. Dhaliwal, S.S.; Nahid, A.-A.; Abbas, R. Effective intrusion detection system using XGBoost. *Information* **2018**, *9*, 149. [CrossRef]
61. Croux, C.; Dehon, C. Influence functions of the Spearman and Kendall correlation measures. *Stat. Methods Appl.* **2010**, *19*, 497–515. [CrossRef]
62. Croux, C.; Dehon, C. Robustness versus efficiency for nonparametric correlation measures. *FBE Res. Rep. KBI_0803* **2008**.
63. Lan, Y.; Huang, Z.; Deng, X.; Zhu, Z.; Huang, H.; Zheng, Z.; Lian, B.; Zeng, G.; Tong, Z. Comparison of machine learning methods for citrus greening detection on UAV multispectral images. *Comput. Electron. Agric.* **2020**, *171*, 105234. [CrossRef]
64. DadrasJavan, F.; Samadzadegan, F.; Seyed Pourazar, S.H.; Fazeli, H. UAV-based multispectral imagery for fast Citrus Greening detection. *J. Plant Dis. Prot.* **2019**, *126*, 307–318. [CrossRef]
65. Su, J.; Liu, C.; Hu, X.; Xu, X.; Guo, L.; Chen, W.-H. Spatio-temporal monitoring of wheat yellow rust using UAV multispectral imagery. *Comput. Electron. Agric.* **2019**, *167*, 105035. [CrossRef]
66. Albetis, J.; Duthoit, S.; Guttler, F.; Jacquin, A.; Goulard, M.; Poilvé, H.; Féret, J.-B.; Dedieu, G. Detection of Flavescence dorée grapevine disease using unmanned aerial vehicle (UAV) multispectral imagery. *Remote Sens.* **2017**, *9*, 308. [CrossRef]
67. West, J.S.; Canning, G.G.; Perryman, S.A.; King, K. Novel Technologies for the detection of Fusarium head blight disease and airborne inoculum. *Trop. Plant Pathol.* **2017**, *42*, 203–209. [CrossRef] [PubMed]
68. Guo, A.; Huang, W.; Dong, Y.; Ye, H.; Ma, H.; Liu, B.; Wu, W.; Ren, Y.; Ruan, C.; Geng, Y. Wheat yellow rust detection using UAV-based hyperspectral technology. *Remote Sens.* **2021**, *13*, 123. [CrossRef]
69. Ye, H.; Huang, W.; Huang, S.; Cui, B.; Dong, Y.; Guo, A.; Ren, Y.; Jin, Y. Recognition of banana fusarium wilt based on UAV remote sensing. *Remote Sens.* **2020**, *12*, 938. [CrossRef]
70. Jain, A.K.; Murty, M.N.; Flynn, P.J. Data clustering: A review. *ACM Comput. Surv.* **1999**, *31*, 264–323. [CrossRef]
71. Xiao, Y.; Dong, Y.; Huang, W.; Liu, L.; Ma, H. Wheat fusarium head blight detection using UAV-based spectral and texture features in optimal window size. *Remote Sens.* **2021**, *13*, 2437. [CrossRef]
72. Hanley, J.A.; McNeil, B.J. The meaning and use of the area under a receiver operating characteristic (ROC) curve. *Radiology* **1982**, *143*, 29–36. [CrossRef]
73. Fernando, W.D.; Oghenekaro, A.O.; Tucker, J.R.; Badea, A. Building on a foundation: Advances in epidemiology, resistance breeding, and forecasting research for reducing the impact of Fusarium head blight in wheat and barley. *Can. J. Plant Pathol.* **2021**, *43*, 495–526. [CrossRef]
74. Filella, I.; Penuelas, J. The red edge position and shape as indicators of plant chlorophyll content, biomass and hydric status. *Int. J. Remote Sens.* **1994**, *15*, 1459–1470. [CrossRef]
75. Fu, Y.; Yang, G.; Wang, J.; Song, X.; Feng, H. Winter wheat biomass estimation based on spectral indices, band depth analysis and partial least squares regression using hyperspectral measurements. *Comput. Electron. Agric.* **2014**, *100*, 51–59. [CrossRef]

Article

Improving Land Use/Cover Classification Accuracy from Random Forest Feature Importance Selection Based on Synergistic Use of Sentinel Data and Digital Elevation Model in Agriculturally Dominated Landscape

Sa'ad Ibrahim

Department of Geography, Adamu Augie College of Education, Argungu PMB 1012, Kebbi State, Nigeria;
saad.ibrahim@aacoeargungu.edu.ng

Abstract: Land use and land cover (LULC) mapping can be of great help in changing land use decisions, but accurate mapping of LULC categories is challenging, especially in semi-arid areas with extensive farming systems and seasonal vegetation phenology. Machine learning algorithms are now widely used for LULC mapping because they provide analytical capabilities for LULC classification. However, the use of machine learning algorithms to improve classification performance is still being explored. The objective of this study is to investigate how to improve the performance of LULC models to reduce prediction errors. To address this question, the study applied a Random Forest (RF) based feature selection approach using Sentinel-1, -2, and Shuttle Radar Topographic Mission (SRTM) data. Results from RF show that the Sentinel-2 data only achieved an out-of-bag overall accuracy of 84.2%, while the Sentinel-1 and SRTM data achieved 83% and 76.44%, respectively. Classification accuracy improved to 89.1% when Sentinel-2, Sentinel-1 backscatter, and SRTM data were combined. This represents a 4.9% improvement in overall accuracy compared to Sentinel-2 alone and a 6.1% and 12.66% improvement compared to Sentinel-1 and SRTM data, respectively. Further independent validation, based on equally sized stratified random samples, consistently found a 5.3% difference between the Sentinel-2 and the combined datasets. This study demonstrates the importance of the synergy between optical, radar, and elevation data in improving the accuracy of LULC maps. In principle, the LULC maps produced in this study could help decision-makers in a wide range of spatial planning applications.

Keywords: land use; land cover; classification; random forest; Sentinel data; SRTM; random forest; feature selection; accuracy; validation

Citation: Ibrahim, S. Improving Land Use/Cover Classification Accuracy from Random Forest Feature Importance Selection Based on Synergistic Use of Sentinel Data and Digital Elevation Model in Agriculturally Dominated Landscape. *Agriculture* **2023**, *13*, 98. https://doi.org/10.3390/agriculture13010098

Academic Editors: Jibo Yue, Chengquan Zhou, Haikuan Feng, Yanjun Yang and Ning Zhang

Received: 4 December 2022
Revised: 21 December 2022
Accepted: 26 December 2022
Published: 29 December 2022

1. Introduction

Earth-observing satellite sensor data can be used for land-cover mapping and monitoring, which is essential for estimating land-cover change. The increase in land use and land cover changes (LULC) in natural ecosystems has adversely affected carbon stocks, climate change, and biodiversity, as well as the global climate over the past few decades [1–4]. It is believed that deforestation due to urbanization and agricultural expansion is one of the most critical threats to the environment in the 21st century [5]. The United Nations (U.N.) sustainable development goal (SDG) 15 has emphasized measures to *"protect, restore, and promote sustainable use of terrestrial ecosystems, sustainably manage forests, combat desertification, and halt and reverse land degradation and biodiversity loss"* [6]. Priority is placed on combating desertification, recovering degraded land and soil, particularly in areas affected by desertification, drought, and floods, and combating land degradation by 2030. Satellite Earth observation data offer one of the most reliable options for monitoring land degradation in the context of the SDGs due to their consistency and repeatability at local and large spatial scales. Information about the land cover of a country is an essential part of the planning and development process. It is useful for environmental reporting [7], assessing the impact of land use on the natural environment [8], conserving biodiversity and habitats [9], mapping population distributions [10], forecasting crops, studying urban heat islands, managing insurance risks, planning telecommunications, and others [11–13].

Even though traditional methods (e.g., field surveying) yield accurate results, they are expensive and inefficient in monitoring large and inaccessible areas. To overcome these limitations, remote sensing scientists have developed analytical tools for detecting, characterizing, parameterizing, and monitoring land variables based on space observations. Remote sensing has experienced rapid advances over the past 40 years. Based on remote sensing technology, data are usually collected across different regions of the electromagnetic spectrum at wide spatiotemporal scales (e.g., the recent Copernicus program/Sentinel missions and the Landsat program/missions- which has been available for over 40 years). Hence, remote sensing provides an interesting option for policymakers to make informed decisions about our environment and also to improve the methodology of assessing ecosystem vulnerability [14,15].

Over the past decades, the scientific community has fully recognized remote sensing/Earth observation data from space for LULC monitoring. These data offer an unparalleled opportunity for large-area measurement and high temporal precision for land cover mapping and monitoring. Today, a large number of global land cover maps are produced (e.g., GLOBCOVER and MODIS land cover products). However, these products have their limitations for regional as well as local assessments due to their low spatial resolution (e.g., 1 km, 250 m), temporal frequency, and inconsistencies in their assigned thematic classes [16]. These limitations primarily occur due to (1) the small number of training data relative to the size of the area being mapped, (2) mismatch definitions/propriety in land cover classification schemes, (3) and the need for a readily and automated algorithm to handle large datasets. In this light, many regional governments have embarked on research projects to provide high and medium-resolution (e.g., 30 m) land cover maps, which are accurate and consistent with their local demands. For example, the operational land cover databases (e.g., the National Land Cover Database for the United States of America (U.S.A.) and the United Kingdom's Land Cover product which is based on the European CORINE land cover mapping scheme [11].

A widespread increase in anthropogenic activities, land use, and land cover changes are occurring at an unprecedented rate, requiring policymakers and stakeholders to pay greater attention to the measures to manage and control environmental degradation. In Nigeria, the threat to environmental sustainability, for example, is encapsulated in the need to ensure the quality of the environment is appropriate for good health and well-being, as well as to protect and utilize the environment and natural resources for the benefit of present and future generations. The policy encourages the compilation of detailed land capability inventories, comprehensive land classifications, assessment of the current land use practices, causes and extent of land degradation, and regulatory framework for sustainable land use [17]. However, despite recent advancements in Earth observation and remote sensing, there is no reliable land LULC for the country. Most of the previous global land cover maps were not also developed based on adequate or training data sets covering Nigeria. And their class labeling and definitions (e.g., International Geosphere-Biosphere Programme) have mixed land cover classes, which are unsuitable for discerning LULC characteristics in Nigeria. Conservation policies in Nigeria have emphasized undertaking land capability classifications based on evolving methods of land evaluation suitable to local conditions.

Land cover monitoring using remotely sensed data involves precise mapping of complex land cover and land use categories, necessitating the employment of strong classification systems [18]. Waske and Braun [19], who compare the ensemble classifiers with approaches such as the maximum likelihood classifier for land cover classification using C-band multi-temporal SAR data, observed that random forest (RF) outperformed maximum likelihood by more than 10%. A comprehensive comparison of machine learning algorithms has been conducted by Lawrence and Moran (2015) using uniform procedures and 30 distinct datasets. Their results showed that RF had the highest classification accuracy of 73.19% than SVM, which had an accuracy of 62.28%. Of the total 30 classifications, RF was the most accurate in 18 classification scenarios. Talukdar et al. [20] reviewed six machine-learning classifiers for LULC classification using satellite observations. Based on overall accuracy, results indicate that RF is the best machine-learning LULC classifier (0.89, RMSE = 0.006), compared to support vector machine (Kappa = 0.86, RMSE = 0.11), artificial neural network (Kappa = 0.87, RMSE = 0.09), fuzzy adaptive resonance theory-supervised predictive mapping (0.85, RMSE = 0.17), spectral angle mapper (Kappa = 0.84, RMSE = 0.23), and Mahalanobis distance (Kappa = 0.82, RMSE = 0.28). This makes the machine learning algorithm suitable for LULC classification. Furthermore, a recent study by Adugna et al. [21], who compare RF and SVM machine learning methods, found that RF outperformed SVM, yielding overall accuracy (OA) of 0.86 and a kappa (k) statistic of 0.83, respectively, which is 1–2% and 3% higher than the best SVM model.

Nowadays, machine learning technology is used for feature selection to assist in mapping LULC categories. The advantage of RF is its capability for feature selection, which has been proven to improve classification accuracy in previous studies [22–24]. A study by Balzter et al. [11], who developed a method for CORINE Land Cover mapping using RFs, demonstrates the importance of variable selection using Sentinel-1A radar backscatter coefficient at HH and HV polarizations (summer acquisitions) and VV and VH polarization (winter acquisitions) and SRTM Digital Elevation Model Data. The classification out-of-bag error rate was 52.5%, and kappa (κ) = 0.38 for the Sentinel-1 variables. When the variables generated from the S.R.T.M. data were added, the quality of the classified map was improved substantially, with an out-of-bag error rate of 31.6% (68.4% accuracy) and κ = 0.63. R.F. clearly describes the benefits of including variable selection in the land cover classification process in a complex environment [25].

The RF technique is well-established in land remote sensing today. Still, it has not been adequately evaluated by the remote sensing community as compared to more traditional pattern recognition algorithms. In addition, there have been observations about how the importance of variables varies depending on the data and ecosystem in question, necessitating further exploration [23,25,26]. To assist decision-makers in a variety of spatial planning applications (e.g., cropland management, irrigated agriculture intensification, flood vulnerability assessment, water management, or human settlement/resettlement planning in floodplains), the thematic LULC classes were created to represent the local characteristics of the semi-arid region, in Nigeria. Specifically, the objectives of this study were; (1) to evaluate the applicability of an RF classification algorithm for LULC mapping using local class definitions and training data sets in an agriculturally dominated landscape in Nigeria; (2) to assess the contribution of an individual satellite band in the RF model; (3) to improve model performance and reduce prediction errors of LULC maps based on RF feature selection. The novelty of this study is the synergistic use of different sources of satellite data to identify the most important variables to reduce prediction error. Therefore, one of the most important contributions of the work is the methodology developed to improve classification performance. The insights gained in this work to improve model performance and reduce prediction errors not only support policymakers in applying accurate LULC maps in spatial planning but also enrich the methodological system of LULC assessment through machine learning.

2. Materials and Methods

2.1. The Study Area

This study was conducted in Kebbi state, the northwestern part of Nigeria (Figure 1a,b). This area is located between latitude 4°27′0″–4°54′0″ N of the equator and longitude 4°19′12″–4°48′0″ E of the Greenwich meridian (Figure 1a). This area falls in the Argungu local government and parts of Augie, Birnin-Kebbi, and Gwandu local government areas. The climate in the area is tropical continental, with two distinct seasons, dry and wet. This is caused by the presence of two contrasting air masses, the tropical continental and the tropical maritime, which originate from the Sahara Desert and the Atlantic Ocean, respectively. The wet season lasts from May to October. The dry season lasts from November to April. The average rainfall is 800 mm. The average temperature is 27 °C which can rise to 40 °C in the summer. Sudan savannah is the predominant vegetation type in the area [27–29]. Geologically, the area is composed of sedimentary rock, primarily undifferentiated sands, gravels, clays (mostly in the upland areas), and floodplains that surround riverine communities [30]. It is, therefore, possible to identify two types of soil in the area: sandy soil for the upland area and clayey and hydromorphic soils for the floodplain area (clay, clay-loam, sandy-loam, loamy sand). The area is mostly characterized by lowland with a few highland areas of up to 344 m, dissected by large flowing rivers (e.g., River Rima) (Figure 1a). The area is characterized by Sudan savannah vegetation type. It includes trees/shrubs (*Pilliostigma reticulatum, Combretum nigricans, Combretum verticellatum, Guira senegalensis Azadirachta indica, Piliostima thonningii, Guira senegalensis* and grass species (*Borroria scabra, Borroria radata, Pennisetum peicellatum, Pennistum peicellatum, Corchorus fascicularis, Digitaria horizontalis, Lam (karangiya), Commelina forskalei, Eragrostis gangetica*, etc.) [28]. These species of plants have different phenological cycles (e.g., leaf flush and senescence period). However, most of these species have their leaf-on up to the end of November.

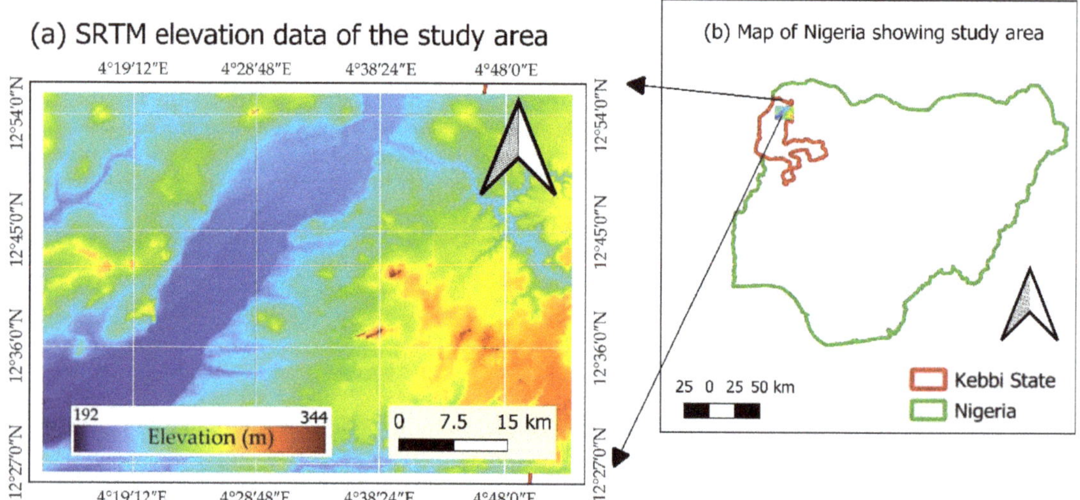

Figure 1. Study area (**a**) study area showing the elevation data based on SRTM data, (**b**) Map of Nigeria showing Kebbi State and the location of the study area.

A large number of subsistence farmers use the floodplain area for irrigation activities where rice, wheat, tomatoes, etc., are being cultivated. But the majority of the farmers engaged in rice cultivation. In the upland areas, crops such as millet, guinea corn, legumes, etc., are usually cultivated. The main harvest time for cereals (such as millet) is late October, except for Guinea corn and legumes, which are usually harvested around mid-November. However, most of the cultivated land in the upland region grows millet. The main harvesting season for rice is November and December, depending on the type and timing of planting. Rice grown in the rainy season is fed to some extent by the rain, as it is strongly supported by irrigation activities. Approximately 75% of the people in the area work as farmers and cultivate crops through rain-fed and irrigation practices in the floodplain area [31]. The area is well-known for its contribution to rice production and fishing in Nigeria. The river Rima in Argungu provides an opportunity for tourism-the famous Argungu International Fishing and Cultural Festival on the one hand, and industrial development- the WACOT rice mill company-employing many thousands of people as well as enhancing rice production in the region. Currently, land use patterns are undergoing various transformations as a result of changing demographic and economic characteristics in the area, creating a wide range of environmental problems. As the land use system continues to undergo rapid changes, there is a need to develop an accurate mapping framework so that an assessment of future land use patterns and the sustainability of land resources may be well-studied.

2.2. Remote Sensing Data

2.2.1. Sentinel-1 Normalized Backscatter

Sentinel-1 is a C-band synthetic aperture radar (SAR) satellite mission of the European Copernicus Program. In this study, the Sentinel-1 Analysis Ready Data (ARD) is one of the remote sensing data used as input variables for feature importance selection with the RF Classifier. The Sentinel-1 data were acquired on 4 October 2020, and were downloaded from the Digital Earth Africa website (https://www.digitalearthafrica.org/, accessed on 15 August 2022. Because the wet season occurs from the end of May to the end of October, the image acquisition period was found suitable to capture the phenology of plant species and crops. The data are available in single polarization (VV) and double polarization (VH). In addition, radiometric terrain correction (RTC) was applied to the normalized backscatter [32]. To increase the number of variables in the RF model, two additional variables were created from these polarizations. The mean and total sum of VV and VH were generated and included in the RF model to assess whether these variables could contribute to model performance. In general, data from SAR, such as those from Sentinel-1, provide different and complementary information than that provided by optical remote sensing. A radar signal can penetrate clouds and provide information about the Earth's surface that optical sensors cannot work due to topography, land cover structure, orientation, and moisture characteristics.

2.2.2. Sentinel-2 Surface Reflectance

In addition to other remote sensing data, this study incorporates the Copernicus sentinel-2 multispectral data to map the LULC of the study area. The Sentinel-2 ARD for 17 October 2020, was downloaded from the Digital Earth Africa website at https://maps.digitalearth.africa/, accessed on 15 August 2022. The acquisition period of the imagery was considered useful in capturing the phenology of woody plants, grasses, and crops. The spectral bands used for this study include blue (band 2), green (band 3), red (band 4), red edge (band 5), red edge (band 6), red edge (band 7), NIR (band 8), NIR (band 8a), SWIR1 (band 11) and SWIR2 (band 12). The spatial resolution of these data is 20 m. The data were pre-processed and atmospherically corrected by the providers. Sentinel-2 has promise in LULC mapping in semi-arid/agriculturally dominant landscapes based on RF feature selection [33,34].

2.2.3. SRTM Digital Model Data

It is a collaboration between the National Geospatial-Intelligence Agency and the National Aeronautics and Space Administration (NASA) to provide elevation data at a global scale to produce the most complete high-resolution digital topographic database of Earth using radar data. The 30 m, ArcGRID format was used in this study and is available at http://www2.jpl.nasa.gov/srtm/index.html (accessed on 15 August 2022). Three variables were created from the data. These include elevation, slope, and aspect. This will be used to represent the surface elevation of the study site. The 30 m DEM was upgraded to 20 m through the nearest neighbor interpolation techniques to make it compatible with Sentinel data.

2.3. Method

2.3.1. Workflow

This study identified the most critical spectral and topographic variables for enhancing model performance using ARD sentinel data (1 and 2) and SRTM DEM. The RF classification method was used to classify LULC types (river, wetland/flooded, irrigated land, barren, built-up area, tree/shrubland, farmland, and grassland) in the area. The flowchart is shown in Figure 2. The figure illustrates a high-level summary of the processes and procedures employed in this study.

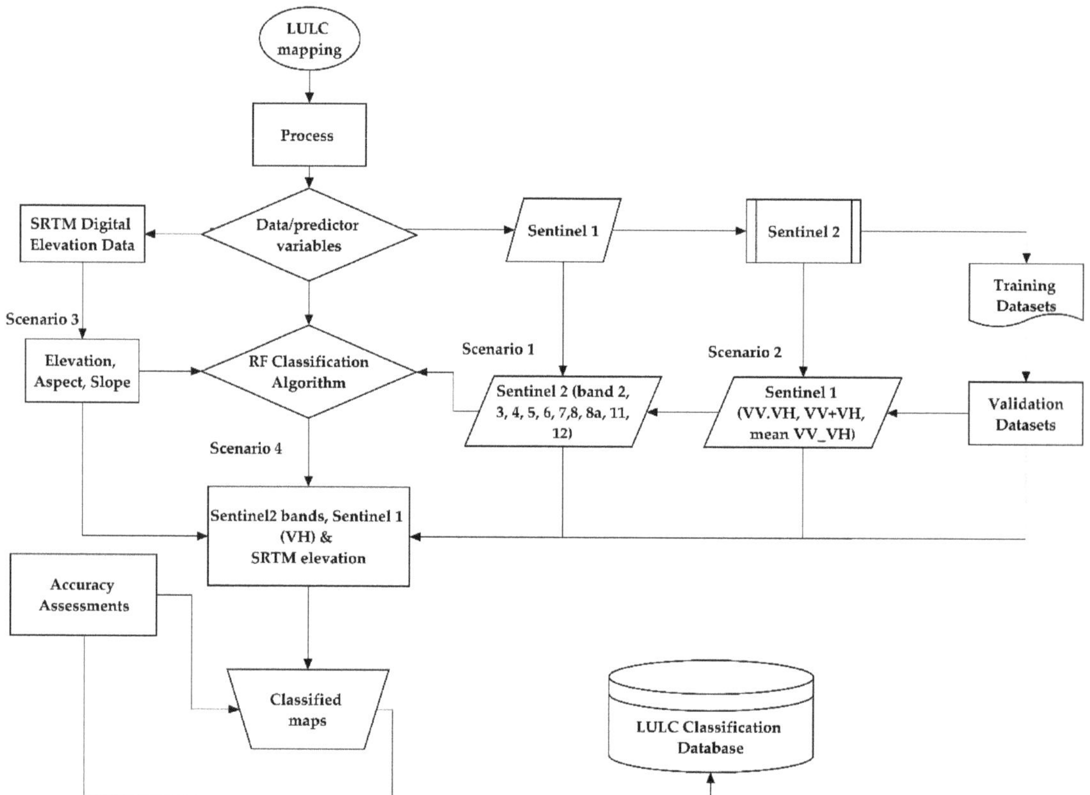

Figure 2. The workflow of the processes and approaches implemented in the study.

2.3.2. Resampling

The data used in this study are not on the same spatial resolution. For example, the ARD Sentinel-2 data is 20 m, Sentinel-1 is 25 m, and STRM DEM is 30 m. This makes it necessary to resample these datasets on the same spatial resolution. However, there are different techniques of resampling that are used depending on the problem. The first one is done as a result of a mismatch between the different raster datasets, while the other type is when a raster dataset is converted into a different coordinate system. In this study, the reason for resampling is due to the mismatch between the different data sets mentioned earlier. Three resampling methods, including bi-linear, nearest neighbor, bicubic or cubic convolutional interpolation, are commonly used for resampling raster data [35]. The nearest neighbor resampling approach was found to be more suitable and was therefore used to resample the normalized sentinel backscatter data and data from DEM to 20 m. The nearest neighbor assigns the DN (or another pixel value, like backscatter) based on the input matrix's nearest pixel. It has the advantage of computational simplicity and does not potentially change input pixel values [35]. Zheng et al. [36], who assessed the effects of different spatial resolution unification schemes and methods on LULC classification, discovered that nearest neighbor interpolation could satisfy the needs of local and regional LULC applications.

2.3.3. Feature Importance Selection

Predictor variables were selected based on an understanding of how spectral reflectance varies across surface features and how it contributes to land surface characterization. The electromagnetic spectrum offers a wide range of options for discriminating among various objects. Even within a land cover category, there are variations in the spectral signatures of different electromagnetic spectrum components. For example, in vegetation, light absorption by leaf pigments dominates in the visible wavelength (400–700 nm), whereas leaf pigments are transparent to NIR (700–1300 nm), and leaf absorption is small [37]. Sentinel-2 data, for instance, has 13 bands, each of which contributes differently to the differentiation of the land surface. A unique characteristic of vegetation is its reflectance signature, which is observed by active sensors such as microwaves (e.g., shortwave or longwave radar data). Whether it is day, night, or cloudy, microwave sensors can image any part of the planet. Through this, radar data complement passive optical data in mapping LULC types. Some variables are more relevant for some phenomena than others, depending on the situation at hand. In Figure 3a, normalized backscatter variability is shown for the 8 LULC types being studied. Based on these spectral variations, the LULC types seem to be distinguished across different polarizations (VV, VH, mean VV & VH, VV+VH). Figure 3b shows the spectral curves of the 8 LULC types from Sentinel-2 multi-spectral data. In general, there is a possibility that these classes could be well distinguished by the classifier based on their emittance behavior (Figure 3a,b). The visible wavelengths, especially the blue and green bands, do not discriminate between these LULC types. The LULC classes of red, NIR, and SWIR 1 and 2, however, possess distinct spectral characteristics (Figure 3b).

The complexity of the environment makes it challenging to easily identify which feature is most useful for predicting land cover categories. This is due to the uncertainty as to which of the features will contribute most to the accuracy of classification. Additionally, auxiliary features such as topographic variables are usually included in an RF feature selection to complement spectral data. The ability to combine numerous variables to enable feature selection to better predict outcomes is provided by the RF machine learning feature selection [38,39]. Mean decrease accuracy (MDA) has been recognized as one of the standard procedures for assessing feature importance, which is based on the OOB estimates of the RF model [40,41]. The higher the value, the more important the variable is. To find the most important features for enhancing model performance, sets of scenarios with various features were established, which assessed these features based on individual datasets and in combinations for this investigation (Table 1).

Table 1. Predictor variables for the RF feature selection.

| S/No | Scenario 1 | Scenario 2 | Scenario 3 | Scenario 4 |
	Sentinel-2 Bands	Sentinel-1 Bands	SRTM Data	Combined (Scenario 1, 2 & 3)
1	Blue	VV	Elevation	Blue
2	Green	VH	Aspect	Green
3	Red	Mean (VV & VH)	Slope	Red
4	NIR_8	VV+VH	-	NIR_8
5	NIR_8a	-	-	NIR_8a
6	SWIR1	-	-	SWIR1
7	SWIR1	-	-	SWIR1
8	Red edge_1	-	-	Red edge_1
9	Red edge_2	-	-	- Red edge_2
10	Red edge_3	-	-	- Red edge_3
11	-	-	-	VH
12	-	-	-	Elevation

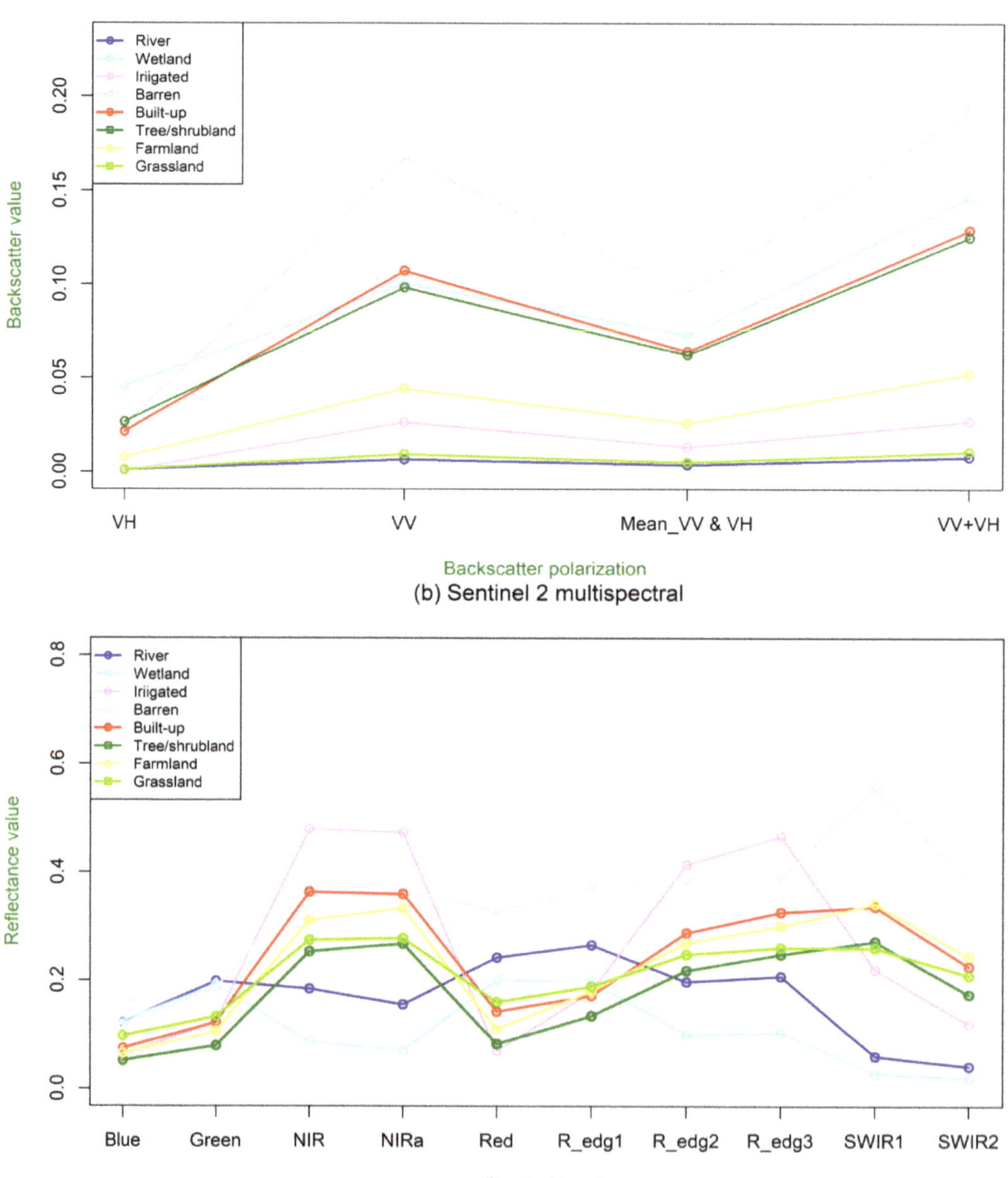

Figure 3. (**a**): Spectral curves of the LULC categories derived from the Sentinel-1 normalized backscatter, (**b**) spectral curves of the LULC categories derived from the Sentinel-2 normalized backscatter.

2.3.4. Training Data

A previous study evaluating the RF method found that classification accuracy increases with increasing training data [40,41]. This means that accurate classification requires many training polygons. Therefore, this study digitized 430 polygons for the seven LULC classes using an RGB composite derived from Sentinel-2 and Google Earth.

2.3.5. RF Classification

Breiman [24] explains that "RF classification is a combination of tree predictors such that each tree depends on the values of a random vector sampled independently and with the same distribution for all trees in the forest". RF classification of images is based on the principles that construct several decision trees. From the large collection of trees, each tree in the RF splits out a class prediction, and model prediction is performed based on the class with the most votes. This method relies on bootstrap and feature randomness when generating each tree [24]. Liaw and Wiener [42] explain the basic steps in the RF classification procedure as follows:

i. First, create n_{tree} bootstrap samples using the original data.
ii. Create an unpruned classification or regression tree for each of the bootstrap samples. At each node, select the best split from a randomly selected subset of the predictors rather than the best among all predictors.
iii. Assemble the predictions of the ntree trees to predict new data (i.e., majority votes for classification, the average for regression).

In this study, the RF classification was implemented in R statistical software by applying the 'RandomForest' package [42] and other packages such as 'raster' [43], 'sp' [44], 'rgdal' [45], 'sf' [46], 'gstat' [47]. As explained earlier, the LULC in the study area was classified into eight classes. For increased classification accuracy, all pixels in the training data were used for each class. Four scenarios were established based on the predictor variables to determine the most important features and accurate results. To evaluate the performance of classifications, various input features were used:

i. In the first scenario, only the Sentinel-2 bands were considered as predictor variables.
ii. In the second scenario, Sentinel-1 normalized backscatter (VV, VH, VV+HH, and the mean of VV and VH) were considered.
iii. In the third scenario, only the DEM variables (elevation, aspect, and slope) were considered.
iv. All of the variables considered in scenario 1 and most variables in the second and third scenarios were utilized in the fourth scenario.

2.3.6. Out-Of-Bag Error Estimates

The accuracy of the classifications was assessed based on the Out-of-bag (OOB) confusion matrix, which is usually computed internally by the model. The training data is divided into 70%, which is used for the classification, while the remaining 30% is used for the OOB estimation. An estimate of the error rate can be obtained, based on the training data, by the following:

1. At each bootstrap iteration, predict the data not in the bootstrap sample (what Breiman (2001) calls "out-of-bag", or OOB, data) using the tree grown with the bootstrap sample.
2. Aggregate the OOB predictions (On average, each data point would be out-of-bag around 30%, which aggregates these predictions). Calculate the error rate and call it the OOB estimate of the error rate.

The OOB confusion matrix, kappa statistics, overall accuracy, and error rate were presented. In addition, class errors were also presented as they can depict LULC type that is more or less accurate and can therefore disentangle the uncertainty associated with overall accuracy based on the classification performance [48].

2.3.7. Independent Validation

Researchers are concerned about the reliability of accuracy assessments. This is even though OOB error calculated by the RF is widely recognized as a standard method of error reporting by the scientific community [49,50], some scholars are still of the view that an independent test is required due to bias nature of the RF accuracy assessment [51,52]. It was proposed that cross-validation can reduce the remaining bias [51]. A well-known phenomenon is RF's preference for predicting classes where the majority of training observations originate [53]. A stratified random sampling of equal size was used for the selection of validation data in this study. The selection of validation shapefile was carried out in R programming software using the 'sp' [44] and 'raster' [43] packages. One hundred points were extracted for each class from the classified maps. However, further confirmation and verification of the individual points were done in QGIS with the help of RGB composite and Google Earth so that the correct class could be assigned to each point data. Similarly, the accuracy assessments of the classified maps were performed in *R* programming software using the validation datasets created earlier. The same R packages were used for accuracy assessments. This cross-validation aims to: (1) complement OOB error estimates of the RF, (2) find out whether two validation results can maintain a consistent pattern, and (3) find out whether sampling the same number of observations in each class could serve as an alternative means of reducing bias.

3. Results

3.1. Feature Importance Selection

Based on the proposed scenarios for evaluating the feature importance, all variables were put into the RF model, and the importance of each variable was calculated by the score of the accuracy of their contribution to the RF classification (4a/d). The RF classification algorithm is robust as it outputs the contribution of different variables in the model. The feature analyses were carried out for each dataset (Sentinel-1, 2, and topographic variables) separately, and the most important features were selected for the last scenario. Based on the random nature of the model, different scores of importance were derived. The Sentinel-2 variables show the lowest out-of-bag error. Therefore, one of the most important features in the second scenario (VH normalized backscatter) and the third scenario (SRTM elevation) were selected to complement the Sentinel-2 data.

Figure 4a,d shows the mean decrease in accuracy of the model for the four scenarios established and implemented in this study. The greater the accuracy, the more influential the variable is for the classification. Figure 4a shows the mean decrease in accuracy of the first scenario, which uses only the Sentinel-2 data. The mean decrease in accuracy between these variables and for this specific scenario. This means that the difference between the most and least important features is substantial. The blue band contributes the most, followed by the SWIR1 band and the NIR band 8a, NIR band 8a, and SWIR2. In this particular scenario, the red edge bands are the least important features (Figure 4a), with mean decrease accuracy ranging from 40–75. In this scenario, the OOB error estimates are less, meaning that all features have yielded the overall accuracy of the model. These results point to the importance of spectral reflectance property variation and the role of the interacting medium.

The feature importance for the second scenario is shown in Figure 4b. Only the Sentinel-1 normalized backscatter was considered in this scenario. In this scenario, VH normalized backscatter appears as the most important feature compared to VV, Mean, and sum of VV and VH. And there is a wide gap between them. And the VH backscatter has the highest contribution to the model. This does not, however, means that the Sentinel-1 data outperformed the Sentinel-2 data when reference is made to the MDA scores. Although the mean decrease accuracy shows the most important feature based on MDA, the scores do not, however, determine the overall accuracy of the model, especially if two different scenarios are being considered. Feature importance in an RF model depends to a large extent on the combinations of the variables in the model. In the third scenario, only topographic variables were assessed. In the third scenario, elevation has the highest scores, followed by slope and the aspect, aspect. The gap between the elevation scores and that of other topographic features is substantial. This suggests that elevation has an important contribution in discriminating land cover/use categories. In the fourth scenario (Figure 4d), 12 features, selected from across the 1st, 2nd, and 3rd scenarios, were combined to optimize the features and therefore ensure an increase in model performance. In this scenario, elevation is the most important feature and therefore has the greatest contribution to the classification, followed by the blue band > VH > NIR_8a > SWIR1 > NIR_8 > SWIR2 > green > red and then the red edge band as the least contributors in that order (Figure 4d).

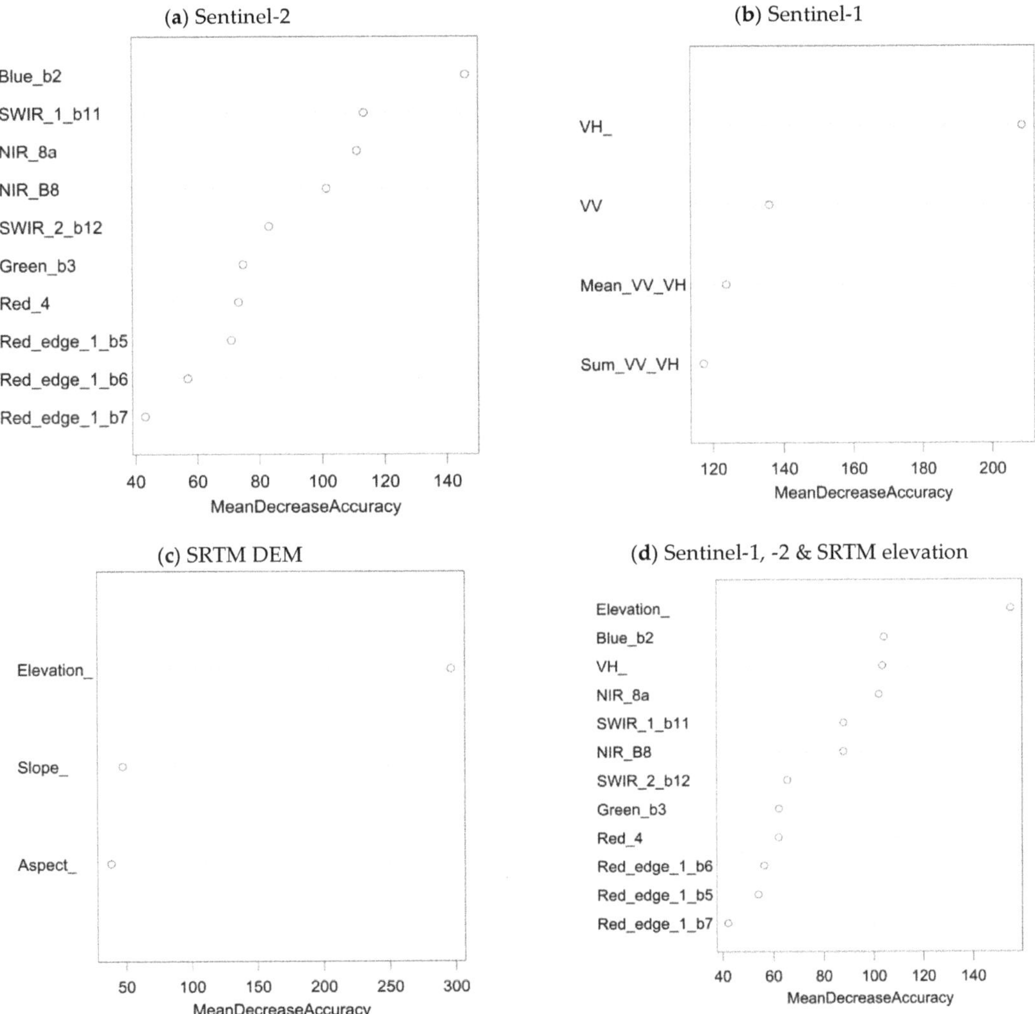

Figure 4. The important contribution of the RF feature importance selection-based MDA, (**a**) Scenario 1 (Sentinel-2 only), (**b**) Scenario 2 (Sentinel-1 only), (**c**) Scenario 3 (SRTM digital elevation model data only), (**d**) Scenario 4 (Sentinel-2 bands, Sentinel-1 (VH backscatter) and SRTM elevation data).

3.2. Out-Of-Bag Error Estimates

Here, the study presented only the two most accurate classifications (based on Sentinel-2 data and based on a combination of Sentinel-1 VH, Sentinel-2 bands, and SRTM elevation data) based on the most important features of the four classification scenarios explained above. As earlier stated, the purpose of different scenarios implemented in this study was to find out the most important feature for model optimization.

3.2.1. Sentinel-2 (Scenario 1) Classification Results

The overall OOB error estimates show that Sentinel-2 bands have an overall accuracy of 84.2%, an OOB error rate of 15.8%, and k = 0.4 (Table 2). Going by the overall accuracy, one can infer that the classification results for these data are highly accurate. However, as expected, and as it is most common to many classification results, there are omission as well as commission errors in the classification results. The RF model provides an error rate for each class of the land cover/use category. Irrigated land has the least class error (5.3%), while grassland has the highest (21%), followed by tree/shrubland (18.30%) and then farmlands (17.49%) (Table 2). This means that there is a probability that pixels classified in these categories may not be the actual land cover on the ground. For example, the spectral signatures of farmlands resembled that of grassland and farmland. This led to confusion and misclassifications of these land categories. The misclassification is confirmed, given that these classes recorded the highest errors. The river and the wetland were overestimated, given the spatial resolution of the datasets. Barren and built-up areas confuse each other with barren land in the floodplain region classified as built-up owing to their spectral similarity.

Table 2. OOB confusion matrix for Sentinel-2 (Scenario 1) classification results. Overall error rate = 15.8%, Overall accuracy = 84.2%, κ = 0.38.

LULC Category	1	2	3	4	5	6	7	8	Row Total	Class Error (%)
River	1145	71	46	11	20	0	0	0	1293	11.45
wetland	84	776	1	6	11	0	1	0	879	11.72
irrigated land	9	0	443	1	0	14	0	1	468	5.34
Built-up	1	3	7	1263	29	0	20	0	1323	4.54
Barren	5	2	0	135	7516	374	187	9	8228	8.65
Tree/shrubland	0	0	108	2	56	1978	208	69	2421	18.30
Farmland	0	2	57	425	581	3702	32,745	2173	39,685	17.49
Grassland	2	0	3	0	0	15	21	147	188	21.81
Column total	1246	854	665	1843	8213	6083	33,182	2399	54,485	

3.2.2. Sentinel-1 (Scenario 2) Classification Results

The overall OOB error estimates show that Sentinel 1 backscatter has an overall accuracy of 83%, an OOB error rate of 17%, and k = 0.22 (Table 3). Going by the overall accuracy, the result is encouraging. However, the RF's class error shows otherwise. Only farmland achieved a class error of less than 5%, while other classes recorded not less than 43%. The Sent1nel backscatter does not separate the different classes proposed in this study.

Table 3. OOB confusion matrix for Sentinel 1 (Scenario 2) classification results. Overall error rate = 17%, Overall accuracy = 83%, κ = 0.22.

LULC Category	1	2	3	4	5	6	7	8	Row Total	Class Error (%)
River	285	118	2	33	48	12	794	1	1293	77.95
wetland	80	496	0	56	7	0	239	0	878	43.50
irrigated land	0	1	109	2	67	5	283	1	468	76.70
Built-up	37	54	0	505	31	2	692	0	1321	61.77
Barren	11	5	14	16	5419	97	2663	3	8228	34.13
Tree/shrubland	8	0	5	3	222	703	1479	1	2421	70.96
Farmland	133	63	37	277	1034	299	37,826	15	39,685	4.68
Grassland	1	0	0	0	28	0	134	25	188	86.70
Column total	555	737	167	892	6856	1118	44,110	46	54,482	

3.2.3. SRTM Elevation Data (Scenario 3) Classification Results

The overall OOB error estimates show that SRTM data (elevation, aspect, and slope) have an overall accuracy of 76.44% and an OOB error rate of 23. 56%, and k = 0.10 (Table 4). Going by the overall accuracy, the result is encouraging. On the contrary, the RF's class error shows otherwise. Only farmland achieved a class error of less than 4%, while other classes recorded not less than 60%. The SRTM data do not separate the different LULC classes proposed in this study. However, it always shows a good result when it is combined with other multi-spectral and radar data.

Table 4. OOB confusion matrix for SRTM data (Scenario 2) classification results. Overall error rate = 23.56%, Overall accuracy = 76.44%, κ = 0.10.

LULC Category	1	2	3	4	5	6	7	8	Row Total	Class Error (%)
River	476	152	34	12	404	1	114	1	1194	60.13
wetland	259	169	52	8	202	0	121	0	811	79.16
irrigated land	75	51	80	6	130	1	96	0	439	81.78
Built-up	62	28	19	125	184	14	830	1	1263	90.10
Barren	92	73	35	24	1881	14	5363	0	7482	74.86
Tree/shrubland	1	5	2	8	109	67	2039	0	2231	97.00
Farmland	39	52	50	46	960	45	36,109	0	37,301	3.20
Grassland	3	0	0	3	11	1	160	0	178	99.00
Column total	1007	530	272	232	3881	143	44,832	2	50,899	

3.2.4. Sentinel-1, 2, VH Backscatter and SRTM Elevation Data (Scenario 4) Classification Results

Table 5 presents the confusion matrix for the classification results in scenario 4. Regarding Sentinel-2 classification results, a consistent pattern has been maintained by the combinations of Sentinel-1, -2, VH backscatter, and SRTM elevation data but with improvement in the classification accuracy (Table 5). The overall accuracy is 89.8% and a κ value of 0.4 (Table 5). This show an increase of 4.9% and 5.3% compared to Sentinel-2 for overall accuracy and kappa statistics, respectively. Similarly, grassland has the highest class error (18.09%), which is still 3% lower than that of Sentinel 2. Grassland was followed by wetland/flooded area (12.19%), farmland (11.8), and tree/shrubland (12.02%). The class error for tree/shrubland is the lowest for this scenario and is 6.28% lower than that obtained in scenario 2 (Table 5). Generally, the addition of the other two features (VH normalized backscatter and elevation data) has improved the overall accuracy of the classification.

Table 5. Out-of-bag confusion matrix of the Sentinel-1, -2, VH backscatter, and SRTM elevation data (Scenario 4) classification. Overall OOB error rate = 10.9%, Overall accuracy = 89.1%, κ = 0.4.

LULC Category	1	2	3	4	5	6	7	8	Row Total	Class Error (%)
River	1178	50	37	11	17	0	0	0	1293	8.89
wetland	88	771	0	1	17	0	1	0	878	12.19
irrigated land	7	0	457	4	0	0	0	0	468	2.35
Built-up	1	3	8	1276	14	0	19	0	1321	3.41
Barren	0	0	1	64	7617	297	242	7	8228	7.43
Tree/shrubland	0	0	60	2	41	2130	156	32	2421	12.02
Farmland	0	0	18	451	682	2143	34,965	1426	39,685	11.89
Grassland	2	0	2	0	0	12	18	154	188	18.09
Column total	1276	824	583	1809	8388	4582	35,401	1619	54,482	

3.3. Independent Validation

To complement the validation results obtained in an RF model (which uses 30% of the training datasets), another independent validation was carried out to compare the two scenarios (Tables 6 and 7).

Table 6. Confusion matrix for Sentinel-2 accuracy assessment (Scenario 2) classification. Overall error rate = 30.1%, Overall accuracy = 69.9%, κ = 0.66.

LULC Category	1	2	3	4	5	6	7	8	Row Total	Accuracy (%)	
										Producer's	User's
River	91	20	16	6	0	0	0	0	133	90.1	68.4
wetland	10	79	1	26	0	1	0	0	117	79	67.5
irrigated land	0	0	72	0	0	16	0	0	88	72	81.8
Built-up	0	0	3	46	2	0	1	6	58	46.9	79.3
Barren	0	0	0	5	90	7	4	6	112	90.9	80.4
Tree/shrubland	0	0	6	0	5	66	20	22	119	64.7	55.5
Farmland	0	1	0	15	2	5	75	25	123	75	61
Grassland	0	0	2	0	0	7	0	39	48	39.8	81.2
Column total	101	100	100	98	99	102	100	98	798		

Table 7. Confusion matrix for Sentinel 2, 1 (VH backscatter) and SRTM elevation data (Scenario 4) classification. Overall error rate = 24.8%, Overall accuracy = 75.2%, κ = 0.71.

LULC Category	1	2	3	4	5	6	7	8	Row Total	Accuracy (%)	
										Producer's	User's
River	93	13	13	6	0	0	0	0	125	92.1	74.4
Wetland	8	86	2	26	0	5	0	0	127	86	67.7
Irrigated land	0	0	75	0	0	7	0	0	82	75	91.5
Built-up	0	0	3	48	2	0	2	7	62	49	77.4
Barren	0	0	0	3	91	5	4	5	108	91.9	84.3
Tree/shrubland	0	0	3	13	5	78	13	12	124	76.5	62.9
Farmland	0	1	1	2	1	3	79	24	111	79	71.2
Grassland	0	0	3	0	0	4	2	50	59	51	84.7
Column total	101	100	100	98	99	102	100	98	798		

3.3.1. Sentinel-2 Accuracy Assessments (Scenario 1)

The confusion matrix for the Sentinel-2 data classification (Scenario 2) is presented in Table 6. In this validation, an equal-size stratified random sampling was used for the selection of validation datasets (800 points, 100 points each for the eight land cover/use categories) were used. The study reports an overall accuracy of 69.9%, an error rate of error, and a κ value of 0.66. Except for grassland and built-up area, all LULC categories achieved producer and user accuracy of more than 55%.

3.3.2. Sentinel-1, -2, VH Backscatter and SRTM Elevation Data (Scenario 4) Accuracy Assessment

Table 7 indicates the accuracy assessment of the RF classification results conducted based on an independent validation for Sentinel-1 and -2, VH backscatter, and SRTM elevation data. In this scenario, the study observed an overall accuracy of 75.2% and a k-value of 0.71. The study noticed an improvement in terms of model performance compared to when Sentinel-2 only was used. In this scenario, all classes recorded the user's accuracy of at least 62%.

3.4. LULC Maps and Area Covered by Each LULC Category

Table 8 shows the area proportion as extracted from the LULC maps obtained from the RF classifications for Sentinel-2 only (Figure 5a) and the combination of Sentinel-1, -2, VH backscatter, and the SRTM elevation (Figure 5b), which are presented in Figure 5a,e, the two most accurate LULC maps. Quantitatively, it is obvious that cultivated areas dominate the landscape, with farmland occupying close to 3000 km^2 of the land. On the other hand, wetlands, rivers, and grassland constitute a smaller proportion. The maps show the types of LULC categories that exist in the area. Visually, the maps show that cropland (upland agriculture) predominates in the area. Despite the predominance of upland agricultural land use, the RF model's ability to discern across classes makes the maps even more intriguing. LULC categories like river, wetland/flooded, irrigated land, and grassland are relatively modest in comparison to other LULC categories, but the amount of specific information that comes from the classification is detailed and relatively accurate. The floodplain area was clearly distinguished from the LULC in upland areas. This has been achieved in both scenarios. Figure 5c,e shows a full-resolution comparison between the two maps based on Sentinel-2 RGB color composite. In comparison to RGB, there is a clear difference between how the two scenarios classified the LULC classes. Sentinel-2 only seems to have observed more barren land than the combined datasets (Figure 5c,f). On the other hand, the map produced from the combined datasets shows a more vegetated area. These differences occur as a result of variations in the spectral reflectance signature of the land categories. But the use of sentinel backscatter and elevation data has helped to adjust the confusion between classes, which led to improved classification performance.

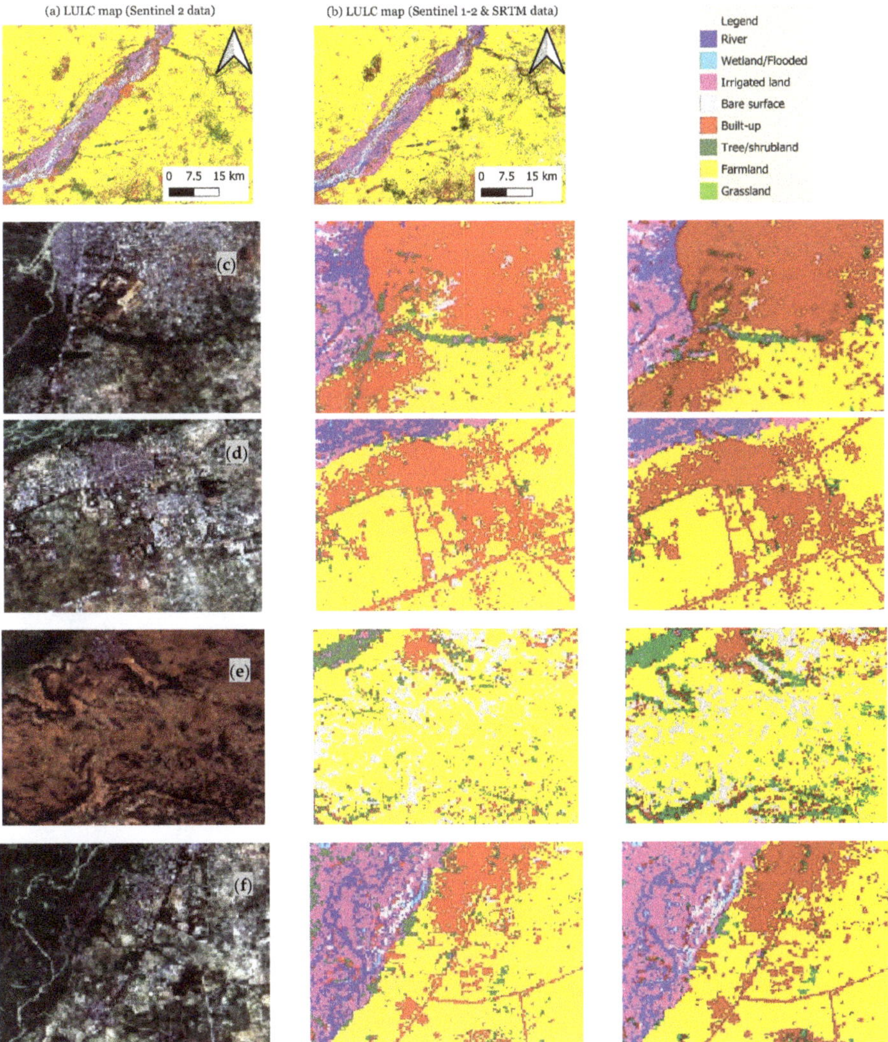

Figure 5. LULC maps (**a**) LULC map derived from Scenario 1 (Sentinel-2) (**b**) Scenario 4 (LULC map derived from Sentinel-1, -2 and SRTM elevation data), (**c–f**) four areas extracted from the RGB composite of Sentinel-2 and their corresponding locations in the two maps presented in (**a**) and (**b**).

Table 8. Area proportion (km^2) for each LULC category.

LULC Category	Sentinel-2	Sentinel-1, -2 & SRTM Elevation
River	58.42	55.38
wetland	22.91	25.34
irrigated land	140.07	183.54
Built-up	110.67	142.28
Barren	481.55	425.56
Tree/shrubland	305.80	310.62
Farmland	2990.68	2964.54
Grassland	3.63	2.27

4. Discussion

This study reports on the production of LULC maps based on random feature selection to evaluate its application in an agriculturally dominated landscape in Nigeria. The potential of using Sentinel-1 optical data, Copernicus Sentinel-1 c radar backscatter, and SRTM topographic variables were investigated to ascertain whether this synergy could improve classification accuracy. The general findings that emerged from this study suggest that: (1) the application of RF classification appears promising in this ecosystem; (2) the use of multiple remote sensing and environmental variables is an important contribution to quantitative remote sensing applications; (3) feature selection methods can improve classification accuracy; however, the evaluation of classification accuracy requires a thorough and critical assessments.

The mapping performed in this study was guided by the RF feature selection procedure based on the ranking of MDA as a function of OOB error estimates. The contribution of each satellite band varies. Some bands make a better contribution than others. What makes these results interesting is the procedure used to test each data set individually and then in combinations. Interestingly, the most important bands also provide the largest spectral differences between classes, except for the normalized backscatter polarizations, where the spread between classes is not very large, but this is similar behavior observed for the Sentinel-2 blue band (Figure 3a,b). Among the Sentinel-2 data variables, the blue, SWIR1, and NIR bands were found to be the most important variables (Figure 4a). Similar behavior for the SWIR1 and blue spectral bands of Sentinel-2 has been observed in previous vegetation, tree species, and crop mapping studies [54,55]. ED Chaves et al. [56] have explained that Sentinel's two SWIR bands are very sensitive to chlorophyll content, which allows them to distinguish different vegetation types and determine classification accuracy for LULC. ED Chaves, CA Picoli, and D. Sanches [56] stated that Sentinel's two SWIR bands are very sensitive to chlorophyll content, allowing them to distinguish different vegetation types and determine classification accuracy for LULC. In addition, visible and shortwave infrared wavelengths are known for their spectral variations, which can explain variations caused by chlorophyll content, soil type, and soil color [57].

Using the normalized backscatter, the VH polarization has the highest rank, which is due to the combinations of the different polarizations (Figure 4b). For the topographic SRTM variables, the elevation data had the highest rank (Figure 4c). The stand-alone classification results for the Sentinel-1 data (Table 3), as well as for the topographic SRTM variables (Table 4), achieved very low accuracy compared to the Sentinel-2 data (Table 2). Therefore, the synergy between VH, elevation data, and Sentinel spectral bands was evaluated to see if the accuracy of the model could be improved. The ranking of the most important variables shows that elevation, blue band, VH, NIR8a, and SWIR1 are the five most important variables (Figure 4d). Elevation makes the largest contribution to the classification. These results are consistent with a recent study by Zhao, Zhu, Wei, Fang, Zhang, Yan, Liu, Zhao, and Wu [57], the only difference being that they do not include radar backscatter as one of their input variables. This study highlights the importance of altitude and radar backscatter data with Sentinel-2 data to improve the classification accuracy of LULC.

The accuracy of the classified maps in this study suggests that it is reasonable to use different remote sensing data for LULC, as has been done in many previous studies. Based on the OOB error estimates, two scenarios were considered the most important, so the comparison is limited to these. The overall OOB classification results for Sentinel-2 data show an overall accuracy of 84.2% and a κ of 0.38, with the lowest and highest class errors for classification at 4.54% and 21.81% for built-up areas and grassland, respectively (Table 2). This level of accuracy is achieved by the Sentinel-2 data alone, further emphasizing their applicability in LULC mapping in this particular ecosystem. However, when the SRTM elevation data and VH backscatter were added to the Sentinel-2 spectral bands, the overall accuracy was 89.1%, and the κ value was 0.4, an increase of 4.9% in overall accuracy (Table 3) compared to the Sentinel-2 data alone. The lowest classification error was found in the irrigated areas, with only 2.15%, while the highest error occurred in the grassland areas (18.09%), which in this case were reduced by 3.72%. The cultivated areas had a class error of 3.41%, which is a further reduction of 1.13% compared to the Sentinel-2 data. For trees and shrubland, the study found a 6.28% difference between the sentinel data and their combination with elevation and VH backscatter data.

This is not independent of the role of the elevation and backscatter data in the overall performance of the model. The topography of the area is heterogeneous, and some of the classes are located in the floodplain, which is typically undulating compared to developed and agricultural areas. Several studies have shown the importance of elevation data to increase the accuracy of the classified map [11,26,40,58]. In the same vein, radar backscatter was found to improve model performance because it can normalize or reduce the effects of the atmosphere, topography, instrument noise, etc., to provide consistent spatial and temporal comparisons [59]. The results from this study are consistent with Meneghini [60], who evaluates the synergy between the Sentinel-1 and Sentinel-2 data for land cover classification. Their results show an overall accuracy of 74% and 78% for Sentinel-2 (Only) and in combination with Sentinel-1 data, respectively. Similarly, several studies have reported the importance of synergy between sentinel-1 and -2 data for increasing model performance for biomass estimation [61], crop type classification [62], irrigation mapping [63], and land cover mapping [64,65].

It has been observed that in a setting in which there is a strong interest in predicting observations from the smaller classes, sampling the same number of observations from each class for validation is a promising alternative [53]. Moreover, one of the objectives of this study was to compare the validation of OOB error estimates of the RF normally performed internally by the model with another independent validation (external) which was performed based on equal-size random stratified sampling using 100 polygons for each LULC category. The overall accuracy of the classification results were 69.9% and 75.2% for Sentinel's 2 data only and the combination of the same data with VH backscatter and elevation data, respectively. The difference between the two is 5.3% which conformed to the OOB estimates of errors even though the overall accuracy obtained from the OOB is higher. The consistency of these two validation results manifested even within the class error. Similar to OOB estimates of error, grassland had the lowest producer's accuracy with an 11.2% difference between the Sentinel's data only and in combination with VH and elevation data based on the independent validation. In this context, the estimates from the OOB are, therefore, reliable since the two validation results have maintained a consistent pattern. The only difference between the two is in kappa statistics, where the external validation shows higher kappa ($k = 0.71$, Table 7) than the estimates from the RF internal validation ($k = 0.4$, Table 3). This is one of the advantages of a balanced setting for applying the equalized stratified random sampling for validation [66], but balancing may not always be possible due to costs or other reasons [4]. But kappa is not a measure of accuracy but of agreement beyond chance, and chance correction is rarely needed [67,68]. The comparison results obtained in this study are consistent with findings by Adelabu et al. [69], who tested the reliability of the internal accuracy assessments of the RF for classifying tree defoliation levels using different validation methods. One of the most important deductions that can be made in this context is that where only the RF approach is applied to the LULC classification, independent validation is not necessary because validation requires a large number of points, and therefore manual class labeling based on external validation is tedious and time-consuming. The findings of this current study provide insights into the reliability and applicability of OOB error estimates.

One of the limitations of this study is the lack of reference ground truth datasets from a field campaign. Although this study relied on RGB composite images and Google Earth data for the selection of training and validation datasets, it should be noted that such datasets are well-acknowledged as a source of training and validation for land cover mapping [70,71]. Furthermore, a comparison of the quantitative and qualitative results showed that the LULC categories are detailed and very accurate (Tables 2, 3 and 6–8 and Figures 4 and 5). The area estimated from the two most accurate results shows that there is extensive agricultural land. The two maps show slight differences for the area of different LULC categories. The study, however, acknowledged the confusion between the barren land and the built-up areas, which occurred primarily due to the presence of settlements in or near the floodplain areas, in addition to the similarity of the spectral reflectance signatures of these LULC classes. Moreover, the difference between the spectral reflectance signatures between the barren land on the upland and in the floodplain probably led to the underestimation of barren land in the upland areas. However, the class error for barren is minimal, as observed for the RF internal validation (7.43%) as well as for independent validation (producer's accuracy = 91.9% and user's accuracy = 84.3%). From these results, it is obvious that further research in this particular ecosystem may require the need to incorporate vegetation (e.g., NDVI), bare soil indices (e.g., modified normalized difference bare-land index), and water indices (e.g., Modified Normalized Difference Water Index) to improve classification performance. The study also noted confusion between the river network and wetlands. Earlier reports indicated that significant flooding occurred in the area on October 1 [72,73]. At this time, the volume of rivers usually increases, and flooding is easily possible when the amount of rainfall is significant, and the dams along these rivers have been opened. These floods have left many people homeless and severely damaged agricultural land and crops. Future research could focus on flood vulnerability assessment based on change detection using sentinel data. In this situation, flood vulnerability mapping can provide critical information to assess flood risk in the region. Policymakers could be well informed about the risk and thus develop appropriate mitigation strategies based on the severity of the impacts [74,75].

Similarly, the study observed confusion between the grassland and farmland. Mapping LULC with Sentinel-2 data in the semi-arid region is quite promising [34] but challenging because most crops are planted during the rainy season, and their growing season is in July and August, during which the cloud cover is high in the area. And the reliance on dry season imagery may not be feasible as there is a transition from cropland to barren land in the area, especially from early November. Since cropland makes up most of the LULC in the area, this is not the most appropriate time for LULC mapping. This study minimized this problem by integrating Sentinel-1 and -2 data in early and mid-October. Van Tricht, Gobin, Gilliams, and Piccard [63] demonstrate the importance of choosing phenological cycles for crop mapping based on the synergy between the sentinel-1 and -2 data using an RF classifier for increasing model performance. Similarly, many studies demonstrated the importance of Sentinel-1 and -2 for rice mapping in a lowland area [76], mapping paddy rice [77], and mapping Maize Areas in heterogeneous agriculture [78] based on RF. By understanding this trade-off, the current study can help in the selection of datasets and periods for LULC classification with specific applications to agricultural landscapes in semi-arid regions. Although cloud cover may result in a lack of cloud-free imagery in this region, a potential area for further research would be to examine crop and vegetation phenological cycles and by incorporating more variables from the Sentinel-1 data during the rainy season to minimize the challenge of cloud cover.

5. Conclusions

This paper proposed LULC mapping by applying an RF classifier to Sentinel-1, -2, and SRTM digital elevation data to evaluate its applicability based on local class definitions and training datasets in an agricultural landscape in Nigeria. The main objective was to develop a methodology to improve model performance and reduce prediction error in LULC classifications. A feature selection method (RF) was implemented to evaluate the contribution of individual bands based on a standard OOB error estimate (MDA). The study showed that the combination of spectral bands, backscatter, and topographic features could improve classification accuracy. The results show that among the variables in the sentinel-2 data, the blue, SWIR1, and NIR bands are the most important variables. Using the normalized backscatter, the VH polarization has the highest rank, which is due to the combination of the different polarizations. For the SRTM topographic variables, the elevation data had the highest rank. The ranking of the most important variables when combining the different data sets shows that height, blue band, VH backscatter, NIR8a, and SWIR1 are the five most important variables.

The overall OOB classification results for Sentinel-2 data show an overall accuracy of 84.2%, with the lowest and highest class errors for classification of 4.54% and 21.81% for built-up areas and grassland, respectively. This level of accuracy is achieved by the Sentinel-2 data alone (scenario 1), further emphasizing its applicability in LULC mapping in this particular ecosystem. On the other hand, the class errors for Sentinel-1 (scenario 2) and SRTM data (scenario 3) show high-class errors. However, when the Sentinel-1, -2, and SRTM elevation data were added to the model, the overall accuracy was 89.1%. This represents a 4.9% improvement in overall accuracy compared to Sentinel-2 alone and a 6.1% and 12.66% improvement compared to Sentinel-1 and SRTM data, respectively. The lowest classification error was found in the irrigated areas at only 2.15%. In comparison, the highest error occurred in the grassland areas (18.09%), which in this case were reduced by 3.72% compared to the Sentinel data alone. According to the study, there was a 6.28% difference between sentinel data and their combination with elevation and VH backscatter data for trees and shrubland. The results of an independent validation based on an equal-size random sampling of 800 points are consistent with OOB error estimates. The study shows how the synergy of optical, radar, and elevation data can significantly improve LULC map accuracy. Based on these results, LULC maps could be used in a broad range of spatial planning applications.

Funding: This research was funded by the Tertiary Education Trust Fund (TETFUND) through the Institution Based Research (IBR).

Data Availability Statement: Not applicable.

Acknowledgments: TETFUND's (IBR) support is gratefully acknowledged. I would also like to give a special thanks to the R Core team for accessing various library packages for data analyses.

Conflicts of Interest: The author declares no conflict of interest.

References

1. Reidsma, P.; Tekelenburg, T.; Berg, M.V.D.; Alkemade, R. Impacts of land-use change on biodiversity: An assessment of agricultural biodiversity in the European Union. *Agric. Ecosyst. Environ.* **2006**, *114*, 86–102. [CrossRef]
2. Newbold, T. Future effects of climate and land-use change on terrestrial vertebrate community diversity under different scenarios. *Proc. R. Soc. B Boil. Sci.* **2018**, *285*, 20180792. [CrossRef] [PubMed]
3. Mahmood, R.; Pielke, R.A.; Hubbard, K.; Niyogi, D.; Bonan, G.; Lawrence, P.; McNider, R.; McAlpine, C.; Etter, A.; Gameda, S.; et al. Impacts of Land Use/Land Cover Change on Climate and Future Research Priorities. *Bull. Am. Me-Teorol. Soc.* **2010**, *91*, 37–46. [CrossRef]
4. Preidl, S.; Lange, M.; Doktor, D. Introducing APiC for regionalised land cover mapping on the national scale using Sentinel-2A imagery. *Remote Sens. Environ.* **2020**, *240*, 111673. [CrossRef]
5. DeFries, R.S.; Rudel, T.; Uriarte, M.; Hansen, M.C. Deforestation driven by urban population growth and agricultural trade in the twenty-first century. *Nat. Geosci.* **2010**, *3*, 178–181. [CrossRef]
6. UN. Envision2030 Goal 15: Life on Land. United Nation. Available online: https://www.un.org/development/desa/disabilities/envision2030-goal15.html (accessed on 20 March 2021).

7. Haines-Young, R. Environmental Accounts for Land Cover: Their Contribution to 'State of the Environment' Reporting. *Trans. Inst. Br. Geogr.* **1999**, *24*, 441–456. [CrossRef]

8. Pérez-Soba, M.; Petit, S.; Jones, L.; Bertrand, N.; Briquel, V.; Omodei-Zorini, L.; Contini, C.; Helming, K.; Farrington, J.H.; Mossello, M.T.; et al. Land Use Functions—A Multifunctionality Approach to Assess the Impact of Land Use Changes on Land Use Sustainability. In *Sustainability Impact Assessment of Land Use Changes*; Springer: Berlin/Heidelberg, Germany, 2008; pp. 375–404.

9. Falcucci, A.; Maiorano, L.; Boitani, L. Changes in land-use/land-cover patterns in Italy and their implications for biodiversity conservation. *Landsc. Ecol.* **2006**, *22*, 617–631. [CrossRef]

10. Tian, Y.; Yue, T.; Zhu, L.; Clinton, N. Modeling population density using land cover data. *Ecol. Model.* **2005**, *189*, 72–88. [CrossRef]

11. Balzter, H.; Cole, B.; Thiel, C.; Schmullius, C. Mapping CORINE Land Cover from Sentinel-1A SAR and SRTM Digital Elevation Model Data Using Random Forests. *Remote Sens.* **2015**, *7*, 14876–14898. [CrossRef]

12. Lambin, E.; Rounsevell, M.; Geist, H. Are agricultural land-use models able to predict changes in land-use intensity? *Agric. Ecosyst. Environ.* **2000**, *82*, 321–331. [CrossRef]

13. Richter, R.; Weingart, U.; Wever, T.; Kahny, U. Urban Land Use Data for the Telecommunications Industry. *Photogramm. Fernerkund. Geoinf.* **2006**, *4*, 297.

14. Peter, N. The use of remote sensing to support the application of multilateral environmental agreements. *Space Policy* **2004**, *20*, 189–195. [CrossRef]

15. Hong, W.; Jiang, R.; Yang, C.; Zhang, F.; Su, M.; Liao, Q. Establishing an ecological vulnerability assessment indicator system for spatial recognition and management of ecologically vulnerable areas in highly urbanized regions: A case study of Shenzhen, China. *Ecol. Indic.* **2016**, *69*, 540–547. [CrossRef]

16. Congalton, R.G.; Gu, J.; Yadav, K.; Thenkabail, P.; Ozdogan, M. Global Land Cover Mapping: A Review and Uncertainty Analysis. *Remote Sens.* **2014**, *6*, 12070–12093. [CrossRef]

17. National Environmental Standards and Regulations Enforcement Agency. *National-Policy-on-Environment*; NESREA: Abuja, Nigeria, 2017.

18. Phan, T.N.; Kuch, V.; Lehnert, L.W. Land Cover Classification using Google Earth Engine and Random Forest Classifier—The Role of Image Composition. *Remote Sens.* **2020**, *12*, 2411. [CrossRef]

19. Waske, B.; Braun, M. Classifier ensembles for land cover mapping using multitemporal SAR imagery. *ISPRS J. Photogramm. Remote Sens.* **2009**, *64*, 450–457. [CrossRef]

20. Talukdar, S.; Singha, P.; Mahato, S.; Shahfahad; Pal, S.; Liou, Y.-A.; Rahman, A. Land-Use Land-Cover Classification by Machine Learning Classifiers for Satellite Observations—A Review. *Remote Sens.* **2020**, *12*, 1135. [CrossRef]

21. Adugna, T.; Xu, W.; Fan, J. Comparison of Random Forest and Support Vector Machine Classifiers for Regional Land Cover Mapping Using Coarse Resolution FY-3C Images. *Remote Sens.* **2022**, *14*, 574. [CrossRef]

22. Li, Y.; Li, C.; Li, M.; Liu, Z. Influence of Variable Selection and Forest Type on Forest Aboveground Biomass Estimation Using Machine Learning Algorithms. *Forests* **2019**, *10*, 1073. [CrossRef]

23. Maxwell, A.E.; Warner, T.A.; Fang, F. Implementation of machine-learning classification in remote sensing: An applied review. *Int. J. Remote Sens.* **2018**, *39*, 2784–2817. [CrossRef]

24. Breiman, L. Random Forests. *Mach. Learn.* **2001**, *45*, 5–32. [CrossRef]

25. Zhang, F.; Yang, X. Improving land cover classification in an urbanized coastal area by random forests: The role of variable selection. *Remote Sens. Environ.* **2020**, *251*, 112105. [CrossRef]

26. Monsalve-Tellez, J.M.; Torres-León, J.L.; Garcés-Gómez, Y.A. Evaluation of SAR and Optical Image Fusion Methods in Oil Palm Crop Cover Classification Using the Random Forest Algorithm. *Agriculture* **2022**, *12*, 955. [CrossRef]

27. Keay, R.W.J. An Example of Sudan Zone Vegetation in Nigeria. *J. Ecol.* **1949**, *37*, 335. [CrossRef]

28. Ibrahim, S.; Kaduk, J.; Tansey, K.; Balzter, H.; Lawal, U.M. Detecting phenological changes in plant functional types over West African savannah dominated landscape. *Int. J. Remote Sens.* **2020**, *42*, 567–594. [CrossRef]

29. Kamba, A.A.; Yelwa, J.F.; Ojomugbokenyode, I.E.; Yaji, I.L. Analysis of the Perceived Effects of Climate Change on Crop Production among Farmers of Argungu Zone of Kebbi State Agricultural Development Programme—Adp. *Int. J. Agric. Ext.* **2022**, *10*, 315–324. [CrossRef]

30. Adelana, S.M.A.; Olasehinde, P.I.; Bale, R.B.; Vrbka, P.; Edet, A.E.; Goni, I.B. An overview of the geology and hydrogeology of Nigeria. In *Applied Groundwater Studies in Africa*; CRC Press: Boca Raton, FL, USA, 2008; pp. 181–208.

31. Sedano, F.; Molini, V.; Azad, M.A.K. A Mapping Framework to Characterize Land Use in the Sudan-Sahel Region from Dense Stacks of Landsat Data. *Remote Sens.* **2019**, *11*, 648. [CrossRef]

32. Yuan, F.; Repse, M.; Leith, A.; Rosenqvist, A.; Milcinski, G.; Moghaddam, N.F.; Dhar, T.; Burton, C.; Hall, L.; Jorand, C.; et al. An Operational Analysis Ready Radar Backscatter Dataset for the African Continent. *Remote Sens.* **2022**, *14*, 351. [CrossRef]

33. Schulz, D.; Yin, H.; Tischbein, B.; Verleysdonk, S.; Adamou, R.; Kumar, N. Land use mapping using Sentinel-1 and Sentinel-2 time series in a heterogeneous landscape in Niger, Sahel. *ISPRS J. Photogramm. Remote Sens.* **2021**, *178*, 97–111. [CrossRef]

34. Abida, K.; Barbouchi, M.; Boudabbous, K.; Toukabri, W.; Saad, K.; Bousnina, H.; Chahed, T.S. Sentinel-2 Data for Land Use Mapping: Comparing Different Supervised Classifications in Semi-Arid Areas. *Agriculture* **2022**, *12*, 1429. [CrossRef]

35. Lillesand, T.; Kiefer, R.W.; Chipman, J. *Remote Sensing and Image Interpretation*, 5th ed.; John Willey and Sons: New York, NY, USA, 2008.

36. Zheng, H.; Du, P.; Chen, J.; Xia, J.; Li, E.; Xu, Z.; Li, X.; Yokoya, N. Performance Evaluation of Downscaling Sentinel-2 Imagery for Land Use and Land Cover Classification by Spectral-Spatial Features. *Remote Sens.* **2017**, *9*, 1274. [CrossRef]

37. Huete, A.R. 11—Remote Sensing for Environmental Monitoring. In *Environmental Monitoring and Characterization*; Artiola, J.F., Pepper, I.L., Brusseau, M.L., Eds.; Academic Press: Burlington, NJ, USA, 2004; pp. 183–206.

38. Xun, L.; Zhang, J.; Cao, D.; Yang, S.; Yao, F. A novel cotton mapping index combining Sentinel-1 SAR and Sentinel-2 multispectral imagery. *ISPRS J. Photogramm. Remote Sens.* **2021**, *181*, 148–166. [CrossRef]

39. Genuer, R.; Poggi, J.-M.; Tuleau-Malot, C. Variable selection using random forests. *Pattern Recognit. Lett.* **2010**, *31*, 2225–2236. [CrossRef]

40. Rodriguez-Galiano, V.F.; Ghimire, B.; Rogan, J.; Chica-Olmo, M.; Rigol-Sanchez, J.P. An assessment of the effectiveness of a random forest classifier for land-cover classification. *ISPRS J. Photogramm. Remote Sens.* **2012**, *67*, 93–104. [CrossRef]

41. Pelletier, C.; Valero, S.; Inglada, J.; Champion, N.; Dedieu, G. Assessing the robustness of Random Forests to map land cover with high resolution satellite image time series over large areas. *Remote Sens. Environ.* **2016**, *187*, 156–168. [CrossRef]

42. Liaw, A.; Wiener, M. Classification and Regression by Randomforest. *R News* **2002**, *2*, 18–22.

43. Hijmans, R.J.; Van Etten, J.; Mattiuzzi, M.; Sumner, M.; Greenberg, J.A.; Lamigueiro, O.P.; Bevan, A.; Racine, E.B.; Shortridge, A. Raster Package in R. Version. 2013. Available online: https://mirrors.sjtug.sjtu.edu.cn/cran/web/packages/raster/raster.pdf (accessed on 18 May 2022).

44. Pebesma, E.; Bivand, R.S. Classes and Methods for Spatial Data: The Sp Package. 2005. Available online: http://cran.nexr.com/web/packages/sp/index.html (accessed on 27 March 2022).

45. Bivand, R.; Keitt, T.; Rowlingson, B.; Pebesma, E.; Sumner, M.; Hijmans, R.; Rouault, E.; Bivand, M.R. Package 'Rgdal'. Bindings for the Geospatial Data Abstraction Library. Available online: https://cran.r-project.org/web/packages/rgdal/index.html (accessed on 15 October 2017).

46. Pebesma, E.J. Simple Features for R: Standardized Support for Spatial Vector Data. *R J.* **2018**, *10*, 439–446. [CrossRef]

47. Pebesma, E. The Meuse Data Set: A Brief Tutorial for the Gstat R Package. ViennaR. 2022. Available online: https://cran.r-project.org/web/packages/gstat/index.html (accessed on 20 August 2022).

48. Shao, G.; Tang, L.; Liao, J. Overselling overall map accuracy misinforms about research reliability. *Landsc. Ecol.* **2019**, *34*, 2487–2492. [CrossRef]

49. Mutanga, O.; Adam, E. High Density Biomass Estimation: Testing the Utility of Vegetation Indices and the Random Forest Regression Algorithm. In Proceedings of the 34th International Symposium for Remote Sensing of the Environment (ISRSE), Sydney, Australia, 10–15 April 2011.

50. Breiman, L.; Cutler, A. State of the Art of Data Mining Using Random Forest. In Proceedings of the Salford Data Mining Conference, San Diego, CA, USA, 21–25 May 2012.

51. Mitchell, M. Bias of the Random Forest out-of-Bag (Oob) Error for Certain Input Parameters. *Open J. Stat.* **2011**, *1*, 205–211. [CrossRef]

52. Tian, Z.; Liu, F.; Liang, Y.; Zhu, X. Mapping soil erodibility in southeast China at 250 m resolution: Using environmental variables and random forest regression with limited samples. *Int. Soil Water Conserv. Res.* **2021**, *10*, 62–74. [CrossRef]

53. Janitza, S.; Hornung, R. On the overestimation of random forest's out-of-bag error. *PLoS ONE* **2018**, *13*, e0201904. [CrossRef] [PubMed]

54. Mohammadpour, P.; Viegas, D.X.; Viegas, C. Vegetation Mapping with Random Forest Using Sentinel 2 and GLCM Texture Feature—A Case Study for Lousã Region, Portugal. *Remote Sens.* **2022**, *14*, 4585. [CrossRef]

55. Immitzer, M.; Vuolo, F.; Atzberger, C. First Experience with Sentinel-2 Data for Crop and Tree Species Classifications in Central Europe. *Remote Sens.* **2016**, *8*, 166. [CrossRef]

56. Chaves, M.E.D.; Picoli, M.C.A.; Sanches, I.D. Sanches. Recent Applications of Landsat 8/Oli and Sentinel-2/Msi for Land Use and Land Cover Mapping: A Systematic Review. *Remote Sens.* **2020**, *12*, 3062. [CrossRef]

57. Zhao, Y.; Zhu, W.; Wei, P.; Fang, P.; Zhang, X.; Yan, N.; Liu, W.; Zhao, H.; Wu, Q. Classification of Zambian grasslands using random forest feature importance selection during the optimal phenological period. *Ecol. Indic.* **2022**, *135*, 108529. [CrossRef]

58. Cui, J.; Zhu, M.; Liang, Y.; Qin, G.; Li, J.; Liu, Y. Land Use/Land Cover Change and Their Driving Factors in the Yellow River Basin of Shandong Province Based on Google Earth Engine from 2000 to 2020. *ISPRS Int. J. Geo-Inf.* **2022**, *11*, 163. [CrossRef]

59. Mladenova, I.E.; Jackson, T.J.; Bindlish, R.; Hensley, S. Incidence angle normalization of radar backscatter data. *IEEE Trans. Geosci. Remote Sens.* **2012**, *51*, 1791–1804. [CrossRef]

60. Meneghini, A. *An Evaluation of Sentinel-1 and Sentinel-2 for Land Cover Classification*; Clark University: Worcester, MA, USA, 2019.

61. Malhi, R.K.M.; Anand, A.; Srivastava, P.K.; Chaudhary, S.K.; Pandey, M.K.; Behera, M.D.; Kumar, A.; Singh, P.; Kiran, G.S. Synergistic evaluation of Sentinel 1 and 2 for biomass estimation in a tropical forest of India. *Adv. Space Res.* **2022**, *69*, 1752–1767. [CrossRef]

62. Orynbaikyzy, A.; Gessner, U.; Conrad, C. Spatial Transferability of Random Forest Models for Crop Type Classification Using Sentinel-1 and Sentinel-2. *Remote Sens.* **2022**, *14*, 1493. [CrossRef]

63. Van Tricht, K.; Gobin, A.; Gilliams, S.; Piccard, I. Synergistic Use of Radar Sentinel-1 and Optical Sentinel-2 Imagery for Crop Mapping: A Case Study for Belgium. *Remote Sens.* **2018**, *10*, 1642. [CrossRef]

64. Gargiulo, M.; Dell'Aglio, D.A.G.; Iodice, A.; Riccio, D.; Ruello, G. Integration of Sentinel-1 and Sentinel-2 Data for Land Cover Mapping Using W-Net. *Sensors* **2020**, *20*, 2969. [CrossRef] [PubMed]

65. Hu, B.; Xu, Y.; Huang, X.; Cheng, Q.; Ding, Q.; Bai, L.; Li, Y. Improving Urban Land Cover Classification with Combined Use of Sentinel-2 and Sentinel-1 Imagery. *ISPRS Int. J. Geo-Inf.* **2021**, *10*, 533. [CrossRef]

66. Stehman, S.V.; Foody, G. Accuracy Assesments. In *Sage Handbook of RemoteSensing*; Nellis, M.D., Warner, T.A., Foody, G.M., Eds.; SAGE: London, UK, 2009.
67. Pontius, R.G., Jr.; Millones, M. Death to Kappa: Birth of quantity disagreement and allocation disagreement for accuracy assessment. *Int. J. Remote Sens.* **2011**, *32*, 4407–4429. [CrossRef]
68. Foody, G.M. Explaining the unsuitability of the kappa coefficient in the assessment and comparison of the accuracy of thematic maps obtained by image classification. *Remote Sens. Environ.* **2020**, *239*, 111630. [CrossRef]
69. Adelabu, S.; Mutanga, O.; Adam, E. Testing the reliability and stability of the internal accuracy assessment of random forest for classifying tree defoliation levels using different validation methods. *Geocarto Int.* **2015**, *30*, 810–821. [CrossRef]
70. Fonte, C.C.; Bastin, L.; See, L.; Foody, G.; Lupia, F. Usability of Vgi for Validation of Land Cover Maps. *Int. J. Geogr. Inf. Sci.* **2015**, *29*, 1269–1291. [CrossRef]
71. Ali, U.; Esau, T.J.; Farooque, A.A.; Zaman, Q.U.; Abbas, F.; Bilodeau, M.F. Limiting the Collection of Ground Truth Data for Land Use and Land Cover Maps with Machine Learning Algorithms. *ISPRS Int. J. Geo-Inf.* **2022**, *11*, 333. [CrossRef]
72. Muktar, A.; Yelwa, S.A.; Bello, M.T.; Elekwachi, W. Geo-Spatial Study of Farmland Affected by 2020 Flooding of River Rima, Northwestern Nigeria. *Nov. Perspect. Eng. Res.* **2021**, *4*, 111–119.
73. Soni, D. *Inside Kebbi's Floods of Fury, Pains and Tears*; Vanguard: Valley Forge, PA, USA, 2020.
74. Le, T.D.H.; Pham, L.H.; Dinh, Q.T.; Nguyen, T.T.H.; Tran, T.A.T. Rapid method for yearly land-use and LULC classification using Random Forest and incorporating time-series NDVI and topography: A case study of Thanh Hoa province, Vietnam. *Geocarto Int.* **2022**, 1–19. [CrossRef]
75. Amani, M.; Kakooei, M.; Ghorbanian, A.; Warren, R.; Mahdavi, S.; Brisco, B.; Moghimi, A.; Bourgeau-Chavez, L.; Toure, S.; Paudel, A.; et al. Forty Years of Wetland Status and Trends Analyses in the Great Lakes Using Landsat Archive Imagery and Google Earth Engine. *Remote Sens.* **2022**, *14*, 3778. [CrossRef]
76. Fiorillo, E.; Di Giuseppe, E.; Fontanelli, G.; Maselli, F. Lowland Rice Mapping in Sédhiou Region (Senegal) Using Sentinel 1 and Sentinel 2 Data and Random Forest. *Remote Sens.* **2020**, *12*, 3403. [CrossRef]
77. Cai, Y.; Lin, H.; Zhang, M. Mapping Paddy Rice by the Object-Based Random Forest Method Using Time Series Sentinel-1/Sentinel-2 Data. *Adv. Space Res.* **2019**, *64*, 2233–2244. [CrossRef]
78. Chen, Y.; Hou, J.; Huang, C.; Zhang, Y.; Li, X. Mapping Maize Area in Heterogeneous Agricultural Landscape with Multi-Temporal Sentinel-1 and Sentinel-2 Images Based on Random Forest. *Remote Sens.* **2021**, *13*, 2988. [CrossRef]

Article

Monitoring of Soybean Maturity Using UAV Remote Sensing and Deep Learning

Shanxin Zhang [1], Hao Feng [1], Shaoyu Han [1,2], Zhengkai Shi [1], Haoran Xu [1], Yang Liu [2,3], Haikuan Feng [2,4], Chengquan Zhou [2,5] and Jibo Yue [1,*]

[1] College of Information and Management Science, Henan Agricultural University, Zhengzhou 450002, China
[2] Key Laboratory of Quantitative Remote Sensing in Agriculture of Ministry of Agriculture, Beijing Research Center for Information Technology in Agriculture, Beijing 100097, China
[3] Key Lab of Smart Agriculture System, Ministry of Education, China Agricultural University, Beijing 100083, China
[4] College of Agriculture, Nanjing Agricultural University, Nanjing 210095, China
[5] Institute of Agricultural Equipment, Zhejiang Academy of Agricultural Sciences (ZAAS), Hangzhou 310000, China
* Correspondence: yuejibo@henau.edu.cn

Abstract: Soybean breeders must develop early-maturing, standard, and late-maturing varieties for planting at different latitudes to ensure that soybean plants fully utilize solar radiation. Therefore, timely monitoring of soybean breeding line maturity is crucial for soybean harvesting management and yield measurement. Currently, the widely used deep learning models focus more on extracting deep image features, whereas shallow image feature information is ignored. In this study, we designed a new convolutional neural network (CNN) architecture, called DS-SoybeanNet, to improve the performance of unmanned aerial vehicle (UAV)-based soybean maturity information monitoring. DS-SoybeanNet can extract and utilize both shallow and deep image features. We used a high-definition digital camera on board a UAV to collect high-definition soybean canopy digital images. A total of 2662 soybean canopy digital images were obtained from two soybean breeding fields (fields F1 and F2). We compared the soybean maturity classification accuracies of (i) conventional machine learning methods (support vector machine (SVM) and random forest (RF)), (ii) current deep learning methods (InceptionResNetV2, MobileNetV2, and ResNet50), and (iii) our proposed DS-SoybeanNet method. Our results show the following: (1) The conventional machine learning methods (SVM and RF) had faster calculation times than the deep learning methods (InceptionResNetV2, MobileNetV2, and ResNet50) and our proposed DS-SoybeanNet method. For example, the computation speed of RF was 0.03 s per 1000 images. However, the conventional machine learning methods had lower overall accuracies (field F2: 63.37–65.38%) than the proposed DS-SoybeanNet (Field F2: 86.26%). (2) The performances of the current deep learning and conventional machine learning methods notably decreased when tested on a new dataset. For example, the overall accuracies of MobileNetV2 for fields F1 and F2 were 97.52% and 52.75%, respectively. (3) The proposed DS-SoybeanNet model can provide high-performance soybean maturity classification results. It showed a computation speed of 11.770 s per 1000 images and overall accuracies for fields F1 and F2 of 99.19% and 86.26%, respectively.

Keywords: unmanned aerial vehicle; soybean; convolutional neural network; deep learning

Citation: Zhang, S.; Feng, H.; Han, S.; Shi, Z.; Xu, H.; Liu, Y.; Feng, H.; Zhou, C.; Yue, J. Monitoring of Soybean Maturity Using UAV Remote Sensing and Deep Learning. *Agriculture* **2023**, *13*, 110. https://doi.org/10.3390/agriculture13010110

Academic Editor: Josef Eitzinger

Received: 22 November 2022
Revised: 27 December 2022
Accepted: 28 December 2022
Published: 30 December 2022

1. Introduction

Soybeans are a high-quality source of plant protein and raw materials for the production of hundreds of chemical products [1,2]. China's soybean-growing areas include the Northeast China Plain [3] and the North China Plain [4] (ranging from the north latitude of 30° to 48°). Soybean breeders must develop early-maturing, standard, and late-maturing varieties for planting at different latitudes to ensure that soybean plants fully utilize solar radiation. Therefore, timely and accurate monitoring of soybean breeding line maturity can facilitate soybean breeding decision-making and agricultural management [5–8].

Traditional methods for measuring field breeding line maturity are time-consuming and labor-intensive [7]. Meanwhile, the expertise and bias of the investigators can affect the accuracy of field surveys. Breeding fields have thousands of breeding lines with different maturation times. Manual surveys cannot quickly provide high-frequency breeding line maturity information to meet harvesting and yield measurement scheduling requirements. Unmanned aerial vehicle (UAV) remote sensing technology can be used to collect high-resolution crop canopy images and has thus been widely used in precision agricultural crop trait monitoring [9–12]. Compared with satellite and airborne remote sensing technologies, UAV remote sensing technology is relatively inexpensive and flexible in its operation, and it requires less space for landing and takeoff [13]. More importantly, the digital images obtained by low-altitude UAVs have a high ground spatial resolution (centimeter-scale or higher); thus, they contain rich crop-canopy surface information for crop phenotypic research [14,15]. In recent years, UAV remote sensing technology has been widely used to collect crop trait information [9–12,16,17]. UAVs equipped with high-definition digital cameras can acquire soybean canopy ultrahigh ground spatial resolution digital images over a field scale [14,15]. Many UAV-based methods have been proposed for monitoring various types of crop trait information, including the leaf area index (LAI) [18], leaf chlorophyll content [18–21], biomass [15,22], and crop height [23].

Machine learning has been successfully applied in several areas, such as image classification, target recognition, and language translation [24–26]. In recent years, machine learning techniques have been widely used to recognize various crop traits based on remote sensing images [27]. Gniewko et al. [28] used an artificial neural network (ANN), growing degree days, and total precipitation to estimate soybean yields. Letícia et al. [29] conducted a study to identify nematode damage to soybeans through the use of UAV remote sensing and a random forest (RF) model. The results obtained by Eugenio et al. [30] and Paulo et al. [31] indicated that machine learning techniques are efficient and flexible for remote sensing monitoring of soybean yields. Abdelbaki et al. [32] conducted a study to predict the soybean LAI and fractional vegetation cover (FVC) based on the RF model and UAV remote sensing. Compared with traditional machine learning methods (e.g., SVM and RF), deep learning methods such as long short-term memory (LSTM) [33,34], deep convolutional neural networks (CNNs) [26,35], and transformers [14] have been applied to image recognition, medical image analysis, climate change, and Weiqi game analysis, where they can provide results with similar or even higher precision than human experts. Deep learning uses multiple layers to extract higher-level features from the raw input. In recent years, deep learning techniques have been widely used to recognize various crop traits in remote sensing images, e.g., in leaf disease identification, weed identification, and crop trait recognition [1,26,33–37]. Wang et al. [34] developed an LSTM model by integrating MODIS LAI data to predict crop yields in China. Khan et al. [37] used a YOLOv4 model to identify apple leaf diseases in digital images captured by mobile phones. Zhang et al. [26] used a YOLOv4 model to identify weeds in digital photos of a peanut field. Khalied et al. [38] proposed a model based on MobileNetV2 for fruit identification and classification. Yonis et al. [39] proposed a CNN model adopting the VGG16 architecture for seed identification and classification. Notably, most of these widely used networks (e.g., YOLOv4 [40], ResNet50 [41], MobileNet [42], VGG16 [39], and InceptionResNetV2 [43]) did not take full advantage of shallow features. Shallow features derived from the shallow layers of CNNs are rich in image details, which are generally used in areas such as fine texture detection or small target detection [44,45]. Fusing the deep and shallow features of CNNs may improve performance in soybean maturity classification [44–46].

The objective of this work was to monitor soybean maturity using UAV remote sensing and deep learning. We designed a new convolutional neural network architecture (DS-SoybeanNet) to extract and utilize both shallow and deep image features to improve the performance of UAV-based soybean maturity information monitoring. We used a high-definition digital camera on board a UAV to collect high-definition soybean canopy digital images from two soybean breeding fields. We compared the UAV-based soybean maturity information monitoring performances of conventional machine learning methods (support vector machine (SVM) and random forest (RF)), current deep learning methods (InceptionResNetV2, MobileNetV2, and ResNet50), and our proposed DS-SoybeanNet method. Our results indicate that the proposed DS-SoybeanNet method can extract both shallow and deep image feature information and can realize high-performance soybean maturity classification.

2. Materials

2.1. Study Area

The study area was located at the Shengfeng Experimental Station (E: 116°22′10″–116°22′20″, N: 35°25′50″–35°26′20″, Figure 1) of the National Center for Soybean Improvement, Jiaxiang County, Jining City, Shandong Province, China. Jiaxiang County is situated on the North China Plain, with a warm continental monsoon climate, concentrated precipitation, and an average annual sunshine duration of 2405.2 h. The average annual temperature is 13.9 °C.

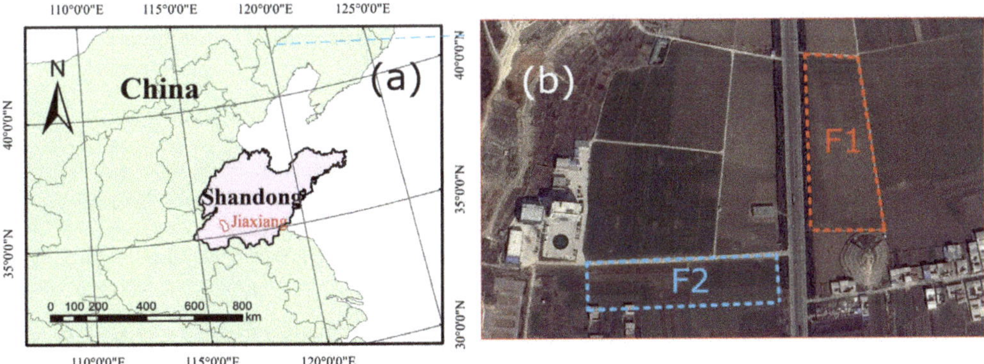

Figure 1. Study area (**a**) and experimental soybean field (**b**).

2.2. UAV Flights and Soybean Canopy Image Collection

We used a high-definition digital camera on board an eight-rotor electric UAV to collect high-resolution soybean canopy remote sensing images (Table 1). In the soybean breeding experimental field, the size of each planting area was approximately 2.5 m × 5 m. As shown in Figure 1, we selected two independent soybean planting fields (fields F1 and F2) in the study area to obtain soybean canopy digital images and maturity information.

Table 1. Parameters of the UAV and digital camera used in this study.

UAV	Parameter	Camera	Parameter
UAV name	DJI S1000	Camera name	SONY DSC-QX100
Flight height	Approximately 50 m	Image size	5472 × 3648
Flight speed	Approximately 8 m/s	Image dpi	350
Flight time	>20 min	Aperture	f/11
		Exposure	1/1250 s
		ISO	ISO-1600
		Focal length	10 mm
		Channels	Red, green, blue
		Ground spatial resolution	0.016 m

For field F1, we conducted five UAV flights (29 July, 13 August, 31 August, 17 September, and 28 September 2015). A total of 2116 soybean canopy digital images and their maturity information were obtained, which were used to calibrate the SoybeanNet model. For field F2, we made only one observation on 30 September, 2015. There were immature, near-mature, mature, and harvested soybean breeding lines in field F2 on 30 September. A total of 546 planting areas were set up in field F2 for the mapping and independent evaluation of the DS-SoybeanNet model.

The soybean image collection and image stitching process mainly included the following three steps:

(1)　Before the UAV took off, we set the flight route information according to the field size; the heading and lateral overlap were set to 80%. Table 1 shows the digital camera exposure parameters.

(2)　During the UAV flight, the soybean canopy images and corresponding position and orientation system (POS) information were collected using the digital camera, inertial measurement unit, and global positioning system device on board the UAV.

(3)　After the UAV flight, we imported the digital images and POS information into PhotoScan software to stitch together the high-definition digital images collected by the UAV. After the image stitching process, five soybean canopy digital orthophoto maps (ground spatial resolution (GSD): 0.016 m) for field F1 and one soybean canopy digital orthophoto map (GSD: 0.016 m) for field F2 were acquired.

2.3. Soybean Canopy Image Labeling

In this study, soybean maturity information was manually labeled. The labeling method was based on the standards of soybean harvesting. The labeling method is described in Table 2. For workers to customize schedules for harvesting soybean planting plots, four categories were used: immature (L0), near-mature (L1), mature (L2), and harvested (L3). L2 plots have the highest harvesting priority and need to be harvested as soon as possible, L1 plots have a high priority because the soybean will mature in less than a week, L0 and L3 plots have a lower priority because L0 plots generally take longer to grow, and no outdoor work is required for L3 plots.

Table 2. Standards used for labeling the soybean plots.

Label	Priority		Description
L0	Immature	Low	All upper canopy leaves are green or there are a few yellow leaves.
L1	Near-mature	High	Approximately half of the upper canopy leaves are yellow.
L2	Mature	Highest	The upper leaves of the canopy are yellow but have yet to be harvested.
L3	Harvested	Low	The soybean planting area has been harvested.

Since different soybean breeding lines have different maturation times, the numbers of images corresponding to the four labels varied between the two fields. Sixty percent of the images of each type in the dataset were randomly chosen to train the model, and the remaining 40% were used to evaluate the model's accuracy. Table 3 shows the numbers of samples used to train and validate the DS-SoybeanNet model. Figure 2 shows the soybean images used for model calibration and validation.

Table 3. Numbers of soybean images for model calibration and validation.

Label	Training Dataset (Field F1)	Validation Dataset (Field F1)	Independent Validation Dataset (Field F2)
L0	542	318	64
L1	257	163	219
L2	70	52	198
L3	400	314	65
Total	1269	847	546
Enhancement	25,380	16,940	-

Figure 2. Examples of the four labels.

2.4. Data Enhancement

In this study, we produced a DOM for the entire area by mosaicking together the digital images collected during each UAV flight. Since an orthoimage has a uniform scale, the ground spatial resolutions and solar angles were the main differences between the five DOMs. We used image rotation (four rotation angles: 0° (i.e., the original image), 90°, 180°, and 270°) and scaling (four scaling factors: 1.0 (i.e., the original image), 1.2, 1.5, 1.8, and 2.0) to enhance the soybean canopy image dataset collected from field F1. Image rotation and magnification helped us to obtain soybean canopy images with different resolutions and angles; in addition, they helped prevent overfitting of the model due to the small number of samples collected in the field.

After data enhancement, the number of original soybean canopy images obtained from field F1 was increased by 20 times. The number of independent validation datasets obtained from field F2 was not increased. In this study we used the Python open-cv and NumPy libraries to extract, rotate, and magnify the soybean canopy images.

3. Methods

3.1. Proposed DS-SoybeanNet

CNNs were originally proposed based on the receptive field mechanism in biology and they are a widely used deep learning technology [47]. CNNs are designed to process images with a lattice-like structure. The multilayer convolution, weight sharing, and rotational-shift invariance of CNNs make them effective in image classification and feature recognition. The deep and complex features extracted by CNNs are often used to effectively describe differences between different image categories and can be used to quickly and accurately complete classification tasks. Currently, widely used networks (e.g., ResNet50 and MobileNetV2) ignore shallow image feature information. We designed a network structure (Figure 3) that considers both shallow and deep image features to enhance the model's generalization ability. The advantage of DS-SoybeanNet is that the shallow and deep features are linked together by means of a concatenation module. Consequently, DS-SoybeanNet can extract and utilize both shallow and deep image features to improve the accuracy of soybean maturity information classifications. Figure 3 shows the architecture of DS-SoybeanNet. DS-SoybeanNet contains five convolutional layers, five flattening modules, one concatenation module, and four fully connected layers. The layers are described as follows:

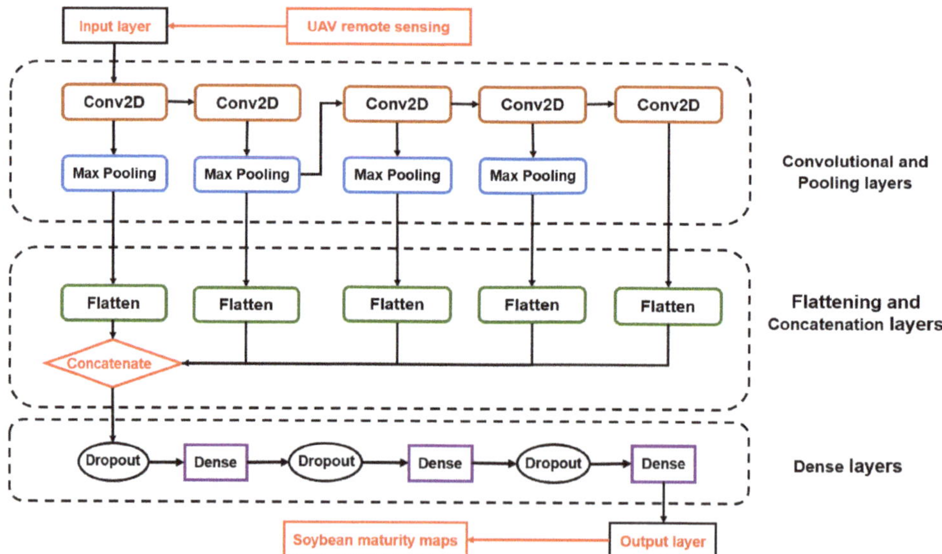

Figure 3. Architecture of DS-SoybeanNet.

(1) Input layer

The input data were collected via UAV remote sensing technology in the form of soybean canopy orthophotos and were then manually labeled and cropped to produce sample data. The sample size was $108 \times 108 \times 3$, and the sample data were divided into four types: immature, near-mature, mature, and harvested.

(2) Convolutional and pooling layers

The purpose of the convolution operation was to extract the different features of the input images. DS-SoybeanNet was designed with five convolutional layers; each convolutional layer was combined with the ReLU activation function to achieve delinearization. The pooling layers can reduce the dimensions of the feature maps by summarizing the presence of features in patches of the feature map.

(3) Flattening and concatenation layers

A flattening layer can reshape the feature maps into the dimensions required for the subsequent layers. A concatenation layer concatenates inputs along a specified dimension.

(4) Fully connected layers and output layer

Four fully connected layers were designed, and dropout layers were attached to the first three layers to prevent overfitting and improve model generalization. The output of the model was soybean maturity information derived from the input images.

3.2. Transfer Learning Based on InceptionResNetV2, MobileNetV2, and ResNet50

Transfer learning is a strategy for solving similar or related tasks using existing methods and data. Many deep learning networks show effective performance in image classification and target recognition from natural images (e.g., InceptionResNetV2 [43], ResNet50 [41], and MobileNetV2 [42]). Using a pretrained model to extract the features of remote sensing images can solve, to a certain extent, the problems involved with training a network for remote sensing image scene classification when there is a lack of training data. In this study, we used InceptionResNetV2, MobileNetV2, and ResNet50 as the pretrained deep learning models for transfer learning and performance comparisons with the proposed DS-SoybeanNet model.

(1) ResNet50: The ResNet50 network contains 49 convolutional layers and a fully connected layer. The core CNN components are the convolutional filter and the pooling layer. ResNet50 is a CNN derivative with a core component skip-connection to circumvent the gradient disappearance problem. The ResNet structure can accelerate training and improve performance (preventing gradient dispersion).

(2) InceptionResNetV2: The Inception module can obtain sparse or nonsparse features in the same layer. InceptionResNetV2 performs very well, but compared with ResNet, InceptionResNetV2 has a more complex network structure.

(3) MobileNetV2: MobileNetV2 is a lightweight CNN model proposed by Google for embedded devices, such as mobile phones, with a focus on optimizing latency while considering the model's size. MobileNetV2 can effectively balance latency and accuracy.

Transfer learning requires a low learning rate for retraining because the feature extraction module of the model already has some ability to extract image feature information after pretraining. An ideal learning rate can promote model convergence, whereas an unsuitable rate can cause training oscillations or even directly lead to the "explosion" of the loss value of the objective function. In addition to transfer learning methods based on InceptionResNetV2, MobileNetV2, and ResNet50, we also tested the performance of the AlexNet [48] and VGG16 [38] models to monitor soybean maturity.

3.3. SVM and RF

We also compared the soybean maturity information classification accuracy of our proposed DS-SoybeanNet with those of conventional machine learning models (SVM and RF). SVM is a generalized linear classifier that performs binary data classification in supervised learning [49]. Its decision boundary is the maximum marginal hyperplane solved for the learned samples, which reduces the classification problem to a convex quadratic programming problem. SVM has a low composition risk, its training is challenging to implement on large samples, and it is not ideal for solving multiclassification problems. RF is based on an integrated learning strategy, which combines multiple decision trees [50]. These decision trees are independent and unrelated to each other. Random forest uses the bagging strategy and repeated sampling to generate multiple trees. Under the bagging and bootstrap aggregation strategy, a subset of the samples are randomly selected from the dataset for training, and voting is conducted to obtain the average value as the resulting output. This strategy significantly avoids incorrect sample data, and thus shows improved accuracy.

3.4. Accuracy Evaluation

Figure 4 shows the experimental methodology used in this work. The canopy images of field F1 were used to calibrate and validate the models, whereas all canopy images of field F2 were used to validate the models.

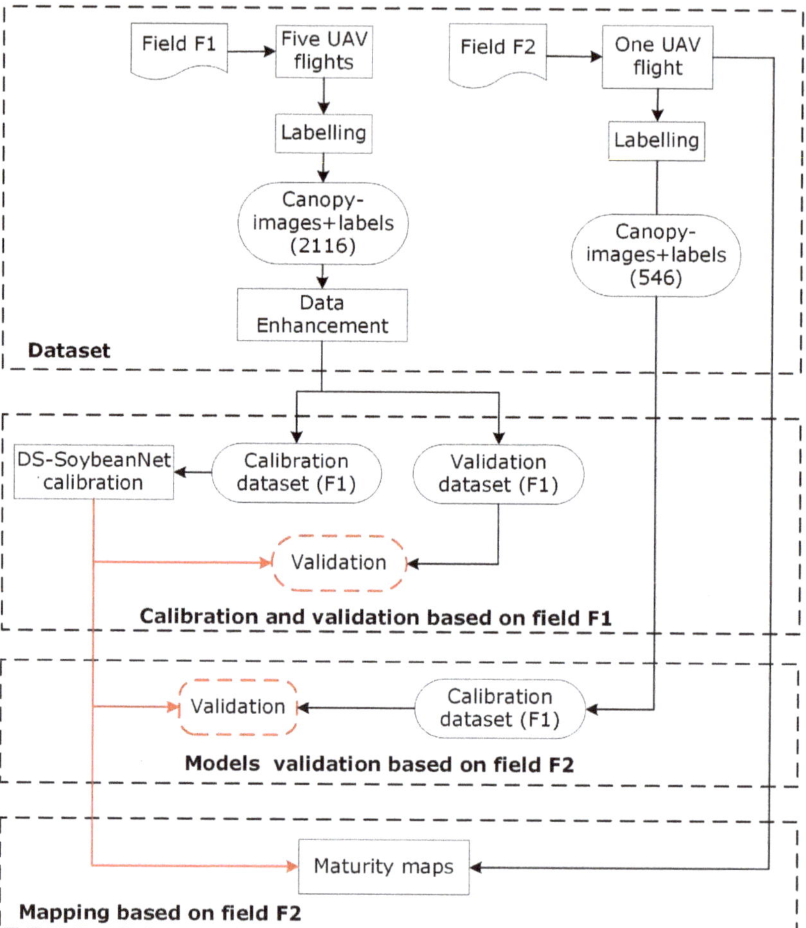

Figure 4. Flowchart of the experimental methodology.

The confusion matrix is a widely used tool for model accuracy evaluations. Table 4 shows the confusion matrix for the binary classification problem. Accuracy and recall can be obtained based on the confusion matrix. Generally, a higher accuracy and recall indicate a higher classification accuracy.

$$Accuracy = \frac{TP + TN}{TP + TN + FP + FN} \tag{1}$$

$$Recall = \frac{TP}{TP + FN} \tag{2}$$

Table 4. Confusion matrix.

Type		Predicted condition	
	Label	Positive (P)	Negative (N)
Actual condition	Positive (T)	True Positive (TP)	False Negative (FN)
	Negative (N)	False Positive (FP)	True Negative (TN)

TP, *TN*, *FP*, and *FN* represent the true-positive, true-negative, false-positive, and false-negative categories, respectively, in the confusion matrix (Table 4). Confusion matrices are not limited to binary classification but can also be used for multiclass classification. In this study, we used the confusion matrix, accuracy, and recall to evaluate the soybean maturity classification accuracy of the proposed DS-SoybeanNet model.

4. Results and Discussion

4.1. Model Calibration and Validation Based on Field F1

We used the calibration dataset of field F1 to train the proposed DS-SoybeanNet, AlexNet, VGG16, InceptionResNetV2, MobileNetV2, ResNet50, SVM, and RF models. Each model was trained and validated three times, and the model with the highest performance was saved. The learning rates were set to 0.0005, 0.0001, and 0.00001 for the transfer learning models (InceptionResNetV2, MobileNetV2, and ResNet50), and the number of epochs was set to 100. For DS-SoybeanNet, we analyzed the model accuracy with different convolution window sizes.

4.1.1. Validation of AlexNet, VGG16, SVM, and RF

We tested the SVM and RF models for monitoring soybean breeding line maturity (Table 5) based on the validation dataset from field F1. The L0, L1, and L3 classification recall values were higher than 99% for the traditional machine learning models (SVM and RF). The classification accuracies of SVM and RF were 92.31% and 94.23%, respectively. We also tested the performance of the AlexNet and VGG16 models (Table 5). The performances of AlexNet (99.44%) and VGG16 (97.99%) were higher than those of SVM (92.31%) and RF (94.23%).

Table 5. Classification results of AlexNet, VGG16, SVM, and RF.

Label	SVM	RF	AlexNet	VGG16
L0	99.69%	99.06%	99.69%	98.74%
L1	100%	100%	99.39%	100%
L2	90.38%	90.38%	98.08%	84.62%
L3	99.04%	99.36%	99.36%	98.41%
Accuracy	92.31%	94.23%	99.44% *	97.99%

Note: * indicates the highest accuracy.

4.1.2. Validation of Transfer Learning Based on InceptionResNetV2, MobileNetV2, and ResNet50

We also tested the performance of the three deep learning models using three learning rates. Table 6 shows the accuracies of the models using different learning rates. The performances of the three deep learning models (InceptionResNetV2, MobileNetV2, and ResNet50) were similar when using different learning rates. Our results indicate that the soybean maturity classification accuracy of traditional machine learning models (RF: 94.23%; SVM: 92.31%) was lower than that of InceptionResNetV2, MobileNetV2, and ResNet50.

There were notable differences in recall among the four labels. For example, the L2 classification recall of InceptionResNetV2 was much lower than those of L0, L1, and L3 when the learning rate was 0.0005. The same was observed for MobileNetV2 and ResNet50, which had L2 classification recalls of 69.23% and 88.46%, respectively.

Table 6. Classification results of transfer learning based on InceptionResNetV2, MobileNetV2, and ResNet50.

Label	InceptionResNetV2			MobileNetV2			ResNet50		
	Rate 1	Rate 2	Rate 3	Rate 1	Rate 2	Rate 3	Rate 1	Rate 2	Rate 3
L0	98.09%	100%	99.69%	100%	100%	99.69%	99.69%	100%	99.69%
L1	96.93%	100%	98.16%	95.09%	96.32%	92.02%	100%	96.93%	98.16%
L2	82.69%	98.08%	98.08%	69.23%	84.62%	82.69%	88.46%	96.15%	94.23%
L3	99.36%	98.73%	99.04%	99.36%	97.77%	97.77%	99.04%	98.73%	99.36%
Accuracy	97.41%	99.49%	99.09%	96.93%	97.52%	96.46%	98.93%	98.77%	98.97%

Note: Rate 1 = 0.0005; Rate 2 = 0.0001; Rate 3 = 0.00001.

4.1.3. Validation of the Proposed DS-SoybeanNet Model

We tested the proposed DS-SoybeanNet model in the monitoring of soybean breeding line maturity. Table 7 shows the classification results of the DS-SoybeanNet model with the convolution kernel size set to $3 \times 3, 5 \times 5, 7 \times 7, 9 \times 9, 11 \times 11, 16 \times 16$, and 21×21. The results indicate that there was little difference in performance among the seven convolution kernel sizes (with classification accuracies ranging from 97.52% to 99.19%). The results suggest that the model had the best soybean maturity classification accuracy when the convolution kernel size was set to 5×5 (99.17%) or 7×7 (99.19%). Figure 5 shows the training accuracy and loss curves of the DS-SoybeanNet with kernel sizes of 5×5 and 7×7. These results indicate that the model reached convergence at about 40 epochs. Training the DS-SoybeanNet (5×5) for about 100 epochs could take about 40 min and 5 s. Tables A1 and A2 show the model architecture and parameter information of DS-SoybeanNet with 5×5 and 7×7 kernels.

Table 7. Classification results of the proposed DS-SoybeanNet.

Label		DS-SoybeanNet						
		3×3	5×5	7×7	9×9	11×11	16×16	21×21
Recall	L0	100%	100%	100%	100%	100%	100%	100%
	L1	96.93%	100%	100%	100%	99.39%	99.39%	99.39%
	L2	92.31%	90.38%	90.38%	78.85%	88.46%	80.77%	75.00%
	L3	99.36%	99.36%	99.68%	99.36%	99.68%	96.50%	97.77%
Accuracy		98.70%	99.17% *	99.19% *	98.47%	99.06%	97.40%	97.52%

Note: * indicates the highest accuracy.

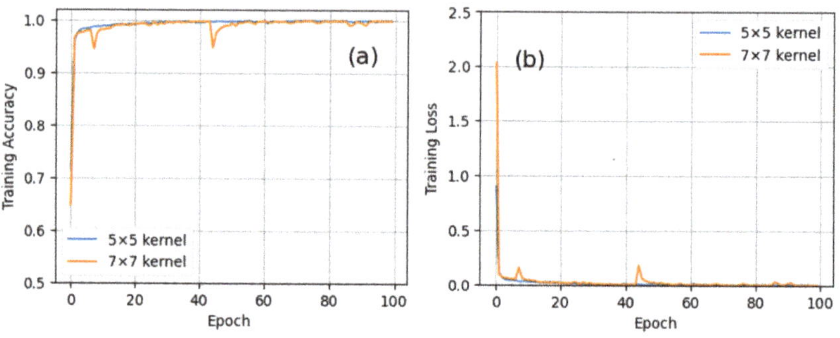

Figure 5. Training accuracy (**a**) and loss (**b**) of the DS-SoybeanNet with kernel sizes of 5×5 and 7×7.

4.2. Performance Comparison Based on Field F2

We used the 546 images from field F2 to test the performance of MobileNetV2, InceptionResNetV2, ResNet50, SVM, RF, and the proposed DS-SoybeanNet model in monitoring soybean maturity. Table 8 shows the confusion matrices of the soybean maturity classifications of the eight models. Table 9 shows the classification results of the eight models using the data from field F2. Our results (Tables 8 and 9) indicated that the proposed DS-SoybeanNet model exhibited a higher classification accuracy than the other machine learning models.

Table 8. Confusion matrices of MobileNetV2 (a), InceptionResNetV2 (b), ResNet50 (c), SVM (d), RF (e), DS-SoybeanNet with kernel sizes of 5 × 5 (f) and 7 × 7 (g), AlexNet (h), and VGG16 (i).

(a) — Actual condition / Predicted Condition

Label	L0	L1	L2	L3
L0	52	0	4	8
L1	20	0	168	31
L2	1	0	17	25
L3	0	0	1	64

(b) — Actual condition / Predicted Condition

Label	L0	L1	L2	L3
L0	39	0	17	8
L1	5	18	184	12
L2	0	1	193	4
L3	0	0	10	55

(c) — Actual condition / Predicted Condition

Label	L0	L1	L2	L3
L0	46	16	1	1
L1	9	97	109	4
L2	0	3	191	4
L3	0	0	5	60

(d) — Actual condition / Predicted Condition

Label	L0	L1	L2	L3
L0	64	0	0	0
L1	98	119	1	1
L2	2	85	102	9
L3	2	0	2	61

(e) — Actual condition / Predicted Condition

Label	L0	L1	L2	L3
L0	64	0	0	0
L1	89	94	33	3
L2	1	40	137	20
L3	0	0	3	62

(f) — Actual condition / Predicted Condition

Label	L0	L1	L2	L3
L0	59	5	0	0
L1	16	185	18	0
L2	0	27	171	0
L3	0	0	9	56

(g) — Actual condition / Predicted Condition (5 × 5)

Label	L0	L1	L2	L3
L0	59	5	0	0
L1	13	179	25	2
L2	0	22	169	7
L3	0	0	12	53

(h) — Actual condition / Predicted Condition

Label	L0	L1	L2	L3
L0	51	11	0	2
L1	5	97	115	2
L2	0	3	192	3
L3	0	0	5	60

(i) — Actual condition / Predicted Condition

Label	L0	L1	L2	L3
L0	59	5	0	0
L1	25	169	19	6
L2	0	28	163	7
L3	0	0	7	58

Table 9. Classification results of eight models from field F2.

Model	Rank	Precision L0	L1	L2	L3	Accuracy
DS-SoybeanNet (5 × 5)	1	92.19%	84.47%	86.36%	86.15%	86.26%
DS-SoybeanNet (7 × 7)	1	92.19%	81.74%	85.35%	81.54%	84.25%
VGG16	2	92.19%	77.17%	82.32%	89.23%	82.23%
AlexNet	3	79.37%	43.89%	96.95%	92.31%	72.89%
ResNet50	4	71.87%	44.29%	96.46%	92.31%	72.16%
RF	5	100%	42.92%	69.19%	95.38%	65.38%
SVM	6	100%	54.34%	51.52%	93.85%	63.37%
InceptionResNetV2	7	60.93%	8.22%	97.47%	84.62%	55.86%
MobileNetV2	8	81.25%	0%	39.53%	98.46%	52.75%

The conventional machine learning models (SVM and RF) exhibited the highest classification recall (100%) in the classification of immature soybeans (L0) (see Table 9). AlexNet (96.95%) showed the highest classification recall for mature soybeans (L2). As shown in Tables 8 and 9, the conventional machine learning models (SVM and RF) and deep learning models (MobileNetV2, InceptionResNetV2, and ResNet50) showed lower recalls for near-mature soybeans (L1), which led to lower overall classification accuracies for these models. DS-SoybeanNet (84.47%) had the highest classification recall for near-mature soybeans (L1) (see Table 9).

As shown in Table 9, the ResNet50 model exhibited a high classification accuracy of 72.16%. The RF (65.38%) and SVM (63.37%) models had similar classification accuracies. The soybean classification accuracies of InceptionResNetV2 (55.86%) and MobileNetV2 (52.75%) were lower than those of the other five models. The accuracies of DS-SoybeanNet based on 5×5 and 7×7 convolution kernels, namely, 86.26% and 84.25%, respectively, were notably higher than those of the other models.

Note that the eight models' performance decreased when using the field F2 dataset to test the models (Tables 5–7 and 9). As shown in Table 9, the top 3 models were DS-SoybeanNet, AlexNet, and VGG16 when monitoring soybean maturity using the field F2 dataset. Recently, the AlexNet [48] and VGG16 [39] models have been used to detect crop maturity by many researchers. Our results show that the new DS-SoybeanNet model performed better than the AlexNet and VGG16 models in the classification of immature (L0) and near-mature soybeans (L1). For the field F1 dataset, the recall of L0 for DS-SoybeanNet was 100%, which is higher than that of AlexNet (99.69%) and VGG16 (98.74%). For the field F2 dataset, the recall of L0 and L1 for DS-SoybeanNet was 92.19% and 84.47%, which was notably higher than that of the AlexNet (L0: 79.37%, L1: 43.89%) model.

To further evaluate the fusion of deep and shallow CNN features and to explore the efficiency of the proposed DS-SoybeanNet model, we set up three ablation experiments for DS-SoybeanNet, as described below. Figure 6 shows the architectures of the CNNs used for experiments 2 and 3. Each model was trained and validated three times, and the model with the highest performance was saved.

- Experiment 1. DS-SoybeanNet (Figure 3);
- Experiment 2. DS-SoybeanNet with only shallow image features (Figure 6a); and
- Experiment 3. DS-SoybeanNet with only deep image features (Figure 6b);

Our results (Table 10) indicate that the soybean maturity classification accuracy in experiment 2 (only shallow image features) and experiment 3 (only deep image features) was lower than that in experiment 1. This further proved that fusing deep and shallow CNN features [44–46] may improve the performance of the model in image classification tasks.

4.3. Soybean Maturity Mapping

For soybean maturity mapping, the following three steps were carried out:

(a) A soybean canopy DOM of field F2 was obtained after the UAV flight and the image stitching process. Then, all soybean breeding line plots (26 rows and 21 columns) were manually labeled, and the soybean plot image coordinates (plot center) were recorded.

(b) The soybean canopy images ($108 \times 108 \times 3$) were extracted automatically using the image coordinates and soybean canopy DOM using a Python script. Then, we used DS-SoybeanNet to classify these soybean canopy images.

(c) We then mapped the soybean maturity based on the soybean maturity information and soybean plot image coordinates.

Figure 7 shows a true-color RGB image and the maturity maps calculated for field F2 using DS-SoybeanNet with 5×5 and 7×7 convolution kernels. Our results indicate that the estimated soybean maturity information for field F2 had a high accuracy. The soybean maturity information obtained from the DS-SoybeanNet model with 5×5 and 7×7 convolution kernels was similar.

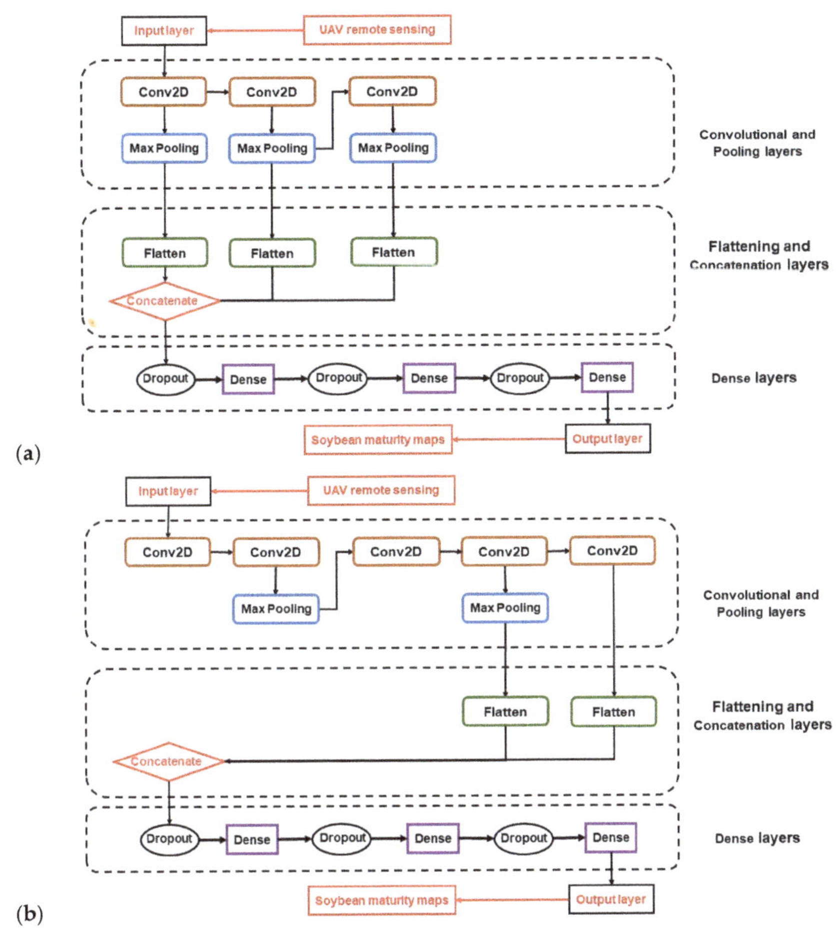

Figure 6. Architecture of CNNs used for experiments 2 (**a**) and 3 (**b**).

Table 10. Classification results of three experiments with 5×5 and 7×7 kernels.

| | | Experiment 1 | | Experiment 2 | | Experiment 3 | |
	Label	Validation Dataset (Field F1)	Independent Validation Dataset (Field F2)	Validation Dataset (Field F1)	Independent Validation Dataset (Field F2)	Validation Dataset (Field F1)	Independent Validation Dataset (Field F2)
Recall	L0	100%	92.19%	100%	98.44%	100%	96.88%
	L1	100%	84.47%	100%	74.89%	99.39%	83.11%
	L2	90.38%	86.36%	84.62%	87.37%	69.23%	71.21%
	L3	99.36%	86.15%	98.09%	75.38%	98.09%	87.69%
Accuracy		**99.17% ***	**86.26% ***	98.35%	82.23%	97.28%	80.95%
Recall	L0	100%	92.19%	100%	75.00%	100%	89.06%
	L1	100%	81.74%	99.39%	78.08%	99.39%	81.74%
	L2	90.38%	85.35%	86.54%	82.83%	78.85%	75.25%
	L3	99.68%	81.54%	98.73%	92.31%	98.41%	81.54%
Accuracy		**99.19% ***	**84.25% ***	98.58%	81.14%	97.99%	80.22%

Note: Bold and * indicate the highest accuracy.

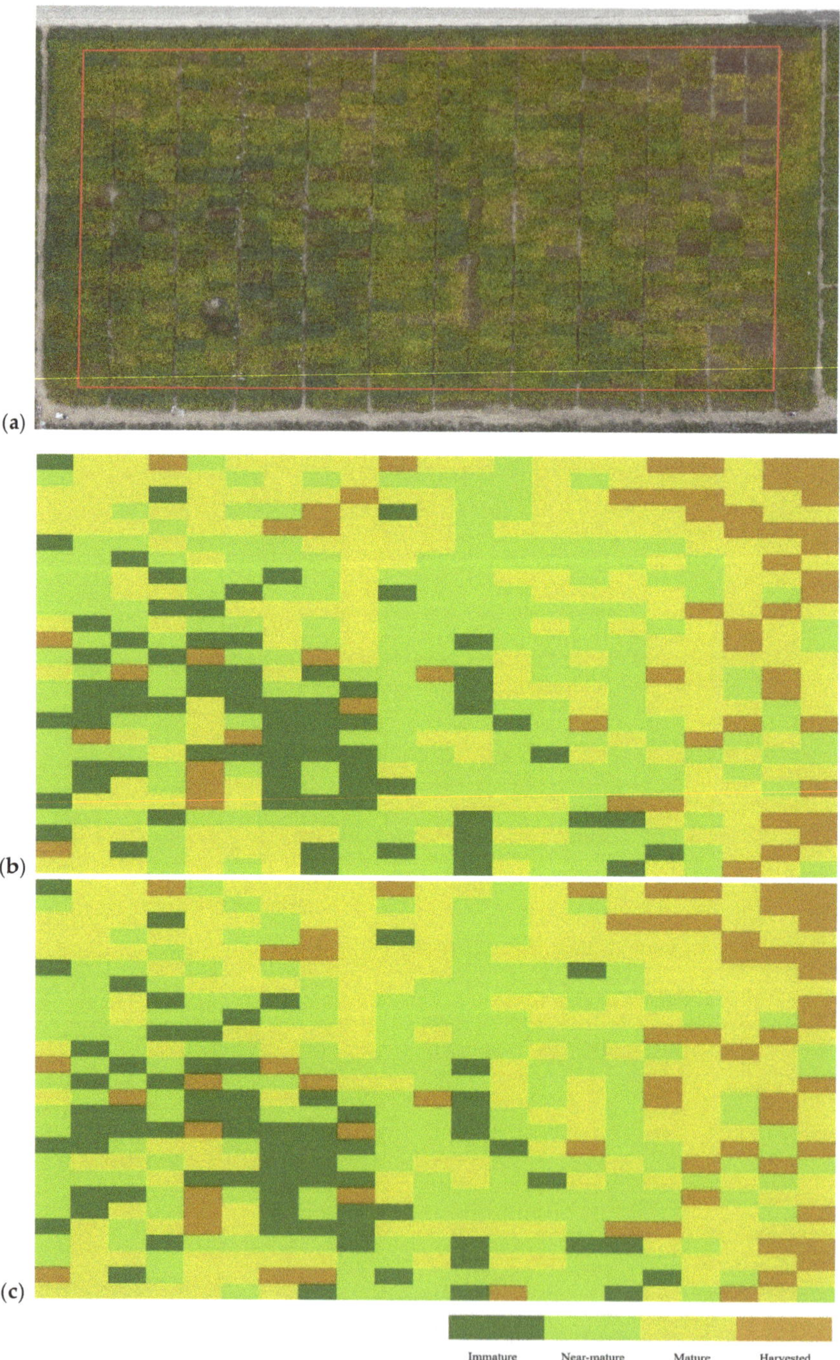

Figure 7. Maturity maps. (**a**) RGB true-color image; (**b**) DS-SoybeanNet (5 × 5); and (**c**) DS-SoybeanNet (7 × 7). Note: The red rectangle indicates the soybean plot region.

4.4. Advantages and Disadvantages of UAV + DS-SoybeanNet

As soybeans mature, the leaf chlorophyll level gradually decreases, contributing to a slow change in the leaves' color from green to yellow [51,52]. Crop leaf chlorophyll variation is asynchronous among layers of leaves [52]. For example, leaves in the top layer of a soybean canopy tend to have a younger leaf age and thus turn yellow later than the leaves in the bottom layer. Consequently, green and yellow leaves appear in the soybean canopy when the soybeans are nearly mature (Figure 2). Breeding fields commonly have thousands of breeding lines with different maturation times. Thus, timely monitoring of soybean breeding line maturity is crucial for soybean harvesting management and yield measurements [5–8]. UAV remote sensing technology can be utilized to collect high-resolution crop canopy images and has been widely used in precision agricultural crop trait monitoring [14,15]. Many studies have evaluated the crop parameter monitoring performance of digital cameras and multispectral sensors on board lightweight UAVs [17–19]. In our study, we attempted to evaluate the potential of using UAV remote sensing to monitor soybean breeding line maturity. We developed DS-SoybeanNet, which can extract and utilize both shallow and deep image features, and which thus helps to provide soybean breeding line maturity monitoring that is more robust than that offered by conventional machine learning methods. DS-SoybeanNet achieved the best accuracy of 86.26% (Table A1), which was notably higher than those of the conventional machine learning models (SVM and RF). However, DS-SoybeanNet has various disadvantages compared with conventional machine learning methods, such as its long elapsed time and large size (Table 11). In machine learning, CNNs have a more complex network structure and higher computational complexity than conventional machine learning models with larger model sizes.

Table 11. Models' elapsed times and sizes.

Model	Time (s)/1000 Samples	Size
RF	0.003	24.1 KB
SVM	0.007	7.70 KB
MobileNetV2	6.607	53.3 MB
DS-SoybeanNet (5 × 5)	11.770	2616 MB
AlexNet	19.011	151 MB
DS-SoybeanNet (7 × 7)	22.955	2616 MB
ResNet50	36.099	306 MB
InceptionResNetV2	44.328	653 MB
VGG16	67.080	623 MB

Table 11 shows the time required to process 1000 samples using each model and the models' sizes. The computation times of the CNN models (ranging from 6.607 s to 67.080 s) were notably higher than those of the conventional machine learning models, SVM and RF (0.003 s and 0.007 s). In addition, a high-performance device is required to calibrate CNN models. As shown in Table 11, the model sizes of DS-SoybeanNet, ResNet50, and InceptionResNetV2 were more than 300 MB. The proposed DS-SoybeanNet model had the largest size (2616 MB) compared to the other models. The DS-SoybeanNet model's large size may mean that it requires large storage when deployed on lightweight platforms (e.g., Raspberry Pi) for stationary observations. Nevertheless, DS-SoybeanNet (5 × 5) had approximately the same calculation speed as MobileNetV2 and a much higher monitoring accuracy than the other deep learning models. Therefore, we consider DS-SoybeanNet a fast and high-performance deep-learning tool for monitoring soybean maturity.

Many previous studies have used AlexNet, VGG16, Inception-V3, and VGG19 in crop maturity classifications. Faisal et al. [53] compared the performance of pre-trained VGG-19 (99.4%), Inception-V3 (99.4%), and NASNet (99.7%) in detecting fruit maturity. Atif et al. [54] used AlexNet and VGG16 to classify the maturity levels of jujube fruits (best: VGG16 = 99.17%). Sahil et al. [55] developed a method that used YOLOv3 to pinpoint the locations of tomatoes (94.67%) and used an AlexNet-like CNN model to classify their maturity levels (90.67%). In this work, we compared the results of conventional machine learning models (SVM (92.31%) and RF (94.23%)) and six CNN machine learning models (DS-SoybeanNet (99.19%), VGG16 (97.99%), AlexNet (99.44%), ResNet50 (98.97%), InceptionResNetV2 (99.49%), and MobileNetV2 (97.52%)) in soybean maturity information monitoring based on UAV remote sensing. The accuracy results reported in this study were close to those of previous studies based on AlexNet, VGG16, Inception-V3, and VGG16. Thus, our results further proved that deep learning is a good tool for crop maturity information monitoring [48,53–56]. The combination of UAV remote sensing and deep learning can be used for high-performance soybean maturity information monitoring. However, our results indicate that selected machine learning models' performance decreased when using the field F2 dataset to test the models (Tables 5–7 and 9). We suspect that changes in the UAV's working environment—for example, varying sunlight intensity over time—led to a direct decline in the models' performance. This is perhaps not surprising because the farmland environment is affected by varying cropland conditions (e.g., irrigation, wind). Thus, future research should be focused on the factors influencing cropland images.

In this study, the performance obtained when using soybean canopy images captured by the UAV's remote sensing digital camera may have been limited by the varying sunlight intensity over time. Since DS-SoybeanNet did not normalize the image differences due to sunlight, a normalization module may improve its performance in soybean maturity classification. Therefore, future studies need to develop a normalization module to weaken the effect of the sun. Thus, more experiments with different varieties and regions of soybeans are needed to improve the generalizability of the DS-SoybeanNet model. In this study, the proposed DS-SoybeanNet was validated using only two breeding fields from a single site; thus, further validation is required from additional fields and study sites.

5. Conclusions

In this study, we designed a network, namely, DS-SoybeanNet, to extract and utilize both shallow and deep image features to improve the performance of UAV-based soybean maturity information monitoring. We compared conventional machine learning methods (SVM and RF), current deep learning methods (AlexNet, VGG16, InceptionResNetV2, MobileNetV2, and ResNet50), and our proposed DS-SoybeanNet model in terms of their soybean maturity classification accuracy. The results were as follows.

(1) The conventional machine learning methods (SVM and RF) had lower calculation times than the deep learning methods (AlexNet, VGG16, InceptionResNetV2, MobileNetV2, and ResNet50) and our proposed DS-SoybeanNet model. For example, the computation speed of RF was 0.03 s per 1000 images. However, the overall accuracies of the conventional machine learning methods were notably lower than those of the deep learning methods and the proposed DS-SoybeanNet model.
(2) The current deep learning methods were outperformed in terms of universality by the DS-SoybeanNet model in the monitoring of soybean maturity. The overall accuracies of MobileNetV2 for fields F1 and F2 were 97.52% and 52.75%, respectively.
(3) The proposed DS-SoybeanNet model was able to provide high-performance soybean maturity classification results. Its computation speed was 11.770 s per 1000 images and its overall accuracies for fields F1 and F2 were 99.19% and 86.26%, respectively.
(4) Furthermore, future studies are needed in order to develop a normalization module to weaken the effect of the sun. Moreover, further validation is required using additional fields and study sites.

Author Contributions: J.Y., H.F. (Haikuan Feng), S.Z. and H.F. (Hao Feng) designed the experiments. J.Y., H.F. (Haikuan Feng), Z.S., H.X. and C.Z. collected the soybean images. J.Y. and S.Z. analyzed the data and wrote the manuscript. Y.L. and S.H. made comments and revised the manuscript. All authors have read and agreed to the published version of the manuscript.

Funding: This research was funded by the National Natural Science Foundation of China (NO. 42101362).

Data Availability Statement: The data that support the findings of this study are available from the corresponding author, J.Y., upon reasonable request.

Acknowledgments: We thank Bo Xu and Guozheng Lu for field management and data collection.

Appendix A

Figure A1 shows the attention regions of different models in the soybean canopy images. Regarding interpretability, the top three models performed differently when their attention regions were visualized by means of the Grad-CAM technique (Figure A1). VGG16 models focused only on luxuriant leaves for all four categories (Figure A1). The AlexNet model showed acceptable attention regions when dealing with L0 and L1 soybean images, whereas it focused only on branches and leaves when analyzing L2 and L3 soybean images (Figure A1). Compared with AlexNet and VGG16 models, DS-SoybeanNet showed acceptable attention regions for the four categories (Figure A1). In most cases, DS-SoybeanNet was able to differentiate among the soybean images accurately based on the leaves, branches, and soil pixels, similarly to farm workers. Tables A1 and A2 show the model architecture and parameter information of DS-SoybeanNet with 5×5 and 7×7 kernels.

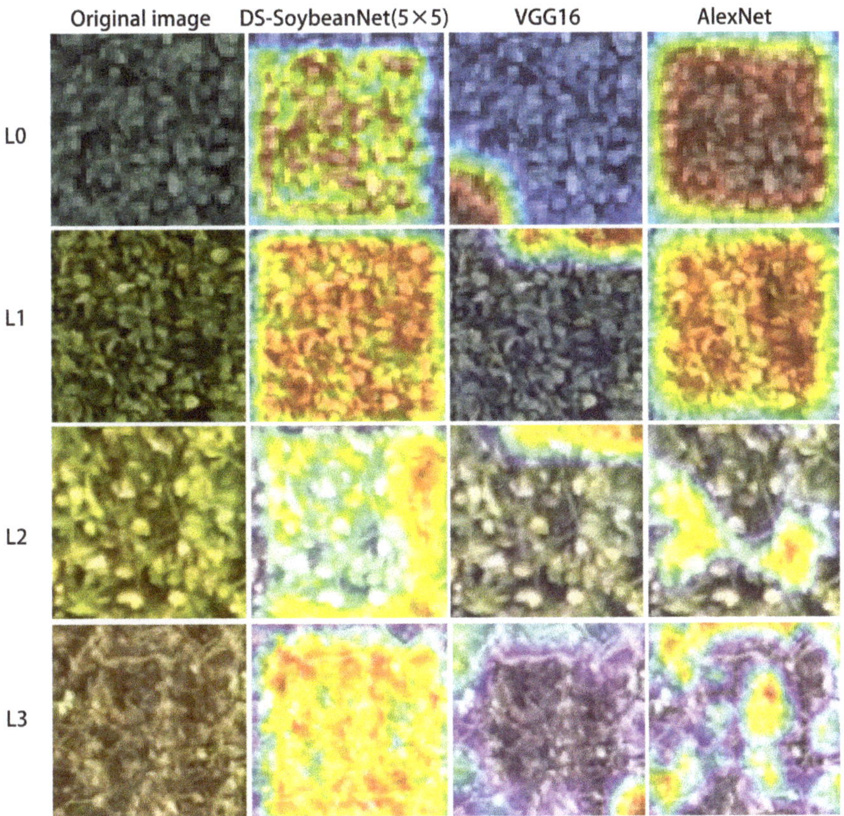

Figure A1. The attention regions of the top 3 (Table 9) models in soybean canopy images.

Table A1. Details of the proposed DS-SoybeanNet with 5 × 5 kernels.

Layer (Type)	Output Shape	Param	Connected to
input_1 (Input Layer)	[(None,108,108,3)	0	
conv2d (Conv2D)	(None,108,108,32)	2432	input_1 [0][0]
conv2d_1 (Conv2D)	(None,108,108,16)	12816	conv2d [0][0]
max_pooling2d_1 (MaxPooling2D)	(None,27,27,16)	0	conv2d_1 [0][0]
conv2d_2 (Conv2D)	(None,27,27,32)	12832	max_pooling2d_1 [0][0]
conv2d_3 (Conv2D)	(None,27,27,16)	12816	conv2d_2 [0][0]
max_pooling2d (MaxPooling2D)	(None,27,27,32)	0	conv2d [0][0]
max_pooling2d_2 (MaxPooling2D)	(None,13,13,32)	0	conv2d_2 [0][0]
max_pooling2d_3 (MaxPooling2D)	(None, 13,13,16)	0	conv2d_3 [0][0]
conv2d_4 (Conv2D)	(None,27,27,16)	6416	conv2d_3 [0][0]
flatten (Flatten)	(None,23328)	0	max_pooling2d [0][0]
flatten_1 (Flatten)	(None,11664)	0	max_pooling2d_1 [0][0]
flatten_2 (Flatten)	(None,5408)	0	max_pooling2d_2 [0][0]
flatten_3 (Flatten)	(None,2704)	0	max_pooling2d_3 [0][0]
flatten_4 (Flatten)	(None,11664)	0	conv2d_4 [0][0]
concatenate (Concatenate)	(None,54768)	0	flatten [0][0] flatten_1 [0][0] flatten_2 [0][0] flatten _3 [0][0] flatten _4 [0][0]
dropout (Dropout)	(None,54768)	0	concatenate [0][0]
dense (Dense)	(None,4096)	224333824	dropout [0][0]
dropout_1 (Dropout)	(None,4096)	0	dense [0][0]
dense_1 (Dense)	(None,512)	4195328	dropout_1 [0][0]
dropout_2 (Dropout)	(None,512)	0	dense_1 [0][0]
dense_2 (Dense)	(None,4)	4100	dropout_2 [0][0]

Total params: 228,580,564
Trainable params: 228,580,564
Non-trainable params: 0

Table A2. Details of the proposed DS-SoybeanNet with 7×7 kernels.

Layer (Type)	Output Shape	Param	Connected to
input_1 (Input Layer)	[(None,108,108,3)	0	
conv2d (Conv2D)	(None,108,108,32)	4736	input_1 [0][0]
conv2d_1 (Conv2D)	(None,108,108,16)	25104	conv2d [0][0]
max_pooling2d_1 (MaxPooling2D)	(None,27,27,16)	0	conv2d_1 [0][0]
conv2d_2 (Conv2D)	(None,27,27,32)	25120	max_pooling2d_1 [0][0]
conv2d_3 (Conv2D)	(None,27,27,16)	25104	conv2d_2 [0][0]
max_pooling2d (MaxPooling2D)	(None,27,27,32)	0	conv2d [0][0]
max_pooling2d_2 (MaxPooling2D)	(None,13,13,32)	0	conv2d_2 [0][0]
max_pooling2d_3 (MaxPooling2D)	(None, 13,13,16)	0	conv2d_3 [0][0]
conv2d_4 (Conv2D)	(None,27,27,16)	12560	conv2d_3 [0][0]
flatten (Flatten)	(None,23328)	0	max_pooling2d [0][0]
flatten_1 (Flatten)	(None,11664)	0	max_pooling2d_1 [0][0]
flatten_2 (Flatten)	(None,5408)	0	max_pooling2d_2 [0][0]
flatten_3 (Flatten)	(None,2704)	0	max_pooling2d_3 [0][0]
flatten_4 (Flatten)	(None,11664)	0	conv2d_4 [0][0]
concatenate (Concatenate)	(None,54768)	0	flatten [0][0] flatten_1 [0][0] flatten_2 [0][0] flatten _3 [0][0] flatten _4 [0][0]
dropout (Dropout)	(None,54768)	0	concatenate [0][0]
dense (Dense)	(None,4096)	224333824	dropout [0][0]
dropout_1 (Dropout)	(None,4096)	0	dense [0][0]
dense_1 (Dense)	(None,512)	4195328	dropout_1 [0][0]
dropout_2 (Dropout)	(None,512)	0	dense_1 [0][0]
dense_2 (Dense)	(None,4)	4100	dropout_2 [0][0]

Total params: 228,625,876
Trainable params: 228,625,876
Non-trainable params: 0

References

1. Maimaitijiang, M.; Sagan, V.; Sidike, P.; Hartling, S.; Esposito, F.; Fritschi, F.B. Soybean Yield Prediction from UAV Using Multimodal Data Fusion and Deep Learning. *Remote Sens. Environ.* **2020**, *237*, 111599. [CrossRef]
2. Qin, P.; Wang, T.; Luo, Y. A Review on Plant-Based Proteins from Soybean: Health Benefits and Soy Product Development. *J. Agric. Food Res.* **2022**, *7*, 100265. [CrossRef]
3. Liu, X.; Jin, J.; Wang, G.; Herbert, S.J. Soybean Yield Physiology and Development of High-Yielding Practices in Northeast China. *Field Crop. Res.* **2008**, *105*, 157–171. [CrossRef]
4. Zhang, Y.M.; Li, Y.; Chen, W.F.; Wang, E.T.; Tian, C.F.; Li, Q.Q.; Zhang, Y.Z.; Sui, X.H.; Chen, W.X. Biodiversity and Biogeography of Rhizobia Associated with Soybean Plants Grown in the North China Plain. *Appl. Environ. Microbiol.* **2011**, *77*, 6331–6342. [CrossRef] [PubMed]
5. Vogel, J.T.; Liu, W.; Olhoft, P.; Crafts-Brandner, S.J.; Pennycooke, J.C.; Christiansen, N. Soybean Yield Formation Physiology—A Foundation for Precision Breeding Based Improvement. *Front. Plant Sci.* **2021**, *12*, 719706. [CrossRef]
6. Maranna, S.; Nataraj, V.; Kumawat, G.; Chandra, S.; Rajesh, V.; Ramteke, R.; Patel, R.M.; Ratnaparkhe, M.B.; Husain, S.M.; Gupta, S.; et al. Breeding for Higher Yield, Early Maturity, Wider Adaptability and Waterlogging Tolerance in Soybean (*Glycine max* L.): A Case Study. *Sci. Rep.* **2021**, *11*, 22853. [CrossRef]
7. Volpato, L.; Dobbels, A.; Borem, A.; Lorenz, A.J. Optimization of Temporal UAS-Based Imagery Analysis to Estimate Plant Maturity Date for Soybean Breeding. *Plant Phenome J.* **2021**, *4*, e20018. [CrossRef]
8. Moeinizade, S.; Pham, H.; Han, Y.; Dobbels, A.; Hu, G. An Applied Deep Learning Approach for Estimating Soybean Relative Maturity from UAV Imagery to Aid Plant Breeding Decisions. *Mach. Learn. Appl.* **2022**, *7*, 100233. [CrossRef]
9. Zhou, J.; Mou, H.; Zhou, J.; Ali, M.L.; Ye, H.; Chen, P.; Nguyen, H.T. Qualification of Soybean Responses to Flooding Stress Using UAV-Based Imagery and Deep Learning. *Plant Phenomics* **2021**, *2021*. [CrossRef]

10. Habibi, L.N.; Watanabe, T.; Matsui, T.; Tanaka, T.S.T. Machine Learning Techniques to Predict Soybean Plant Density Using UAV and Satellite-Based Remote Sensing. *Remote Sens.* **2021**, *13*, 2548. [CrossRef]

11. Luo, S.; Liu, W.; Zhang, Y.; Wang, C.; Xi, X.; Nie, S.; Ma, D.; Lin, Y.; Zhou, G. Maize and Soybean Heights Estimation from Unmanned Aerial Vehicle (UAV) LiDAR Data. *Comput. Electron. Agric.* **2021**, *182*, 106005. [CrossRef]

12. Fukano, Y.; Guo, W.; Aoki, N.; Ootsuka, S.; Noshita, K.; Uchida, K.; Kato, Y.; Sasaki, K.; Kamikawa, S.; Kubota, H. GIS-Based Analysis for UAV-Supported Field Experiments Reveals Soybean Traits Associated with Rotational Benefit. *Front. Plant Sci.* **2021**, *12*, 637694. [CrossRef] [PubMed]

13. Yang, G.; Li, C.; Wang, Y.; Yuan, H.; Feng, H.; Xu, B.; Yang, X. The DOM Generation and Precise Radiometric Calibration of a UAV-Mounted Miniature Snapshot Hyperspectral Imager. *Remote Sens.* **2017**, *9*, 642. [CrossRef]

14. Zhou, C.; Ye, H.; Sun, D.; Yue, J.; Yang, G.; Hu, J. An Automated, High-Performance Approach for Detecting and Characterizing Broccoli Based on UAV Remote-Sensing and Transformers: A Case Study from Haining, China. *Int. J. Appl. Earth Obs. Geoinf.* **2022**, *114*, 103055. [CrossRef]

15. Yue, J.; Yang, G.; Tian, Q.; Feng, H.; Xu, K.; Zhou, C. Estimate of Winter-Wheat above-Ground Biomass Based on UAV Ultrahigh-Ground-Resolution Image Textures and Vegetation Indices. *ISPRS J. Photogramm. Remote Sens.* **2019**, *150*, 226–244. [CrossRef]

16. Haghighattalab, A.; González Pérez, L.; Mondal, S.; Singh, D.; Schinstock, D.; Rutkoski, J.; Ortiz-Monasterio, I.; Singh, R.P.; Goodin, D.; Poland, J. Application of Unmanned Aerial Systems for High Throughput Phenotyping of Large Wheat Breeding Nurseries. *Plant Methods* **2016**, *12*, 35. [CrossRef] [PubMed]

17. Singhal, G.; Bansod, B.; Mathew, L.; Goswami, J.; Choudhury, B.U.; Raju, P.L.N. Chlorophyll Estimation Using Multi-Spectral Unmanned Aerial System Based on Machine Learning Techniques. *Remote Sens. Appl. Soc. Environ.* **2019**, *15*, 100235. [CrossRef]

18. Roosjen, P.P.J.; Brede, B.; Suomalainen, J.M.; Bartholomeus, H.M.; Kooistra, L.; Clevers, J.G.P.W. Improved Estimation of Leaf Area Index and Leaf Chlorophyll Content of a Potato Crop Using Multi-Angle Spectral Data—Potential of Unmanned Aerial Vehicle Imagery. *Int. J. Appl. Earth Obs. Geoinf.* **2018**, *66*, 14–26. [CrossRef]

19. Yue, J.; Feng, H.; Tian, Q.; Zhou, C. A Robust Spectral Angle Index for Remotely Assessing Soybean Canopy Chlorophyll Content in Different Growing Stages. *Plant Methods* **2020**, *16*, 104. [CrossRef]

20. Wang, W.; Gao, X.; Cheng, Y.; Ren, Y.; Zhang, Z.; Wang, R.; Cao, J.; Geng, H. QTL Mapping of Leaf Area Index and Chlorophyll Content Based on UAV Remote Sensing in Wheat. *Agriculture* **2022**, *12*, 595. [CrossRef]

21. Wójcik-Gront, E.; Gozdowski, D.; Stępień, W. UAV-Derived Spectral Indices for the Evaluation of the Condition of Rye in Long-Term Field Experiments. *Agriculture* **2022**, *12*, 1671. [CrossRef]

22. Yue, J.; Feng, H.; Li, Z.; Zhou, C.; Xu, K. Mapping Winter-Wheat Biomass and Grain Yield Based on a Crop Model and UAV Remote Sensing. *Int. J. Remote Sens.* **2021**, *42*, 1577–1601. [CrossRef]

23. Han, L.; Yang, G.; Yang, H.; Xu, B.; Li, Z.; Yang, X. Clustering Field-Based Maize Phenotyping of Plant-Height Growth and Canopy Spectral Dynamics Using a UAV Remote-Sensing Approach. *Front. Plant Sci.* **2018**, *9*, 1638. [CrossRef] [PubMed]

24. Ofer, D.; Brandes, N.; Linial, M. The Language of Proteins: NLP, Machine Learning & Protein Sequences. *Comput. Struct. Biotechnol. J.* **2021**, *19*, 1750–1758. [CrossRef] [PubMed]

25. Janiesch, C.; Zschech, P.; Heinrich, K. Machine Learning and Deep Learning. *Electron. Mark.* **2021**, *31*, 685–695. [CrossRef]

26. Zhang, H.; Wang, Z.; Guo, Y.; Ma, Y.; Cao, W.; Chen, D.; Yang, S.; Gao, R. Weed Detection in Peanut Fields Based on Machine Vision. *Agriculture* **2022**, *12*, 1541. [CrossRef]

27. Yue, J.; Feng, H.; Yang, G.; Li, Z. A Comparison of Regression Techniques for Estimation of Above-Ground Winter Wheat Biomass Using Near-Surface Spectroscopy. *Remote Sens.* **2018**, *10*, 66. [CrossRef]

28. Niedbała, G.; Kurasiak-Popowska, D.; Piekutowska, M.; Wojciechowski, T.; Kwiatek, M.; Nawracała, J. Application of Artificial Neural Network Sensitivity Analysis to Identify Key Determinants of Harvesting Date and Yield of Soybean (*Glycine max* [L.] Merrill) Cultivar Augusta. *Agriculture* **2022**, *12*, 754. [CrossRef]

29. Santos, L.B.; Bastos, L.M.; de Oliveira, M.F.; Soares, P.L.M.; Ciampitti, I.A.; da Silva, R.P. Identifying Nematode Damage on Soybean through Remote Sensing and Machine Learning Techniques. *Agronomy* **2022**, *12*, 2404. [CrossRef]

30. Eugenio, F.C.; Grohs, M.; Venancio, L.P.; Schuh, M.; Bottega, E.L.; Ruoso, R.; Schons, C.; Mallmann, C.L.; Badin, T.L.; Fernandes, P. Estimation of Soybean Yield from Machine Learning Techniques and Multispectral RPAS Imagery. *Remote Sens. Appl. Soc. Environ.* **2020**, *20*, 100397. [CrossRef]

31. Teodoro, P.E.; Teodoro, L.P.R.; Baio, F.H.R.; da Silva Junior, C.A.; Dos Santos, R.G.; Ramos, A.P.M.; Pinheiro, M.M.F.; Osco, L.P.; Gonçalves, W.N.; Carneiro, A.M.; et al. Predicting Days to Maturity, Plant Height, and Grain Yield in Soybean: A Machine and Deep Learning Approach Using Multispectral Data. *Remote Sens.* **2021**, *13*, 4632. [CrossRef]

32. Abdelbaki, A.; Schlerf, M.; Retzlaff, R.; Machwitz, M.; Verrelst, J.; Udelhoven, T. Comparison of Crop Trait Retrieval Strategies Using UAV-Based VNIR Hyperspectral Imaging. *Remote Sens.* **2021**, *13*, 1748. [CrossRef] [PubMed]

33. Sun, J.; Di, L.; Sun, Z.; Shen, Y.; Lai, Z. County-Level Soybean Yield Prediction Using Deep CNN-LSTM Model. *Sensors* **2019**, *19*, 4363. [CrossRef]

34. Wang, J.; Si, H.; Gao, Z.; Shi, L. Winter Wheat Yield Prediction Using an LSTM Model from MODIS LAI Products. *Agriculture* **2022**, *12*, 1707. [CrossRef]

35. Tian, H.; Wang, P.; Tansey, K.; Han, D.; Zhang, J.; Zhang, S.; Li, H. A Deep Learning Framework under Attention Mechanism for Wheat Yield Estimation Using Remotely Sensed Indices in the Guanzhong Plain, PR China. *Int. J. Appl. Earth Obs. Geoinf.* **2021**, *102*, 102375. [CrossRef]

36. Khaki, S.; Wang, L. Crop Yield Prediction Using Deep Neural Networks. *Front. Plant Sci.* **2019**, *10*, 621. [CrossRef]
37. Khan, A.I.; Quadri, S.M.K.; Banday, S.; Latief Shah, J. Deep Diagnosis: A Real-Time Apple Leaf Disease Detection System Based on Deep Learning. *Comput. Electron. Agric.* **2022**, *198*, 107093. [CrossRef]
38. Albarrak, K.; Gulzar, Y.; Hamid, Y.; Mehmood, A.; Soomro, A.B. A Deep Learning-Based Model for Date Fruit Classification. *Sustainability* **2022**, *14*, 6339. [CrossRef]
39. Gulzar, Y.; Hamid, Y.; Soomro, A.B.; Alwan, A.A.; Journaux, L. A Convolution Neural Network-Based Seed Classification System. *Symmetry* **2020**, *12*, 2018. [CrossRef]
40. Bochkovskiy, A.; Wang, C.-Y.; Liao, H.-Y.M. YOLOv4: Optimal Speed and Accuracy of Object Detection. *arXiv* **2020**, arXiv:2004.10934.
41. Sangeetha, V.; Prasad, K.J.R. Syntheses of Novel Derivatives of 2-Acetylfuro[2,3-a]Carbazoles, Benzo[1,2-b]-1,4-Thiazepino[2,3-a]Carbazoles and 1-Acetyloxycarbazole-2- Carbaldehydes. *Indian J. Chem. Sect. B Org. Med. Chem.* **2006**, *45*, 1951–1954. [CrossRef]
42. Howard, A.G.; Zhu, M.; Chen, B.; Kalenichenko, D.; Wang, W.; Weyand, T.; Andreetto, M.; Adam, H. MobileNets: Efficient Convolutional Neural Networks for Mobile Vision Applications. *arXiv* **2017**, arXiv:1704.04861.
43. Szegedy, C.; Ioffe, S.; Vanhoucke, V.; Alemi, A.A. Inception-v4, Inception-ResNet and the Impact of Residual Connections on Learning. In Proceedings of the 31st AAAI Conference on Artificial Intelligence (AAAI-17), San Francisco, CA, USA, 4–9 February 2017; pp. 4278–4284. [CrossRef]
44. Miao, Y.; Lin, Z.; Ding, G.; Han, J. Shallow Feature Based Dense Attention Network for Crowd Counting. In Proceedings of the 34th AAAI Conference on Artificial Intelligence (AAAI-20), New York, NY, USA, 7–12 February 2020; pp. 11765–11772. [CrossRef]
45. Wei, J.; Wang, Q.; Li, Z.; Wang, S.; Zhou, S.K.; Cui, S. Shallow Feature Matters for Weakly Supervised Object Localization. *Proc. IEEE Comput. Soc. Conf. Comput. Vis. Pattern Recognit.* **2021**, *1*, 5989–5997. [CrossRef]
46. Bougourzi, F.; Dornaika, F.; Mokrani, K.; Taleb-Ahmed, A.; Ruichek, Y. Fusing Transformed Deep and Shallow Features (FTDS) for Image-Based Facial Expression Recognition. *Expert Syst. Appl.* **2020**, *156*, 113459. [CrossRef]
47. Lecun, Y.; Bengio, Y.; Hinton, G. Deep Learning. *Nature* **2015**, *521*, 436–444. [CrossRef] [PubMed]
48. Behera, S.K.; Rath, A.K.; Sethy, P.K. Maturity Status Classification of Papaya Fruits Based on Machine Learning and Transfer Learning Approach. *Inf. Process. Agric.* **2021**, *8*, 244–250. [CrossRef]
49. Hosseini, M.; McNairn, H.; Mitchell, S.; Robertson, L.D.; Davidson, A.; Ahmadian, N.; Bhattacharya, A.; Borg, E.; Conrad, C.; Dabrowska-Zielinska, K.; et al. A Comparison between Support Vector Machine and Water Cloud Model for Estimating Crop Leaf Area Index. *Remote Sens.* **2021**, *13*, 1348. [CrossRef]
50. Breiman, L. Random Forests. *Mach. Learn.* **2001**, *45*, 5–32. [CrossRef]
51. Huang, W.; Wang, Z.; Huang, L.; Lamb, D.W.; Ma, Z.; Zhang, J.; Wang, J.; Zhao, C. Estimation of Vertical Distribution of Chlorophyll Concentration by Bi-Directional Canopy Reflectance Spectra in Winter Wheat. *Precis. Agric.* **2011**, *12*, 165–178. [CrossRef]
52. Wang, J.; Zhao, C.; Huang, W. *Fundamental and Application of Quantitative Remote Sensing in Agriculture*; Science China Press: Beijing, China, 2008.
53. Faisal, M.; Alsulaiman, M.; Arafah, M.; Mekhtiche, M.A. IHDS: Intelligent Harvesting Decision System for Date Fruit Based on Maturity Stage Using Deep Learning and Computer Vision. *IEEE Access* **2020**, *8*, 167985–167997. [CrossRef]
54. Mahmood, A.; Singh, S.K.; Tiwari, A.K. Pre-Trained Deep Learning-Based Classification of Jujube Fruits According to Their Maturity Level. *Neural Comput. Appl.* **2022**, *34*, 13925–13935. [CrossRef]
55. Mutha, S.A.; Shah, A.M.; Ahmed, M.Z. Maturity Detection of Tomatoes Using Deep Learning. *SN Comput. Sci.* **2021**, *2*, 441. [CrossRef]
56. Zhou, X.; Lee, W.S.; Ampatzidis, Y.; Chen, Y.; Peres, N.; Fraisse, C. Strawberry Maturity Classification from UAV and Near-Ground Imaging Using Deep Learning. *Smart Agric. Technol.* **2021**, *1*, 100001. [CrossRef]

Article

A Synthetic Angle Normalization Model of Vegetation Canopy Reflectance for Geostationary Satellite Remote Sensing Data

Yinghao Lin [1,2,3], Qingjiu Tian [2], Baojun Qiao [1,*], Yu Wu [4,*], Xianyu Zuo [1], Yi Xie [1] and Yang Lian [5]

[1] Henan Key Laboratory of Big Data Analysis and Processing, School of Computer and Information Engineering, Henan University, Kaifeng 475001, China
[2] International Institute for Earth System Science, School of Geographic and Oceanographic Sciences, Nanjing University, Nanjing 210023, China
[3] Zhong Ke Langfang Institute of Spatial Information Applications, Langfang 065000, China
[4] School of Earth System Science, Tianjin University, Tianjin 300072, China
[5] Henan Yellow River Administration Bureau, Zhengzhou 450053, China
* Correspondence: qbj@henu.edu.cn (B.Q.); wu_yu@tju.edu.cn (Y.W.)

Citation: Lin, Y.; Tian, Q.; Qiao, B.; Wu, Y.; Zuo, X.; Xie, Y.; Lian, Y. A Synthetic Angle Normalization Model of Vegetation Canopy Reflectance for Geostationary Satellite Remote Sensing Data. *Agriculture* **2022**, *12*, 1658. https://doi.org/10.3390/agriculture12101658

Academic Editor: Josef Eitzinger

Received: 7 September 2022
Accepted: 7 October 2022
Published: 10 October 2022

Abstract: High-frequency imaging characteristics allow a geostationary satellite (GSS) to capture the diurnal variation in vegetation canopy reflectance spectra, which is of very important practical significance for monitoring vegetation via remote sensing (RS). However, the observation angle and solar angle of high-frequency GSS RS data usually differ, and the differences in bidirectional reflectance from the reflectance spectra of the vegetation canopy are significant, which makes it necessary to normalize angles for GSS RS data. The BRDF (Bidirectional Reflectance Distribution Function) prototype library is effective for the angle normalization of RS data. However, its spatiotemporal applicability and error propagation are currently unclear. To resolve this problem, we herein propose a synthetic angle normalization model (SANM) for RS vegetation canopy reflectance; this model exploits the GSS imaging characteristics, whereby each pixel has a fixed observation angle. The established model references a topographic correction method for vegetation canopies based on path-length correction, solar zenith angle normalization, and the Minnaert model. It also considers the characteristics of diurnal variations in vegetation canopy reflectance spectra by setting the time window. Experiments were carried out on the eight Geostationary Ocean Color Imager (GOCI) images obtained on 22 April 2015 to validate the performance of the proposed SANM. The results show that SANM significantly improves the phase-to-phase correlation of the GOCI band reflectance in the morning time window and retains the instability of vegetation canopy spectra in the noon time window. The SANM provides a preliminary solution for normalizing the angles for the GSS RS data and makes the quantitative comparison of spatiotemporal RS data possible.

Keywords: angle normalization; vegetation canopy reflectance; geostationary satellite; path length correction; Minnaert model; GOCI

1. Introduction

A geostationary satellite (GSS) is characterized by a wide coverage area and strong maneuverability. It can realize minute-level high-frequency observations of specific areas, which greatly improves the efficiency of remote sensing (RS) data acquisition in cloudy and rainy areas [1]. Imaging sensors deployed on the traditional GSSs only have a single channel with a wide band range in the visible and near-infrared range (VNIR), and the spatial resolution is usually less than 1 km (e.g., the Fengyun-2 satellites [2] and the GOES (Geostationary Operational Environmental Satellite) generations before the GOES-R series). In recent years, imaging sensors deployed on GSSs have developed capabilities with multiple channels in the VNIR, and spatial resolutions have increased to 50–500 m (e.g., the COMS (Communication, Ocean, and Meteorological Satellite) [3], the Gaofen-4 satellite [4], the Fengyun-4 satellites [5], the Himawari-8 satellite [6], the GOES-R series, the INSAT (Indian National Satellite System) satellite [7], the ELECTRO-L satellite [8], and the MTG (Meteosat Third Generation) satellite [9]). The optimized design of GSSs extends its application area from traditional meteorological, communications, and broadcasting to land-surface and ocean-water-color RS monitoring.

Vegetation is an important part of the Earth's ecosystem, and vegetation monitoring is the most complex part of land-surface RS monitoring. Vegetation has typical spectral characteristics and has a different canopy morphology due to differences in organizational structure, seasonal phase, and ecological conditions. Changes in canopy morphological features such as LAI (Leaf Area Index) and LAD (Leaf Angle Distribution) lead to changes in canopy porosity and extinction of cross-sectional size [10]. Therefore, it strongly influences the reflection and scattering characteristics in the optical and microwave bands, and this influence is perturbed by the terrain, illumination conditions, and observation geometry. Consequently, angle normalization should be urgently applied to RS monitoring of vegetation, which is better applied to monitoring land-surface phenology [11], biomass estimation [12], and surface vegetation patterns [13]. However, taking LAI and LAD as input parameters will reduce the usability of the angle normalization model: it is difficult to obtain ground observation of these features for large areas; remote sensing inversion products are obtained using remote sensing reflectance, and these products will introduce iteration errors. Therefore, it is necessary to use a simplified representation of BRDF (Bidirectional Reflectance Distribution Function).

The angle normalization of RS data, and of reflectivity in particular, consists of normalizing a uniform solar zenith angle and observation zenith angle, usually involving topographic correction (TC), solar angle correction or normalization (SAC), and detector angle correction or normalization (DAC). The digital elevation model (DEM)-based TC methods are the most widely used in the existing TC methods [14–16]. In recent years, many scholars have introduced non-Lambertian models and vegetation canopy structure parameters into TC methods to improve the accuracy of vegetation-canopy spectral topographic correction [17,18]. The existing SAC models use the cosine of the solar zenith angle as the main correction factor [19]. More complex algorithms introduced the intercept and slope for SAC models to solve the problem involving ground radiation signals in the presence of atmospheric scattering and refraction from the adjacent background, but no direct sunlight [20]. As for the DAC, only the 16-day synthetic products of MODIS (Moderate Resolution Imaging Spectroradiometer)/VIIRS (Visible Infrared Imaging Radiometer Suite) involving albedo and BRDF are currently widely recognized and applied. The spatial resolution of these products is 500 m, and the core of the production algorithm is the solution of a kernel-driven model [21,22].

The viewing angle on a per-pixel basis is constant, while the sun angle of GSS RS data changes from hour to hour, unlike those from sun-synchronous satellite sensors, and wide-field imaging characteristics magnify this difference [1], so it is urgent to normalize angles in the quantitative vegetation applications of GSS RS data. The operational BRDF and albedo algorithm uses a multi-day period of cloud-free angular surface reflectance that adequately samples the viewing geometry (at least seven observations) to fit an appropriate kernel-driven, RossThick-LiSparse-Reciprocal semi-empirical bidirectional reflectance model for the given surface location. However, the MODIS/VIIRS and sentinel-2A BRDF products have a lower spatial or temporal resolution, their applications are faced with the problem of spatial and temporal adaptability. Therefore, the research on angle normalization of RS data remains a hot topic and is the focus of this paper.

In this paper, the high-frequency and wide-field imaging characteristics of GSS sensors are fully exploited to propose a synthetic angle normalization model (SANM) for RS vegetation canopy reflectance. The GOCI (Geostationary, Ocean Color Imager) data obtained from GSS COMS were used to construct and verify the proposed SANM while considering the characteristics of diurnal variations in vegetation canopy spectra. The proposed SANM can provide a reference for the production of angle-normalization products for GSS RS data and optimize the temporal resolution of angle-normalization products for RS vegetation canopy reflectance, which has important applications and practical significance.

2. Materials and Methods

2.1. Synthetic Angle Normalization Model Overview

The SANM proposed herein is based on the definition of angle normalization for RS data, using GSS RS data to get the normalized reflectance with the terrain slope, solar, and detector zenith angle are all $0°$. The framework of the proposed model is presented schematically in Figure 1; the order of three core steps (TC, SAC, and DAC) was designed to satisfy the SAC and DAC models' assumption that the ground objects are aligned horizontally. Based on the literature research and comparison, the TC step uses the path-length correction (PLC) model, the SAC step uses the cosine of the solar zenith angle as the correction factor, and the DAC combines the imaging geometric coordinate rotation and the Minnaert model.

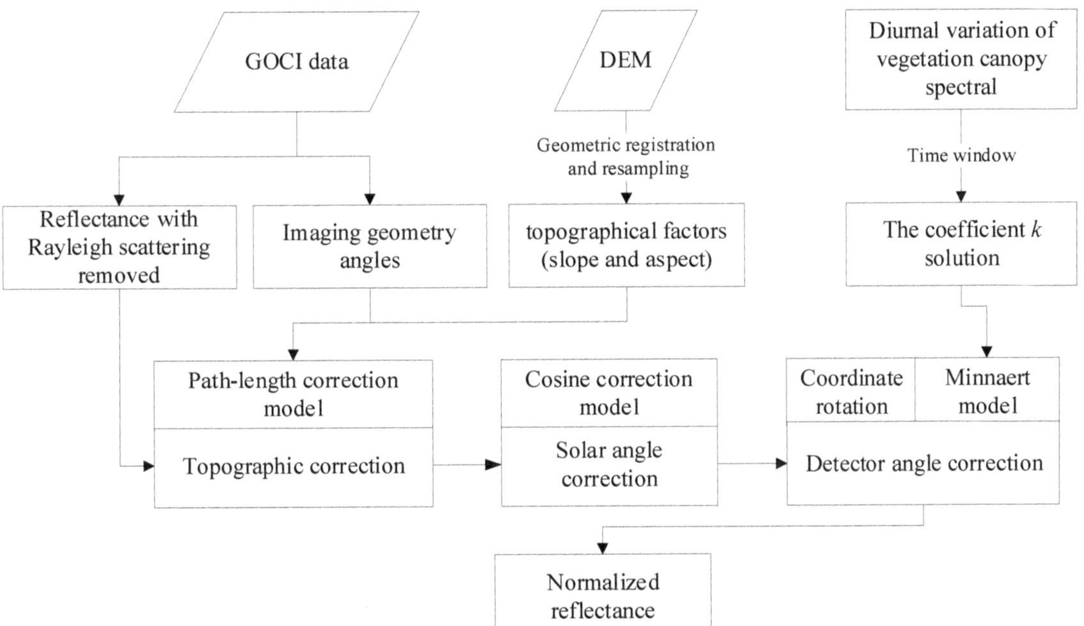

Figure 1. Schematic showing the workflow of the proposed method.

These three core steps are described in detail in the following three subsections. The angles and reflectance symbols used in each step and model application are described in Table 1.

The cosine of the angle between any two directions $\cos(\theta_{1-2})$ can be calculated as:

$$\cos(\theta_{1-2}) = \cos(\theta_1)\cos(\theta_2) + \sin(\theta_1)\sin(\theta_2)\cos(\varphi_1 - \varphi_2) \tag{1}$$

where θ_1 and θ_2 are zenith angles, and φ_1 and φ_2 are azimuth angles.

Table 1. Symbols used in the SANM.

Symbol	Explanation
θ_S	Solar zenith angle
φ_S	Solar azimuth angle
θ_D	Detector zenith angle
φ_D	Detector azimuth angle
θ_T	Slope
φ_T	Slope aspect
θ_{D-S}	Angle from observation direction to the solar incidence direction; derived from Equation (1)
θ_{S-T}	Angle from solar incidence direction to ground surface normal (solar incidence angle); derived from Equation (1)
θ_{D-T}	Angle from observation direction to ground surface normal; derived from Equation (1)
ρ_t	Vegetation canopy reflectance observed by sensor
ρ_{PLC}	Vegetation canopy reflectance after PLC model processing
ρ_{pre}	Vegetation canopy reflectance after PLC model and SACM processing
$\rho_{Minnaert}$	Vegetation canopy reflectance after Minnaert model processing
ρ_{nom}	Vegetation canopy reflectance after SANM processing

2.2. Topographic Correction for Vegetation Canopies-PLC

Vegetation grows geotropically; the terrain affects only the angle of the vegetation relative to the surface rather than the geometric relationship between the sun and the vegetation [23]. The TC method for vegetation canopies based on PLC [18] satisfies Assumption I, in which the radiance collected by the sensor is only from single scattering from leaves (i.e., the contributions from soil reflectance and from multiple scattering from leaves are negligible). In order to reduce the influence of mixed pixels and meet this assumption as far as possible, we select the mountainous area and field crop with full vegetation coverage to verify the algorithm. The relationship between ρ_t and ρ_{PLC} can be formulated as follows [18]:

$$\rho_{PLC} = \rho_t \frac{S_t(\varphi_S) + S_t(\varphi_D)}{S(\varphi_S) + S(\varphi_D)} \tag{2}$$

where $S(\varphi_S)$ and $S(\varphi_D)$ are the path lengths along the solar and viewing directions over flat terrain, respectively, and $S_t(\varphi_S)$ and $S_t(\varphi_D)$ are their counterparts over sloping terrain.

The path length along the direction of gravity is unity under any terrain conditions. The geometry of the extinction path at different angles is shown in Figure 2.

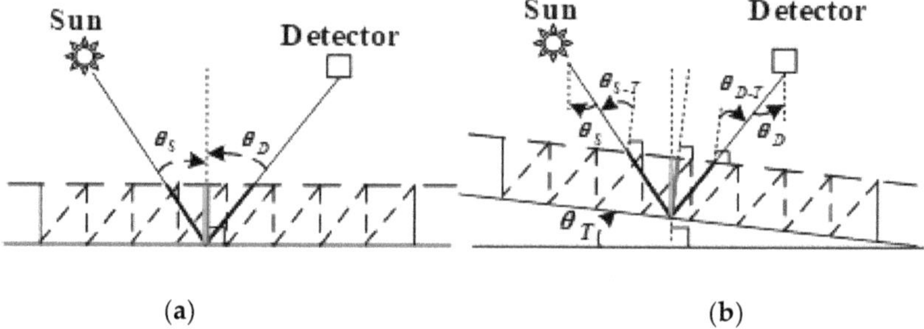

Figure 2. Path length of a (solar) beam through a canopy: (**a**) canopy on a horizontal surface; (**b**) canopy on an inclined surface. Green bold lines represent the path length along the zenith direction; it has unit magnitude. Red bold lines represent the path length along the direction normal to the vegetation canopy; its magnitude is $\cos(\theta_T)$. Black bold lines represent the path length (S) along an arbitrary direction in the vegetation canopy.

The path length in an arbitrary direction can be calculated as:

$$S(\theta_1, \varphi_1, \theta_T, \varphi_T) = \frac{\cos(\theta_T)}{\cos(\theta_{1-T})} = \frac{1}{\cos(\theta_1)[1 + \tan(\theta_1)\tan(\theta_T)\cos(\varphi_1 - \varphi_T)]} \tag{3}$$

where θ_1 is θ_D or θ_S, φ_1 is φ_D or φ_S, and θ_{1-T} is θ_{S-T} or θ_{D-T}.

2.3. Correction of Solar Angle

The solar angle includes the θ_S and the φ_S. The θ_S strongly influences the surface solar irradiance, whereas the φ_S only affects the image detail [24]. Therefore, the existing SAC models only involves the θ_S. Considering the BRDF characteristics of the land objects, we use the φ_S to calculate θ_{D-S} as a comprehensive angle to carry out the alternative correction, see section "Correction of Detector Angle" for details.

The classical SAC model (SACM) formula is usually expressed as [25]:

$$\rho_{\text{pre}} = \rho_{\text{PLC}} / \cos(\theta_S) \tag{4}$$

2.4. Correction of Detector Angle

After the TC and SAC steps, ρ_{pre} corrects for the influence of terrain and solar zenith angle, it does not take into account the difference in BRDF caused by imaging geometric differences between different phases. We rotated the coordinate to create an equivalent condition where the observation zenith angle is $0°$ (see Figure 3). Specifically, each pixel is simplified into a point object to ensure the BRDF character is unchanged; and finally, the four imaging geometric angles are converted to θ_{D-S} in DAC.

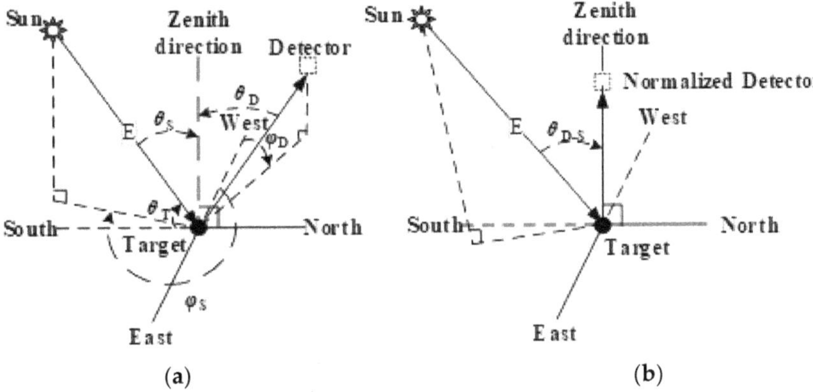

Figure 3. Schematic diagram of (a) the real imaging geometry and (b) the equivalent imaging geometry.

The Minnaert function was proposed for the TC of non-Lambertian albedo [26], where the k coefficient of the Minnaert function is the simplified representation of BRDF, and it is a constant in a given area. Note that the traditional k coefficient was found by simply applying a linear regression analysis with all types of objects in the Minnaert model early used for a single RS datum, and the optimized k coefficient was solved by applying a polynomial fit in the slope grading strategy in the modified Minnaert model to better represent the terrain change. However, the above Minnaert model ignores the influence of the ground object on the k coefficient. In this paper, the k coefficient is solved pixel by pixel using the high-frequency imaging feature of GSS RS data.

The DAC formula based on coordinate rotation and the Minnaert model can be expressed as:

$$\rho_{\text{Minnaert}} = \rho_{\text{pre}} \cos(\theta_T) / [\cos(\theta_T)\cos(\theta_{D-S})]^k \tag{5}$$

The θ_T of each pixel has been corrected to $0°$ after the application of Equation (4), so Equation (5) can be further reduced to:

$$\rho_{nom} = \rho_{pre}/[\cos(\theta_{D-S})]^k \tag{6}$$

Note that the diurnal variation in the vegetation canopy spectra based on field experiments [27] and related studies [28] shows that the local time period before 11:00 (called the morning time window) and after 13:30 (called the afternoon time window) are the periods when the vegetation canopy spectrum itself is relatively stable; whereas the local time period from 11:00 to 13:30 (called the noon time window) is when the vegetation canopy spectrum changes drastically. Thus, to ensure that the vegetation canopy spectrum itself is relatively stable for data screening, we find the k coefficient as a function of the time window for each pixel.

2.5. Study Area and Data

The study area was located at the junction point of Jiangsu province and Anhui province of China (117°57′43″ E~118°38′13″ E, 32°09′43″ N~32°24′14″ N) (see Figure 4a). The study area spans in altitude from −52 to 392 m (see Figure 4b), and its slope ranges from $0°$ to $30°$. The conventional crops include wheat, rice, rapeseed, soybean, etc., and forests include poplar, Masson pine, etc.

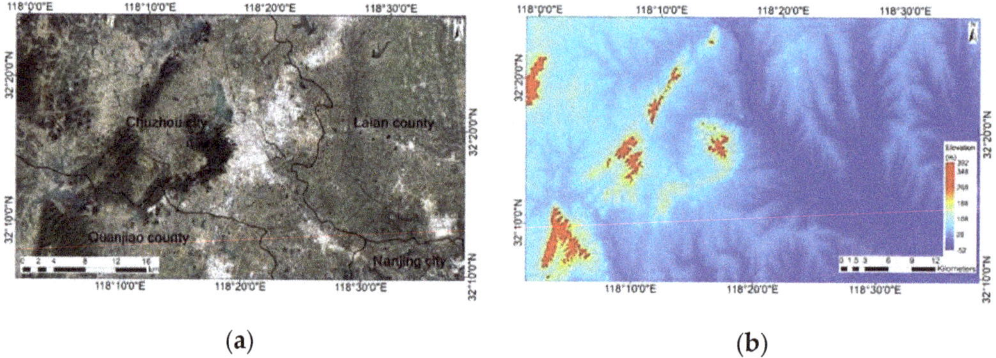

(a)   **(b)**

Figure 4. (a) Geographic location and (b) DEM of the study area.

The GOCIs onboard the Communication, Ocean, and Meteorological Satellite (COMS), observation area of 2500 × 2500 km is centered on the Korean Peninsula (130° E, 36° N) and supports a spatial resolution of 500 m; the spectral features are shown in Table 2. The GOCI is capable of producing images at hourly intervals and receives images eight times a day from 08:15 to 15:45 CST (China Standard Time UT + 8:00).

Table 2. GOCI satellite band parameter information.

	B1	B2	B3	B4	B5	B6	B7	B8
Band length (nm)	412	443	488	555	660	680	745	865
Band width (nm)	20	20	20	20	20	10	20	40

The GOCI images acquired on 22 April 2015 were used for model verification because of the advantageous winter wheat growth cycle and the good spatial distribution of the cloud coverage for the GOCI images. In the study area, 22 April 2015 was during the jointing stage of winter wheat; the crops appeared to be growing well with full ground coverage. However, the GOCI images received after 14:00 CST on 22 April 2015 suffered from thin cloud coverage in the study area, so these two images were not used in the data processing and analysis. Subsequently, each image is represented by the imaging hour in CST.

We first used the GDPS (the GOCI data-processing system) to process GOCI L1B data to obtain the Rayleigh-corrected reflectance, the latitudes and longitudes of the four corner points and the center point, the solar angles and observation angles of each pixel, etc. We then subset the reflectance products according to the coordinate range of the study area. Furthermore, for comprehensive considerations of the synchronous ground observation experiment on winter wheat [29] and the GOCI pixel NDVI covering the samples, we took 0.6 as the NDVI threshold and used LAND_NDVI products to screen the ground object type, and the model was applied only to the 5292 selected vegetation pixels. Finally, after projection conversion and resembling operations, the 90 m Chinese resolution digital elevation data product was used to calculate the topographical factors (slope and aspect) of each pixel of the GOCI reflectance products after geometric registration and resampling.

2.6. Method Evaluation Strategies

Numerous strategies have been used to assess the performance of topographic correction methods and solar normalization methods [18,30]. To obtain an objective evaluation, we used three different methods:

(i)   Correlation analysis between reflectance in different imaging periods. Because the vegetation canopy spectrum is relatively stable in the morning time window, the effective angle normalization model should strengthen the reflectance correlation of different imaging phases in the morning time window and make the slope of the linear regression equation closer to unity. Conversely, the vegetation canopy spectrum changes drastically in the noon time window, so the effective angle normalization model should weaken the reflectance correlation of different imaging phases and make the slope of the linear regression equation further depart from unity.

(ii) Analysis of the correlation between the cosine of the imaging geometry angles and reflectance. This is one of the most widely used quantitative evaluation methods. The efficiency of the normalization methods can be quantified by using R^2 and the imaging geometry angles of the corresponding linear regression. The ideal normalization method should make R^2 approach zero [31].

(iii) Radiometric stability. Theoretically, the maximum (minimum) reflectance in the original image before correction should appear in the sunny (shady) slope and will decrease (increase) after topographic correction. Consequently, a successful correction method will reduce the reflectance range. Moreover, the median reflectance is relatively stable and invariable after correction [30].

3. Results

According to the typical vegetation spectral characteristics, the bands 400–730 nm and 730–900 nm are two typical spectral bands in the winter wheat canopy spectrum [29]. GOCI band 5 (650–670 nm) and band 8 (845–885 nm) are used to produce NDVI (Normalized Difference Vegetation Index) products and were selected for model application analysis.

3.1. Correlation between Different Imaging Phases

To comprehensively compare how normalizing the angles affects the treatment of the models of the GOCI reflectance bands, Table 3 shows the detailed regression results for the band 5 reflectance and band 8 reflectance for different imaging hours.

Table 3 shows that the correlations for the band 8 reflectance between different imaging phases are significantly better than for the band 5 reflectance in the corresponding phases, which is consistent with the diurnal variation in the field-measured reflectance spectra of the vegetation canopy [29]. The slope of the linear fit and R^2 in Table 3 further indicates that the normalization has no effect on the results of the PLC model for the GOCI reflectance correlation between different imaging phases, the SACM suffers from over-correction, and the SANM not only significantly reduces the over-correction of the SACM but also preserves the instability of the vegetation canopy reflectance spectra in the noon time window.

Table 3. Slope and R^2 of fit for GOCI band 5 reflectance and GOCI band 8 reflectance between different imaging times.

	Imaging Hour	Linear Fit	Band 5 Reflectance				Band 8 Reflectance			
			Ori	PLC	SACM	SANM	Ori	PLC	SACM	SANM
Morning window	08–09	Slope	1.014	0.998	0.802	0.976	0.924	0.917	0.731	1.021
		R^2	0.768	0.767	0.771	0.889	0.901	0.900	0.901	0.951
	08–10	Slope	1.144	1.116	0.785	1.008	0.857	0.848	0.589	0.994
		R^2	0.844	0.832	0.847	0.993	0.862	0.856	0.860	0.997
	09–10	Slope	0.958	0.957	0.831	0.891	0.904	0.903	0.785	0.914
		R^2	0.792	0.794	0.793	0.833	0.906	0.906	0.906	0.925
Noon window	11–12	Slope	0.804	0.808	0.797	0.786	0.877	0.878	0.871	0.814
		R^2	0.705	0.708	0.705	0.718	0.871	0.872	0.871	0.855
	11–13	Slope	0.847	0.849	0.887	0.801	0.859	0.859	0.902	0.689
		R^2	0.324	0.327	0.328	0.368	0.798	0.799	0.798	0.703
	12–13	Slope	0.848	0.849	0.890	0.846	0.962	0.962	1.016	0.860
		R^2	0.298	0.302	0.298	0.353	0.883	0.884	0.884	0.850

3.2. Sensitivity to Imaging Geometry Angles

Band 8 normalization has a consistent effect with band 5, but with higher reflectance, so we take band 5 reflectance from 08:15 CST as an example; Figure 5 compares the cosine of the imaging geometry angle with the reflectance before and after each normalization model (i.e., the PLC model, the SACM, and the proposed SANM).

The correlation is extremely weak between the original band 5 reflectance with $\cos(\theta_{S-T})$, $\cos(\theta_{D-T})$, and $\cos(\text{slope})$: R^2 for the linear fit is 2.88×10^{-4} (see Figure 5a), 0.001 (see Figure 5e), and 0.011 (see Figure 5i). These results are attributed to the small difference in imaging geometry when the study area is small. The use of the PLC model significantly improves the correlation between the band 5 reflectance and $\cos(\theta_{S-T})$ and $\cos(\theta_{D-T})$: R^2 for the linear fit increased to 0.038 (see Figure 5b) and 0.03 (see Figure 5f). The use of the SACM significantly reduced the correlation between band 5 reflectance and $\cos(\theta_{S-T})$ and $\cos(\text{slope})$: R^2 for the linear fit decreased to 1.402×10^{-4} (see Figure 5c) and 0.01 (see Figure 5k). The use of the SANM significantly improved the correlation between band 5 reflectance and $\cos(\theta_{D-T})$: R^2 for the linear fit increased to 0.015 (see Figure 5h) from the original 0.001 (see Figure 5e); whereas the correlation is significantly reduced between band 5 reflectance and $\cos(\theta_{S-T})$ and $\cos(\text{slope})$: R^2 for the linear fit decreased to 1.6×10^{-4} (see Figure 5c) and 0.004 (see Figure 5k). These results indicate that the normalization by SANM proposed herein has a better effect on the solar angle of incidence and slope (i.e., a lower R^2); however, it presents a poor normalization effect on θ_{D-T}.

3.3. Radiometric Stability

Theoretically, after correction, the reflectance ranges should be contained in their counterparts before correction [30]. Box plots of band 5 reflectance and band 8 reflectance from the uncorrected and corrected images shows that each angle normalization for a given model has the same effect on the reflectance of bands 5 and 8, and the reflectance distribution is more concentrated when the mean reflectance is lower (see Figure 6). Figure 6 also shows that the PLC model did not change the distribution of the GOCI band reflectance and the variations in the imaging phases: the SACM suffered from over-correction, which increased with the solar zenith angle, and the SANM significantly improved the over-correction problem of the SACM. The band reflectances were stable in the morning time window and decreased in the noon time window after SANM processing, which is consistent with the intraday variation of the field-measured reflectance spectra of the vegetation canopy [28].

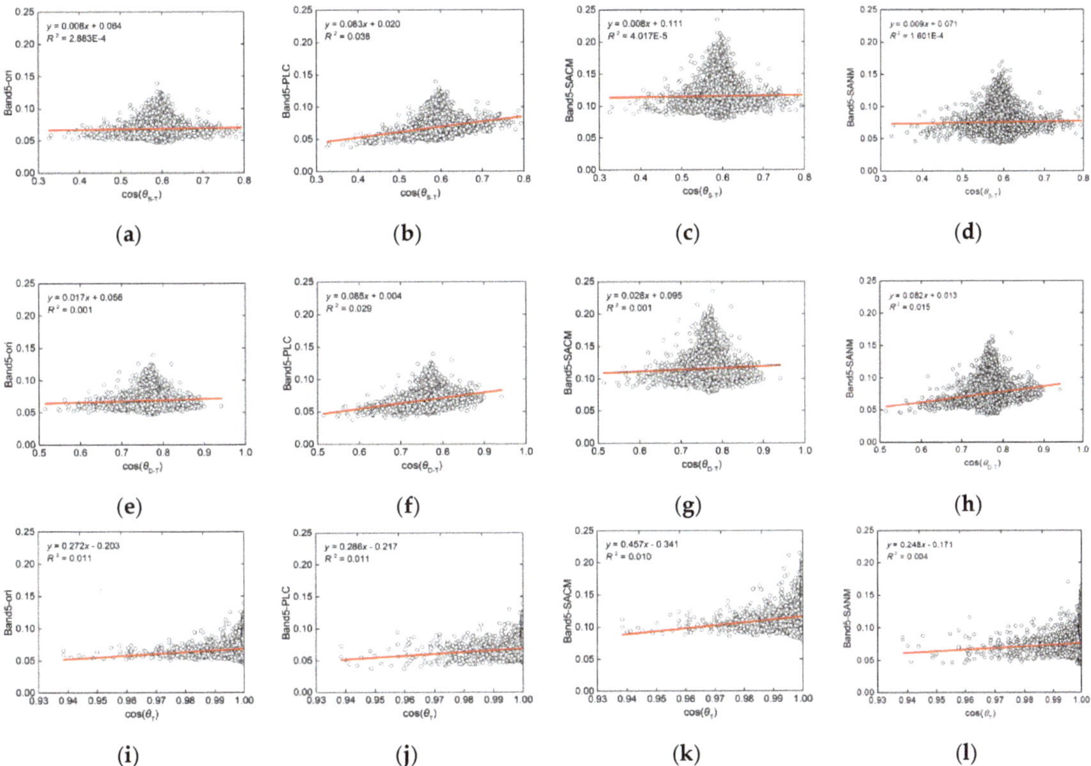

Figure 5. (**a**) Density scatter plots between the original Rayleigh-corrected GOCI band 5 reflectance (Band5-ori) at 08:15 CST and $\cos(\theta_{S-T})$, the red line is linearly fit to data; (**b**) Same as (**a**) except using the GOCI band 5 reflectance after PLC correction (Band5-PLC); (**c**) Same as (**a**) except between the GOCI band 5 reflectance after ASCM correction with Band5-ori as ρ_t (Band5-SACM); (**d**) Same as (**a**) except between the GOCI band 5 reflectance after SANM normalization (Band5-SANM); (**e**) Same as (**a**) except using the $\cos(\theta_{D-T})$; (**f–h**) Same as (**e**) except using Band5-PLC, Band5-SACM, and Band5-SANM, respectively; (**i**) Same as (**a**) except using the $\cos(\theta_T)$; (**j–l**) Same as (**i**) except using Band5-PLC, Band5-SACM, and Band5-SANM, respectively.

Figure 6a shows that the mean original reflectance of band 5 increases from 08:15 to 11:15 CST, after which it decreases. After PLC processing, band 5 reflectance underwent no significant change in range or distribution (Figure 6b) compared with Figure 6a and retained the variations of band 5 reflectance for the various imaging phases. Figure 6c shows that, because of the over-correction problem, large differences exist in the band 5 reflectance after SACM processing. The mean reflectance of band 5 after SACM processing decreased along the imaging phase and increased up to the 04 phase. Figure 6d shows that the band 5 reflectance in the 00 phase to the 03 phase are more similar after SANM processing, and the band 5 reflectance in the 04 phase and 05 phase are lower than those for the other phases after SANM processing. Since all selected pixels found almost no shadows, the reduction in the reflectance range is inconspicuous.

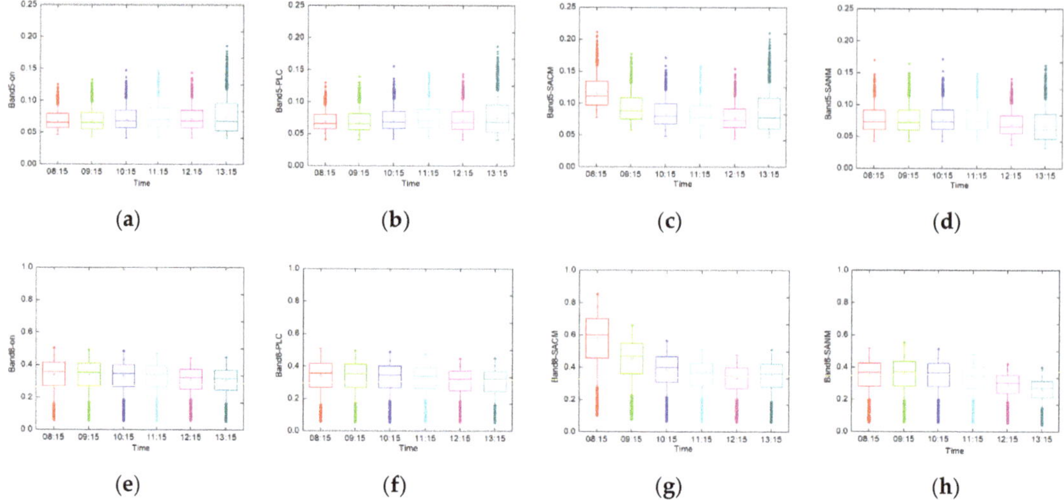

Figure 6. Box plots of (**a**) the original Rayleigh-corrected GOCI band 5 reflectance; (**b**) the GOCI band 5 reflectance after PLC correction (Band5-PLC); (**c**) the GOCI band 5 reflectance after ASCM correction with Band5-ori as ρ_t (Band5-SACM); (**d**) the GOCI band 5 reflectance after SANM normalization (Band5-SANM) in each imaging time for the morning time window and the double-peak time window; (**e**–**h**) same as (**a**–**d**) correspondingly except using band 8 reflectance.

4. Discussion

With the help of GSS RS data, the proposed SANM can improve the time resolution of angle-normalized products for RS reflectance from a vegetation canopy to the hourly level. However, the following problems exist in the model-construction process:

In the TC step, the extinction path-length formula is derived as a hypothetical condition for a dense canopy without considering the effects of a sparse canopy [32]. However, the actual vegetation canopy structure usually has daily, quarterly, and annual variations and regional differences, which strongly impact the BRDF and biomass retrieval [33]. In the subsequent model optimization, we propose to introduce vegetation cover factor variables to distinguish how a dense canopy versus a sparse canopy affects the reflectance spectrum from a vegetation canopy [34,35].

In the SAC step, we used the simplest cosine correction model and did not consider whether the fit to the reflectance and cosine of the solar zenith angle passes through the origin, which depends on atmospheric scattering and refraction from adjacent pixels [36]. Subsequent research should introduce the intercept and slope into the SAC step for optimization. However, the difficulty is the determination of the intercept, especially in the case of large changes in solar angle caused by the wide field and high frequency of GSSs.

In the DAC step, we set the time window when solving for the Minnaert model k coefficient. The given time window only considered few a diurnal variations of the reflectance spectra from the vegetation canopy, which limited the effective data used for calculating the k coefficient. In the follow-up study, a database will be created of the diurnal variations of reflectance spectra from canopies of different types of vegetation through literature research and field measurements so as to screen data to more accurately solve for the k coefficient. In addition, the distribution of the k coefficient obtained herein exceeded the conventional range of 0–1 (see Figure 7), which may be due to the GSS imaging regions that are located in the backscattering area [26]. Subsequent research should study how the scattering orientation affects the k coefficient and determine the range of the k coefficient under the conditions of fixed observation angle.

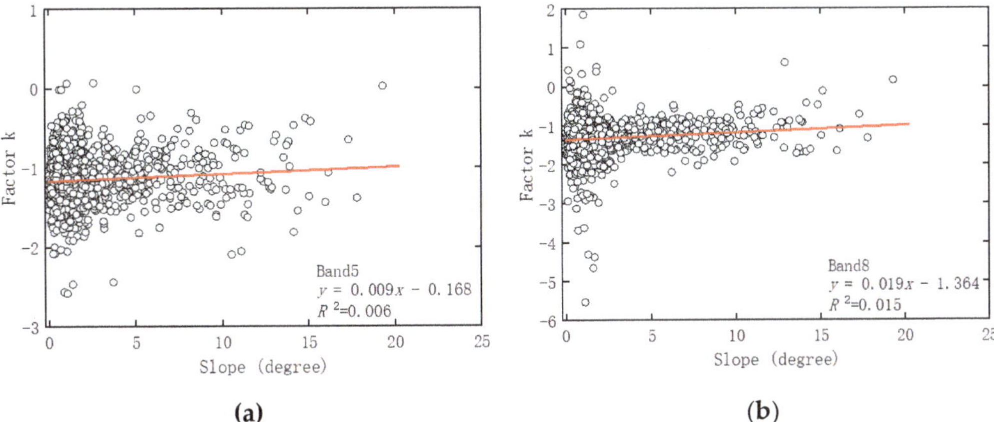

Figure 7. Density scatterplots between the slope and factor k in SANM for GOCI images band 5 (**a**) and band 8 (**b**) ground object reflectance.

5. Conclusions

By using the PLC model, the cosine model for solar-angle normalization, and the Minnaert model, we herein establish a SANM for the reflectance of the GSS RS vegetation canopy. GOCI images were used to test the SANM, and a multi-criteria analysis was used in the evaluation. With the PLC model, normalization has no effect on the correlation of the GOCI reflectance between different imaging phases. However, the correlation is significantly improved between the band reflectance with the cosine of the solar angle of incidence and the cosine of the angle from the observation direction to the ground surface normal. The SACM significantly reduced the correlation between band 5 reflectance with the cosine of the solar angle of incidence and slope, but it suffered from over-correction. Whereas the SANM significantly improved the over-correction problem for the SACM and also preserved the instability of the vegetation canopy spectra in the noon time window. The use of the SANM significantly reduced the correlation between the band reflectance with the cosine of the solar angle of incidence and the slope. For normalizing the angle of the high-frequency GSS RS, the SANM outperformed all other methods, which indicates that it has a strong potential for applications and for monitoring land-surface phenology, estimating biomass, etc.

Author Contributions: Conceptualization, Y.L. (Yinghao Lin) and Q.T.; methodology, Y.L. (Yinghao Lin) and B.Q.; software, Y.W. and Y.X.; validation, Y.L. (Yinghao Lin), X.Z. and Y.W.; writing—original draft preparation, Y.L. (Yinghao Lin) and Q.T.; writing—review and editing, Y.L. (Yinghao Lin) and Y.L. (Yang Lian); visualization, Y.L. (Yinghao Lin) and Y.X.; funding acquisition, Y.L. (Yinghao Lin) and Y.W. All authors have read and agreed to the published version of the manuscript.

Funding: This research was funded by Key R&D and Promotion Projects of Henan Province, grant number 222102320163; China High-resolution Earth Observation System, grant number 80-Y50G19-9001-22/23; Major Project of Science and Technology of Henan Province, grant number 201400210300; National Natural Science Foundation of China, grant number 42071318; National Defense Basic Research Projects of China, grant number JCKY2020908B001; National Basic Research Program of China, grant number 2019YFE0126600 and Kaifeng science and technology development plan, grant number 2002001.

Institutional Review Board Statement: Not applicable.

Informed Consent Statement: Not applicable.

Data Availability Statement: The GOCI data and GDPS can be downloaded from: http://kosc.kiost.ac.kr/index.nm?menuCd=54&lang=en (accessed on 10 June 2022).

Acknowledgments: The author would like to thank all contributors to this study.

Conflicts of Interest: The authors declare no conflict of interest.

References

1. Hashimoto, H.; Wang, W.; Dungan, J.L.; Li, S.; Michaelis, A.R.; Takenaka, H.; Higuchi, A.; Myneni, R.B.; Nemani, R.R. New generation geostationary satellite observations support seasonality in greenness of the Amazon evergreen forests. *Nat. Commun.* **2021**, *12*, 684. [CrossRef] [PubMed]
2. Yang, J.; Jiang, L.; Shi, J.; Wu, S.; Sun, R.; Yang, H. Monitoring snow cover using Chinese meteorological satellite data over China. *Remote Sens. Environ.* **2014**, *143*, 192–203. [CrossRef]
3. Wang, M.; Ahn, J.-H.; Jiang, L.; Shi, W.; Son, S.; Park, Y.-J.; Ryu, J.-H. Ocean color products from the Korean Geostationary Ocean Color Imager (GOCI). *Opt. Express* **2013**, *21*, 3835. [CrossRef] [PubMed]
4. Wang, M.; Cheng, Y.; Chang, X.; Jin, S.; Zhu, Y. On-orbit geometric calibration and geometric quality assessment for the high-resolution geostationary optical satellite GaoFen4. *ISPRS J. Photogramm. Remote Sens.* **2017**, *125*, 63–77. [CrossRef]
5. Min, M.; Wu, C.; Li, C.; Liu, H.; Xu, N.; Wu, X.; Chen, L.; Wang, F.; Sun, F.; Qin, D.; et al. Developing the science product algorithm testbed for Chinese next-generation geostationary meteorological satellites: Fengyun-4 series. *J. Meteorol. Res.* **2017**, *31*, 708–719. [CrossRef]
6. Yumimoto, K.; Nagao, T.M.; Kikuchi, M.; Sekiyama, T.T.; Murakami, H.; Tanaka, T.Y.; Ogi, A.; Irie, H.; Khatri, P.; Okumura, H.; et al. Aerosol data assimilation using data from Himawari-8, a next-generation geostationary meteorological satellite. *Geophys. Res. Lett.* **2016**, *43*, 5886–5894. [CrossRef]
7. Di, A.; Xue, Y.; Yang, X.; Leys, J.; Guang, J.; Mei, L.; Wang, J.; She, L.; Hu, Y.; He, X.; et al. Dust aerosol optical depth retrieval and dust storm detection for Xinjiang Region using Indian national satellite observations. *Remote Sens.* **2016**, *8*, 702. [CrossRef]
8. Bloshchinskiy, V.D.; Kuchma, M.O.; Andreev, A.I.; Sorokin, A.A. Snow and cloud detection using a convolutional neural network and low-resolution data from the Electro-L No. 2 Satellite. *J. Appl. Remote Sens.* **2020**, *14*, 1. [CrossRef]
9. Aminou, D.M.; Lamarre, D.; Stark, H.; Van Den Braembussche, P.; Blythe, P.; Fowler, G.; Gigli, S.; Stuhlmann, R.; Rota, S. Meteosat Third Generation (MTG) status of space segment definition. In *Sensors, Systems, and Next-Generation Satellites XIII*; SPIE: Berlin, Germany, 2009; Volume 7474, p. 747406.
10. Du, H.; Liu, Q.; Li, J.; Yang, L. Retrieving crop leaf area index by combining optical and microwave vegetation indices: A feasibility analysis. *Yaogan Xuebao/J. Remote Sens.* **2013**, *17*, 1587–1611. [CrossRef]
11. Babcock, C.; Finley, A.O.; Looker, N. A Bayesian model to estimate land surface phenology parameters with harmonized Landsat 8 and Sentinel-2 images. *Remote Sens. Environ.* **2021**, *261*, 112471. [CrossRef]
12. Puliti, S.; Breidenbach, J.; Schumacher, J.; Hauglin, M.; Klingenberg, T.F.; Astrup, R. Above-ground biomass change estimation using national forest inventory data with Sentinel-2 and Landsat. *Remote Sens. Environ.* **2021**, *265*, 112644. [CrossRef]
13. Novillo, C.J.; Arrogante-Funes, P.; Romero-Calcerrada, R. Improving land cover classifications with multiangular data: MISR data in mainland Spain. *Remote Sens.* **2018**, *10*, 1717. [CrossRef]
14. Teillet, P.M.; Guindon, B.; Goodenough, D.G. On the slope-aspect correction of multispectral scanner data. *Can. J. Remote Sens.* **1982**, *8*, 84–106. [CrossRef]
15. Blesius, L.; Weirich, F. The use of the Minnaert correction for land-cover classification in mountainous terrain. *Int. J. Remote Sens.* **2005**, *26*, 3831–3851. [CrossRef]
16. Ekstrand, S. Landsat TM-based forest damage assessment: Correction for topographic effects. *Photogramm. Eng. Remote Sens.* **1996**, *62*, 151–161.
17. Li, H.; Xu, L.; Shen, H.; Zhang, L. A general variational framework considering cast shadows for the topographic correction of remote sensing imagery. *ISPRS J. Photogramm. Remote Sens.* **2016**, *117*, 161–171. [CrossRef]
18. Yin, G.; Li, A.; Wu, S.; Fan, W.; Zeng, Y.; Yan, K.; Xu, B.; Li, J.; Liu, Q. PLC: A simple and semi-physical topographic correction method for vegetation canopies based on path length correction. *Remote Sens. Environ.* **2018**, *215*, 184–198. [CrossRef]
19. Kowalik, W.S.; Marsh, S.E.; Lyon, R.J.P. A relation between landsat digital numbers, surface reflectance, and the cosine of the solar zenith angle. *Remote Sens. Environ.* **1982**, *12*, 39–55. [CrossRef]
20. Li, L. *The Influence of the Satellite Observation and Sunshine Direction on Vegetation-Shade—A Case Study of Qinghai-Tibet Railway*; China University of Geosciences: Beijing, China, 2016.
21. Liu, Y.; Wang, Z.; Sun, Q.; Erb, A.M.; Li, Z.; Schaaf, C.B.; Zhang, X.; Román, M.O.; Scott, R.L.; Zhang, Q.; et al. Evaluation of the VIIRS BRDF, Albedo and NBAR products suite and an assessment of continuity with the long term MODIS record. *Remote Sens. Environ.* **2017**, *201*, 256–274. [CrossRef]
22. Roy, D.P.; Li, J.; Zhang, H.K.; Yan, L.; Huang, H.; Li, Z. Examination of Sentinel-2A multi-spectral instrument (MSI) reflectance anisotropy and the suitability of a general method to normalize MSI reflectance to nadir BRDF adjusted reflectance. *Remote Sens. Environ.* **2017**, *199*, 25–38. [CrossRef]
23. Chance, C.M.; Hermosilla, T.; Coops, N.C.; Wulder, M.A.; White, J.C. Effect of topographic correction on forest change detection using spectral trend analysis of Landsat pixel-based composites. *Int. J. Appl. Earth Obs. Geoinf.* **2016**, *44*, 186–194. [CrossRef]
24. Li, L.; Hu, Y.; Gong, C.; He, H. Solar elevation angle's effect on image energy and its correction. *J. Atmos. Environ. Opt.* **2013**, *8*, 11–17.

25. Wolfe, R.E.; Roy, D.P.; Vermote, E. MODIS land data storage, gridding, and compositing methodology: Level 2 grid. *IEEE Trans. Geosci. Remote Sens.* **1998**, *36*, 1324–1338. [CrossRef]
26. Gao, M.; Gong, H.; Zhao, W.; Chen, B.; Chen, Z.; Shi, M. An improved topographic correction model based on Minnaert. *GISci. Remote Sens.* **2016**, *53*, 247–264. [CrossRef]
27. Lin, Y.; Shen, H.; Tian, Q.; Gu, X. Improving leaf area index retrieval using spectral characteristic parameters and data splitting. *Int. J. Remote Sens.* **2020**, *41*, 1741–1759. [CrossRef]
28. Guo, J.; Wang, Q.; Tong, Y.; Fei, D.; Liu, J. Effect of solar radiation intensity and observation angle on canopy reflectance hyperspectra for winter wheat. *Nongye Gongcheng Xuebao/Trans. Chin. Soc. Agric. Eng.* **2016**, *32*, 157–163. [CrossRef]
29. Lin, Y.; Shen, H.; Tian, Q.; Gu, X.; Yang, R.; Qiao, B. Mechanisms underlying diurnal variations in the canopy spectral reflectance of winter wheat in the jointing stage. *Curr. Sci.* **2020**, *118*, 1401–1406. [CrossRef]
30. Sola, I.; González-Audícana, M.; Álvarez-Mozos, J. Multi-criteria evaluation of topographic correction methods. *Remote Sens. Environ.* **2016**, *184*, 247–262. [CrossRef]
31. Yin, G.; Li, A.; Zhao, W.; Jin, H.; Bian, J.; Wu, S. Modeling Canopy Reflectance over Sloping Terrain Based on Path Length Correction. *IEEE Trans. Geosci. Remote Sens.* **2017**, *55*, 4597–4609. [CrossRef]
32. Kane, V.R.; Gillespie, A.R.; McGaughey, R.; Lutz, J.A.; Ceder, K.; Franklin, J.F. Interpretation and topographic compensation of conifer canopy self-shadowing. *Remote Sens. Environ.* **2008**, *112*, 3820–3832. [CrossRef]
33. Yang, G.; Pu, R.; Zhang, J.; Zhao, C.; Feng, H.; Wang, J. Remote sensing of seasonal variability of fractional vegetation cover and its object-based spatial pattern analysis over mountain areas. *ISPRS J. Photogramm. Remote Sens.* **2013**, *77*, 79–93. [CrossRef]
34. Wen, J.; Liu, G.; Gong, Z.; Pang, Y.; Cai, Z.; Xua, J. Aquatic Vegetation Canopy Spectral Characteristics under Different Coverage Percentages. *J. Appl. Spectrosc.* **2018**, *85*, 885–890. [CrossRef]
35. Gao, R.; Xie, Y.; Gu, X.; Han, J.; Sun, Y.; Liu, J. A model of topographic radiance correction in view of fractional vegetation cover. *Sci. Surv. Mapp.* **2016**, *41*, 132–138.
36. Schott, J. *Remote Sensing: The Image Chain Approach*; Oxford University Press: New York, USA, 2007; Volume 45, ISBN 9780195178173.

 *agriculture*

Article

Winter Wheat Yield Prediction Using an LSTM Model from MODIS LAI Products

Jian Wang [1], Haiping Si [1], Zhao Gao [2] and Lei Shi [1,*]

[1] College of Information and Management Science, Henan Agricultural University, Zhengzhou 450002, China
[2] The First Geodetic Survey Team of the Ministry of Natural Resources, Shaanxi Bureau of Surveying, Mapping and Geoinformation, Xi'an 710054, China
* Correspondence: shilei@henau.edu.cn; Tel.: +86-0371-56990030

Abstract: Yield estimation using remote sensing data is a research priority in modern agriculture. The rapid and accurate estimation of winter wheat yields over large areas is an important prerequisite for food security policy formulation and implementation. In most county-level yield estimation processes, multiple input data are used for yield prediction as much as possible, however, in some regions, data are more difficult to obtain, so we used the single-leaf area index (LAI) as input data for the model for yield prediction. In this study, the effects of different time steps as well as the LAI time series on the estimation results were analyzed for the properties of long short-term memory (LSTM), and multiple machine learning methods were compared with yield estimation models constructed by the LSTM networks. The results show that the accuracy of the yield estimation results using LSTM did not show an increasing trend with the increasing step size and data volume, while the yield estimation results of the LSTM were generally better than those of conventional machine learning methods, with the best R^2 and RMSE results of 0.87 and 522.3 kg/ha, respectively, in the comparison between predicted and actual yields. Although the use of LAI as a single input factor may cause yield uncertainty in some extreme years, it is a reliable and promising method for improving the yield estimation, which has important implications for crop yield forecasting, agricultural disaster monitoring, food trade policy, and food security early warning.

Keywords: winter wheat; yield estimation; LSTM; LAI; deep learning

Citation: Wang, J.; Si, H.; Gao, Z.; Shi, L. Winter Wheat Yield Prediction Using an LSTM Model from MODIS LAI Products. *Agriculture* **2022**, *12*, 1707. https://doi.org/10.3390/agriculture12101707

Academic Editor: Koki Homma

Received: 25 August 2022
Accepted: 12 October 2022
Published: 17 October 2022

Publisher's Note: MDPI stays neutral with regard to jurisdictional claims in published maps and institutional affiliations.

1. Introduction

Wheat is an important crop in China, and its yield is directly related to the development of the national economy. Timely, accurate, and wide-ranging monitoring and forecasting of wheat yields is of great practical significance for national economic development and food policy formulation [1,2]. Due to its large coverage area and short detection period, satellite remote sensing provides a new technical tool for large-scale crop estimation and is rapidly becoming the most widely used technology in crop estimation.

At present, the methods of crop yield estimation using remote sensing technology can be broadly classified into three categories according to the characteristics of the models used: (1) the empirical modeling method; (2) the mechanistic modeling method; and (3) the semi-empirical (semi-mechanistic) modeling method. The empirical model directly uses spectral vegetation indices or canopy remote sensing inversion parameters to establish relationships with crop yields, which are characterized by their simplicity and ease, involving fewer crop yield formation mechanisms, and the relationships are generally established using conventional machine learning methods, such as support vector machine and random forest, with NDVI or leaf area index (LAI) as input parameters [3–8]. Such relationships are usually localized and difficult to generalize to other agricultural areas. Semi-empirical models and semi-mechanical models are also known as parametric models, among which the light energy utilization efficiency model is the most widely used [6,9–11], but some parameters are difficult to quantitatively simulate. Mechanical models fully consider the mechanism of crop yield formation, but their solution process is complex and requires more input parameters, and some necessary parameters in the operation are difficult to obtain at a regional scale; meanwhile, most of these models estimate crop yields at the field scale [12–15]. Although the models work well for archiving yields, the accuracy may be reduced when scaled to the national level. Many a priori parameters are required in regional estimation. Due to the heterogeneity of the ground surface, the accuracy of the ground parameters is generally low, especially in the case of small farmland in China, resulting in low regional accuracy. Moreover, the computational process is complicated and requires many parameters, which can be limited in practical use.

In recent years, deep learning has been successfully applied to several fields, such as image recognition and language translation [11,16–22]. Compared with traditional machine learning methods, deep learning techniques often achieve better performance. CNN and recurrent neural network (RNN) are more widely used models in neural networks and have also been applied to crop yield estimation and prediction [19,23–29]. LSTM is a special kind of RNN [30,31], due to its recursive structure and gating mechanism that regulates the entry and exit of information into and out of cells, as well as its processing of sequential data. The LSTM has feedback connections and can handle the input sequences of arbitrary length and is often preferred in the classification, processing, and prediction based on time series data. Several studies used LSTM for crop yield prediction with impressive results. LSTM not only captures trends in the data but also describes the dependencies of the time series data. Tian et al. built an LSTM model by integrating the meteorological data and two remote sensing indices (vegetation temperature condition index (VTCI) and LAI) to estimate wheat yield in Guanzhong Plain [32]. Jeong et al. used water-related indices and the maximum temperature as inputs for rice yield prediction using an LSTM model, which showed reliable early prediction accuracy [16]. Sun et al. used a CNN-LSTM model to predict the end-of-season and in-season yields of soybean in the county. The input data for the model included meteorological data and MODIS surface temperature (LST) [24]. The LSTM model was shown to have high prediction accuracy for crop yield estimation, but all of the above methods for estimating crop yield use multiple data as input parameters, and it is difficult to obtain non-remote sensing data in some regions [33–35]. There are a large number of mature remote sensing products for LAI data, so this study mainly considered using LAI remote sensing products as single model input data and what kind of accuracy could be achieved when using deep learning algorithms for yield estimation.

In this study, LSTM was used to estimate the winter wheat yield at the county scale based on the relationship between time series LAI products and winter wheat yield. Considering the simplicity of obtaining LAI data, the model input parameters were only the leaf area index to verify the accuracy that could be achieved under the influence of a single factor. Moreover, the time step of the input remote sensing data of the model was considered so as to determine the accuracy in different time data.

2. Materials and Methods

2.1. Study Area

Henan Province is located in eastern central China (31°23′~36°22′ N, 110°21′~116°39′ E). Figure 1 is a schematic diagram of the location of the study area. Henan Province is located between the warm temperate zone and the subtropical zone. The terrain of Henan Province is high in the west and low in the east, with mountains above 1000 m above sea level in the west and plains below 100 m in the east. Mountains and hills account for 44.3% of the total area, and plains account for 55.7%. The total wheat output of Henan Province ranks first in China, accounting for more than 28% of the country's total wheat output. The sowing time of wheat varies from north to south by nearly two weeks, and there is a large gap in yield in different regions. The topography of Henan Province is complex and diverse, the topography is low in the east and high in the west, with significant differences; the surface morphology is complex and diverse. Due to the influence of landforms and monsoons, Henan Province has a wide variety of soil types and large differences in climate resources, resulting in significant differences in crop yields in different regions. According to the production conditions of producing areas and the climatic characteristics of the wheat growth period, the wheat-growing areas in Henan Province can be divided into five regions.

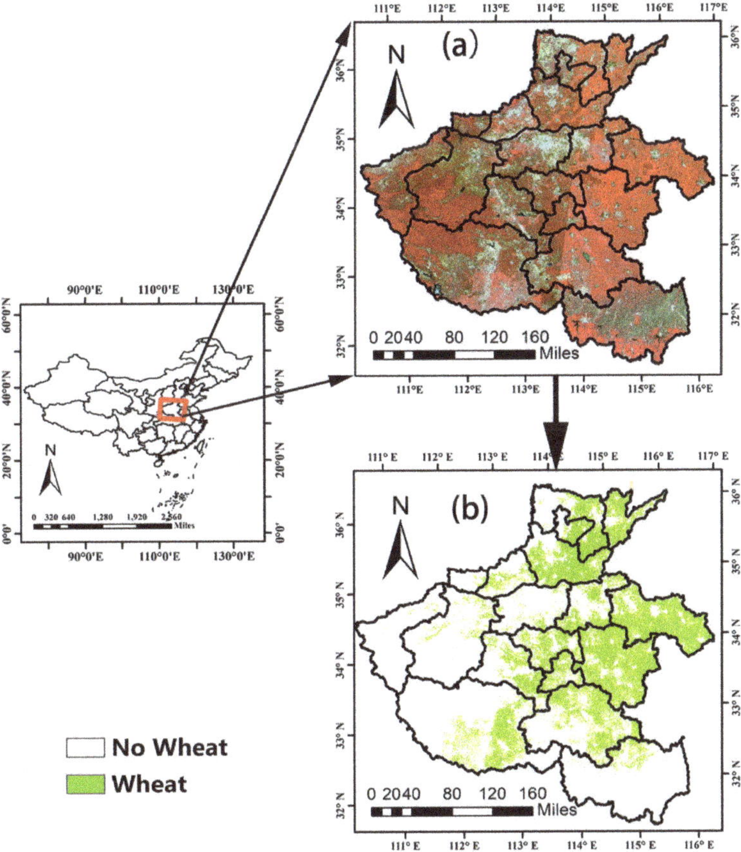

Figure 1. The geographic location of the study area and classification results (The red box on the left is the location of Henan Province. (**a**)—false color composite with the MODIS09 data; and (**b**)—land classification, green represents the winter wheat region).

1. The wheat area in Nanyang Basin, including Nanyang City and Biyang Countyin Zhumadian, is a typical rainfed and semi-rainfed area due to the relatively poor field supporting projects and irrigation conditions.
2. Rice stubble and wheat areas in southern Henan, including Xinyang, southern Zhumadian, and Nanyang Tongbai. The soil in this area is heavy, and the precipitation during the wheat growth period is relatively high.
3. In western Henan, southwestern Henan, and northern Henan, dry wheat areas, including Luoyang, Sanmenxia, Jiyuan, Pingdingshan, Anyang, and other shallow hilly areas, drought, winter, and spring freezing damage, rust, and yellow dwarf disease are the main factors affecting wheat yield.
4. The wheat area in north-central Henan Province, including Xuchang, Zhengzhou, Luoyang, and the irrigated land north of the Yellow River, has good production conditions and high production levels.

5. The wheat area in the central and eastern part of Henan Province includes the irrigated land in the middle- and high-yield wheat areas in the north-central part of Zhumadian, Luohe, Zhoukou, Shangqiu, and Pingdingshan Mountain.

The topography and climate within each of the above production areas are relatively consistent; so, we developed a winter wheat yield model for each production area. The actual yield data for winter wheat in this study were provided by the Henan Provincial Bureau of Statistics.

2.2. MODIS LAI

LAI is defined as the area of unilateral green leaves per unit of ground area in a broadleaf canopy and half of the total needle surface area per unit ground area in a coniferous canopy. The LAI product selected for this study was MCD15A2H. MCD15A2H is an 8-day composite product with a total of 46 scenes per year and a spatial resolution of 500 m. The inversion algorithm for the Moderate Resolution Imaging Spectroradiometer (MODIS) LAI product was a look-up table constructed based on a three-dimensional radiative transfer model. When the main algorithm failed, a backup algorithm using an empirical relationship between NDVI and the canopy LAI was triggered to estimate the LAI for each pixel, and the look-up table was used to compare whether the observed and simulated canopy top BRFs were within a given biologically relevant range. All canopy/soil patterns and corresponding LAI values that differed between the modeled and observed BRFs within a given level of uncertainty were considered acceptable solutions. The product used for this study was version 6, and the final result was the true LAI.

All satellite data are archived in HDF-EOS format, and the MODIS Reprojection Tool (MRT) software provided by NASA enables the user to read the HDF-EOS format. This software supports the performing geo-transformations to different coordinate systems or cartographic projections and writing the output to other file formats (GeoTIFF). All data are initially projected onto an integer sine wave (ISIN) mapping grid. These data were corrected to UTM coordinates using MRT and resampled using the nearest neighbor algorithm. Non-wheat fields were masked using a land cover classification, and then the corresponding areas were cut out of the remotely sensed imagery using SHP data based on the extent of each city. The MODIS LAI products include the quality control (QC) information designed to help users make the best use of these data. Each QC layer has a large amount of quality information associated with each pixel, whether the pixel is labeled as cloudy, clear, or in cloud shadow. In a subsequent study, MODIS LAI selected high-quality pixels derived using the master algorithm under cloud-free conditions. Figure 2 shows the time series results of the MODIS LAI for the entire winter wheat growing region in Henan, and the maximum value of the LAI was approximately 2. Considering that LAI is at a low level in the early and late stages of wheat growth, and that these growth stages do not have a particularly strong influence on yield formation, we used LAI from flowering to maturity each year as input data.

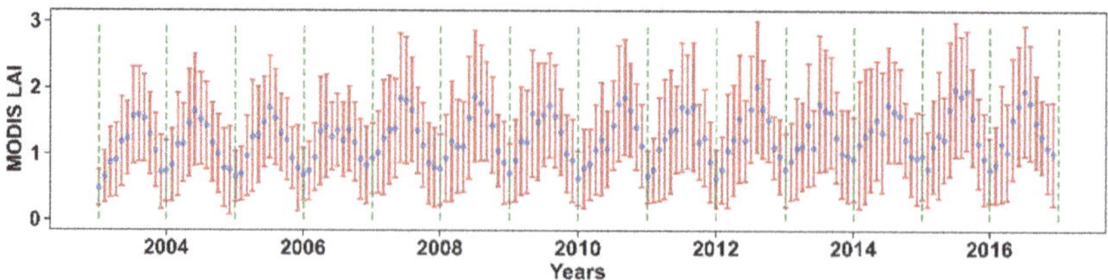

Figure 2. The mean curve of MODIS LAI in Henan Province from 2003 to 2016. The red line is the error bars plotted using standard deviation; blue dots represent the mean; between the two green dotted lines is the LAI for one year.

2.3. Mapping of Wheat Distribution

This study focused on winter wheat, which first required the extraction of winter wheat distribution areas from remote sensing images. We used the product produced by Dong [36]. The product was produced by first synthesizing monthly NDVI maxima and using FROM-GLC as a priori knowledge to obtain crop distribution information; then, we used the standard seasonal growth curve of winter wheat combined with the time-weighted dynamic time warping (TWDTW) to determine the area of winter wheat. The accuracy of the final product was higher than 89.30% and 90.59% for producers and users, respectively. To maintain consistency with the resolution of MODIS data, the classification results were finally resampled to 500 m using the mode resampling method. Mode resampling selects the value with the highest frequency of occurrence among all sampling points, and the results maintain the real state of the ground surface to some extent. The classification results are shown in Figure 1b.

2.4. LSTM

The recurrent neural network (RNN) is a type of neural network with short-term memory capabilities. In a cyclic neural network, a neuron cannot only receive information from other neurons but also its own information, forming a network structure with loops. Compared with feedforward neural networks, recurrent neural networks are more in line with the structure of biological neural networks. Recurrent neural networks have been widely used in tasks such as speech recognition, language modeling, and natural language generation. The parameter learning of recurrent neural networks can be learned by the backpropagation algorithm over time. The backpropagation algorithm with time transmits the error information step by step in the reverse order of time. When the input sequence is relatively long, there will be the problem of gradient explosion and disappearance. In order to solve this problem, people have made many improvements to the cyclic neural network. The most effective means of improvement is to introduce a gating mechanism, one of which is called a long short-term memory network (LSTM). LSTM is a variant of a cyclic neural network, which can effectively solve the problem of gradient explosion or the disappearance of a simple cyclic neural network.

The ingenuity of LSTM is that, by increasing the input threshold, the forgetting threshold, and output threshold, the weight of the self-loop is changed. As such, when the model parameters are fixed, the integration scale at different times can be dynamically changed, thereby avoiding the problem of gradient disappearance or gradient expansion.

The LSTM network introduces a gating mechanism to control the path of information transmission. The three "gates" are the input gate i_t, the forget gate f_t, and the output gate o_t. The functions of these three gates are as follows:

(1) The forgetting gate f_t controls how much information needs to be forgotten to control the internal state c_{t-1} at the last moment;

(2) The enter gate i_t controls the candidate state $\tilde{c}_t$ at the current moment and how much information needs to be saved;

(3) The output gate o_t controls how much information of the internal state c_t at the current moment needs to be output to the external state h_t.

When $f_t = 0$ and $i_t = 1$, the memory unit clears the historical information and writes the candidate state vector $\tilde{c}_t$. However, the memory unit c_t is still related to the historical information at the previous moment. When $f_t = 1$ and $i_t = 0$, the memory unit will copy the content of the previous moment without writing new information.

Figure 3 shows the cyclic unit structure of the LSTM network. The calculation process is: (1) first use the external state h_{t-1} at the previous moment and the input x_t at the current moment to calculate the three gates and the candidate state $\tilde{c}_t$; (2) combine the forget gate f_t and the input gate i_t to update the memory unit c_t; and (3) combine the output gate o_t to transfer the information of the internal state to the external state h_t.

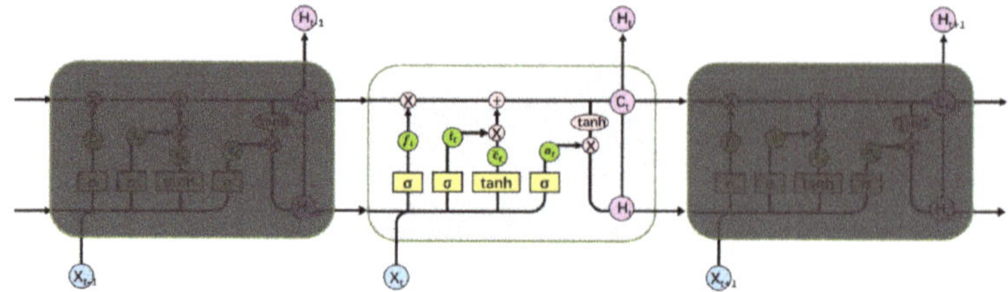

Figure 3. Cyclic cell structure of the LSTM network.

By means of LSTM cyclic units, the whole network can be built up with long-distance temporal dependencies. It can be succinctly described as

$$\begin{bmatrix} \widetilde{c}_t \\ o_t \\ i_t \\ f_t \end{bmatrix} = \begin{bmatrix} tanh \\ \sigma \\ \sigma \\ \sigma \end{bmatrix} \left(w \begin{bmatrix} x_t \\ h_{t-1} \end{bmatrix} + b \right) \tag{1}$$

$$c_t = f_t \odot c_{t-1} + i_t \odot \widetilde{c}_t \tag{2}$$

$$h_t = o_t \odot \tan h(c_t) \tag{3}$$

In the LSTM network, a memory unit c can capture a key piece of information at a certain moment and can store this key information for a certain time interval. The lifetime of information stored in memory unit c is longer than that of short-term memory h but much shorter than that of long-term memory. Figure 4 shows the workflow from LAI data processing to LSTM model building and yield estimation.

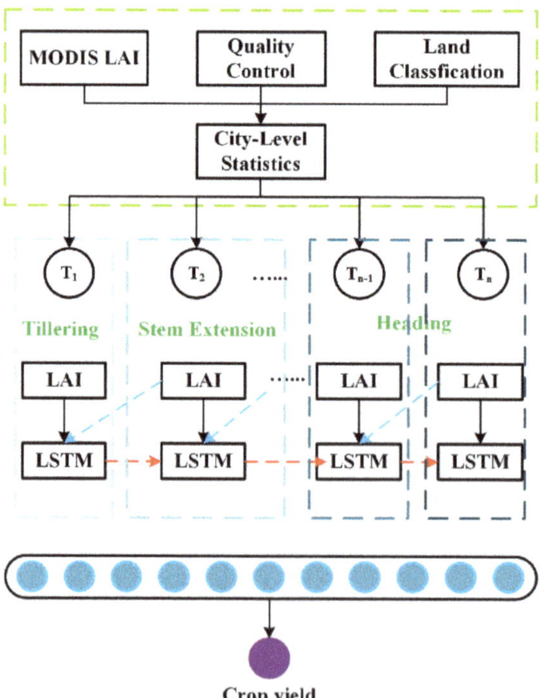

Figure 4. Overall structure of the LSTM model for wheat yield estimation.

2.5. Model Evaluation Metrics

We chose root mean square error (RMSE), coefficient of determination (R^2), and mean absolute percentage error (MAPE), the three metrics in Equations (4)–(6). To evaluate the performance of the model, the difference between the predicted and actual statistical values of the model was calculated. A smaller RMSE indicates a better performance of the model, and a larger R^2 indicates a higher regression accuracy of the model. The lower the value of MAPE and RMSE, the higher the accuracy of the obtained predictive model. MAPE measures the error in percentage and specifies the average percentage deviation between the forecast value and the actual implementation [37]. Usually, the fit of the model is perfect when the MAPE value is below 10% and when it is in the range from 10% to 20%, the model fit is good. In the range of 20–30%, the error level is acceptable and when it exceeds 30%, the model is a poor fit and should be rejected [38].

$$\text{RMSE} = \sqrt{\frac{1}{n}\sum_{i=1}^{n}(\hat{y}_i - y_i)^2} \tag{4}$$

$$R^2 = 1 - \frac{\sum_{i=1}^{n}(\hat{y}_i - y_i)^2}{\sum_{i=1}^{n}(y_i - \hat{y}_i)^2}, \tag{5}$$

$$\text{MAPE} = \frac{100\%}{n}\sum_{i=1}^{n}\left|\frac{\hat{y}_i - y_i}{y_i}\right|, \tag{6}$$

where y_i represents the actual yield, $\hat{y}_i$ is the predicted yield, and n is the sample size.

3. Results and Discussion

3.1. Yield Estimation under Different Time Steps and Input Combinations

We used winter wheat LAI time series data from 2003 to 2015 as training samples and modeled them with LSTM to obtain the winter wheat yield prediction results for 2016 and compared the predicted yield with the statistical yield in 2016 to verify the prediction accuracy of the model. Moreover, considering the characteristics of the LSTM model and the requirements of the yield prediction task, the input data required for training were processed as subsequently described, and the winter wheat growth data from March to the end of May each year were selected so that the input data for one year had 12 LAI values containing data from the flowering stage to the maturity stage. The fitted data of the model changed at different input steps, which also had an impact on the accuracy of the prediction results. To obtain the input step with the highest accuracy, the input steps from 1 to 6 were compared. The overall RMSE for Henan Province under the six step-size scenarios is shown in Figure 5. The solid red line in the figure is the 1:1 line, and the estimation results were mostly evenly distributed on both sides with the solid line as the center, indicating that the model had a good prediction for the yield. However, when the yield exceeded 7000 kg/ha, the estimation accuracy of the model tended to decrease, and the yield prediction tended to be underestimated, which is consistent with the systematic trend of underestimation at high yields in previous studies, mainly due to the relatively low proportion of data from high production areas in the sample. We first constructed LSTM models for five winter wheat production areas and finally obtained the winter wheat yield estimation results for the whole of Henan Province. A–E in Table 1 show the RMSE, R^2 and MAPE of the five grain-producing regions at different step sizes, and F shows the statistical results for the whole Henan Province. Figure 6 shows the RMSE histograms of the LSTM model for the five regions and the whole of Henan Province under different step lengths. When the step size was short, the predicted yields of the LSTM model for different production areas had great instability compared with the statistical yields in the field, but when the step size increased to 3 and 4, the accuracy started to improve and stabilized. However, the accuracy decreased again when the step size was 6. The LSTM model achieved optimal values of R^2 and RMSE at a step size of 4, with results of 88% and 532.16 kg/ha, respectively, while the value of MAPE was 4.44%, which was at a very high level of fit. Although the problem of

LSTM was effectively solved in terms of gradient disappearance or explosion compared with the general RNN networks, the winter wheat yields were not closely related to yields from many years ago, and to some extent, were only strongly correlated with historical yields from the last three or four years, and there may be a loss of accuracy if the current results are fitted with data from a longer period of time ago.

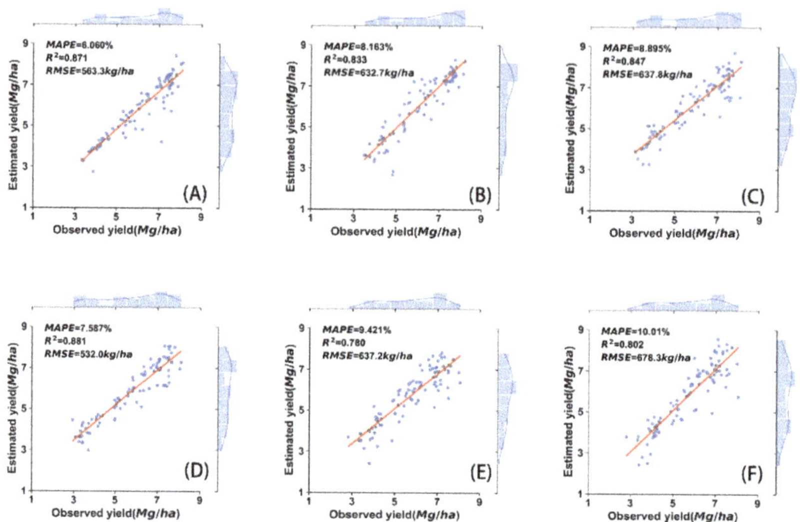

Figure 5. Comparison of yield estimation results for LSTM models based on different time steps: (**A**) one time step; (**B**) two time steps; (**C**) three time steps; (**D**) four time steps; (**E**) five time steps; and (**F**) six time steps. The probability distributions of the statistical and estimated yields are plotted on the right and top of the Y axis, respectively.

Table 1. The accuracy evaluation of the LSTM prediction model was compared between different time steps when 12 LAI data were input.

Time Steps	RMSE (kg/ha)						R^2						MAPE (%)				
	A	B	C	D	E	F	A	B	C	D	E	F	A	B	C	D	E
1	1039.93	436.63	350.59	386.97	687.70	563.38	0.40	0.79	0.95	0.92	0.01	0.87	12.33	5.01	4.28	5.19	7.29
2	468.29	548.72	831.34	699.47	219.07	632.77	0.86	0.73	0.74	0.81	0.27	0.83	7.82	7.72	10.57	12.40	2.31
3	604.58	589.19	635.35	767.26	583.36	637.88	0.86	0.78	0.87	0.82	0.03	0.85	10.17	9.29	8.79	12.63	5.90
4	530.79	461.56	568.10	667.68	396.56	532.08	0.83	0.78	0.87	0.90	0.38	0.88	7.56	7.24	8.02	11.75	4.44
5	393.49	549.11	635.34	699.27	713.08	637.27	0.90	0.74	0.80	0.70	0.00	0.78	6.99	8.70	9.40	13.01	8.45
6	795.51	414.61	770.62	548.72	674.49	678.36	0.86	0.80	0.78	0.81	0.24	0.80	13.41	6.96	12.04	9.17	7.95

The respective highest accuracy estimation results in different regions differed significantly in time steps. For the whole of Henan Province, the overall accuracy may not be optimal if the same step was used for yield estimation. The lowest value of RMSE can be seen in Figure 6 in the southwestern production area, with an RMSE of approximately 400, and the highest in time steps was in southwestern Henan, with approximately 700. There was a correlation between the winter wheat yield and time series LAI data, but the interannual correlation was to some extent not a more accurate result by modeling with more data, and there are many factors affecting the wheat yield, including variation in variety, temperature, and topography, which can all cause yield fluctuations, although all these factors can affect the LAI time series to some extent. However, when using time series LAI data alone as input parameters, it is not better to use more data, and it is not better to set a higher time step; these should be quantitatively and individually tested for different regions and not generalized.

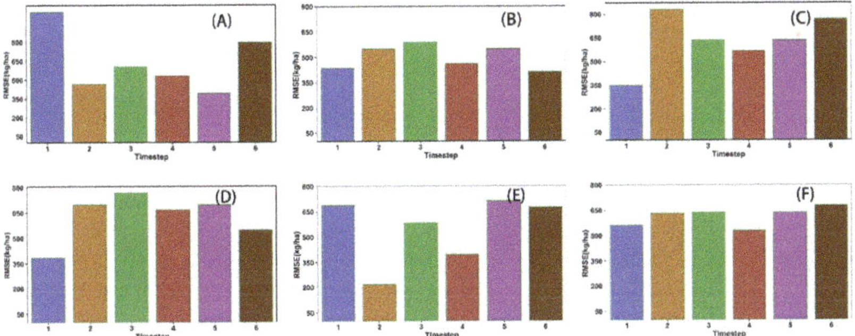

Figure 6. The model performance (predicted RMSE) using different time steps for the whole growing season. (**A**) one time step; (**B**) two time steps; (**C**) three time steps; (**D**) four time steps; (**E**) five time steps; and (**F**) six time steps.

3.2. Yield Estimation with Different Input Time Series Data

In the process of winter wheat yield estimation, it is always desirable to predict the yield as early as possible. Therefore, we considered shortening the time range of the LAI data used and the used data from the plucking stage to the filling stage for modeling so that their yield could be predicted 16 days before maturity. The LAI data used in this study were only six per year, which halved the amount of data compared to previous studies, making it easier to obtain the data, especially the high spatial resolution satellite data. The results of the comparison between the predicted yield and the actual statistical yield are shown in Figure 7, from which there was no significant decrease in the prediction accuracy for all of Henan compared to the 12 data, and the accuracy of individual time steps increased. The best time step for the overall accuracy occurred at 2, where the R^2 and RMSE were 87% and 522.32 kg/ha, respectively, while MAPE was 5.67%. A–E in Table 2 shows the RMSE, R^2 and MAPE of the five grain-producing regions at different step sizes, and F shows the statistical results for the whole Henan Province; Figure 8 shows the histogram of RMSE. However, it can also be seen that the value of RMSE gradually increased with increasing step size. This may be due to the fact that LAI data were strongly correlated only in the last two years, and the larger the time difference, the lower the correlation between LAI and yield; moreover, the interannual LAI was not strongly correlated with early and late yield formation, so in the process of yield estimation using time series LAI, a higher accuracy was obtained by using data from the nodulation to filling stage. It has been shown that the accumulation of dry matter in winter wheat is mainly concentrated at the nodulation and gestation stages, and the LAI in this period was closely correlated with the yield formation of winter wheat, which is also more consistent with the results of this study. It should also be noted that the overall RMSE accuracy was the best except for the case of step size 2. The RMSE of the remaining steps was significantly lower compared to the model with 12 data; therefore, using as many growing period data as possible would also make the model more robust when using time-series LAI for yield estimation.

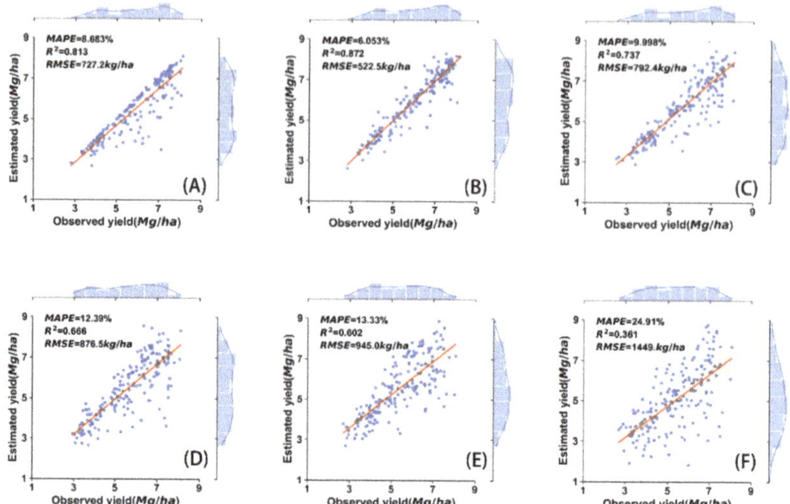

Figure 7. Comparison of the yield estimation results for LSTM models based on different time steps (**A**) one time step; (**B**) two time steps; (**C**) three time steps; (**D**) four time steps; (**E**) five time steps; and (**F**) six time steps. The probability distributions of the statistical and estimated yields are plotted on the right and top of the Y axis, respectively.

Table 2. The accuracy evaluation of the LSTM prediction model was compared between different time steps when 6 LAI data were input.

Time Steps	RMSE (kg/ha)						R^2						MAPE (%)				
	A	B	C	D	E	F	A	B	C	D	E	F	A	B	C	D	E
1	1180.61	534.22	798.75	338.77	656.31	727.11	0.17	0.70	0.80	0.96	0.19	0.81	17.63	5.22	10.29	5.67	6.64
2	372.66	552.27	534.55	411.21	598.78	522.32	0.87	0.67	0.86	0.93	0.18	0.87	6.03	7.41	5.61	6.53	5.67
3	704.62	362.19	867.82	739.41	901.01	792.37	0.53	0.85	0.75	0.67	0.01	0.74	9.78	4.63	11.11	12.23	9.83
4	1010.14	789.28	800.46	850.57	978.56	876.48	0.18	0.31	0.74	0.63	0.00	0.67	14.36	12.82	11.30	12.37	12.99
5	743.15	591.05	928.28	743.79	1255.65	945.34	0.44	0.61	0.63	0.67	0.00	0.60	11.69	10.62	12.73	12.17	17.03
6	1052.22	1572.15	1170.50	1167.80	1954.85	1448.93	0.28	0.02	0.50	0.36	0.02	0.36	20.81	45.16	17.92	21.91	27.96

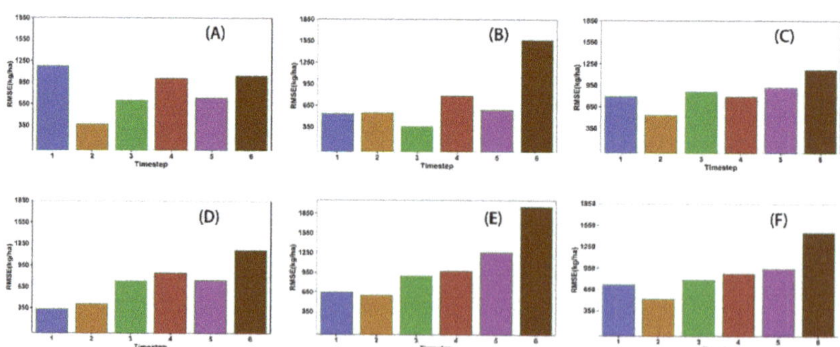

Figure 8. The model performance (predicted RMSE) using different time steps for only one period of the growing season. (**A**) one time step; (**B**) two time steps; (**C**) three time steps; (**D**) four time steps; (**E**) five time steps; and (**F**) six time steps.

3.3. Performance Comparison with Machine Learning Methods

Here, four machine learning methods (random forest, support vector regression, partial least squares regression, and XGBoost) were used to construct county-level wheat yield models for each agroecological zone. Random forest (RF) is a supervised machine learning algorithm based on integration learning [39]. Different subsets are randomly drawn from the provided data and used to build several different decision trees and integrate the results of one decision tree according to Bagging's rules. Support vector regression (SVR) is a regression algorithm that is a variant of SVM in regression analysis [40]. SVR also considers maximization intervals but considers points within the decision boundary so that as many sample points as possible lie within the interval. The partial least squares regression (PLSR) [41] algorithm is a regression modeling method for multiple dependent variables Y on multiple independent variables X. The algorithm considers extracting as many principal components as possible from Y and X in building the regression and maximizing the correlation between the principal components extracted from X and Y, respectively. XGBoost [42] is a scalable machine learning system that adds to the objective function of each iteration regular term to further reduce the risk of overfitting, XGBoost is an all-in-one machine learning algorithm.

The yield prediction of winter wheat in Henan Province was constructed using the four methods mentioned above. The prediction model first used winter wheat LAI time series data from 2003 to 2015, followed by yield prediction for 2016, and the R^2 and RMSE of the prediction results are shown in Table 3. For the whole of Henan Province, the best performance among the four methods was the SVR with R2, RMSE and MAPE of 0.76, 725.8 kg/ha and 6.33%, respectively, and the worst was PLSR with R2, RMSE and MAPE of 0.7 and 809.1 kg/ha and 7.74%, respectively. Compared with these machine learning methods, the prediction results of the LSTM had better accuracy and performance both for individual wheat growing areas and for the whole of Henan province.

Table 3. Accuracy evaluations comparison among different methods.

Model	RMSE (kg/ha)						R^2						MAPE (%)				
	A	B	C	D	E	F	A	B	C	D	E	F	A	B	C	D	E
RF	566.2	705.8	946.5	693.3	656.6	774.5	0.66	0.37	0.62	0.73	0.35	0.72	9.31	9.48	9.05	11.24	7.04
SVR	605.0	558.9	980.0	545.2	505.4	725.8	0.60	0.56	0.60	0.83	0.61	0.76	10.26	8.53	10.26	9.31	6.33
PLSR	627.7	680.4	1051.4	558.2	676.8	809.1	0.57	0.37	0.55	0.82	0.31	0.70	11.03	11.18	11.28	9.06	7.74
XGBOOST	579.2	659.4	976.4	737.8	638.1	785.8	0.62	0.45	0.60	0.69	0.38	0.72	8.62	10.40	10.12	11.60	6.98

The prediction accuracy of the four machine learning methods was the lowest in the southwest Henan region, which is a mountainous and hilly area with complex topography, a small winter wheat growing area, and large yield variation, and there may be a lack of yield accuracy in this region using conventional methods to construct the model. Compared with these, using LSTM model for winter wheat yield prediction had a superior performance. Compared to algorithms such as SVR and RF, the accuracy of estimation was not sufficient, because the ability to analyze complex nonlinear relationships between long time series variables is not as good as LSTM, resulting in poor model performance. These machine learning methods do not consider the time correlation between winter wheat yields in the modeling process, and the estimation for each year's yield is conducted independently, while LSTM takes into account the time series correlation of yields and also can better handle the nonlinear relationship, so it has higher accuracy compared to machine learning. Overall, a better performance capability can be obtained using LSTM models for forecasting time series data.

4. Conclusions

In this study, considering the complexity of data collection, we used LAI as a single input variable and five models, including four machine learning models (RF, SVR, PLSR, and XGBOOST) and one deep learning model (LSTM) to predict the winter wheat yield in Henan Province in 2016. In general, the LSTM model had superior performance compared with the machine learning models. Moreover, considering the characteristics of the LSTM, the time step of the modeled data as well as the growth period data were analyzed, and the time step needs to be analyzed for different growth regions, while using only the necessary growth period data can also obtain a high prediction accuracy; however, for the robustness of the model, the more growth period data are used accordingly, the better. To date, winter wheat yield prediction based on remote sensing images has been carried out at the county level. However, the determination of crop yield remains a challenge because the variability and uncertainty within the region is unknown. The results of our study on winter wheat yield prediction at a regional scale using publicly available data, using LAI as an input variable for determining crop yield, can potentially be applied to crop yield estimation in regions with sparse observational data and worldwide.

Author Contributions: Conceptualization, J.W.; resources, Z.G.; writing—original draft preparation, J.W.; writing—review and editing, J.W., L.S. and H.S. All authors have read and agreed to the published version of the manuscript.

Funding: This research was funded by the National Natural Science Foundation of China (NO.42101362, 31501225); the Natural Science Foundation of Henan Province of China (NO.222300420463).

Institutional Review Board Statement: Not applicable.

Data Availability Statement: Not applicable.

Conflicts of Interest: The authors declare no conflict of interest.

References

1. Becker-Reshef, I.; Justice, C.; Sullivan, M.; Vermote, E.; Tucker, C.; Anyamba, A.; Small, J.; Pak, E.; Masuoka, E.; Schmaltz, J. Monitoring global croplands with coarse resolution earth observations: The Global Agriculture Monitoring (GLAM) project. *Remote Sens.* **2010**, *2*, 1589–1609. [CrossRef]
2. Tilman, D.; Balzer, C.; Hill, J.; Befort, B.L. Global food demand and the sustainable intensification of agriculture. *Proc. Natl. Acad. Sci. USA* **2011**, *108*, 20260–20264. [CrossRef] [PubMed]
3. Becker-Reshef, I.; Vermote, E.; Lindeman, M.; Justice, C. A generalized regression-based model for forecasting winter wheat yields in Kansas and Ukraine using MODIS data. *Remote Sens. Environ.* **2010**, *114*, 1312–1323. [CrossRef]
4. Xie, Y.; Wang, P.; Bai, X.; Khan, J.; Zhang, S.; Li, L.; Wang, L. Assimilation of the leaf area index and vegetation temperature condition index for winter wheat yield estimation using Landsat imagery and the CERES-Wheat model. *Agric. For. Meteorol.* **2017**, *246*, 194–206. [CrossRef]
5. Mkhabela, M.; Bullock, P.; Raj, S.; Wang, S.; Yang, Y. Crop yield forecasting on the Canadian Prairies using MODIS NDVI data. *Agric. For. Meteorol.* **2011**, *151*, 385–393. [CrossRef]
6. Lai, Y.; Pringle, M.; Kopittke, P.M.; Menzies, N.W.; Orton, T.G.; Dang, Y.P. An empirical model for prediction of wheat yield, using time-integrated Landsat NDVI. *Int. J. Appl. Earth Obs. Geoinf. ITC J.* **2018**, *72*, 99–108. [CrossRef]
7. Moriondo, M.; Maselli, F.; Bindi, M. A simple model of regional wheat yield based on NDVI data. *Eur. J. Agron.* **2007**, *26*, 266–274. [CrossRef]
8. Wall, L.; Larocque, D.; Léger, P. The early explanatory power of NDVI in crop yield modelling. *Int. J. Remote Sens.* **2008**, *29*, 2211–2225. [CrossRef]
9. Chen, Y.; Zhang, Z.; Tao, F. Improving regional winter wheat yield estimation through assimilation of phenology and leaf area index from remote sensing data. *Eur. J. Agron.* **2018**, *101*, 163–173. [CrossRef]
10. Franch, B.; Vermote, E.; Becker-Reshef, I.; Claverie, M.; Huang, J.; Zhang, J.; Justice, C.; Sobrino, J.A. Improving the timeliness of winter wheat production forecast in the United States of America, Ukraine and China using MODIS data and NCAR Growing Degree Day information. *Remote Sens. Environ.* **2015**, *161*, 131–148. [CrossRef]
11. Pantazi, X.E.; Moshou, D.; Alexandridis, T.; Whetton, R.L.; Mouazen, A.M. Wheat yield prediction using machine learning and advanced sensing techniques. *Comput. Electron. Agric.* **2016**, *121*, 57–65. [CrossRef]
12. Huang, J.; Ma, H.; Sedano, F.; Lewis, P.; Liang, S.; Wu, Q.; Su, W.; Zhang, X.; Zhu, D. Evaluation of regional estimates of winter wheat yield by assimilating three remotely sensed reflectance datasets into the coupled WOFOST–PROSAIL model. *Eur. J. Agron.* **2019**, *102*, 1–13. [CrossRef]
13. Huang, J.; Sedano, F.; Huang, Y.; Ma, H.; Li, X.; Liang, S.; Tian, L.; Zhang, X.; Fan, J.; Wu, W.; et al. Assimilating a synthetic Kalman filter leaf area index series into the WOFOST model to improve regional winter wheat yield estimation. *Agric. For. Meteorol.* **2016**, *216*, 188–202. [CrossRef]
14. Huang, J.; Tian, L.; Liang, S.; Ma, H.; Becker-Reshef, I.; Huang, Y.; Su, W.; Zhang, X.; Zhu, D.; Wu, W.; et al. Improving winter wheat yield estimation by assimilation of the leaf area index from Landsat TM and MODIS data into the WOFOST model. *Agric. For. Meteorol.* **2015**, *204*, 106–121. [CrossRef]

15. Zhuo, W.; Huang, J.; Li, L.; Zhang, X.; Ma, H.; Gao, X.; Huang, H.; Xu, B.; Xiao, X. Assimilating soil moisture retrieved from Sentinel-1 and Sentinel-2 data into WOFOST model to improve winter wheat yield estimation. *Remote Sens.* **2019**, *11*, 1618. [CrossRef]
16. Jeong, S.; Ko, J.; Yeom, J.-M. Predicting rice yield at pixel scale through synthetic use of crop and deep learning models with satellite data in South and North Korea. *Sci. Total Environ.* **2022**, *802*, 149726. [CrossRef] [PubMed]
17. Bolton, D.K.; Friedl, M.A. Forecasting crop yield using remotely sensed vegetation indices and crop phenology metrics. *Agric. For. Meteorol.* **2013**, *173*, 74–84. [CrossRef]
18. Krizhevsky, A.; Sutskever, I.; Hinton, G.E. Imagenet classification with deep convolutional neural networks. *Commun. ACM* **2017**, *60*, 84–90. [CrossRef]
19. Koirala, A.; Walsh, K.B.; Wang, Z.; McCarthy, C. Deep learning–Method overview and review of use for fruit detection and yield estimation. *Comput. Electron. Agric.* **2019**, *162*, 219–234. [CrossRef]
20. Fischer, T.; Krauss, C. Deep learning with long short-term memory networks for financial market predictions. *Eur. J. Oper. Res.* **2018**, *270*, 654–669. [CrossRef]
21. LeCun, Y.; Bengio, Y.; Hinton, G. Deep learning. *Nature* **2015**, *521*, 436–444. [CrossRef] [PubMed]
22. Schmidhuber, J. Deep learning in neural networks: An overview. *Neural Netw.* **2015**, *61*, 85–117. [CrossRef] [PubMed]
23. Khaki, S.; Wang, L.; Archontoulis, S.V. A cnn-rnn framework for crop yield prediction. *Front. Plant Sci.* **2020**, *10*, 1750. [CrossRef] [PubMed]
24. Sun, J.; Di, L.; Sun, Z.; Shen, Y.; Lai, Z. County-level soybean yield prediction using deep CNN-LSTM model. *Sensors* **2019**, *19*, 4363. [CrossRef]
25. Khaki, S.; Wang, L. Crop yield prediction using deep neural networks. *Front. Plant Sci.* **2019**, *10*, 621. [CrossRef]
26. Maimaitijiang, M.; Sagan, V.; Sidike, P.; Hartling, S.; Esposito, F.; Fritschi, F.B. Soybean yield prediction from UAV using multimodal data fusion and deep learning. *Remote Sens. Environ.* **2020**, *237*, 111599. [CrossRef]
27. Cao, J.; Zhang, Z.; Luo, Y.; Zhang, L.; Zhang, J.; Li, Z.; Tao, F. Wheat yield predictions at a county and field scale with deep learning, machine learning, and google earth engine. *Eur. J. Agron.* **2021**, *123*, 126204. [CrossRef]
28. Nevavuori, P.; Narra, N.; Linna, P.; Lipping, T. Crop yield prediction using multitemporal UAV data and spatio-temporal deep learning models. *Remote Sens.* **2020**, *12*, 4000. [CrossRef]
29. Hara, P.; Piekutowska, M.; Niedbała, G. Selection of independent variables for crop yield prediction using artificial neural network models with remote sensing data. *Land* **2021**, *10*, 609. [CrossRef]
30. Hochreiter, S.; Schmidhuber, J. Long short-term memory. *Neural Comput.* **1997**, *9*, 1735–1780. [CrossRef]
31. Graves, A. Generating sequences with recurrent neural networks. *arXiv Prepr.* **2013**, arXiv:1308.0850.
32. Tian, H.; Wang, P.; Tansey, K.; Han, D.; Zhang, J.; Zhang, S.; Li, H. A deep learning framework under attention mechanism for wheat yield estimation using remotely sensed indices in the Guanzhong Plain, PR China. *Int. J. Appl. Earth Obs. Geoinf. ITC J.* **2021**, *102*, 102375. [CrossRef]
33. Liu, Y.; Wang, S.; Wang, X.; Chen, B.; Chen, J.; Wang, J.; Huang, M.; Wang, Z.; Ma, L.; Wang, P.; et al. Exploring the superiority of solar-induced chlorophyll fluorescence data in predicting wheat yield using machine learning and deep learning methods. *Comput. Electron. Agric.* **2022**, *192*, 106612. [CrossRef]
34. Zhang, L.; Zhang, Z.; Luo, Y.; Cao, J.; Xie, R.; Li, S. Integrating satellite-derived climatic and vegetation indices to predict smallholder maize yield using deep learning. *Agric. For. Meteorol.* **2021**, *311*, 108666. [CrossRef]
35. Cai, Y.; Guan, K.; Lobell, D.; Potgieter, A.B.; Wang, S.; Peng, J.; Xu, T.; Asseng, S.; Zhang, Y.; You, L.; et al. Integrating satellite and climate data to predict wheat yield in Australia using machine learning approaches. *Agric. For. Meteorol.* **2019**, *274*, 144–159. [CrossRef]
36. Dong, J.; Fu, Y.; Wang, J.; Tian, H.; Fu, S.; Niu, Z.; Han, W.; Zheng, Y.; Huang, J.; Yuan, W. Early-season mapping of winter wheat in China based on Landsat and Sentinel images. *Earth Syst. Sci. Data* **2020**, *12*, 3081–3095. [CrossRef]
37. Sharma, L.K.; Singh, T.N. Regression-based models for the prediction of unconfined compressive strength of artificially structured soil. *Eng. Comput.* **2018**, *34*, 175–186. [CrossRef]
38. Peng, J.; Kim, M.; Kim, Y.; Jo, M.; Kim, B.; Sung, K.; Lv, S. Constructing Italian ryegrass yield prediction model based on climatic data by locations in South Korea. *Grassl. Sci.* **2017**, *63*, 184–195. [CrossRef]
39. Breiman, L. Random forests. *Mach. Learn.* **2001**, *45*, 5–32. [CrossRef]
40. Chang, C.-C.; Lin, C.-J. LIBSVM: A library for support vector machines. *ACM Trans. Intell. Syst. Technol.* **2011**, *2*, 1–27. [CrossRef]
41. Wegelin, J.A. *A Survey of Partial Least Squares (PLS) Methods, with Emphasis on the Two-Block Case*; Technical Report; University of Washington: Seattle, DC, USA, 2000.
42. Chen, T.; Guestrin, C. Xgboost: A scalable tree boosting system. In Proceedings of the 22nd ACM SIGKDD International Conference on Knowledge Discovery and Data Mining, San Francisco, CA, USA, 13–17 August 2016; pp. 785–794.

Article

Remotely Sensed Prediction of Rice Yield at Different Growth Durations Using UAV Multispectral Imagery

Shanjun Luo [1], Xueqin Jiang [2,*], Weihua Jiao [3], Kaili Yang [1], Yuanjin Li [1] and Shenghui Fang [1,*]

[1] School of Remote Sensing and Information Engineering, Wuhan University, Wuhan 430079, China
[2] School of Information Science and Engineering, Shandong Agriculture University, Tai'an 271001, China
[3] Center for Agricultural and Rural Economic Research, Shandong University of Finance and Economics, Jinan 250014, China
* Correspondence: xueqinjiang@whu.edu.cn (X.J.); shfang@whu.edu.cn (S.F.)

Abstract: A precise forecast of rice yields at the plot scale is essential for both food security and precision agriculture. In this work, we developed a novel technique to integrate UAV-based vegetation indices (VIs) with brightness, greenness, and moisture information obtained via tasseled cap transformation (TCT) to improve the precision of rice-yield estimates and eliminate saturation. Eight nitrogen gradients of rice were cultivated to acquire measurements on the ground, as well as six-band UAV images during the booting and heading periods. Several plot-level VIs were then computed based on the canopy reflectance derived from the UAV images. Meanwhile, the TCT-based retrieval of the plot brightness (B), greenness (G), and a third component (T) indicating the state of the rice growing and environmental information, was performed. The findings indicate that ground measurements are solely applicable to estimating rice yields at the booting stage. Furthermore, the VIs in conjunction with the TCT parameters exhibited a greater ability to predict the rice yields than the VIs alone. The final simulation models showed the highest accuracy at the booting stage, but with varying degrees of saturation. The yield-prediction models at the heading stage satisfied the requirement of high precision, without any obvious saturation phenomenon. The product of the VIs and the difference between the T and G (T − G) and the quotient of the T and B (T/B) was the optimum parameter for predicting the rice yield at the heading stage, with an estimation error below 7%. This study offers a guide and reference for rice-yield estimation and precision agriculture.

Keywords: yield estimation; rice; unmanned aerial vehicle (UAV); tasseled cap transformation; precision agriculture

Citation: Luo, S.; Jiang, X.; Jiao, W.; Yang, K.; Li, Y.; Fang, S. Remotely Sensed Prediction of Rice Yield at Different Growth Durations Using UAV Multispectral Imagery. *Agriculture* **2022**, *12*, 1447. https://doi.org/10.3390/agriculture12091447

Academic Editors: Jibo Yue, Chengquan Zhou, Haikuan Feng, Yanjun Yang and Ning Zhang

Received: 10 August 2022
Accepted: 9 September 2022
Published: 13 September 2022

Publisher's Note: MDPI stays neutral with regard to jurisdictional claims in published maps and institutional affiliations.

1. Introduction

As the largest grain crop in the world and a staple food for over half of the global population, the research on rice is of crucial importance for agricultural systems and food production [1]. Rice-yield data are vital reference indicators for species selection and breeding, determined by the combination of genes and the growth environment. The accurate prediction of rice yields, and especially at the regional level, is of great relevance to guaranteeing food security and sustainable agricultural development, and it is concerned with the elaboration of major policies for national livelihoods [2].

The conventional methods for crop-yield estimates include field sampling [3] and the crop-growth model [4]. The field survey is a devastating assessment method. Although the accuracy of the results can be maintained through a comprehensive investigation, it is undoubtedly a laborious and lengthy task [5]. Crop-growth models incorporate multiple data sources and approaches, which greatly compound their complexity due to the many model input parameters [6]. The remote estimation of yields is a technology that can be used to develop a connection between crop spectra and yield data. Remote sensing (RS) provides a convenient way to efficiently acquire spectral data of vegetation canopies in a nondestructive manner, which carries considerable valuable information regarding the interaction between the canopy and solar radiation, such as the vegetation absorption and scattering [7]. Vegetation-canopy spectra are intimately associated with crop growth, and especially in the visible ranges affected by pigmentation and the near-infrared (NIR) bands subject to the cell tissue and canopy structure [8]. Therefore, the vegetation indices (VIs) derived from these bands have been frequently adopted to estimate the vegetation phenotypic parameters, such as the leaf-area index (LAI), biomass, chlorophyll content, and nitrogen content [9,10]. In general, remote estimates of crop yields using VI-based methods have become mainstream [11].

As for the data source, it is an influential factor in crop-yield estimations. Yield estimations using ground-based measurement spectra are hardly adequate for large areas, and the real-time forecasting requirements and performance are regionally limited [12]. Satellite imagery can be appropriate and cost-effective data for crop monitoring at the regional scale [13]. However, cloud coverage is pervasive during the pivotal crop-growing season, and thus, sufficient spatial- and temporal-resolution data may not be available for precision agriculture. The emergence and development of unmanned aerial vehicles (UAVs) and lightweight sensors can be complementary between satellites and ground-based sensors [14]. Because UAVs have easy access to dynamic data, they have enormous potential to solve and refine strategies for the challenges encountered in agriculture. Despite some shortcomings, such as the flight time, load capacity, and weather situation, UAVs are expected to be applied with high frequency in agriculture from now on due to the valued information gained and effective implementation [15]. The spectral information gained from UAV-based multispectral or hyperspectral data has been broadly applied for crop-growth monitoring and parameter estimation [16]. Moreover, multispectral images, free from the information redundancy and complicated processing of hyperspectral data, have a red-edge band that RGB digital images lack, and their centimeter-level spatial resolutions make multispectral sensors preferred devices in precision agriculture [17].

The reproductive stages of rice can be divided into the tillering, jointing, booting, heading, filling, and maturity stages. During the booting and heading stages, the rice plant progressively accomplishes the conversion from nutritional to reproductive growth, and the appropriate parameters for the yield estimation differ at different stages [18]. In practical terms, it is imperative to access early and precise rice-yield data prior to harvest for market decisions and policymaking. In the early stages of rice growth, the leaves are not yet fully grown, and the variations in the later growth process make it difficult to estimate the yield accurately. However, it may be too late to use data collected at the later stages for yield estimation, as some effective measures need to be scheduled in advance. Moreover, the appearance of rice spikes during mid-to-late growth can interfere with the spectral characteristics of rice, as the color of the spikes eventually turns yellow, causing the overall spectral pattern of the rice to deviate from the normal green vegetation. Zhou et al. revealed that the presence of spikes increased the challenge of yield prediction in the late reproductive stage of rice [19]. Duan et al. also noted that the reduced predictive ability during the heading stage may be related to the uneven penetration of spikes into the sensor field [20]. Hence, the booting and heading stages are suitable for rice-yield estimation, but the heading stage needs to overcome the effect of the spikes.

The tasseled cap transformation (TCT), a viable pioneer of feature-detection algorithms, is a linear-conversion technique that is commonly utilized in the areas of vegetation, soil, and land-cover mapping [21]. The vast majority of the variations in the spectra of a single scene can be interpreted in terms of the brightness, greenness, and humidity retrieved from TCT [22]. Therefore, TCT was exploited to extract the brightness, greenness, and moisture components of the rice fields to provide potentially valuable variables for yield estimation.

In our experiment, the canopy spectral data of the paddy field was remotely measured from both the ground and UAV-mounted platforms, which had quite high spatial resolution, and thus, well-reflecting variations of field. Meanwhile, the LAI and chlorophyll-content data (SPAD) in the same period were obtained. Unlike previous studies, we compared the yield-estimation performance of the ground and UAV-based parameters at different periods, and we combined the TCT parameters to improve the accuracy of the rice-yield estimation without saturation. With rice grown under different nitrogen-fertilizer treatments, our objectives were: (1) to compare the ability of the rice-yield estimation at the booting and heading stages; (2) to accurately estimate the rice yield by ground measurements and UAV data; (3) to explore improving the VI-based approach for rice-yield estimation by integrating the brightness, greenness, and wetness retrieval from TCT.

2. Materials and Methods

2.1. Study Area and Experimental Design

The study area was located in Wuxue City, Hubei Province, China (Figure 1a). It has a humid subtropical monsoon climate with long plant production cycles and abundant rainfall. It is suitable for the comprehensive development of agriculture, forestry, animal husbandry, and fishing. As shown in Figure 1b, the identical rice variety was cultivated in 24 plots with different N-fertilizer-application levels, with a whole area of about 480 m^2 and a total of 7920 rice plants. The ridges between each plot were covered with white plastic film to isolate the mixing of water in the field. There were eight N-fertilizer gradients, and three replications, applying N_0, N_3, $N_{5.5}$, $N_{8.5}$, N_{11}, N_{14}, $N_{16.5}$, and $N_{19.5}$ (unit: kg/ha). Two important growth periods (the booting and heading stages) were selected for the UAV flight experiments. In the former period, no spikes appeared, and in the later period, almost all spikes were clearly present. The conditions were strictly identically controlled, except for differences in the amount of the nitrogen-fertilizer application. Field maintenance, including weeding and pest control, was performed by professionals throughout the growing season.

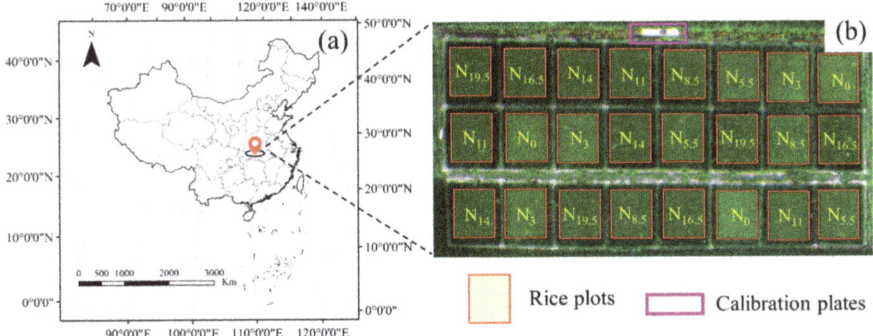

Figure 1. Study area and rice-plot settings: (**a**) experimental-area location; (**b**) nitrogen-gradient layout of rice plots.

2.2. Ground-Data Acquisition

The LAI, canopy chlorophyll content (CCC), and canopy height (CH) are significant indicators for characterizing crop yields [23,24]. Therefore, the SunScan canopy analysis system (Delta Inc., Cambridge, UK) was applied to measure the LAI of each plot. The five-point sampling method was performed at the four corners and center of each plot, and the average value was taken as the canopy LAI of each plot. At each location where the LAI was measured, three rice plants were selected, and the leaf chlorophyll was measured in the upper, middle, and lower parts of the plant using the SPAD-502 chlorophyll meter (Konica, Minolts Sensing Inc., Osaka, Japan); the mean value was recorded as the leaf chlorophyll content of each plot. The CCC is generally expressed using the product of the LAI and SPAD [25]. At each location where the LAI and SPAD were observed, three rice plants were randomly selected, and the height of the rice was measured using a millimeter ruler; the final CH was the mean value of all the readings. Rice seeds were collected by hand harvesting when the rice was fully mature. The seeds were then separated and left to dry in the sun until there were no variations in their weights. All the sun-dried seeds in each plot were weighed independently to obtain the rice yield of each plot.

2.3. Canopy Reflectance Derived from UAV Images

The UAV flights were implemented before the ground measurements. As shown in Figure 2, a multirotor UAV (S1000, SZ DJI Technology Co., Ltd., Shenzhen, China) equipped with a six-band MCA camera (Mini-MCA 6, Tetracam, Inc., Chatsworth, CA, USA) was employed to collect multispectral images of the rice at the booting (13 August 2015) and heading (29 August 2015) stages. The drone flights were conducted between 11:00 a.m. and 1:00 p.m. local time, thus ensuring minimal variation in the solar zenith angle. The multispectral camera has a center band of 490@10, 550@10, 670@10, 720@10, 800@20, or 900@20 nm. The UAV is also equipped with a three-axis gimbal to ensure that the camera is always shooting vertically downward. Four gray plates (reflectances of 6%, 12%, 24%, and 48%) were placed in the camera field of view for the radiometric calibrations to obtain the reflectance data. To prevent reflectance errors caused by solar-illumination variations, panoramic photographs of the entire study area were taken with a UAV flight altitude of 60 m and an image spatial resolution of approximately 0.03 m.

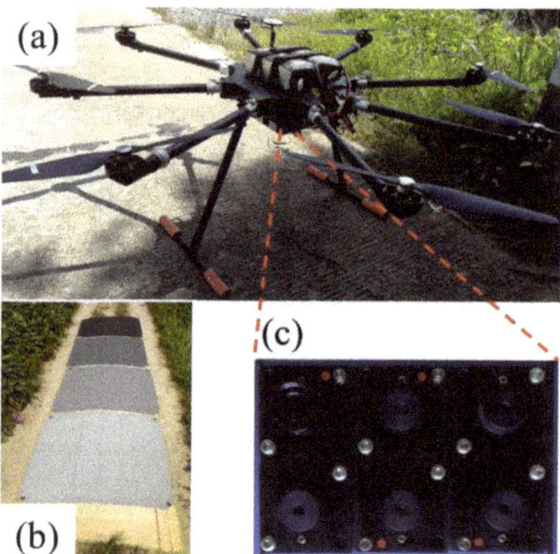

Figure 2. UAV reflectance-data-acquisition system: (**a**) UAV; (**b**) fixed reflectance grey plates for radiometric calibration; (**c**) Mini-MCA 6 multispectral camera.

The classical linear-radiometric-calibration method was utilized to transform the DN values of the multispectral images into reflectances to ensure the comparability of the data from different periods [26]. The reflectance was computed as follows:

$$R_i = DN_i \times G_i + O_i \ (i = 490, 550, 670, 720, 800, \text{and } 900),\tag{1}$$

$$\begin{pmatrix} 0.06 \\ 0.12 \\ 0.24 \\ 0.48 \end{pmatrix} = \begin{pmatrix} DN_{0.06} \\ DN_{0.12} \\ DN_{0.24} \\ DN_{0.48} \end{pmatrix} \times G_i + O_i \tag{2}$$

where R_i represents the calculated reflectance of the ith band, DN_i is the digital number of the ith band in the original multispectral images, and G_i and O_i represent the gain and offset values of the ith band, respectively.

For the 24 rice plots, we determined the maximum region of interest (ROI) suitable for each plot (equal to 10,000 pixels), and then the plot-level reflectance was the average of all the pixels in that rectangle.

2.4. VI Calculation Based on UAV Data

Several commonly available VIs obtained using combinations of visible, red-edge, and NIR bands are shown in Table 1. These VIs were selected for their good performance in crop-yield estimation and inversion of the phenotypic parameters.

Table 1. The common spectral indices selected in this paper.

Vegetation Indices	Formulas	References
Normalized Difference Vegetation Index (NDVI)	$(R_{800} - R_{670})/(R_{800} + R_{670})$	[27]
Red-Edge Chlorophyll Index ($CI_{red\ edge}$)	$R_{800}/R_{720} - 1$	[28]
Green-Edge Chlorophyll Index (CI_{green})	$R_{800}/R_{550} - 1$	[28]
Two-Band Enhanced Vegetation Index (EVI2)	$2.5(R_{800} - R_{670})/(1 + R_{800} + 2.4R_{670})$	[29]
Normalized Difference Red Edge (NDRE)	$(R_{800} - R_{720})/(R_{800} + R_{720})$	[30]
Wide-Dynamic-Range Vegetation Index (WDRVI)	$(\alpha R_{800} - \rho_{670})/(\alpha R_{800} + R_{670}),\ \alpha = 2$	[31]
MERIS Terrestrial Chlorophyll Index (MTCI)	$(R_{800} - R_{720})/(R_{720} - R_{670})$	[32]
Soil-Adjusted Vegetation Index (SAVI)	$(1 + L)(R_{800} - R_{670})/(R_{800} + R_{670} + L),\ L = 0.5$	[33]

2.5. Tasseled Cap Transformation

TCT, which is a quadrature conversion, provides the projection of feature messages, such as the soil and vegetation in the spectral domain, into the tasseled-cap space, following the structural characteristics of the distribution of ground information in multispectral remote sensing [34]. After the TCT was performed, the spectral dimensions could be reduced, and the information was concentrated in a few feature spaces. Its defining equation is given in Equation (3):

$$y = Ax + b, \tag{3}$$

where y is the vector after the TCT; A is the unit quadrature matrix and the coefficient matrix of the TCT; x is the gray value of the image, or the apparent reflectance of the sensor; b is served as an offset vector to avoid negative values after the transformation.

When TCT was performed on six-band UAV images, the results were composed of three factors: the brightness (B), greenness (G), and third component (T). All three of these variables are intimately associated with the surface landscape. The B component represents the variation information of the reflectance, which is a weighted sum of six bands and reflects the overall brightness variation of the surface object. The G component is vertical to the B component, and it also shows the contrast between the visible band (especially the red band) and NIR band, showing the variation in the greenness of the ground vegetation, which is closely related to the ground-vegetation cover, LAI, and biomass. The T variable reflects the moisture characteristics of the soil and vegetation. Similar to calculating the plot-level reflectance, the plot-level TCT parameters were obtained by defining a rectangle (ROI).

2.6. Accuracy Evaluation Using Leave-One-Out Cross-Validation

The yield-prediction model was assessed by employing the leave-one-out cross-validation (LOO-CV) method to reduce the reliance on a single random fraction of the calibration and validation dataset [35]. In this paper, the iterative process was repeated 22 times to ensure that each piece of data was engaged in the validation (the yields of two plots were removed as a result of serious problems during harvesting). The adjusted R^2, RMSE, and MRE were selected as the final accuracy metrics [11].

3. Results

3.1. Rice-Yield Estimation using Ground Measurements at Different Stages

In this study, the rice yield at the booting and heading stages was predicted based on the plot-level LAI, CH, and CCC measured on the ground. Rice-yield data from 24 plots were compared and analyzed, two of which were removed due to obvious errors, and the remaining 22 yield data, along with the corresponding ground data, were applied for modeling analysis. The Shapiro–Wilk test was chosen to check the normality of the data before modeling the rice yield.

In Table 2, the ground measurements (yield, LAI, CH, and CCC) approximately followed a normal distribution ($p > 0.05$). Then, the phenotypic parameters and yield of the rice measured on the ground at the booting and heading stages were fitted by least-squares regression (Figure 3). It was found that the LAI was a good fit for the yield at the booting stage ($R^2 = 0.569$), but relatively poor at the heading stage ($R^2 = 0.468$).

Table 2. Data description and normality test.

Variable	Growth Stage	Min	Max	Mean	*p*-Value	CV
Yield	–	2.70	4.34	3.57	0.89	11.17%
LAI	Booting stage	2.70	6.20	4.53	0.14	15.24%
	Heading stage	2.50	6.40	4.66	0.31	17.13%
CH	Booting stage	0.70	1.03	0.91	0.08	12.36%
	Heading stage	1.03	1.25	1.16	0.06	21.66%
CCC	Booting stage	87.66	201.74	148.61	0.07	23.14%
	Heading stage	86.74	233.92	163.50	0.06	28.63%
Brightness	Booting stage	0.34	0.49	0.44	0.17	9.44%
	Heading stage	0.33	0.53	0.44	0.29	13.95%
Greenness	Booting stage	0.07	0.11	0.09	0.08	14.67%
	Heading stage	0.05	0.13	0.10	0.49	21.14%
Third Component	Booting stage	0.31	0.59	0.49	0.17	14.81%
	Heading stage	0.30	0.59	0.46	0.98	16.73%
T – G	Booting stage	0.21	0.51	0.40	0.08	20.04%
	Heading stage	0.20	0.51	0.37	0.43	22.62%
T/B	Booting stage	0.91	1.19	1.11	0.00	7.30%
	Heading stage	0.85	1.27	1.06	0.26	11.63%

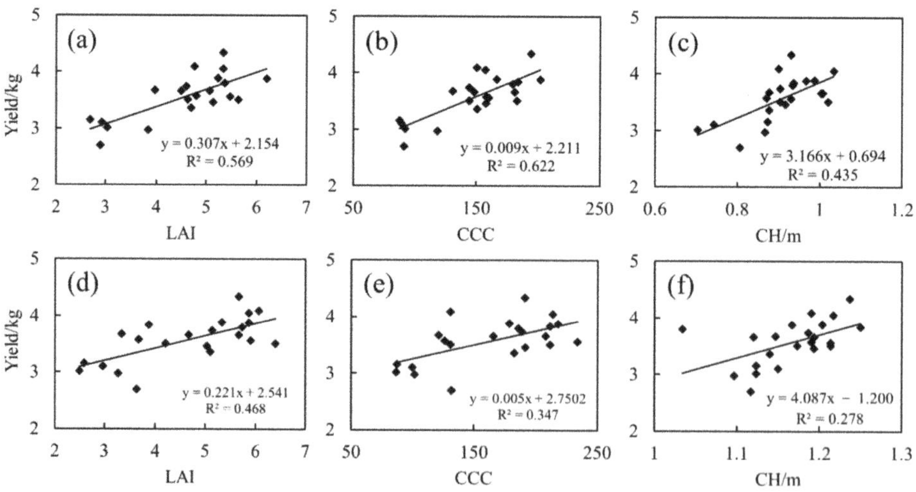

Figure 3. Fitting of ground data to yield at different growth stages: (**a–c**) booting stage; (**d–f**) heading stage.

By comparing the yield-estimation performance of the CH, LAI, and CCC, the results of a Pearson correlation analysis showed that the correlation (r) between the LAI and yield was improved by integrating SPAD data at the booting stage (Table 3), and the yield fit was significantly improved ($R^2 = 0.622$ vs. 0.569) (Figure 3). However, at the heading stage, the correlation decreased after combining SPAD data ($R^2 = 0.468$ vs. 0.347). The yield estimation with the CH was the poorest of the two periods, and especially at the heading stage. Therefore, the ground-measured LAI and SPAD data (CCC) were not suitable for predicting the yield of rice at the heading stage, but they had a good fit at the booting stage.

Table 3. Accuracy comparison of different regression models.

Growth Stage	LAI	CH	CCC (LAI × SPAD)	Brightness	Greenness	Third Component	T − G	T/B
Booting stage	0.754 **	0.659 **	0.789 **	0.585 **	−0.648 **	0.750 **	0.787 **	0.815 **
Heading stage	0.684 **	0.527 **	0.589 **	0.343 **	−0.407 **	0.739 **	0.794 **	0.702 **

** indicates that the correlation is significant at the 0.01 level (two-tailed).

3.2. Rice-Yield Estimation Using TCT Parameters

The change in rice from booting to heading is a process from nutritional to reproductive growth. Spikes basically do not appear on the surface of the rice field during the former period. In contrast, spikes progressively emerge after about two weeks. In addition to the different apparent information, there is also significant spectral diversity in rice at these two stages (Figure 4). The spectral characteristics of the rice at the booting stage were consistent with those of typical green vegetation, but they changed significantly at the heading stage. The reflectance of the rice canopy during the heading stage was significantly higher in the visible–NIR range. The appearance of rice spikes has a great influence on the spectral properties of rice.

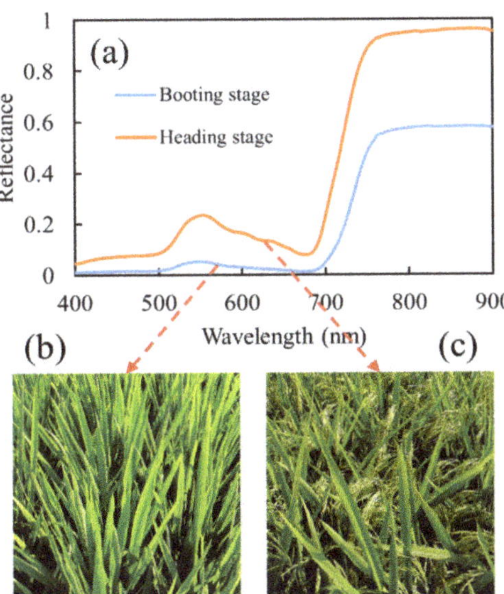

Figure 4. Spectra and field photos of rice at different stages: (**a**) spectral curves of rice at different stages; (**b**) actual view of rice at booting stage; (**c**) actual view of rice at heading stage.

Given the obvious change in the color and texture of the rice canopy caused by the appearance of panicles, paddy-field images at the booting and heading stages were obtained through a UAV equipped with a six-band Mini-MCA camera. Subsequently, the brightness, greenness, and third-component maps of the spectral-dimension reduction were retrieved by TCT (Figure 5).

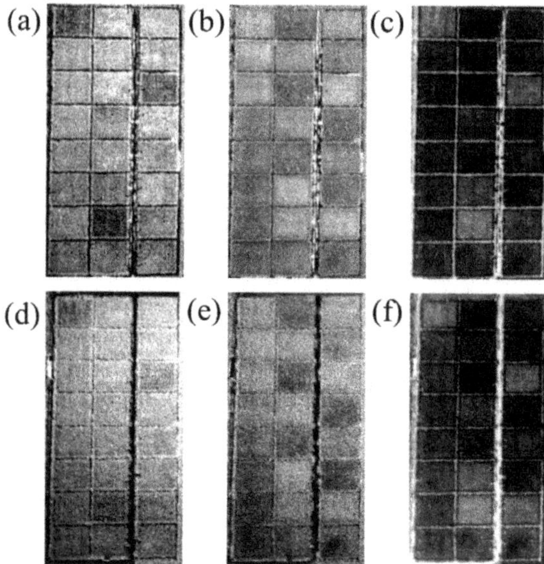

Figure 5. TCT-component images: (**a**) brightness at booting stage; (**b**) greenness at booting stage; (**c**) third component at booting stage; (**d**) brightness at heading stage; (**e**) greenness at heading stage; (**f**) third component at heading stage.

It can be noted that the TCT-component diagrams at the booting and heading stages showed a similar variation pattern. The brightness-component maps showed the overall variation in the rice reflectance throughout the experimental area, which was remarkably brighter than the greenness and third-component ones. In a single growth stage, the brightness of each plot varied with the different nitrogen-gradient conditions: the less nitrogen fertilizer applied, the darker the image. The gray distributions of the greenness and third-component images were contrary to that of the brightness-component image, which showed that the more nitrogen application applied, the darker the image. On account of the lighter color of the panicle compared with the leaf, the uneven occurrence of panicles was reflected by different greenness performances during the same growth period. For the third-component maps, the color at the booting stage was darker than that at the heading stage, reflecting the water status of the paddy fields and rice.

After completing the normality test (Table 2), a strong correlation ($r > 0.5$) was shown between the TCT parameters and rice yield during the booting period (Table 3). However, in the latter stage, the correlation between the yield and the brightness and greenness components was significantly lower. A linear fit of the yield and TCT parameters revealed a satisfactory result for the third component at both stages (R^2 values more than 0.5), but saturation was present at the booting stage (Figure 6). Meanwhile, no saturation was observed when using the brightness and greenness to predict the yield, but the performance was poor (R^2 values below 0.5 at the booting stage, and below 0.2 at the heading stage).

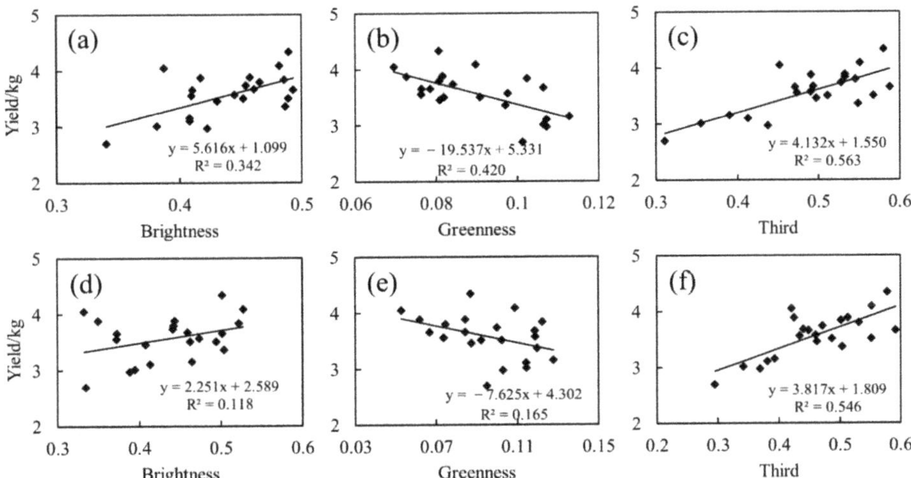

Figure 6. Fitting of TCT parameters to yield at different growth stages: (**a–c**) booting stage; (**d–f**) heading stage.

3.3. Rice-Yield Estimation Combining TCT Parameters and VIs

The correlation analysis of the yield vs. UAV-based VIs and TCT-based parameters (VI × Brightness, VI × Greenness, and VI × Third Component) was performed to compare the precision of the yield prediction at different growth stages (Figure 7). The results suggested that there was a strong correlation between the VIs and the yield at the booting stage ($r > 0.7$), while at the heading stage, except for the EVI2, NDRE, and SAVI, the correlation of the VIs vs. the yield decreased to some extent, and especially the CI_{green} vs. yield. Multiplied by the TCT parameters, some of the VIs had a stronger correlation with the yield, which was more obvious at the heading stage. At the heading stage, the brightness improved the correlation between the $CI_{red\ edge}$, CI_{green}, NDRE, WDRVI, and MTCI and the yield. The greenness only improved the correlation of the $CI_{red\ edge}$ and CI_{green} vs. the yield, and the third component basically improved the correlation between all the listed VIs and the yield. Based on the correlation analysis, the yield estimation of the rice was carried out on 22 samples at the booting and heading stages: (1) yield vs. Vis; (2) yield vs. VI × Brightness; (3) yield vs. VI × Greenness; (4) yield vs. VI × Third Component. The adjusted R^2 and RMSE were used to evaluate the performance of the yield prediction.

The yield-estimation results of the rice at the booting stage are shown in Table 4. The best-estimated yield parameter in the VIs was the WDRVI, with an adjusted R^2 of 0.634. The prediction results of the VI × Brightness and VI × Third Component were improved to a certain degree. In contrast, the performance of the VI × Greenness was worse. After combining the TCT parameters, the optimal yield-estimation variable was the CI_{green}. Several models with the best fitting effect were selected for analysis (shown in Figure 8). Except for the CI_{green} × Brightness, the other models (WDRVI, CI_{green} × Greenness, CI_{green} × Third Component) were saturated with different degrees, of which the WDRVI was the most obvious one.

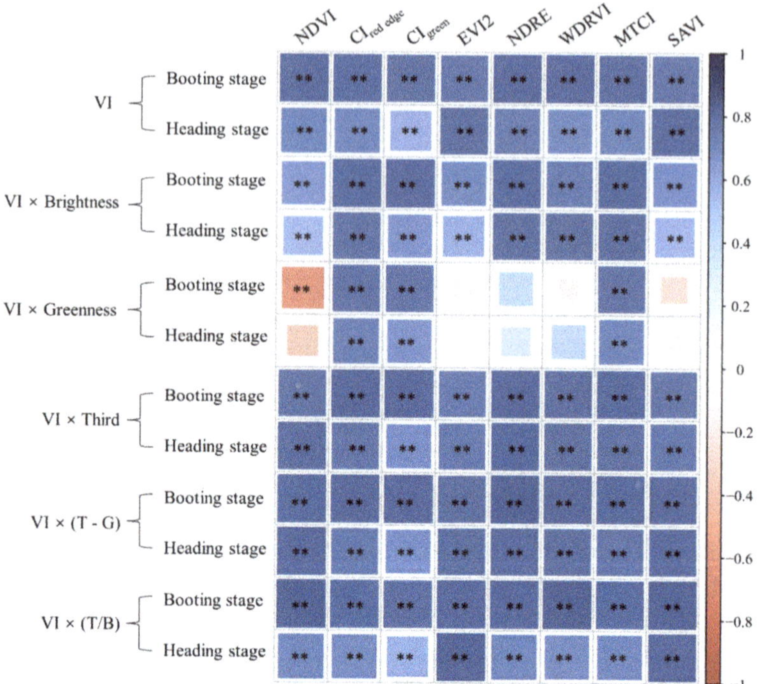

Figure 7. Correlation coefficients between parameters of VI and TCT combinations and yield (** indicates that the correlation is significant at the 0.01 level).

Table 4. Yield-estimation models incorporating TCT parameters and VIs at the booting stage.

Evaluating Indicators	Parameters	NDVI	CI$_{\text{red edge}}$	CI$_{\text{green}}$	EVI2	NDRE	WDRVI	MTCI	SAVI
	VI	0.628	0.614	0.591	0.553	0.624	0.634	0.606	0.558
	VI × Brightness	0.406	0.622	0.638	0.449	0.614	0.532	0.620	0.441
Adjusted R^2	VI × Greenness	0.345	0.562	0.568	0.016	0.152	0.019	0.565	0.062
	VI × Third Component	0.575	0.624	0.637	0.545	0.633	0.604	0.622	0.550
	VI × (T − G)	0.622	0.623	0.636	0.584	0.639	0.631	0.621	0.592
	VI × (T/B)	0.665	0.620	0.603	0.635	0.637	0.662	0.614	0.645
	VI	0.254	0.265	0.273	0.283	0.261	0.254	0.268	0.281
	VI × Brightness	0.334	0.264	0.257	0.321	0.265	0.290	0.265	0.323
RMSE	VI × Greenness	0.342	0.281	0.278	0.426	0.407	0.428	0.280	0.411
	VI × Third Component	0.277	0.264	0.258	0.289	0.259	0.266	0.265	0.286
	VI × (T − G)	0.262	0.265	0.259	0.276	0.257	0.258	0.266	0.273
	VI × (T/B)	0.245	0.263	0.269	0.256	0.256	0.245	0.266	0.252

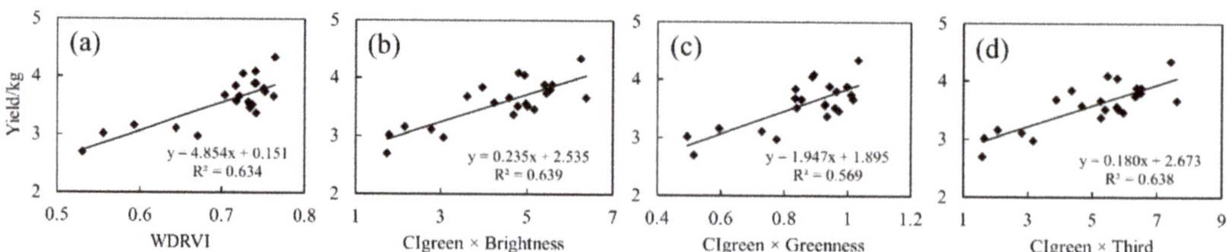

Figure 8. Well-performing yield-estimation models incorporating TCT parameters and VIs at the booting stage.

The rice-yield-prediction results at the heading stage are shown in Table 5. Compared with the booting stage, the parameters with the best fitting performance changed, indicating that the sensitivity of the VIs to the yield varied after the emergence of panicles. The VI with the best fitting performance in this period was the SAVI (adjusted R^2 = 0.600). Multiplied by the TCT parameters, the estimated result of the VI $\times$ Third Component had a certain degree of improvement. However, the fitting effects of the VI $\times$ Brightness and VI $\times$ Greenness were worse. Similarly, the models with good fitting effects were selected for analysis (Figure 9). Except for the SAVI, the other models (WDRVI $\times$ Brightness, $CI_{red\ edge} \times$ Greenness, NDRE $\times$ Third Component) did not show significant saturation. Therefore, the most appropriate parameter to predict the rice yield at the heading stage was the NDRE $\times$ Third Component (Adjusted R^2 = 0.612, RMSE = 0.272).

Table 5. Yield-estimation models incorporating TCT parameters and VIs at the heading stage.

Evaluating Indicators	Parameters	NDVI	$CI_{red\ edge}$	CI_{green}	EVI2	NDRE	WDRVI	MTCI	SAVI
Adjusted R^2	VI	0.477	0.472	0.305	0.585	0.506	0.460	0.485	0.600
	VI $\times$ Brightness	0.273	0.574	0.409	0.324	0.582	0.583	0.580	0.302
	VI $\times$ Greenness	0.105	0.497	0.409	0.003	0.088	0.158	0.466	0.010
	VI $\times$ Third Component	0.597	0.555	0.436	0.533	0.612	0.589	0.567	0.542
	VI $\times$ (T $-$ G)	0.634	0.546	0.436	0.583	0.604	0.581	0.558	0.595
	VI $\times$ (T/B)	0.484	0.459	0.314	0.640	0.488	0.454	0.472	0.633
RMSE	VI	0.308	0.310	0.352	0.276	0.299	0.312	0.307	0.269
	VI $\times$ Brightness	0.379	0.287	0.329	0.363	0.281	0.279	0.285	0.369
	VI $\times$ Greenness	0.396	0.306	0.329	0.432	0.436	0.417	0.317	0.426
	VI $\times$ Third Component	0.275	0.292	0.320	0.298	0.272	0.277	0.289	0.295
	VI $\times$ (T $-$ G)	0.264	0.295	0.320	0.283	0.275	0.280	0.291	0.279
	VI $\times$ (T/B)	0.305	0.313	0.349	0.258	0.304	0.314	0.310	0.260

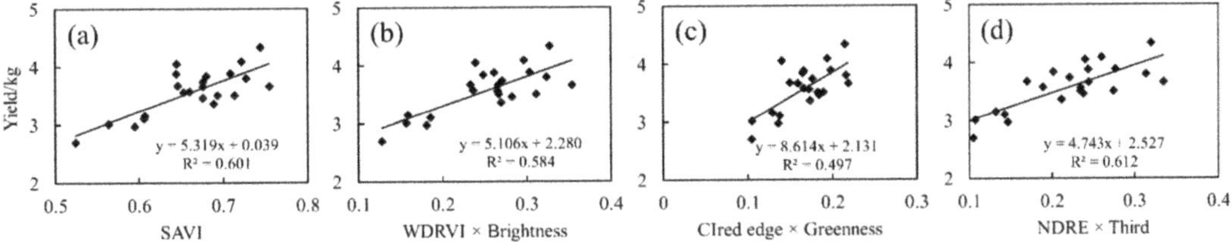

Figure 9. Well-performing yield-estimation models incorporating TCT parameters and VIs at the heading stage.

The TCT parameters were transformed to further improve the accuracy of the rice-yield estimation and reduce the model saturation. On the one hand, Figure 6 reveals that the third component was the best fit for the yield at the booting and heading stages, while the brightness and greenness components were poorly fitted for the yield. On the other hand, the yield was positively correlated with the brightness and third component, and negatively correlated with the greenness component. In addition, the fitting model of the third component and the yield had an obvious saturation phenomenon at the booting stage, which did not exist in the other components. Therefore, two new parameters of the difference between the third component and greenness (T $-$ G) and the quotient of the third component and brightness (T/B) were constructed to fuse the various features of the TCT parameters. The correlation between the new parameters and the yield was significantly enhanced at the booting and heading stages (Figure 7). The rice-yield-prediction results of the VIs incorporating the new TCT parameters are shown in Tables 4 and 5. At the booting stage, the VIs incorporating T $-$ G and T/B had high yield-estimation accuracy (RMSE < 0.276 in the VI $\times$ (T $-$ G) model, and RMSE < 0.269 in the VI $\times$ (T/B) model). Figure 10 shows the yield-simulation models of the VIs incorporating the newly constructed parameters, and there was high accuracy in all the models at both periods (R^2 values are more than 0.6). Nevertheless, all the models at the booting stage had obvious saturation, but none at the heading stage. This indicated that the VIs combining the information of the brightness, greenness, and wetness had good suitability for estimating the rice yield at the heading stage: high accuracy and low saturation.

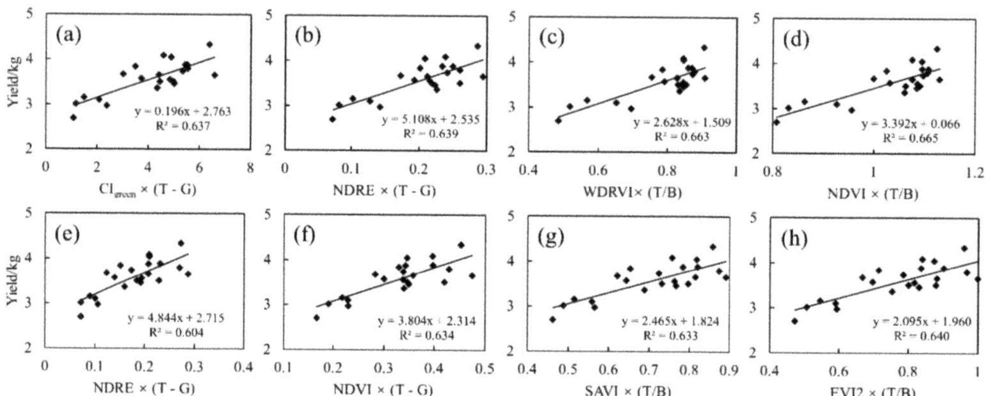

Figure 10. Well-performing yield-estimation models incorporating TCT combination parameters and VIs at different stages: (**a–d**) booting stage; (**e–h**) heading stage.

At length, the LOO-CV method was used to verify the model of the rice-yield estimation at the heading stage, and the results are shown in Figure 11. The model estimation errors of the NDRE × (T − G), NDVI × (T − G), SAVI × (T/B), and EVI2 × (T/B) were less than 7%.

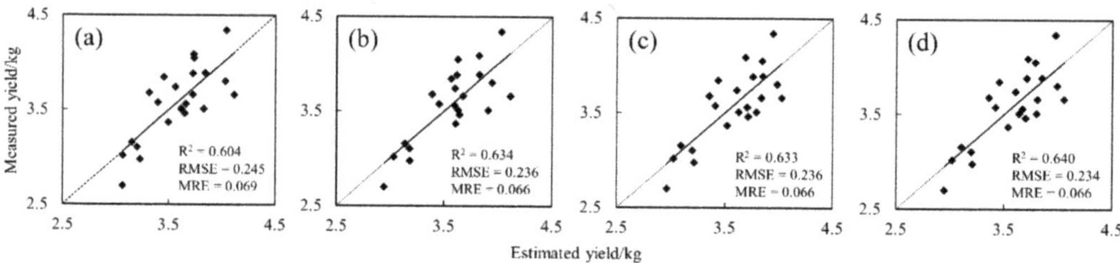

Figure 11. Accuracy-assessment results of the VI × (T − G) and VI × (T/B) models at the heading stage: (**a**) NDRE × (T − G); (**b**) NDVI × (T − G); (**c**) SAVI × (T/B); (**d**) EVI2 × (T/B).

4. Discussion

The main purpose of this paper is to improve the accuracy of rice-yield estimation and reduce the saturation of the models by using the information on the brightness, greenness, and wetness obtained from TCT and combining the UAV-based VIs. The results demonstrated that the VIs incorporating the TCT parameters had good potential to solve these two problems. In crop-yield-estimation studies, an increasing number of parameters have been used in conjunction with VIs. For example, variables such as the canopy texture [36], canopy height [24,37], canopy coverage [24], and temperature [38] are frequently fused by machine-learning methods to improve the crop-yield-estimation accuracy [39]. However, this approach is too complex, and the models have limited robustness. In this paper, we combine the advantages of the VIs and TCT parameters in a simple way through a quadratic operation, which is both easy and has significant accuracy improvement.

The reason for selecting the research period of rice in this paper was that the morphology was not completely stable at the tillering stage and jointing stage, and the leaves and stems changed greatly in a short time. Furthermore, the filling stage and ripening stage were close to the harvesting stage, and thus the yield data obtained was of little value. At the booting and heading stages, the rice gradually completed the transition from vegetative growth to reproductive growth. At the booting stage, there were almost no panicles in the rice canopy, while at the heading stage, with the continuous growth of the rice, the panicles gradually appeared until they covered the whole canopy. Other than that, there was no significant change in this stage relative to the booting stage (Figure 4b,c). Wang et al. proposed that the single-growth-stage model (RNDVI) $_{(880,712)}$ at the booting stage was most suitable for the yield estimation of rice, with an R^2 of 0.75 [18]. Duan et al. pointed out a new method integrating UAV-based VIs and abundance information retrieved from spectral mixture analysis to improve the yield-estimation precision of rice at the heading stage [11]. Zhang et al. put forward that the estimation of the grain yield during the early to mid-growth stages was significant for the initial diagnosis of rice and the quantitative regulation of topdressing [40]. Kawamura et al. demonstrated that the booting stage might be the optimum time for in-season rice-grain assessment [41]. Zhou et al. held that the booting stage was determined as the optimal period for grain-yield estimation using VIs at a single stage for both digital images and multispectral images [19]. Therefore, based on the principle of prediction possibility and time advance, the optimum growth period for rice-yield simulation was determined to be the booting stage, but the heading stage also had great potential for high-precision estimation.

We tried to compare the effects of different data sources on the rice-yield estimation in various stages by collecting the ground data (CH, LAI, and CCC) and UAV remote-sensing images at the booting and heading stages. Peng et al. remotely predicted the yield of oilseed rape based on LAI estimation, with good performance [42]. Hence, the rice yield was first estimated by LAI data. The results showed that the predictive ability of the LAI at the booting stage was significantly better than that at the heading stage (R^2 = 0.569 vs. 0.468) (Figure 3). Liu et al. utilized the LAI integrated with SPAD (LAI × SPAD) data to demonstrate the potential of estimating rice yields [25]. The LAI × SPAD data and rice yield in this paper were also used for regression analysis, and the results showed that they significantly enhanced the ability to predict the rice yield at the booting stage, with an obvious improvement compared with the LAI (Figure 3). However, at the heading stage, the CCC reduced the prediction ability of the rice yield, even worse than the simulation ability of the LAI, which indicated that the appearance of panicles at the heading stage weakened the predicted potential of the LAI and SPAD. Liu et al. also deemed that the CCC completely derived from the green leaves of rice had a good correlation with the yield [25]. Consequently, it was reasonable to speculate that the main reason for the decline in the yield-estimation ability at the heading stage was the emergence of panicles because the SunScan canopy analysis system was used in the LAI measurement. According to its measuring principle, panicles and stems were also a part of the LAI output information, which was probably unrelated to the yield estimation.

With the improvements in RS technology, more crop-canopy images with different spatial scales can be obtained, including multispectral and hyperspectral images [17,43]. VIs calculated by the combination of different bands is one of the most used methods for yield estimation [44]. The eight plot-level VIs (NDVI, $CI_{red\ edge}$, CI_{green}, EVI2, NDRE, WDRVI, MTCI, and SAVI) were extracted from the multispectral images at the booting and heading stages of rice. Then, the VIs and yield data were fitted by the least-squares method. The results showed a good performance at the booting stage, with a minimum RMSE of the WDRVI of 0.254 (Table 4). The simulation results of the other VIs listed at this stage were also satisfactory (RMSE < 0.283). However, there was a most prominent problem of vulnerability to saturation in the VI-based simulation models. The apparent saturation phenomenon exhibited that most of the WDRVI values were concentrated around 0.75 (Figure 8). At the heading stage, the ability of the VIs to predict the yield decreased significantly. Except for the SAVI and EVI2, the simulation accuracies of the rest of the VIs were very poor (RMSE > 0.3) (Table 5). The CI_{green} of the fitting model, in particular, had an adjustment R^2 of 0.305. The calculation of the CI_{green} combined with the reflectance of the green band, and the appearance of panicles, largely reflected the green characteristics of the rice, thus affecting the correlation with the yield. According to the simulation results of the SAVI, the saturation phenomenon was still very distinct (Figure 9). In general, whether it was the booting stage or heading stage, the rice yield simulated by the VIs was inevitably saturated.

The appearance of panicles at the heading stage will lead to changes in the rice-canopy color and other characteristics, which, in turn, have a direct impact on the canopy reflectance. Therefore, the TCT method was used to extract the brightness, greenness, and wetness information of the rice at the booting and heading stages to improve the yield-estimation accuracy and eliminate the saturation. It was found from the TCT-component maps (Figure 5) that the brightness image at the booting stage was darker than that at the heading stage, while the greenness and wetness images at the heading stage were darker than those at the booting stage. This is because the reflectance of the rice canopy at the heading stage was significantly higher than that at the booting stage, and the brightness map was a direct mirror of the reflectance. Due to the appearance of panicles, the greenness of the rice-canopy leaves at the heading stage was replaced by the light color of some of the panicles, resulting in a decrease in the greenness. According to the water requirement of rice, it is necessary to irrigate enough water at the booting stage, and at the heading stage, the water in the paddy fields should be drained off irregularly. Thereby, the wetness of rice fields at the heading stage would decrease. A correlation analysis and regression analysis were performed on the TCT parameters and yield data—(Table 3 and Figure 6). The results showed that the brightness and greenness had poor simulation effects on the yield, while the wetness had a better effect. However, the saturation appeared in the simulation model of the wetness at the booting stage, but it did not exist at the heading stage. In a word, the direct use of the brightness, greenness, and wetness information was not enough to accurately simulate the rice yield.

Combining the different advantages of the VIs and TCT parameters to simulate the rice yield (high precision and low saturation), the method of VIs multiplied by TCT components was employed in this paper. Although some of the parameters were well simulated, the simulation accuracy of the VI × Greenness models at the booting stage, the VI × Brightness, and the VI × Greenness models at the heading stage were lower than those of the VI models. Moreover, there were different degrees of saturation in the fitting models at the booting stage, but none at the heading stage. In combination with the characteristics of different TCT parameters (correlation and saturation), VI × (T − G) and VI × (T/B) were established to estimate the yield of rice at different stages. The models at the booting stage were still saturated, while the simulation models at the heading stage showed high precision with no obvious saturation, and estimation errors below 7%. Consequently, the VIs, which combined the information of the brightness, greenness, and wetness, were suitable for estimating the rice yield at the heading stage.

In this paper, we developed a new approach to estimating rice yields at the booting and heading stages using the integration of VIs and the brightness, greenness, and wetness information retrieved from UAV multispectral images. This method is simple and feasible, but it has crucial reference significance for the yield estimation of rice and similar crops. Moreover, the theoretical and technical support was provided for the crop-yield estimation with evident changes in the canopy over time. In the future, we will further set up more dense nitrogen-fertilizer gradients to explore the best nitrogen-application amount for rice. In terms of the LAI measurement, some new instruments (for example, the LI-3100C table leaf-area meter, LI-COR, USA) will be used to avoid the impact of the panicles and stems on the output of the LAI, and more realistic LAI data will be obtained to improve and validate the accuracy of rice-yield estimations by ground-measurement data. Concurrently, this method will be applied to satellite data and other crops to enable the rapid, nondestructive, and high-precision estimation of crop yields over larger areas.

5. Conclusions

In this study, we developed a technique to improve the estimation of rice yields at the booting and heading stages using UAV-based VIs and TCT-based parameter data. The ground-measurement data could only be used to predict the rice yield at the booting stage, and the prediction ability was lost at the heading stage due to the uneven occurrence of panicles. The UAV-based VIs had similar prediction performances to the ground measurements. Although the accuracy was high at the booting stage, the yield-estimation models were seriously saturated. To improve the prediction accuracy and reduce the saturation of the models, TCT was applied to eliminate the effect of the panicle emergence at the heading stage on the yield estimation. The TCT-component images at the booting and heading stages of the paddy fields were produced based on the six-band UAV images, including the brightness, greenness, and third component (wetness). It was more accurate to use the integration of the plot-level VIs and TCT-parameter information to estimate the rice yield than using VIs alone. Among all the parameters, the $CI_{green} \times (T - G)$, $NDRE \times (T - G)$, $WDRVI \times (T/B)$, and $NDVI \times (T/B)$ at the booting stage, and $NDRE \times (T - G)$, $NDVI \times (T - G)$, $SAVI \times (T/B)$, and $EVI2 \times (T/B)$ at the heading stage, were the most accurate indicators for the rice-yield estimation under different nitrogen-fertilizer treatments, with estimation errors below 7%. The VIs, which combined the brightness, greenness, and third component, were more suitable for estimating the rice yield at the heading stage, with their advantages of high accuracy and low saturation. This paper can provide theoretical and technical support for crop-phenotype-parameter extraction and precision agriculture.

Author Contributions: Conceptualization, S.L. and X.J.; methodology, S.L.; software, W.J.; validation, S.F.; writing—original draft preparation, S.L.; writing—review and editing, S.L., K.Y. and Y.L.; funding acquisition, S.F. All authors have read and agreed to the published version of the manuscript.

Funding: This research was funded by the National Key Research and Development Project (2016YFD0101105), the Phenomics Research and New Variety Creation of Hybrid Rice Based on UAV Remote Sensing (2020BBB058), and the National High-tech Research and Development Program (863 Program) (2013AA102401).

Institutional Review Board Statement: Not applicable.

Informed Consent Statement: Not applicable.

Data Availability Statement: Not applicable.

Conflicts of Interest: The authors declare no conflict of interest.

References

1. Zhu, Y.G.; Williams, P.N.; Meharg, A.A. Exposure to inorganic arsenic from rice: A global health issue? *Environ. Pollut.* **2008**, *154*, 169–171. [CrossRef]
2. Zhang, J.T.; Feng, L.P.; Zou, H.P.; Liu, D.L. Using ORYZA2000 to model cold rice yield response to climate change in the Heilongjiang province, China. *Crop J.* **2015**, *3*, 317–327. [CrossRef]
3. Yang, G.J.; Liu, J.G.; Zhao, C.J.; Li, Z.H.; Huang, Y.B.; Yu, H.Y.; Xu, B.; Yang, X.D.; Zhu, D.M.; Zhang, X.Y.; et al. Unmanned aerial vehicle remote sensing for field-based crop phenotyping: Current status and perspectives. *Front. Plant Sci.* **2017**, *8*, 26. [CrossRef]
4. Zhang, J.T.; Tian, H.Q.; Yang, J.; Pan, S.F. Improving representation of crop growth and yield in the dynamic land ecosystem model and its application to China. *J. Adv. Model. Earth Syst.* **2018**, *10*, 1680–1707. [CrossRef]
5. Luo, S.; He, Y.B.; Li, Q.; Jiao, W.H.; Zhu, Y.Q.; Zhao, X.H. Nondestructive estimation of potato yield using relative variables derived from multi-period LAI and hyperspectral data based on weighted growth stage. *Plant Methods* **2020**, *16*, 14. [CrossRef]
6. Palosuo, T.; Kersebaum, K.C.; Angulo, C.; Hlavinka, P.; Moriondo, M.; Olesen, J.E.; Patil, R.H.; Ruget, F.; Rumbaur, C.; Takac, J.; et al. Simulation of winter wheat yield and its variability in different climates of Europe: A comparison of eight crop growth models. *Eur. J. Agron.* **2011**, *35*, 103–114. [CrossRef]
7. Liu, N.F.; Budkewitsch, P.; Treitz, P. Examining spectral reflectance features related to Arctic percent vegetation cover: Implications for hyperspectral remote sensing of Arctic tundra. *Remote Sens. Environ.* **2017**, *192*, 58–72. [CrossRef]
8. Vilfan, N.; van der Tol, C.; Muller, O.; Rascher, U.; Verhoef, W. Fluspect-B: A model for leaf fluorescence, reflectance and transmittance spectra. *Remote Sens. Environ.* **2016**, *186*, 596–615. [CrossRef]
9. Kross, A.; McNairn, H.; Lapen, D.; Sunohara, M.; Champagne, C. Assessment of RapidEye vegetation indices for estimation of leaf area index and biomass in corn and soybean crops. *Int. J. Appl. Earth Obs. Geoinf.* **2015**, *34*, 235–248. [CrossRef]

10. Moharana, S.; Dutta, S. Spatial variability of chlorophyll and nitrogen content of rice from hyperspectral imagery. *ISPRS-J. Photogramm. Remote Sens.* **2016**, *122*, 17–29. [CrossRef]

11. Duan, B.; Fang, S.H.; Zhu, R.S.; Wu, X.T.; Wang, S.Q.; Gong, Y.; Peng, Y. Remote estimation of rice yield with unmanned aerial vehicle (UAV) data and spectral mixture analysis. *Front. Plant Sci.* **2019**, *10*, 14. [CrossRef] [PubMed]

12. Lobell, D.B.; Azzari, G.; Burke, M.; Gourlay, S.; Jin, Z.; Kilic, T.; Murray, S. Eyes in the sky, boots on the ground: Assessing satellite- and ground-based approaches to crop yield measurement and analysis. *Am. J. Agr. Econ.* **2020**, *102*, 202–219. [CrossRef]

13. Schwalbert, R.A.; Amado, T.J.C.; Nieto, L.; Varela, S.; Corassa, G.M.; Horbe, T.A.N.; Rice, C.W.; Peralta, N.R.; Ciampitti, I.A. Forecasting maize yield at field scale based on high-resolution satellite imagery. *Biosyst. Eng.* **2018**, *171*, 179–192. [CrossRef]

14. Cao, H.T.; Gu, X.F.; Sun, Y.; Gao, H.L.; Tao, Z.; Shi, S.Y. Comparing, validating and improving the performance of reflectance obtention method for UAV-Remote sensing. *Int. J. Appl. Earth Obs. Geoinf.* **2021**, *102*, 15. [CrossRef]

15. Aslan, M.F.; Durdu, A.; Sabanci, K.; Ropelewska, E.; Gueltekin, S.S. A comprehensive survey of the recent studies with UAV for precision agriculture in open fields and greenhouses. *Appl. Sci.* **2022**, *12*, 29. [CrossRef]

16. Liu, S.S.; Li, L.T.; Gao, W.H.; Zhang, Y.K.; Liu, Y.N.; Wang, S.Q.; Lu, J.W. Diagnosis of nitrogen status in winter oilseed rape (*Brassica napus* L.) using in-situ hyperspectral data and unmanned aerial vehicle (UAV) multispectral images. *Comput. Electron. Agric.* **2018**, *151*, 185–195. [CrossRef]

17. Deng, L.; Mao, Z.H.; Li, X.J.; Hu, Z.W.; Duan, F.Z.; Yan, Y.N. UAV-based multispectral remote sensing for precision agriculture: A comparison between different cameras. *ISPRS-J. Photogramm. Remote Sens.* **2018**, *146*, 124–136. [CrossRef]

18. Wang, F.L.; Wang, F.M.; Zhang, Y.; Hu, J.H.; Huang, J.F.; Xie, J.K. Rice yield estimation using parcel-level relative spectra variables from UAV-based hyperspectral imagery. *Front. Plant Sci.* **2019**, *10*, 12. [CrossRef]

19. Zhou, X.; Zheng, H.B.; Xu, X.Q.; He, J.Y.; Ge, X.K.; Yao, X.; Cheng, T.; Zhu, Y.; Cao, W.X.; Tian, Y.C. Predicting grain yield in rice using multi-temporal vegetation indices from UAV-based multispectral and digital imagery. *ISPRS-J. Photogramm. Remote Sens.* **2017**, *130*, 246–255. [CrossRef]

20. Duan, B.; Fang, S.H.; Gong, Y.; Peng, Y.; Wu, X.T.; Zhu, R.S. Remote estimation of grain yield based on UAV data in different rice cultivars under contrasting climatic zone. *Field Crop. Res.* **2021**, *267*, 11. [CrossRef]

21. Joshi, P.P.; Wynne, R.H.; Thomas, V.A. Cloud detection algorithm using SVM with SWIR2 and tasseled cap applied to Landsat 8. *Int. J. Appl. Earth Obs. Geoinf.* **2019**, *82*, 10. [CrossRef]

22. Mostafiz, C.; Chang, N.B. Tasseled cap transformation for assessing hurricane landfall impact on a coastal watershed. *Int. J. Appl. Earth Obs. Geoinf.* **2018**, *73*, 736–745. [CrossRef]

23. Wang, Z.L.; Chen, J.X.; Zhang, J.W.; Fan, Y.F.; Cheng, Y.J.; Wang, B.B.; Wu, X.L.; Tan, X.M.; Tan, T.T.; Li, S.L.; et al. Predicting grain yield and protein content using canopy reflectance in maize grown under different water and nitrogen levels. *Field Crop. Res.* **2021**, *260*, 15. [CrossRef]

24. Wan, L.; Cen, H.Y.; Zhu, J.P.; Zhang, J.F.; Zhu, Y.M.; Sun, D.W.; Du, X.Y.; Zhai, L.; Weng, H.Y.; Li, Y.J.; et al. Grain yield prediction of rice using multi-temporal UAV-based RGB and multispectral images and model transfer–A case study of small farmlands in the South of China. *Agric. For. Meteorol.* **2020**, *291*, 15. [CrossRef]

25. Liu, X.J.; Zhang, K.; Zhang, Z.Y.; Cao, Q.; Lv, Z.F.; Yuan, Z.F.; Tian, Y.C.; Cao, W.X.; Zhu, Y. Canopy chlorophyll density based index for estimating nitrogen status and predicting grain yield in rice. *Front. Plant Sci.* **2017**, *8*, 12. [CrossRef]

26. Smith, G.M.; Milton, E.J. The use of the empirical line method to calibrate remotely sensed data to reflectance. *Int. J. Remote Sens.* **1999**, *20*, 2653–2662. [CrossRef]

27. Tucker, C.J. Red and photographic infrared linear combinations for monitoring vegetation. *Remote Sens. Environ.* **1979**, *8*, 127–150. [CrossRef]

28. Gitelson, A.A.; Gritz, Y.; Merzlyak, M.N. Relationships between leaf chlorophyll content and spectral reflectance and algorithms for non-destructive chlorophyll assessment in higher plant leaves. *J. Plant Physiol.* **2003**, *160*, 271–282. [CrossRef]

29. Jiang, Z.Y.; Huete, A.R.; Didan, K.; Miura, T. Development of a two-band enhanced vegetation index without a blue band. *Remote Sens. Environ.* **2008**, *112*, 3833–3845. [CrossRef]

30. Gitelson, A.; Merzlyak, M.N. Quantitative estimation of chlorophyll-a using reflectance spectra: Experiments with autumn chestnut and maple leaves. *J. Photochem. Photobiol. B-Biol.* **1994**, *22*, 247–252. [CrossRef]

31. Gitelson, A.A. Wide dynamic range vegetation index for remote quantification of biophysical characteristics of vegetation. *J. Plant Physiol.* **2004**, *161*, 165–173. [CrossRef] [PubMed]

32. Dash, J.; Curran, P.J. The MERIS terrestrial chlorophyll index. *Int. J. Remote Sens.* **2004**, *25*, 5403–5413. [CrossRef]

33. Huete, A.R. A soil-adjusted vegetation index (SAVI). *Remote Sens. Environ.* **1988**, *25*, 295–309. [CrossRef]

34. Crist, E.P.; Cicone, R.C. A physically-based transformation of thematic mapper data—The TM tasseled cap. *IEEE Trans. Geosci. Remote Sensing* **1984**, *22*, 256–263. [CrossRef]

35. Wong, T.T. Performance evaluation of classification algorithms by k-fold and leave-one-out cross validation. *Pattern Recognit.* **2015**, *48*, 2839–2846. [CrossRef]

36. Ma, Y.R.; Ma, L.L.; Zhang, Q.; Huang, C.P.; Yi, X.; Chen, X.Y.; Hou, T.Y.; Lv, X.; Zhang, Z. Cotton yield estimation based on vegetation indices and texture features derived from RGB image. *Front. Plant Sci.* **2022**, *13*, 17. [CrossRef]

37. Li, B.; Xu, X.M.; Zhang, L.; Han, J.W.; Bian, C.S.; Li, G.C.; Liu, J.G.; Jin, L.P. Above-ground biomass estimation and yield prediction in potato by using UAV-based RGB and hyperspectral imaging. *ISPRS-J. Photogramm. Remote Sens.* **2020**, *162*, 161–172. [CrossRef]

38. Fei, S.P.; Hassan, M.A.; Xiao, Y.G.; Su, X.; Chen, Z.; Cheng, Q.; Duan, F.Y.; Chen, R.Q.; Ma, Y.T. UAV-based multi-sensor data fusion and machine learning algorithm for yield prediction in wheat. *Precis. Agric.* **2022**, *23*, 26. [CrossRef]

39. Ashapure, A.; Jung, J.H.; Chang, A.J.; Oh, S.; Yeom, J.; Maeda, M.; Maeda, A.; Dube, N.; Landivar, J.; Hague, S.; et al. Developing a machine learning based cotton yield estimation framework using multi-temporal UAS data. *ISPRS-J. Photogramm. Remote Sens.* **2020**, *169*, 180–194. [CrossRef]

40. Zhang, K.; Ge, X.K.; Shen, P.C.; Li, W.Y.; Liu, X.J.; Cao, Q.; Zhu, Y.; Cao, W.X.; Tian, Y.C. Predicting rice grain yield based on dynamic changes in vegetation indexes during early to mid-growth stages. *Remote Sens.* **2019**, *11*, 24. [CrossRef]

41. Kawamura, K.; Ikeura, H.; Phongchanmaixay, S.; Khanthavong, P. Canopy hyperspectral sensing of paddy fields at the booting stage and PLS regression can assess grain yield. *Remote Sens.* **2018**, *10*, 15. [CrossRef]

42. Peng, Y.; Zhu, T.E.; Li, Y.C.; Dai, C.; Fang, S.H.; Gong, Y.; Wu, X.T.; Zhu, R.S.; Liu, K. Remote prediction of yield based on LAI estimation in oilseed rape under different planting methods and nitrogen fertilizer applications. *Agric. For. Meteorol.* **2019**, *271*, 116–125. [CrossRef]

43. Feng, W.; Guo, B.B.; Zhang, H.Y.; He, L.; Zhang, Y.S.; Wang, Y.H.; Zhu, Y.J.; Guo, T.C. Remote estimation of above ground nitrogen uptake during vegetative growth in winter wheat using hyperspectral red-edge ratio data. *Field Crop. Res.* **2015**, *180*, 197–206. [CrossRef]

44. Hatfield, J.L.; Gitelson, A.A.; Schepers, J.S.; Walthall, C.L. Application of spectral remote sensing for agronomic decisions. *Agron. J.* **2008**, *100*, S117–S131. [CrossRef]

Article

Weed Detection in Peanut Fields Based on Machine Vision

Hui Zhang, Zhi Wang, Yufeng Guo, Ye Ma, Wenkai Cao, Dexin Chen, Shangbin Yang and Rui Gao *

College of Information and Management Science, Henan Agricultural University, Zhengzhou 450046, China
* Correspondence: ronnierrui@henau.edu.cn; Tel.: +86-13673626972

Abstract: The accurate identification of weeds in peanut fields can significantly reduce the use of herbicides in the weed control process. To address the identification difficulties caused by the cross-growth of peanuts and weeds and by the variety of weed species, this paper proposes a weed identification model named EM-YOLOv4-Tiny incorporating multiscale detection and attention mechanisms based on YOLOv4-Tiny. Firstly, an Efficient Channel Attention (ECA) module is added to the Feature Pyramid Network (FPN) of YOLOv4-Tiny to improve the recognition of small target weeds by using the detailed information of shallow features. Secondly, the soft Non-Maximum Suppression (soft-NMS) is used in the output prediction layer to filter the best prediction frames to avoid the problem of missed weed detection caused by overlapping anchor frames. Finally, the Complete Intersection over Union (CIoU) loss is used to replace the original Intersection over Union (IoU) loss so that the model can reach the convergence state faster. The experimental results show that the EM-YOLOv4-Tiny network is 28.7 M in size and takes 10.4 ms to detect a single image, which meets the requirement of real-time weed detection. Meanwhile, the mAP on the test dataset reached 94.54%, which is 6.83%, 4.78%, 6.76%, 4.84%, and 9.64% higher compared with YOLOv4-Tiny, YOLOv4, YOLOv5s, Swin-Transformer, and Faster-RCNN, respectively. The method has much reference value for solving the problem of fast and accurate weed identification in peanut fields.

Keywords: weed identification; YOLOv4-Tiny; attention mechanism; multiscale detection; precision agriculture

Citation: Zhang, H.; Wang, Z.; Guo, Y.; Ma, Y.; Cao, W.; Chen, D.; Yang, S.; Gao, R. Weed Detection in Peanut Fields Based on Machine Vision. *Agriculture* **2022**, *12*, 1541. https://doi.org/10.3390/agriculture12101541

Academic Editor: Michele Pisante

Received: 6 August 2022
Accepted: 22 September 2022
Published: 24 September 2022

Publisher's Note: MDPI stays neutral with regard to jurisdictional claims in published maps and institutional affiliations.

1. Introduction

Peanut is one of the leading oil crops in the world and is vital to global oil production. However, weed competition [1] is an essential factor restricting peanut production, reducing peanut production by 5–15% owing to annual grass damage. Research has shown that peanut production in farmlands with 20 weeds per square meter is 48.31% less than a no-weed control group. In addition, weeds facilitate the breeding and spread of diseases and insect pests, resulting in the frequent emergence of peanut diseases and insect pests [2]. The conventional weeding method of spraying pesticides incurs a significant amount of pesticide waste and causes irreversible pollution to the farmland. Owing to the development of precision agriculture [3], the investigation of site-specific weed management [4] for weed prevention and control has intensified gradually. An efficient detection and identification method for peanuts and weeds is necessary to achieve accurate weed control and management in the farmland.

Currently, many methods are proposed for weed detection, including remote sensing analysis [5], spectral identification [6], and machine vision identification [7]. The equipment required for remote sensing analysis and spectral identification methods is expensive and difficult to promote in agricultural production. The machine vision identification method has been widely used in weed identification because of its low cost and high portability. Bakhshish et al. [8] used Fourier descriptors and invariant moment features to form a shape feature set and implemented weed detection based on artificial neural networks. Rojas et al. [9] extracted the texture features of weeds using the gray-level co-occurrence matrix. They used principal component analysis to reduce the dimensionality of the features and finally used a support vector machine algorithm to complete the classification. Although these methods achieve the identification of crops and weeds, they rely excessively on the manual design and selection of image features, are susceptible to environmental factors such as lighting, and have poor stability and low recognition accuracy.

The development of deep learning technology [10] has enabled convolutional neural networks to reveal deeper features in images, which possess stronger generalization ability than manually selected features. Gai et al. [11] proposed an improved YOLOv4 model for fast and accurate detection of cherry fruit in complex environments. Khan et al. [12] established a weed identification system for pea and strawberry fields based on an improved Faster-RCNN, whose maximum average accuracy for weed recognition was 94.73%. Sun et al. [13] used YOLOv3 to identify Chinese cabbages in a vegetable field. They employed image processing methods to tag plants around Chinese cabbages as weeds. To detect weeds in a carrot field, Ying et al. [14] incorporated deep separable convolutions and an inverted residual block structure into YOLOv4 and replaced its backbone network with MobileNetV3-Small, which improved the recognition speed of the model; however, the average recognition accuracy was only 86.62%. The studies mentioned above indicate that although deep learning can solve the problem of manual feature design in conventional image processing methods, the following issues remain: 1) although using a deep-seated network model for weed detection improves the recognition accuracy, the recognition speed cannot satisfy real-time requirements owing to its large volume; 2) improving the recognition speed by trimming the model network renders the model insensitive to smaller target recognition and reduces its recognition accuracy.

In this study, peanuts and six types of weeds were used as recognition objects, and a weed recognition model based on the improved YOLOv4-Tiny [15] was developed to address the issues above. First, based on YOLOv4-Tiny, CSPDarkNet53-Tiny [16] was used as the backbone network of the model to ensure real-time detection performance; next, a multiscale detection model was implemented by introducing the detailed information of shallow-layer features in an FPN [17] to improve the ability of smaller target recognition. In addition, an ECA [18] module was used to calibrate the effective feature layer to enhance key information pertaining to weeds in the image. Finally, the soft-NMS [19] function was used in the output prediction layer to replace the NMS [20] function to filter the prediction box.

2. Materials and Methods

2.1. Materials

2.1.1. Data Acquisition

The weed images used in this study were obtained from peanut fields in more than 20 areas in Henan Province, China. A Fuji Finepixs4500 camera was used to capture artificial images with a resolution of 2017 × 2155 in JPG format; 855 images were obtained, including those of a single weed, sparsely distributed weeds, and overgrown weeds. The images were captured at 7:00, 13:00, and 17:00 via high-angle overhead shots from approximately 70 cm relative to the ground. Based on investigation and screening, the weed types selected were Portulaca oleracea, Eleusine indica, Chenopodium album, Amaranth blitum, Abutilon theophrasti, and Calystegia. No imbalance was indicated between any two types of weeds. The shape and color of the six weeds are shown in Figure 1.

Figure 1. Shape and color of six weeds. (**a**) Portulaca oleracea, (**b**) Eleusine indica, (**c**) Chenopodium album, (**d**) Amaranth blitum, (**e**) Abutilon thophrasti, (**f**) Calystegia hederacea.

2.1.2. Data Enhancement and Annotation

Overfitting in the training set caused by excessively small data sizes was prevented using the following methods: image horizontal and vertical flip, brightness increase and decrease (randomly increase or decrease the original brightness by 10%–20%), and Gaussian noise addition (variance $\sigma = 0.05$) for random image enhancement [21]. Figure 2 shows an example of the effect of data enhancement. The data enhancement method was only used in the training set. The expanded dataset contained 3355 images. Information regarding weeds and peanuts in the image was annotated using the LabelImg software. The annotation format was Pascal VOC2007, and the file type was .xml. The dataset was categorized into training and test sets. The number of pictures in each dataset is shown in Table 1.

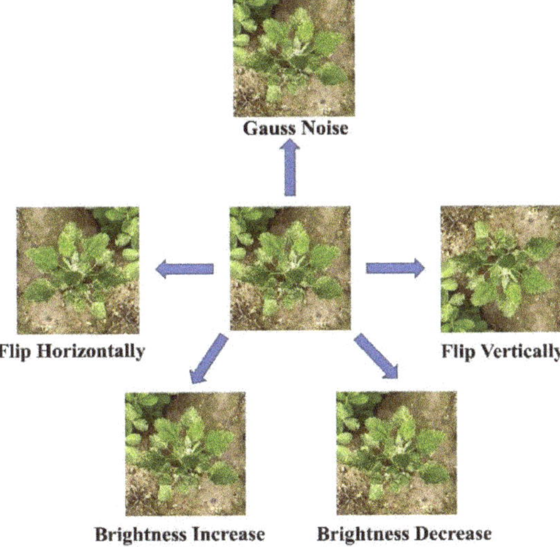

Figure 2. Data enhancement.

Table 1. Dataset after data enhancement.

Dataset	Train	Test	Total
Original Images	700	155	855
Flip Horizontally	500	0	500
Flip Vertically	500	0	500
Brightness Increase	500	0	500
Brightness Decrease	500	0	500
Gauss Noise	500	0	500
Total Number	3200	155	3355

2.2. Methods

2.2.1. EM-YOLOv4-Tiny Network

YOLOv4-Tiny comprises four components: an input layer, a backbone network, an FPN, and an output prediction layer. The images received were uniformly scaled to a size of 416×416. The features were extracted from CSPDarkNet53-Tiny and then sent to the FPN for feature fusion. The location and category information of the target was obtained in the output prediction layer. CSPDarkNet53-Tiny primarily comprises a CBL module and a cross-stage partial (CSP) module [22]. The CBL module comprises a convolutional layer, batch normalization, and a Leaky Relu [23] activation function in series. It is the smallest module in the overall network structure and is used for feature control splicing and sampling. The CSP module is an improved residual network structure that can segment the input feature map into two components: the main component stacks the residual, and the other is fused in series with the main component after some processing. CSPDarkNet53-Tiny contains three CSP modules: CSP1, CSP2, and CSP3. As the dimensions of the output feature map are reduced, the location information in the CSP module becomes increasingly vague, the detailed information becomes increasingly scarce, and the ability to detect smaller targets is gradually weakened. To solve these problems, a path connected to the CSP2 layer in the FPN was added, while the output characteristics of the CSP2 layer were fused with the upsampling results in the channel dimension to form an output focused on the detection of smaller targets. The EM-YOLOv4-Tiny network structure is shown in Figure 3.

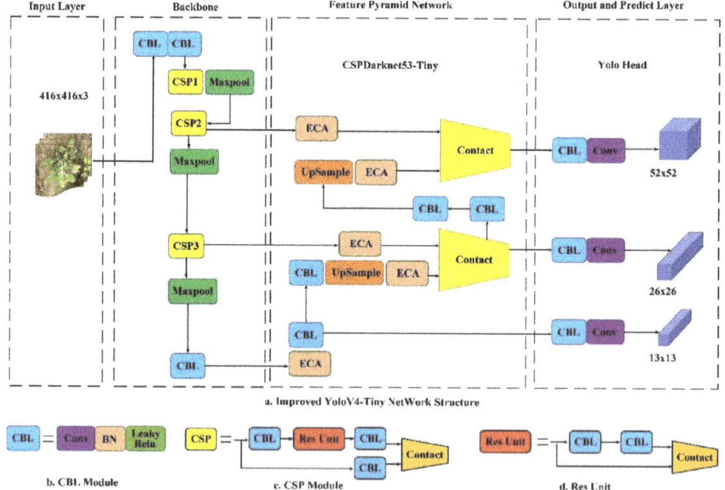

Figure 3. EM-YOLOv4-Tiny network Structure, where Conv is convolution, BN is batch normalization, Leak Relu is activation function, Maxpool is maximum pooling, ResUnit is the residual unit, Upsample is upsampling, ECA is efficient channel attention module, Contact is the feature fusion method of adding channel numbers, Yolo Head is the prediction anchor, CBL is series fusion module of Conv, BN, and Leak Relu, and CSP is cross-stage partial module.

To further improve the detection accuracy, the ECA module was used repetitively to process the effective features in the FPN. The attention module suppressed the background information in the image and enhanced the key information through weight calibration [24]. Regarding the predicted output, the EM-YOLOv4-Tiny network yielded three outputs of different scales, namely 13 × 13, 26 × 26, and 52 × 52.

2.2.2. ECA Attention Mechanisms

Multiscale prediction for hierarchical detection was utilized in this study to detect smaller targets. Although shallow features have smaller receptive fields, which can enable better detection of smaller targets, they result in considerable irrelevant noises, thus affecting the network's ability to assess the importance of information obtained from an image. By introducing the ECA attention module into the neck section of YOLOv4-Tiny, the weed features in the image could be further enhanced while irrelevant background weights were suppressed.

In the ECA network, the input features were first pooled globally, and a single numerical value was used to represent the characteristics of each channel. Next, a fast one-dimensional convolution [25] of size k was performed to assign weights for each channel to realize information exchange between channels. Finally, the weight proportion of each channel was generated using the sigmoid function [26], and features with channel attention were obtained by merging with the original input features. More details about the ECA network can be found in Appendix A.

2.2.3. Use of Complete Intersection over Union Loss

Owing to the scale invariance and non-negativity of the IoU [27], the latter is typically set as the bounding box loss function in conventional target detection networks. Specifically, IoU refers to calculating the ratio of the prediction box and the real box, which can better reflect the quality of the regression box. However, using IoU as the loss function still has some problems. On the one hand, when the positions of two bounding boxes do not intersect (IoU = 0), the loss function will become non-differentiable. On the other hand, when the overlap rate of prediction frames is the same, IoU cannot accurately reflect the location information of both.

Therefore, the CIoU [28] was used in this study as the loss function for training. Additionally, the overlap degree and the distance between the prediction and real boxes were considered comprehensively, and the aspect ratio of the prediction box was added as a penalty term to stabilize the regression results. More details about CIoU loss can be found in Appendix B.

2.2.4. Soft-NMS Algorithm for Filtering Prediction Boxes

For the output and prediction of YOLOv4-Tiny, the NMS algorithm filters redundant prediction boxes around the target to be detected. The NMS algorithm deletes prediction boxes whose confidence is below the preset threshold, filters boxes that belong to the same category, and obtains the highest score in a specific area; hence, it effectively eliminates redundant bounding boxes. However, in cases involving dense weed growth or severe mutual occlusion between weeds and peanuts, the NMS algorithm deletes prediction boxes that belong to other targets, thus resulting in missed detections. To solve this issue, the soft-NMS instead of the original NMS was used in this study. When multiple prediction boxes appeared around a weed, their scores were multiplied by a weighting function to weaken those that overlapped with the box with the highest score. In this regard, the Gaussian [29] function was used as the weighting function, and the calculation is as follows:

$$\text{Score}_i = \text{Score}_i \cdot e^{\frac{-\text{IoU}(C_i, B)}{\sigma}} \tag{1}$$

where Score_i represents the score of the current box, C_i represents the current bounding box, and B represents the prediction box with the highest score. The greater the overlap between the prediction box and the box with the highest score, the stronger the weakening ability of the weighting function and the lower the score assigned to it.

2.2.5. Model Performance Evaluation Indices

In this study, indices typically used in multiclass target detection models, such as precision, recall rate, mean average precision (mAP), and F1 value, were used to evaluate the model performance.

Precision indicates the proportion of correct detections in all the prediction boxes, and Recall indicates the proportion of correctly detected label boxes in all label boxes.

$$\text{Precision} = \frac{TP}{TP + FP}, \tag{2}$$

$$\text{Recall} = \frac{TP}{TP + FN}, \tag{3}$$

where TP represents the number of correctly detected weeds; FP represents the number of incorrectly detected weeds; and FN represents the number of missed detections of weeds.

AP represents the average precision of a class of detected objects, and mAP is the mean average value of AP for all classes.

$$AP = \int_1^0 \text{Precision d Recall}, \tag{4}$$

$$\text{mAP} = \frac{1}{N} \sum_1^N AP(k) \tag{5}$$

The F1 value can be regarded as a harmonic mean of Precision and Recall, as follows:

$$F1 = 2 \times \frac{Precision \times Recall}{Precision + Recall} \tag{6}$$

The evaluation indices selected in this study were calculated based on a threshold of 0.5. In the follow-up experiments, the mAP was used as the primary performance evaluation index of the model.

2.2.6. Model Training

The software and hardware environment of model training and testing are shown in Table 2. In order to further improve the recognition accuracy of the model, this study used a transfer learning method to initialize the weights of the model. Before model training, the EM-YOLOv4-Tiny network was pretrained with the Pascal VOC dataset, and the weight file with the highest map in the training results was used as the pretraining weight to initialize the model. Meanwhile, the K-means [30] algorithm was used to cluster the anchor boxes in the dataset, and a total of 9 anchor boxes with different sizes were obtained: (19, 31), (56, 62), (90, 82), (103, 158), (149, 125), (175, 217), (250, 171), (241, 291), and (320, 335). This makes the true size of the anchor frame closer to the size of the weed to be detected. During training, the number of samples in each batch was set to 16, and the loading of the entire training set was considered an iteration. The adaptive moment estimation algorithm was used to optimize the model, the initial learning rate was set to 0.001, and the cosine annealing algorithm was employed for attenuation. After 150 iterations, the model converged.

Table 2. Training and test environment configuration table.

Configuration	Parameter
Operating System	Ubuntu 18.04.1 LTS
CPU	Intel(R) Xeon(R) Silver 4114 CPU @ 2.20GHz
GPU	NVIDIA Tesla T4
Accelerate Environment	CUDA10.2 CuDNN7.6.5
Pytorch	1.2
Python	3.6.2

3. Results

3.1. Performance Evaluation of EM-YOLOv4-Tiny

Based on the standard of the MS COCO dataset provided by Microsoft, weeds with a resolution lower than 32×32 were defined as smaller targets. Several types of weeds exist in peanut fields, with some being smaller in morphological appearance than others. The standard YOLOv4-Tiny network tends to misdetect when identifying smaller targets. Based on the comparison results of EM-YOLOv4-Tiny and YOLOv4-Tiny using the same test set as shown in Table 3, the recognition precision rates of the EM-YOLOv4-Tiny for smaller targets and all targets were 89.65% and 94.54%, respectively, which surpassed the precision rates of the original network by 10.12% and 6.83%, respectively. The improved network combined the location and detailed information of the shallow-layer feature and improved the ability to identify smaller weeds via the addition of a channel attention mechanism, which suppresses the abundant noise in smaller receptive fields. The recognition performances before and after the network improvement are shown in Figure 4. The EM-YOLOV4-Tiny network included a new scale output in the neck section while the backbone network structure of the model remained unchanged, and the average inference time of each image increased to only 4.4 ms, indicating that the proposed network maintained a high inference speed while improving the recognition precision.

Table 3. Comparison of detection results of YOLOv4-Tiny and EM-YOLOv4-Tiny.

Models	mAP/%		Volume/MB	Time/ms
	Small Targets	All Targets		
YOLOv4-Tiny	79.53	87.71	22.4	6
EM-YOLOv4-Tiny	89.65	94.54	28.7	10.4

Figure 4. Comparison of detection results of YOLOv4-Tiny and EM-YOLOv4-Tiny, where (**a–c**) represent the recognition effect of the YOLOv4-Tiny model, and (**d–f**) represent the recognition effect of the EM-YOLOv4-Tiny model.

3.2. Performance Comparison of Improved Methods

To further demonstrate the effectiveness of the improved method in enhancing the model performance, different modules were benchmarked against the original YOLOv4-Tiny target detection network. The results are shown in Table 4.

Table 4. Influence of different improved modules on YOLOv4-Tiny network.

Method	Precision/%	Recall/%	mAP/%	F1/%	Time/ms
YOLOv4-Tiny	87.60	75.60	87.71	0.80	6.0
YOLOv4-Tiny + K-Means	91.80	74.80	88.90	0.82	6.0
YOLOv4-Tiny + K-Means+ Soft-NMS	88.16	84.91	90.37	0.86	6.0
YOLOv4-Tiny + K-Means+ Soft-NMS + scale3	95.40	82.90	93.72	0.89	9.0
YOLOv4-Tiny + K-Means+ Soft-NMS + scale3 + ECA(EM-YOLOv4-Tiny)	96.7	85.90	94.54	0.90	10.4

scale3 represents an improved strategy for employing multiscale detection in the network.

After obtaining the anchor box using the K-means clustering algorithm, the mAP and F1 values of the model were 1.2% and 2% higher than the original values, respectively, indicating a better match in size between the anchor box and the target to be detected. When using the soft-NMS algorithm to filter the prediction box, the recognition precision decreased. Still, the recall rate increased by approximately 10%, indicating the effectiveness of soft-NMS in improving missed detections. When a new functional layer was added to focus on detecting smaller targets, the detection time increased slightly, but the mAP and F1 values increased by approximately 3%. When the ECA attention mechanism was introduced into the network, the noise caused by shallow features was reduced, and Recall increased by 3%. In general, the proposed methods improved the weed detection performance of the network.

3.3. Performance Comparison of Different Attention Mechanisms

To further verify the advantages of the channel attention mechanism used in this study, under the same experimental conditions, the SE attention mechanism and CBAM attention machine were used as controls at the same location as the network. The experimental results are shown in Table 5.

Table 5. Performance comparison after using different attention modules.

Method	Precision/%	Recall/%	mAP/%	F1/%	Time/ms
Base-SE	96.3	79.6	92.32	0.87	11
Base-CBAM	97.5	80.8	93.15	0.88	12
Base-ECA(EM-YOLOv4-Tiny)	96.7	85.9	94.54	0.90	10.4

Base represents the combined model obtained by using methods of K-Means, multiscale strategy, and soft-NMS, and its result can be found in Table 4.

Compared with the ECA attention network, the SE network uses a full connection to realize information exchange between channels, which increases the computational load and causes feature loss due to dimensionality reductions. The CBAM network is a convolutional block attention module that introduces location information in the channel dimension using the global maximum pool. However, it is limited to local range information instead of long-range dependent information. As shown in Table 5, after different attention mechanisms were added, each performance index improved compared with those of the original model. Among them, the ECA attention module outperformed the others; its mAP was higher than those of the other two attention modules by 2.22% and 1.39%, respectively, implying that the ECA network is more suitable for the model used in this study.

Similarly, in order to further explore the impact of the attention module on the weed detection model, the grad cam method was used in this study to visually analyze the features of the networks before and after adding the attention mechanism. From the detection results in Figure 5, we know that when the attention mechanism module is not added, the network will appear to pay attention to the background information when performing the detection. In contrast, the network incorporating the attention mechanism pays more attention to the information of the object to be detected through the recalibration of the weights. Comparing the feature visualization results of the three attention networks, the ECA network used in this study shades darker on the small target weeds in the images, indicating more attention to the information of small targets.

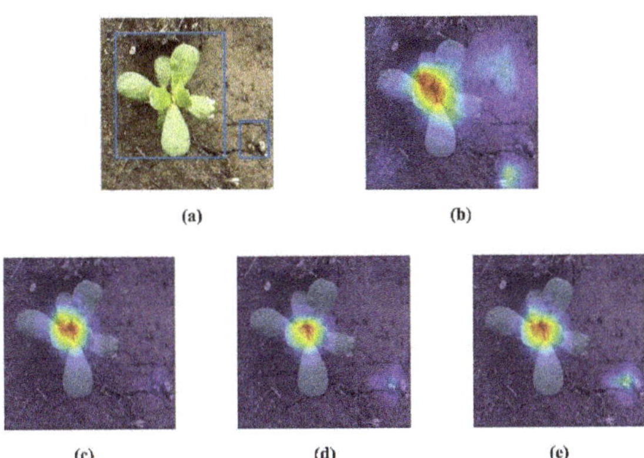

Figure 5. Visual heat map of attentional mechanisms, where (**a**) represents the original image; (**b**) represents the results of using the base model; (**c**) represents the results of using the base model and the SE attention mechanism; (**d**) represents the results of using the base model and the CBAM attention mechanism; (**e**) represents the results of using the base model and the ECA attention mechanism.

3.4. Comparison of Performance with Different Network Models

To verify the efficiency and practicability of the proposed model, several classical target detection models, such as YOLOv4, YOLOv5s, and the Faster-RCNN, were used to test the efficiency of weed detection. In the comparison experiments, strict control was exerted over the parameters. Specifically, 416×416 images were used uniformly as the input to the training network, and identical training and test sets were used throughout the experiments. The results are shown in Table 6.

Table 6. Performance comparison results of multiple target detection networks.

Model	mAP/%	F1/%	Time/ms	Volume/MB	Parameter/$\times 10^6$
Faster-RCNN	84.90	0.78	121	111.4	28.3
YOLOv4	89.76	0.80	25.2	234	64.0
YOLOv5s	87.78	0.86	15	27.1	7.1
Swin-Transformer	89.70	0.89	20.4	117.8	30.8
DETR	95.3	0.92	32.7	158.9	41
EM-YOLOv4-Tiny	94.54	0.90	10.4	27.8	6.8

As shown in Table 6, the average recognition accuracy of all types of networks for weeds exceeded 85%. The mAP of the EM-YOLOv4-Tiny network proposed herein was 94.54%, and its F1 value was 0.9, which is higher than those of the other four target detection networks. Because the test set contained a few smaller target weeds, the Faster-RCNN network did not construct an image pyramid and was insensitive to the detection of smaller targets, resulting in a low Recall and a mAP of only 87.71%. Compared with YOLOv4, the proposed network introduced multiscale detection and the attention mechanism based on

YOLOv4-Tiny, whose mAP and F1 were 4.78% and 10% higher than those of the YOLOv4 network, respectively. Moreover, the volume and number of parameters of the proposed model were much smaller than those of the original YOLOv4 network, indicating that the improved network preserved the merit of lightness. The lightweight YOLOv5s and EM-YOLOv4-Tiny exhibited similar model volumes and testing times; however, the mAP of YOLOv5s was only 87.78%, which was similar to that of the original YOLOv4-Tiny. Although the lightweight network had a simple structure, it was susceptible to overlooking occluded and smaller targets during detection.

Transformer-based target detection networks like Swin-Transformer and DETR were also trained and tested on the dataset in this study. The recognition accuracy is generally better than that of the CNN-based network. Still, the size of the model and the slow detection speed is not conducive to the deployment and development of embedded devices. It is worth mentioning that the Transformer structure is on an unstoppable trend to overtake the CNN structure in the existing studies. In future research, this study will also consider incorporating the Transformer structure into EM-YOLOv4-Tiny, working to improve the accuracy of the model further.

3.5. Comparison of Performances under Different Scenarios

To evaluate the robustness of the model in different scenarios, three different datasets were prepared based on the different growth densities of peanuts and weeds: single weed, sparsely distributed weeds, and overgrown weeds. The test results obtained using the proposed network on the three datasets are shown in Table 7 and Figure 6. The experimental results show that the proposed model performed favorably in terms of weed detection under different growing conditions and accurately located peanuts and various weeds via boundary box regression. The average recognition accuracies of the three datasets mentioned above were 98.48%, 98.16%, and 94.3%, respectively, with a mean value of 96.98%. When the density of peanuts and weeds was high, the model accurately identified occluded weeds while demonstrating excellent recognition of small target weeds.

Table 7. Performance comparison results of models in different scenarios.

Scenarios	Precision/%	Recall/%	mAP/%	F1/%
Single Weed	94.67	96.03	98.48	0.95
Sparsely Distributed	95.97	93.21	98.16	0.94
Vigorous Growth	90.24	89.52	94.30	0.90
Mean	93.62	93.01	96.98	0.93

Figure 6. Effect of model recognition under different scenarios.

4. Discussion

4.1. Deep Learning for Weed Detection

In this study, the target detection technology based on deep learning was used to detect weeds in peanut fields and achieved good results. In similar weed detection work, many researchers [31,32] used unmanned aerial vehicles (UAVs) with intelligent sensors to detect weeds in the field. The UAV can cover a large area in a short time and generate a weed map of the field to guide the weeding device to the designated area for weeding. However, producing a weed map is very challenging due to the similarity of the crops and the weeds. In contrast, deep learning technology can automatically learn the discriminant characteristics between crops and weeds through a deep convolution neural network, which can better solve the problem of weed detection in a complex environment. Hussain [33] used the improved YOLOv3-Tiny network model to detect two kinds of weeds in the wild blueberry field, and the F1 values of the two kinds were 0.97 and 0.90, respectively. This also shows the great potential of the deep learning method in the field of weed detection. However, the actual agricultural production environment is often changeable and uncontrollable. The proposed method may also have certain limitations when the application scenario changes, such as a large increase in weed species and extreme weather. Although deep learning technology has a strong learning and adaptive ability, it must be combined with many other technologies to contribute to agricultural development.

4.2. Challenge of Small Target Detection

Small target detection has always been a research hotspot in the field of target detection. Multiscale detection and feature fusion are the most commonly used methods to solve the problem of small target detection. In this study, the idea of multiscale detection and the attention mechanism were introduced into YOLOv4-Tiny, which improved the recognition ability of the model for small target weeds. The multiscale feature learning method improves the sensitivity of the original network to small target detection by fusing the details of shallow features. The attention module recalibrates the input features with weights, which makes up for the defect that the receptive field of shallow features is small and easily produces noise. However, the existing feature fusion methods, such as concatenation, cannot fully take into account the feature information of the context, which also leads to the model missing or falsely detecting weeds on some small targets. In agricultural production, many application scenarios for small target detection will also exist. The pests are too small and mostly have protective colors, making pest detection a challenge in the pest control process. The accurate identification and positioning of small fruits and vegetables is also key to fruit and vegetable picking. Therefore, small target detection remains a more significant challenge in agriculture. Fortunately, the detection regarding small targets has been ongoing. Wei et al. [34] used a Path Aggregation Feature Pyramid Network (PAFPN) structure to fuse the multiscale features obtained by the Attention Mechanism Network to get high-level multiscale semantic features. The global feature fusion method, like PAFPN, is better than the local feature fusion method in small target detection. Therefore, in subsequent research we will consider adding appropriate feature fusion algorithms to our own networks to further improve the recognition ability of the model for small targets.

4.3. Limitations and Shortcomings

Although the network proposed in this study can better identify weeds in peanut fields, some noteworthy problems still need further research. First of all, the data in this study only include weeds in the peanut seedling stage, and the collected area is only in Henan Province, China. Future research will focus on collecting weed data in peanuts in other growth stages and will cover as many regions as possible. Secondly, although the network in this paper improves the recognition accuracy of the model compared with the original YOLOv4-Tiny network, it also increases the volume of the model to a certain extent. Zhang et al. [35] used the deep separation convolutional network to replace the original convolutional network, which not only improved the accuracy of the model but also reduced the number of parameters and calculations of the network. In subsequent research, we plan to introduce this method into the network of this paper. Finally, the improvement strategy of the multiscale detection and the attention mechanism has been proved to be highly practical in this study. Still, other advanced research continues, such as on the Transformer [36], the Generative Adversarial Network [37], and so on, which have attracted more and more attention. It is worth further exploring the introduction of these technologies into our own network and improving the detection performance of the model.

5. Conclusions

To rapidly and accurately identify various types of weeds in peanut fields, a weed recognition method named EM-YOLOv4-Tiny was proposed. Based on YOLOv4-Tiny, multiscale detection and the attention mechanism were introduced, the CIoU was used as the loss function for training, and the soft-NMS method was used to screen the prediction box to improve the model performance in identifying small targets. The proposed model shows better recognition accuracy than Faster-RCNN, YOLOv5s, YOLOv4, and Swin-Transformer. In addition, the volume of the EM-YOLOv4-Tiny model was 28.7 M, and the single detection time was 10.9 ms, which rendered the model suitable for the embedded development of intelligent weeding robots.

In future work, this research will transplant the constructed model to a suitable embedded device for testing and select an intelligent spraying device to complete the precise weeding in the peanut field. In addition, the model will also be used in applications on smartphones so that farmers can better understand field information and make timely decisions.

Author Contributions: Conceptualization, H.Z.; methodology, H.Z. and Z.W.; software, Y.G.; validation, H.Z., Z.W., and Y.M.; formal analysis, Y.G.; investigation, Y.M.; resources, Y.G.; data curation, Z.W., Y.M., D.C., and W.C.; writing—original draft preparation, Z.W.; writing—review and editing, H.Z. and Z.W.; visualization, S.Y.; supervision, H.Z.; project administration, R.G.; funding acquisition, H.Z. All authors have read and agreed to the published version of the manuscript.

Funding: This research was funded by the Key Research and Development program of Henan Province (No. 212102110028) and Henan Provincial Science and Technology Research and Development Plan Joint Fund (No. 22103810025).

Institutional Review Board Statement: Not applicable.

Data Availability Statement: The datasets used and/or analyzed during the current study are available from the corresponding author upon reasonable request.

Acknowledgments: The author would like to thank all contributors to this study.

Conflicts of Interest: The authors declare no conflict of interest.

Appendix A

The ECA network structure is shown in Figure A1. In the ECA network, a fast one-dimensional convolution with a convolution kernel k was performed to realize local cross-channel interactions, which reduced the computational workload and complexity of the entire connection layer. A positive interaction occurred between the channel dimension C and the convolution kernel size k, i.e., a larger C resulted in a larger k. The relationship between the two can be expressed as follows:

$$C = \varnothing(k) \tag{A1}$$

C is typically measured in an exponential multiple of 2. Therefore, the relationship between the two can be more reasonably expressed as follows:

$$C = \varnothing(k) = 2^{(\gamma \times k - b)}, \tag{A2}$$

Here,

$$k = \varphi(C) = \left| \frac{\log_2(C)}{r} + \frac{b}{\gamma} \right|_{odd}, \tag{A3}$$

where $|n|_{odd}$ represents the odd number closest to n, with γ and b being 2 and 1, respectively.

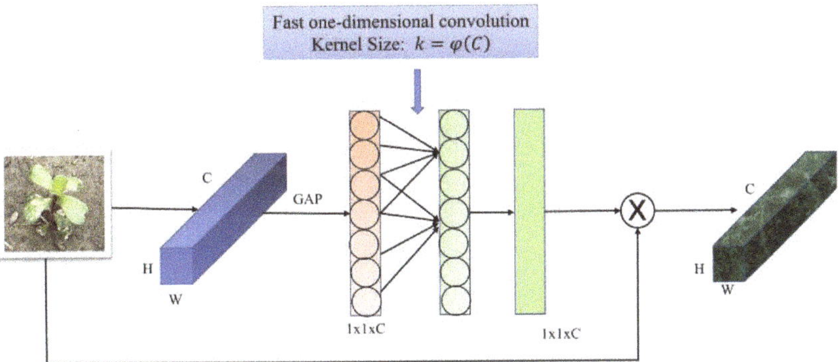

Figure A1. ECA network structure, where C is the channel dimension of the input data, H is the height of the input data, and W is the width of the input data. GAP denotes global average pooling, and k denotes the size of the convolution kernel using fast one-dimensional convolution.

Appendix B

As shown in Figure A2, the CIoU bounding box regression loss function directly minimizes the normalized distance between the predicted box and the real target box, taking into account the overlapping area of the detection box as well as the distance from the center point of the detection box. The measurement parameter of the consistency of the aspect ratio between the detection frame and the real target frame is also added to make the model more inclined to optimize in the direction of the dense overlapping area.

Figure A2. CIoU diagram, where r represents the center point distance d of the two detection boxes, and d represents the distance between the diagonals of the smallest rectangle containing the two detection boxes.

The loss function of the CIoU is calculated as follows:

$$\text{CIoU}_{Loss} = 1 - \text{CIoU} = 1 - \text{IoU} + \frac{\rho^2(b,c)}{d^2} + av \tag{A4}$$

where d represents the distance between the diagonals of the smallest rectangle containing the two boxes; b and c represent the coordinates of the central points of the real and prediction boxes, respectively; $\rho^2(b, c)$ is the function for solving the Euclidean distance between the two mentioned points; and av is the penalty term for border scale.

The a in Equation (7) is the parameter used to balance the ratio, and v is the parameter that measures whether the ratio of the true frame is consistent with the predicted frame. The calculation of both is as follows:

$$v = \frac{4}{\pi^2}\left\{\arctan\frac{w^c}{h^c} - \arctan\frac{w^b}{h^b}\right\}^2 \tag{A5}$$

$$a = \begin{cases} 0, & if\ IoU < 0.5 \\ \frac{v}{(1-IoU)+v}, & if\ IoU \geq 0.5' \end{cases} \tag{A6}$$

where w^c and h^c represent the width and height of the prediction box, and w^b and h^b represent the width and height of the real box.

References

1. Renton, M.; Chauhan, B.S. Modelling crop-weed competition: Why, what, how and what lies ahead? *Crop Prot.* **2017**, *95*, 101–108. [CrossRef]
2. Zhuang, J.; Li, X.; Bagavathiannan, M.; Jin, X.; Yang, J.; Meng, W.; Li, T.; Li, L.; Wang, Y.; Chen, Y.; et al. Evaluation of different deep convolutional neural networks for detection of broadleaf weed seedlings in wheat. *Pest Manag. Sci.* **2022**, *78*, 521–529. [CrossRef] [PubMed]
3. Kanagasingham, S.; Ekpanyapong, M.; Chaihan, R. Integrating machine vision-based row guidance with GPS and compass-based routing to achieve autonomous navigation for a rice field weeding robot. *Precis. Agric.* **2020**, *21*, 831–855. [CrossRef]
4. Wang, A.; Zhang, W.; Wei, X. A review on weed detection using ground-based machine vision and image processing techniques. *Comput. Electron. Agric.* **2019**, *158*, 226–240. [CrossRef]
5. Reedha, R.; Dericquebourg, E.; Canals, R.; Hafiane, A. Transformer Neural Network for Weed and Crop Classification of High Resolution UAV Images. *Remote Sens.* **2022**, *14*, 592. [CrossRef]
6. Peteinatos, G.G.; Weis, M.; Andújar, D.; Ayala, V.R.; Gerhards, R. Potential use of ground-based sensor technologies for weed detection. *Pest Manag. Sci.* **2014**, *70*, 190–199. [CrossRef]
7. García-Santillán, I.D.; Pajares, G. On-line crop/weed discrimination through the Mahalanobis distance from images in maize fields. *Biosyst. Eng.* **2018**, *166*, 28–43. [CrossRef]
8. Bakhshipour, A.; Jafari, A. Evaluation of support vector machine and artificial neural networks in weed detection using shape features. *Comput. Electron. Agric.* **2018**, *145*, 153–160. [CrossRef]
9. Pulido, C.; Solaque, L.; Velasco, N. Weed recognition by SVM texture feature classification in outdoor vegetable crop images. *Ing. E Investig.* **2017**, *37*, 68–74. [CrossRef]
10. Kamilaris, A.; Prenafeta-Boldú, F.X. Deep learning in agriculture: A survey. *Comput. Electron. Agric.* **2018**, *147*, 70–90. [CrossRef]
11. Gai, R.; Chen, N.; Yuan, H. A detection algorithm for cherry fruits based on the improved YOLO-v4 model. *Neural Comput. Appl.* **2021**, 1–12. [CrossRef]
12. Khan, S.; Tufail, M.; Khan, M.T.; Khan, Z.A.; Anwar, S. Deep learning-based identification system of weeds and crops in strawberry and pea fields for a precision agriculture sprayer. *Precis. Agric.* **2021**, *22*, 1711–1727. [CrossRef]
13. Jin, X.; Sun, Y.; Che, J.; Bagavathiannan, M.; Yu, J.; Chen, Y. A novel deep learning-based method for detection of weeds in vegetables. *Pest Manag. Sci.* **2022**, *78*, 1861–1869. [CrossRef]
14. Ying, B.; Xu, Y.; Zhang, S.; Shi, Y.; Liu, L. Weed detection in images of carrot fields based on improved YOLO v4. *Traitement Du Signal* **2021**, *38*, 341–348. [CrossRef]
15. Li, X.; Pan, J.; Xie, F.; Zeng, J.; Li, Q.; Huang, X.; Liu, D.; Wang, X. Fast and accurate green pepper detection in complex backgrounds via an improved YOLOv4-tiny model. *Comput. Electron. Agric.* **2021**, *191*, 106503. [CrossRef]
16. Li, H.; Li, C.; Li, G.; Chen, L. A real-time table grape detection method based on improved YOLOv4-tiny network in complex background. *Biosyst. Eng.* **2021**, *212*, 347–359. [CrossRef]
17. Lin, T.Y.; Dollár, P.; Girshick, R.; He, K.; Hariharan, B.; Belongie, S. Feature pyramid networks for object detection. In Proceedings of the IEEE Conference on Computer Vision and Pattern Recognition, Honolulu, HI, USA, 21–26 July 2017; pp. 2117–2125.
18. Gao, C.; Cai, Q.; Ming, S. YOLOv4 object detection algorithm with efficient channel attention mechanism. In Proceedings of the 2020 5th International Conference on Mechanical, Control and Computer Engineering (ICMCCE), Harbin, China, 25–27 December 2020; IEEE: Piscataway Township, NJ, USA, 2020; pp. 1764–1770.
19. Bodla, N.; Singh, B.; Chellappa, R.; Davis, L.S. Soft-NMS–improving object detection with one line of code. In Proceedings of the IEEE international conference on computer vision, Venice, Italy, 22–29 October 2017; pp. 5561–5569.

20. Neubeck, A.; Van Gool, L. Efficient non-maximum suppression. In Proceedings of the 18th International Conference on Pattern Recognition (ICPR'06), Hong Kong, China, 20–24 August 2006; IEEE: Piscataway Township, NJ, USA, 2006; Volume 3, pp. 850–855.

21. Wu, D.; Lv, S.; Jiang, M.; Song, H. Using channel pruning-based YOLO v4 deep learning algorithm for the real-time and accurate detection of apple flowers in natural environments. *Comput. Electron. Agric.* **2020**, *178*, 105742. [CrossRef]

22. Wang, L.; Qin, M.; Lei, J.; Wang, X.; Tan, K. Blueberry maturity recognition method based on improved YOLOv4-Tiny. *Nongye Gongcheng Xuebao/Trans. Chin. Soc. Agric. Eng.* **2021**, *37*, 170–178.

23. Xu, J.; Li, Z.; Du, B.; Zhang, M.; Liu, J. Reluplex made more practical: Leaky ReLU. In Proceedings of the 2020 IEEE Symposium on Computers and communications (ISCC), Rennes, France, 7–10 July 2020; IEEE: Piscataway Township, NJ, USA, 2020; pp. 1–7.

24. Chen, Z.; Tian, S.; Yu, L.; Zhang, L.; Zhang, X. An object detection network based on YOLOv4 and improved spatial attention mechanism. *J. Intell. Fuzzy Syst.* **2022**, *42*, 2359–2368. [CrossRef]

25. Choi, E.; Bahadori, M.T.; Sun, J.; Kulas, J.; Schuetz, A.; Stewart, W. Retain: An interpretable predictive model for healthcare using reverse time attention mechanism. *arXiv* **2016**, arXiv:1608.05745.

26. Schmidt-Hieber, J. Nonparametric regression using deep neural networks with ReLU activation function. *Ann. Stat.* **2020**, *48*, 1875–1897.

27. Zheng, Z.; Wang, P.; Liu, W.; Li, J.; Ye, R.; Ren, D. Distance-IoU loss: Faster and better learning for bounding box regression. In Proceedings of the AAAI conference on artificial intelligence, New York, NY, USA, 7–12 February 2020; Volume 34, pp. 12993–13000.

28. Zhou, T.; Fu, H.; Gong, C.; Shen, J.; Shao, L.; Porikli, F. Multi-mutual consistency induced transfer subspace learning for human motion segmentation. In Proceedings of the IEEE/CVF Conference on Computer Vision and Pattern Recognition, Seattle, WA, USA, 13–19 June 2020; pp. 10277–10286.

29. Zhong, S.; Chen, D.; Xu, Q.; Chen, T. Optimizing the Gaussian kernel function with the formulated kernel target alignment criterion for two-class pattern classification. *Pattern Recognit.* **2013**, *46*, 2045–2054. [CrossRef]

30. Ismkhan, H. Ik-means−+: An iterative clustering algorithm based on an enhanced version of the k-means. *Pattern Recognit.* **2018**, *79*, 402–413. [CrossRef]

31. Eide, A.; Koparan, C.; Zhang, Y.; Ostlie, M.; Howatt, K.; Sun, X. UAV-Assisted Thermal Infrared and Multispectral Imaging of Weed Canopies for Glyphosate Resistance Detection. *Remote Sens.* **2021**, *13*, 4606. [CrossRef]

32. De Castro, A.I.; Torres-Sánchez, J.; Peña, J.M.; Jiménez-Brenes, F.M.; Csillik, O.; López-Granados, F. An automatic random forest-OBIA algorithm for early weed mapping between and within crop rows using UAV imagery. *Remote Sens.* **2018**, *10*, 285. [CrossRef]

33. Hussain, N.; Farooque, A.A.; Schumann, A.W.; McKenzie-Gopsill, A.; Esau, T.; Abbas, F.; Acharya, B.; Zaman, Q. Design and development of a smart variable rate sprayer using deep learning. *Remote Sens.* **2020**, *12*, 4091. [CrossRef]

34. Wei, H.; Zhang, Q.; Qian, Y.; Xu, Z.; Han, J. MTSDet: Multi-scale traffic sign detection with attention and path aggregation. *Appl. Intell.* **2022**, 1–13. [CrossRef]

35. Zhang, M.; Xu, S.; Song, W.; He, Q.; Wei, Q. Lightweight underwater object detection based on yolo v4 and multi-scale attentional feature fusion. *Remote Sens.* **2021**, *13*, 4706. [CrossRef]

36. Kitaev, N.; Kaiser, Ł.; Levskaya, A. Reformer: The efficient transformer. *arXiv* **2020**, arXiv:2001.04451.

37. Goodfellow, I.; Pouget-Abadie, J.; Mirza, M.; Xu, B.; Warde-Farley, D.; Ozair, S.; Courville, A.; Bengio, Y. Generative adversarial nets. *Commun. ACM* **2020**, *63*, 139–144. [CrossRef]

 *agriculture*

Article

Hyperspectral Estimates of Soil Moisture Content Incorporating Harmonic Indicators and Machine Learning

Xueqin Jiang [1,2,†], Shanjun Luo [2,†], Qin Ye [1,*], Xican Li [3] and Weihua Jiao [4]

[1] College of Surveying and Geo-Informatics, Tongji University, Shanghai 200092, China
[2] School of Remote Sensing and Information Engineering, Wuhan University, Wuhan 430079, China
[3] School of Information Science and Engineering, Shandong Agriculture University, Tai'an 271001, China
[4] Center for Agricultural and Rural Economic Research, Shandong University of Finance and Economics, Jinan 250014, China
* Correspondence: yeqin@tongji.edu.cn; Tel.: +86-139-1786-6679
† These authors contributed equally to this work.

Abstract: Soil is one of the most significant natural resources in the world, and its health is closely related to food security, ecological security, and water security. It is the basic task of soil environmental quality assessment to monitor the temporal and spatial variation of soil properties scientifically and reasonably. Soil moisture content (SMC) is an important soil property, which plays an important role in agricultural practice, hydrological process, and ecological balance. In this paper, a hyperspectral SMC estimation method for mixed soil types was proposed combining some spectral processing technologies and principal component analysis (PCA). The original spectra were processed by wavelet packet transform (WPT), first-order differential (FOD), and harmonic decomposition (HD) successively, and then PCA dimensionality reduction was used to obtain two groups of characteristic variables: WPT-FOD-PCA (WFP) and WPT-FOD-HD-PCA (WFHP). On this basis, three regression models of principal component regression (PCR), partial least squares regression (PLSR), and back propagation (BP) neural network were applied to compare the SMC predictive ability of different parameters. Meanwhile, we also compared the results with the estimates of conventional spectral indices. The results indicate that the estimation results based on spectral indices have significant errors. Moreover, the BP models (WFP-BP and WFHP-BP) show more accurate results when the same variables are selected. For the same regression model, the choice of variables is more important. The three models based on WFHP (WFHP-PCR, WFHP-PLSR, and WFHP-BP) all show high accuracy and maintain good consistency in the prediction of high and low SMC values. The optimal model was determined to be WFHP-BP with an R^2 of 0.932 and a prediction error below 2%. This study can provide information on farm entropy before planting crops on arable land as well as a technical reference for estimating SMC from hyperspectral images (satellite and UAV, etc.).

Keywords: soil moisture content; spectral processing technology; hyperspectral; principal component analysis; feature parameters extraction

Citation: Jiang, X.; Luo, S.; Ye, Q.; Li, X.; Jiao, W. Hyperspectral Estimates of Soil Moisture Content Incorporating Harmonic Indicators and Machine Learning. *Agriculture* **2022**, *12*, 1188. https://doi.org/10.3390/agriculture12081188

Academic Editor: Jose L. Gabriel

Received: 25 July 2022
Accepted: 8 August 2022
Published: 10 August 2022

1. Introduction

Soil moisture content (SMC) is the carrier of material and energy cycle in the soil system, which has an important influence on soil characteristics, vegetation growth and distribution, and the regional ecosystem [1,2]. Meanwhile, the SMC is related to soil nutrient contents by facilitating organic matter decomposition [3], enhancing carbon sequestration [4], and resulting in an increase in crop yield [5]. In agriculture, a timely and effective grasp of the distribution and future trend of soil moisture in the field is of great significance to effectively save water resources, improve the utilization efficiency of agricultural water and sustainable utilization of water and soil resources, and effectively monitor and control farmland drought in real time [6,7].

The traditional artificial SMC measurement method, which is based on point and laboratory measurement, has high precision but the limited scope, a large workload, low efficiency, and high cost, and is difficult to meet the actual needs of SMC monitoring [8,9]. Remote sensing and satellite data have been widely used in monitoring soil and crop systems, such as soil organic matter [10], crop evapotranspiration [11], water stress [12], and yield monitoring [13]. In the case of soil moisture, researchers have reported that hyperspectral imagery has more advantages over regular satellite-based multi-spectral imagery owing to the higher information level stored in the hyperspectral images [14]. Accordingly, hyperspectral remote sensing (HRS) technology has been widely used in SMC monitoring research due to its advantages of large area, non-contact, and timeliness, making up for the shortcomings of traditional methods [15]. HRS can be used for large-scale non-destructive monitoring by analyzing the spectral variation characteristics of different soil properties, which is more suitable for assessing and mapping the spatial variation of soil properties [16]. As a robust stoichiometric means, soil spectroscopy has been proven to be an effective alternative to wet chemistry in soil environmental quality monitoring [17]. However, there are obvious spectral noise and serious scattering phenomena in the original soil spectral data obtained by HRS [18]. There is inevitably noise unrelated to SMC in the soil hyperspectral, which will increase the detection difficulty of SMC. In addition, HRS contains huge amounts of data. Therefore, more thorough denoising and variable optimization become the key to establishing a model with higher accuracy [19].

In the aspect of hyperspectral data preprocessing, many studies have been carried out, such as reciprocal, logarithm, and first differential studies [20–22]. Because the soil spectral curve is the comprehensive expression of the interaction and superposition of various substances, the determination of characteristic bands is not only difficult, but also has a high degree of uncertainty and weak denoising. Subsequently, scholars used spectral denoising methods to process hyperspectral data, such as Savitzky–Golay filtering, median operation, moving average, etc. However, for white noise, especially random and low-frequency signals, these methods are difficult to remove noise without affecting the effective signal [23]. The wavelet packet transform (WPT) can compress the signal while retaining the original information and has been gradually used in the estimation of soil properties and achieved certain results. For example, Gu et al. found that the high-frequency coefficient generated by wavelet transform and random forest algorithm can be used to invert soil organic matter content [24]. Given the above spectral pretreatment technologies, some new methods for estimating SMC still need to be explored.

In the study of SMC estimation, the estimation accuracy of SMC depends on the selection of characteristic variables and the estimation model. At present, there are two kinds of models for estimating soil composition based on soil spectral properties: the physical model based on mechanism information and the statistical model based on experience. In the mechanism model method, the quantitative change mechanism of soil reflectance caused by different water content is very complex, and its inversion effect and adaptability of results are limited [8]. However, the widely used statistical model has the advantages of being simple and direct and can obtain accurate and stable results. At present, the estimation of soil characteristics by soil spectra mostly adopts stepwise multilinear regression [25,26], principal component regression [27], neural network regression [16,28], support vector machine regression [17,29], and partial least squares regression [30,31]. The relationship between SMC and soil hyperspectral is complex and has great nonlinearity and randomness. Its spectral characteristics are difficult to be explained by several bands. Therefore, the simple regression model has certain deficiencies in dealing with nonlinear, heteroscedasticity, multicollinearity, and other complex problems, and it is difficult to obtain good estimation accuracy [32]. In SMC estimation, these methods inevitably lead to missing or redundant information, which directly affects the results. There is a need to explore approaches that can overcome these obstacles, such as machine learning. The neural network model has a strong nonlinear approximation ability, can effectively establish the global nonlinear mapping relationship between input and output [33–35], and has advantages in data fitting, function approximation, and other aspects [36,37]. Good results have been achieved by using the neural network model to estimate soil composition. For example, Pellegrini et al. obtained satisfactory results by using the artificial neural network in estimating soil microbial biomass [16].

In this paper, the hyperspectral data of different types of soils were measured to analyze the variation trend of reflectance with different SMC. Meanwhile, some spectral processing technologies and PCA were employed to extract characteristics variables for estimating SMC of mixed soils. This non-destructive estimation technique is simple, fast, and time efficient. Finally, the PCR, PLSR, and back propagation (BP) regression models were constructed and compared with the spectral-index models. The use of machine learning makes full use of its nonlinear learning characteristics to achieve accurate estimation of SMC under different conditions. Our objectives are (1) to compare the role of characteristic parameters obtained by different spectral processing techniques in estimating SMC, (2) to compare the performance of different regression algorithms in estimating SMC, and (3) to compare the importance of the selection of characteristic variables with the selection of regression models and to construct the SMC high-precision prediction model suitable for mixed soil scenarios.

2. Materials and Methods

2.1. Study Area

The study area is located in Hengshan County, Northern Shaanxi Province, China. As shown in Figure 1, the sampling areas are located in the Loess Plateau of Northern Shaanxi, adjacent to the Mu Us Desert in the north and the Loess hill in the south. The region has a temperate semi-arid continental monsoon climate with a year-round average daily temperature of 8.6°, and the general characteristics of temperature and rainfall are large inter-annual and inter-monthly variations. The soil types mainly include sandy and loessial soil (SS and LS). The sampling points of different soil types are evenly distributed in the whole study area as far as possible. The main tributaries in the area include the Wuding River, Dali River, etc. Due to these geographical factors, the experimental area is not only rich in soil types, but also has great differences in SMC, which is of great significance for the study of SMC estimation.

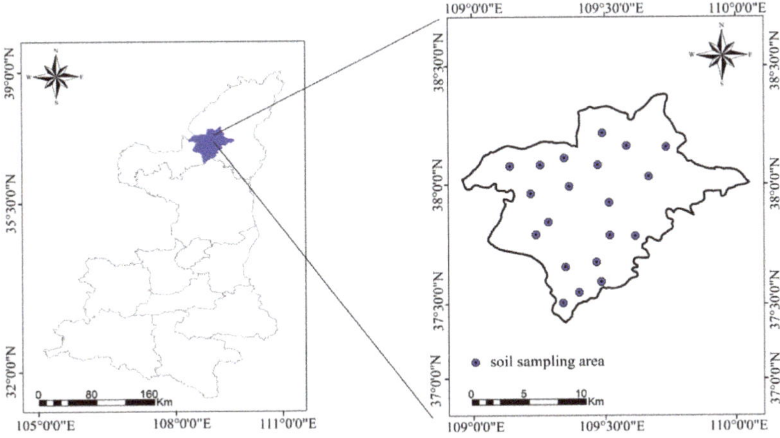

Figure 1. Study region and soil sampling area (the blue dots show the sampling areas).

2.2. Soil Spectral Measurement

The collected soil samples are quickly measured for spectral data in the laboratory. The soil spectral reflectance was measured using the ASD Field Spec FR spectrometer (Analytical Spectral Devices, Inc., Boulder, CO, USA), with a wavelength range of 350–2500 nm.

The soil samples were placed in a black vessel (with a diameter of 8 cm and a depth of 2 cm) in turn, and their surface was scraped flat. A 50 W halogen lamp was used as the light source, and the distance between the light source and the experimental sample is 0.5 m. The distance between the spectrometer probe and the soil sample was 0.2 m. Before each spectral measurement, the diffuse reflection standard reference plate was used for calibration. Four spectral curves were collected for each soil sample, and their arithmetic mean value was taken as the spectral data of the soil sample.

2.3. Determination of SMC

To obtain more accurate and regionally representative SMC data, the destructive sampling approach was recommended [38]. The areas with flat terrain, exposed surface, and no shelter were selected as the sampling areas. About 20 sampling points were determined in total in the sampling areas (Figure 1). In addition, different soil types were considered in sampling, and a total of 84 soil samples were collected. The soil samples were collected from the surface soil with a depth of 0.2 m. They were brought back to the laboratory through aluminum boxes to avoid water evaporation. The soil samples placed in the aluminum box were dried in the oven (105 °C) until the weight did not change, and the SMC was measured by the drying method. The calculation formula is as follows:

$$\text{SMC} = \frac{M_1 - M_2}{M_2 - M_3} \times 100\%,\tag{1}$$

where M_1 is the total weight of the aluminum box and soil before drying, M_2 is the total weight of the aluminum box and soil sample after drying, and M_3 is the weight of each aluminum box after drying.

2.4. Spectral Indices Construction

Since the strong absorption of water leads to changes in reflectance, spectral indices with some physical significance calculated from the reflectance of different bands have been proposed for predicting SMC. Due to the unambiguous physical significance, some spectral indices have been proposed to predict SMC. However, these parameters inevitably remain somewhat regional and generalized. To compare with the method presented in this study, we selected some common two- and three-band spectral indices (Table 1).

Table 1. The common spectral indices selected in this paper.

Spectral Indices	Formula	Reference
EVI	$\frac{2.5(R_{1828}-R_{630})}{R_{1828}+6R_{630}-7.5R_{450}+1}$	[39]
TVI	$0.5[120(R_{666} - R_{834}) - 200(R_{794} - R_{834})]$	[38]
DSI	$R_{1760} - R_{1715}$	[40]
NDMI	$\frac{R_{2027}-R_{1878}}{R_{2027}+R_{1878}}$	[41]
SARVI	$\frac{1.5(R_{1820}-R_{670})}{R_{1820}+R_{670}+0.5}$	[39]

2.5. Spectral Processing Technologies

Spectral preprocessing is very useful for feature extraction and noise removal [30]. For example, WPT can perform a more detailed decomposition and reconstruction of high and low-frequency information of hyperspectral data [19]. This information processing result has no redundancy or omission, which is more conducive to spectral information noise reduction and original information retention, so it is widely used. In this research, the decomposition and reconstruction of the spectral data by WPT were performed according to the following steps.

(i)   Wavelet packet analysis. The wavelet master function used in the study was Db10 [42], by which the noise-bearing spectra were decomposed.

(ii) Determination of the optimal wavelet packet basis. The calculation of the optimal wavelet packet basis was based on the least-cost principle.

(iii) Wavelet packet coefficient thresholding. This process required quantization of the wavelet packet coefficients, which was based on a soft threshold "s" of good continuation.

(iv) Spectral reconstruction. The results in (ii) and (iii) were applied to reconstruct the spectral information, and finally, the noise-reduced spectra were obtained.

Spectral measurements are susceptible to factors, such as observation angle and illumination, making the signal-to-noise ratio of spectral data comparatively poor. After differential processing, not only can the influence of changes in illumination conditions on the target spectra be reduced, but also the background can be partially eliminated, thus better strengthening the spectral variance and highlighting the target characteristics. The first-order differential (FOD) treatment can improve the spectral sensitivity and eliminate the influence of the partial environmental background to reveal the spectral characteristics of the soil interior. The FOD was calculated as follows.

$$\mathrm{Ref}'(\lambda_i) = [\mathrm{Ref}(\lambda_{i+1}) - \mathrm{Ref}(\lambda_{i-1})]/(\lambda_{i+1} - \lambda_{i-1}), \tag{2}$$

where λ_{i-1}, λ_i, and λ_{i+1} are the wavelengths of adjacent bands and Ref is the first-order differential value.

However, none of these traditional methods can obtain robust and noiseless characteristic variables. Harmonic decomposition (HD) transforms hyperspectral data from the time domain to the frequency domain in the form of sine and cosine phase superposition, and finally obtains parameters such as residual term, amplitude, and phase. The calculation method is shown in Figure 2. These variables can reveal the average value and variation of the energy, and the position of the maximum value in different bands of the spectra.

Harmonic decomposition algorithm

Input: Hyperspectral data $R = [r_1, r_2, \ldots, r_N]$:

  N = number of bands; r_k = reflectance of each band

Output: Reconstructed data in the spectral domain $\tilde{R}$, the characteristic variables R' obtained

  by harmonic decomposition in the frequency domain

for each r_k in R

  Calculate the remainder: $A_0/2 = \frac{1}{N}(r_1 + r_2 + \ldots + r_N)$;

  for h = 1 to N

    Calculate the harmonic coefficient:

    $$A_h = \frac{2}{N}(r_1 \cos(2\pi h/N) + r_2 \cos(2\pi h/N) + \ldots + r_N \cos(2\pi h));$$

    $$B_h = \frac{2}{N}(r_1 \sin(2\pi h/N) + r_2 \sin(2\pi h/N) + \ldots + r_N \sin(2\pi h));$$

    $$C_h = \sqrt{(A_h^2 + B_h^2)};$$

    $$\varphi_h = arctan(A_h/B_h);$$

  end for

  Obtain transformed data:

  $$\tilde{R}_k = \frac{A_0}{2} + C_1 \sin(2\pi k/N + \varphi_1) + C_2 \sin(4\pi k/N + \varphi_2) + \ldots + C_N \sin(2\pi k + \varphi_N);$$

  Obtain harmonic components:

  $$R'_i = [A_0/2, C_1, \ldots, C_N, \varphi_1, \ldots, \varphi_N];$$

  end for

  return $\tilde{R}$ and R'

Figure 2. Illustration of the harmonic decomposition algorithm in pseudo-code.

2.6. Model Construction and Validation

After the correlated characteristic variables (WF and WFH) were obtained by spectral processing technologies (WPT, FOD, and HD), they need to be dimensionally reduced to remove redundancy. The principal component analysis (PCA) method can recombine original variables into a group of new comprehensive variables unrelated to each other to achieve feature extraction and dimension reduction [43]. When performing PCA, the components whose cumulative variance contribution rate exceeds 95% of the variable is taken as the new characteristic variable in this study.

It is very important to determine the regression model based on the relationship between independent and dependent variables for accurate estimation of SMC. Principal component regression (PCR) is one of the common methods to solve the problem of collinearity in logistic regression analysis [44]. It integrates the information of variables with high correlation into the principal component with low correlation through principal component transformation and then replaces the original variable to participate in regression calculation. Partial least squares regression (PLSR) is more commonly used as a linear multiple regression analysis method [45]. By analyzing the relationship between the prediction matrix X (independent variable) and the response matrix Y (dependent variable), the initial input data are projected into a potential space, and then many potential variables are extracted by using orthogonal structure, and the linear relationship between these new variables and Y is found. This method does not directly consider the regression modeling of the dependent variable and independent variable, but comprehensively screens the information in the variable system, and selects several new components with the best explanatory ability for the system for regression modeling. Through such information screening, the noise that has no explanatory effect on the dependent variable is eliminated. Backpropagation (BP) neural network is a widely used nonlinear modeling method in the artificial neural network, which is suitable for data prediction [46]. The learning process is composed of forwarding propagation and backpropagation. In the forward propagation process, input data are gradually processed from the input layer to the output layer through the hidden layer. If the data error obtained by the output layer is not within the allowed range, the error is backpropagated and the weight of each neuron is adjusted layer by layer by the gradient descent method. Until the error meets the specified requirements, it has a better estimation effect for complex nonlinear prediction. In this paper, we choose these three methods to conduct regression modeling for spectral characteristic parameters and SMC and compare their advantages and disadvantages.

Hyperspectral estimation of SMC based on spectral processing technologies and PCA mainly includes the following four steps (Figure 3):

(i)   Data collection: preliminary investigation, spatial layout planning of soil sampling sites, and laboratory spectroscopy and SMC measurements were included.

(ii)   Data processing: the original hyperspectral data were processed by WPT, FOD, and HD, and the characteristic variables were obtained by PCA dimensionality reduction.

(iii)   Data set partitioning: 54 groups were randomly selected from 84 groups of sample data as training samples, and the other 30 groups were used as validation data to form the training and validation datasets. The SMC data description is shown in Table 2.

(iv)   Modeling and validation: PCR, PLSR, and BP were used to construct the estimation models of SMC. The coefficient of determination (R2), root mean square error (RMSE), and mean absolute error (MAE) were used to evaluate the model accuracy. Related calculations are shown as follows.

$$\text{RMSE} = \sqrt{\sum\nolimits_{i=1}^{n} \frac{(y_i - \hat{y}_i)^2}{n}}, \tag{3}$$

$$\text{MAE} = \frac{1}{n}\sum_{i=1}^{n} \left| y_i - \hat{y}_i \right|, \tag{4}$$

where y_i is the true value, $\hat{y}_i$ is the predicted value, and n is the number of samples.

Table 2. Descriptive statistics of SMC in soils.

Soil Types	Samples	SMC (%)				
		Min	Max	Mean	SD	CV(%)
Loessial soil	51	3.36	58.43	9.65	8.05	83.40
Sandy soil	33	0.46	38.65	12.09	11.03	91.18
Training data	54	2.09	58.43	10.99	10.02	91.14
Validation data	30	0.46	34.83	10.72	8.87	82.74
Mixed soil	84	0.46	58.43	10.89	9.62	88.34

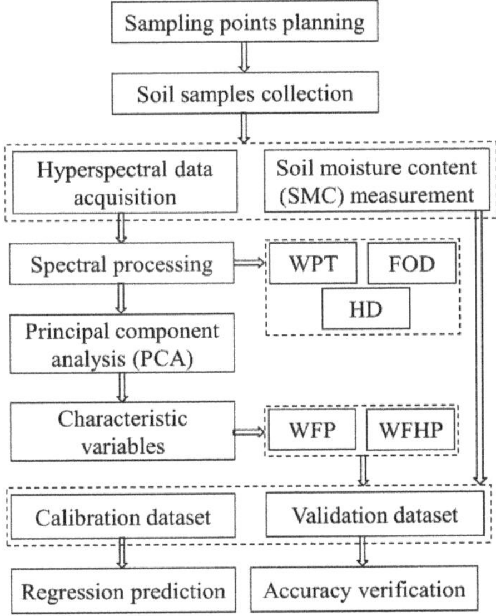

Figure 3. Flowchart indicating experimental methodology (WPT: wavelet packet transform; FOD: first-order differential; HD: harmonic decomposition; WFP: WPT-FOD-PCA; WFHP: WPT-FOD-HD-PCA).

3. Results

3.1. Comparison of Hyperspectral Characteristics of Soils with Different SMC

Some spectral curves over the whole moisture content range were randomly selected for comparison. Hyperspectral curves of different soil types (LS, SS, and MS) are shown in Figure 4. The spectral curves of different soil types have similar shapes and the absorption characteristics of water at 1450 nm and 1960 nm dominate the spectral characteristic curves of soil. For LS, the reflectance of all observation bands generally decreases with the increase of SMC (Figure 4a). However, for SS and MS, the variation of spectral reflectance with SMC does not show a consistent variation law (Figure 4b,c). For these three different soil types, the sensitivity of spectral reflectance to SMC is low in visible and near-infrared bands, and the change is more obvious in other bands. Moreover, the characteristic of mineral absorption at 2200 nm is obvious when SMC is low but disappears gradually with the increase of SMC.

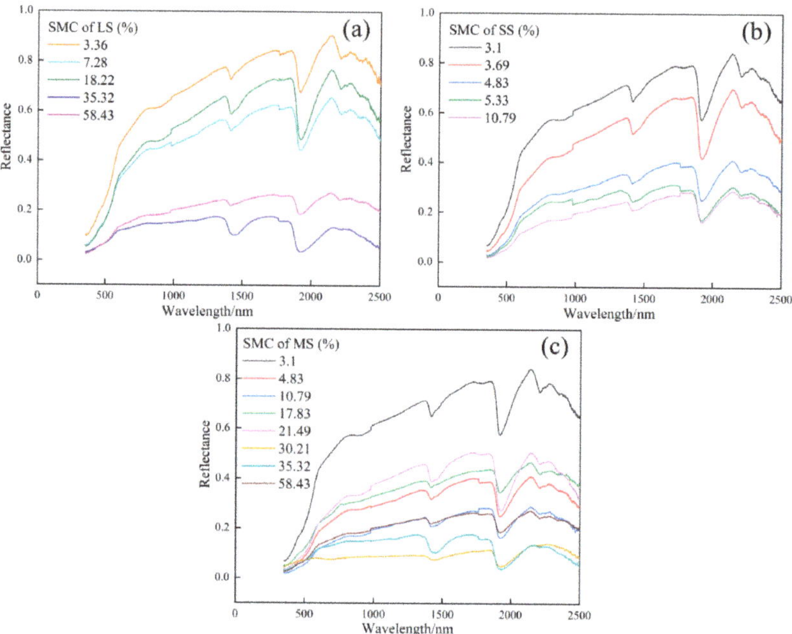

Figure 4. Hyperspectral curves of different soil types: (**a**) LS; (**b**) SS; (**c**) MS.

3.2. Estimation of SMC by Conventional Spectral Indices

The SMC data in the calibration set were adopted as the dependent variables, and five commonly available spectral indices (EVI, TVI, DSI, NDMI, and SARVI) were applied as independent variables to construct the inverse models using linear regression and the PLSR method, and the validation results were shown in Figure 5. The results showed that the selected spectral indices had limited accuracy in predicting the SMC of mixed soil types. Except for TVI, the remaining four indices exhibited varying degrees of overestimation or underestimation at different SMC. Although TVI did not demonstrate overestimation or underestimation (the regression line was close to the 1:1 line), the model errors were large and the points deviating from the 1:1 line were more clustered. Compared with the individual spectral indices inversion results, the PLSR model based on five indices had a higher accuracy (R^2 over 0.8 and error below 4%). In addition, the model did not exhibit local overestimation or underestimation.

3.3. Correlation Analysis between Spectral Data and SMC

The correlation analysis between the original spectral data and the processed data of the original spectra (including WPT, FOD, and HD) and SMC was performed. The results are shown in Figure 6. SS, LS, and MS indicate the Pearson correlation coefficient (r) between the original spectra of different soil types (LS, SS, and MS) and the corresponding SMC. WO and WF represent the correlation between the WPT of original spectral data and FOD after WPT and SMC, respectively. The original spectral reflectance of SS is highly correlated with SMC except for the visible bands ($|r| > 0.6$, $p < 0.01$). The correlation between LS and SMC becomes much weaker ($|r| < 0.5$, $p < 0.01$). For MS, the correlation is between SS and LS (about 0.5, $p < 0.01$). Therefore, for the estimation of SMC of MS, parameters with higher correlation need to be extracted. Compared with the original spectra, WO does not significantly improve the correlation with SMC. Although WF cannot improve the correlation with SMC in all bands, it can produce parameters with a strong correlation in many characteristic bands. Finally, 180 characteristic bands were selected from WF data with $|r| > 0.6$ to estimate SMC.

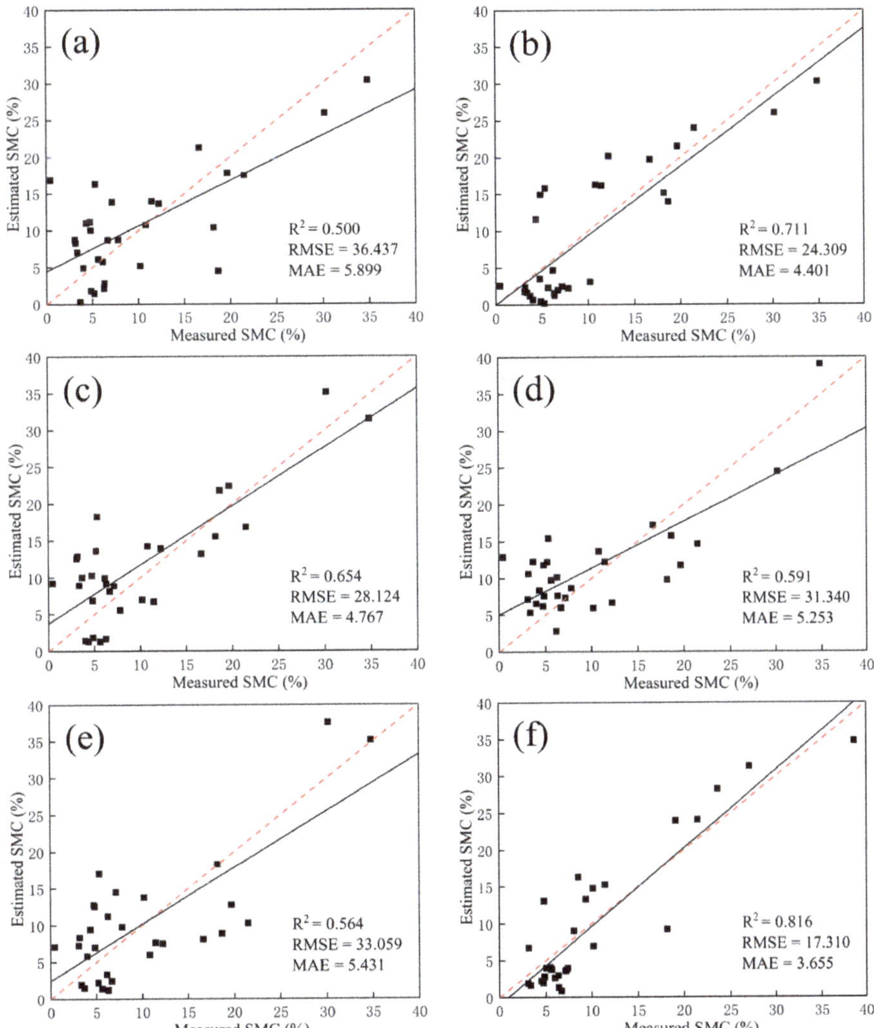

Figure 5. The comparison of measured and estimated SMC: (**a**) EVI; (**b**) TVI; (**c**) DSI; (**d**) NDMI; (**e**) SARVI; (**f**) PLSR model of five spectral indices.

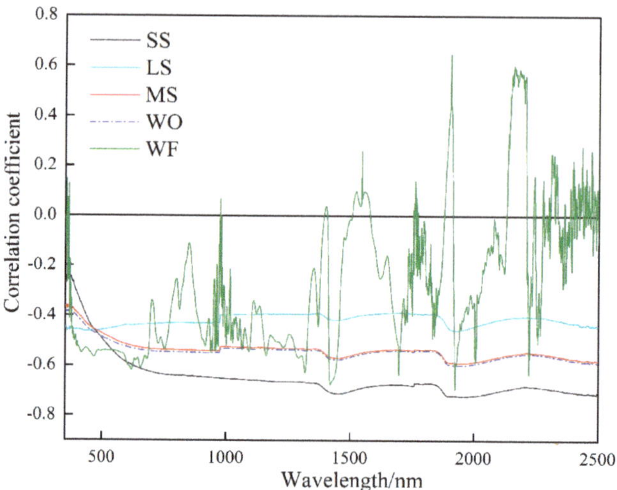

Figure 6. The Pearson correlation coefficient between spectra and SMC.

3.4. Harmonic Characteristic Parameter Acquisition

The feature parameters of harmonic spectra (WFH: remainder, amplitude, and phase) were acquired by decomposing the selected WF data of MS. The correlation between these extracted components and SMC was computed. To keep consistent with the number of characteristic parameters of the selected WF data, the number of harmonic decompositions was determined to be 180. Figure 7 demonstrates the correlation between harmonic characteristic parameters and SMC of MS.

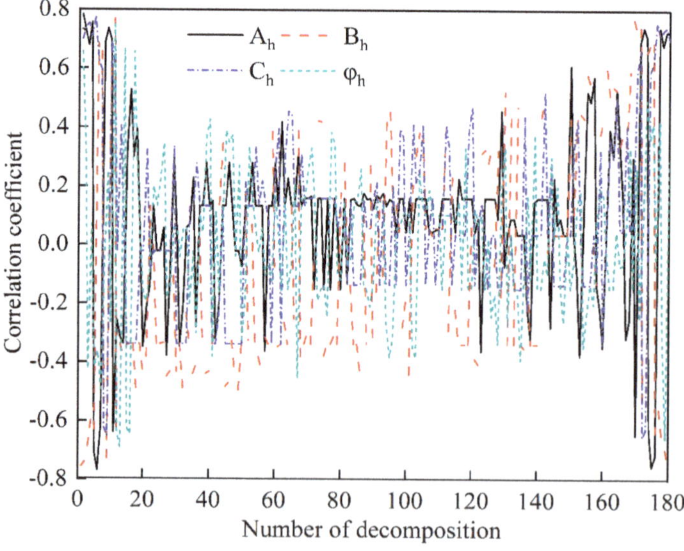

Figure 7. The Pearson correlation coefficient between harmonic characteristic parameters and SMC.

The result of correlation analysis reveals that the variables at the beginning and end of the decomposition numbers have a strong correlation with SMC (| r | close to 0.8, $p < 0.01$). The figure is roughly symmetrical in the center. Furthermore, the correlation coefficient shows a periodic change of alternating positive and negative values. Except for the beginning and the end, the correlation between other characteristic parameters close to the middle and SMC is weak ($|r| < 0.5$, $p < 0.01$). Since the correlation of characteristic parameters is periodic, half of the parameters ($A_0/2$, $A_{h=1,2,4}$, $B_{h=1,2,3}$, $C_{h=1,2,3}$, and $\varphi_{h=1}$) with high correlation with SMC ($|r| > 0.7$) were selected.

3.5. Dimension Reduction of Characteristic Parameters Based on PCA

After extracting the characteristic parameters through a series of spectral processing techniques (including WPT, FOD, and HD), WF and WFH data were obtained. Since many relevant characteristic parameters are included (180 of WF and 11 of WFH), it is necessary to simplify these parameters. To reduce the redundancy of variables and the input of the models, WF and WFH were processed by the PCA method, and the first five variables of the PCA results (PCA1-5) were chosen as the input characteristic variables of the SMC estimation models. The results of PCA are shown in Table 3.

Table 3. The PCA results in eigenvalue and variance contribution rate.

PCA	Eigenvalue		Variance Contribution (%)		Accumulative Contribution (%)	
	WF	**WFH**	**WF**	**WFH**	**WF**	**WFH**
PCA1	927.6×10^{-8}	0.0756	89.742	94.279	89.742	94.279
PCA2	40.8×10^{-8}	4.613×10^{-8}	3.216	3.457	92.958	97.736
PCA3	16.55×10^{-8}	1.572×10^{-9}	1.762	1.253	94.720	98.989
PCA4	10.17×10^{-8}	1.396×10^{-10}	0.965	0.102	95.685	99.091
PCA5	9.36×10^{-8}	1.631×10^{-10}	0.230	0.056	95.915	99.147

It turns out that the contribution rates of cumulative variance of the first five principal components of WF and WFH were 95.915% and 99.147%, respectively. The PCA performance of WFH data is better than that of WF data. PCA1-5 of WFH data roughly includes the harmonic characteristic variable information before processing, which not only retains a large amount of original data information, but also effectively compresses the original data. According to all PCA results, two characteristic variables were established: WFP (PCA of WF) and WFHP (PCA of WFH).

3.6. SMC Estimation and Model Validation Using Spectral Processing Technologies and Harmonic Indicators

Three regression estimation models (PCR, BP, and PLSR) were selected to explore the validity of characteristic variables and the accuracy of the soil moisture estimation models. Based on the modeling of WFP and WFHP, six SMC prediction models were established: WFP-PCR, WFHP-PCR, WFP-BP, WFHP-BP, WFP-PLSR, and WFHP-PLSR. For the BP neural network model, the topology of the model was finally determined as 5-3-1 after several debugging. That is, the number of nodes in the input layer is 5, the number of hidden layers is 5, and the output result layer is 1. Meanwhile, the times of iterations, adaptive learning rate, momentum factor, and the learning error were set as 3000, 0.01, 0.9, and 0.001, respectively. The precision and error of the modeling set and validation set are shown in Table 4. The WFHP has better performance than WFP for the PCR, PLSR, and BP models in calibration and validation datasets. For the same regression model, BP neural network has the highest accuracy than PCR and PLSR. In all similar models, the accuracy of the validation set is slightly lower than that of the modeling set.

To further observe the effect of different variables and different methods on the estimation of different SMC, the scatter diagram of the estimated and measured value of SMC in the validation dataset is shown in Figure 8. Each row represents different regression models of similar characteristic variables (WFP or WFHP), and each column represents the same regression model of different characteristic variables (PCR, PLSR, or BP). The red dotted line indicates the 1:1 line. It can be found that the WFP-based models are prone to underestimation when the SMC exceeds 10% (below the 1:1 line), while the WFHP-based models can accurately estimate SMC in the whole range (almost overlaps with the 1:1 line). For the same characteristic variable, the effect of PLSR and BP is significantly better than that of PCR (closer to the 1:1 line).

Table 4. Accuracy comparison of different regression models.

Model	Calibration			Validation		
	R^2	RMSE (%)	MAE (%)	R^2	RMSE (%)	MAE (%)
WFP-PCR	0.812	3.693	3.363	0.763	4.261	3.771
WFHP-PCR	0.851	3.279	2.819	0.836	3.523	2.902
WFP-PLSR	0.882	2.977	2.632	0.863	3.086	2.759
WFHP-PLSR	0.902	2.673	2.601	0.907	2.826	2.583
WFP-BP	0.917	2.504	2.132	0.909	2.626	2.286
WFHP-BP	0.945	2.115	1.653	0.932	2.311	1.834

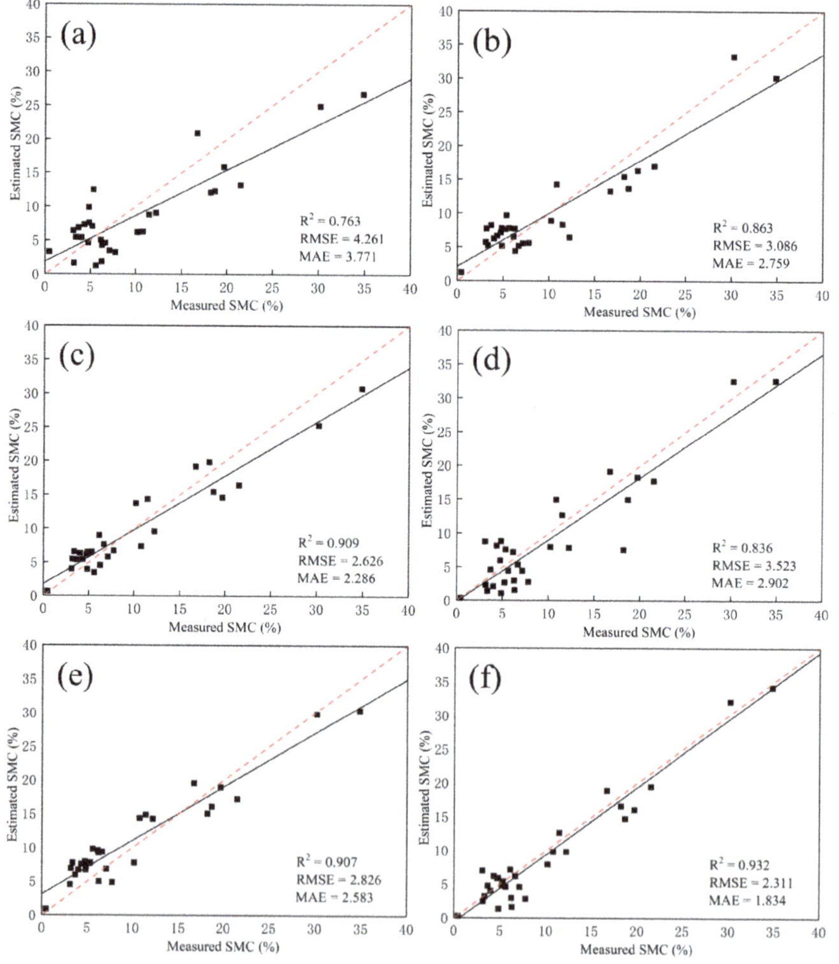

Figure 8. The comparison of measured and estimated SMC: (**a**) WFP-PCR; (**b**) WFP-PLSR; (**c**) WFP-BP; (**d**) WFHP-PCR; (**e**) WFHP-PLSR; (**f**) WFHP-BP.

Compared with the traditional spectral indices prediction results (Figure 5), the validation accuracy of all models, except the WFP-PCR model, was higher with an error below 3%

(Table 4 and Figure 8). This indicated that there was a great potential for spectral variables based on spectral processing techniques upon SMC estimation for mixed soil types.

4. Discussion

Traditional soil moisture measurements using neutron scattering, drying method, and resistance method have been part of many agricultural studies [47–49]. While these measurements provide accurate results, they are tedious, time consuming, and laborious, making it difficult to scale in large areas [50]. Compared with traditional soil moisture monitoring methods, remote sensing has incomparable advantages such as a large area and being a macroscopic, real-time, and dynamic method [30]. The hyperspectral sensor can detect the subtle changes in surface characteristics, and hyperspectral quantitative inversion provides an effective technical means for dynamic monitoring of regional SMC [9,19]. However, obtaining the best characteristic variables of SMC estimation of mixed soil types has always been difficult in modeling. In SMC estimation, the original soil spectral reflectance data contain much noise and a lot of redundant information, which cannot be used directly to estimate SMC. There are many differences in spectral characteristics of different soil types. For example, in SS spectral analysis, the reflectance of all bands decreases with the increase of SMC overall (Figure 4), showing a strong negative correlation (Figure 5). In LS, except for SMC, the variation rule of reflectance is not obvious due to the difference in organic matter content, grain size distribution, mineral composition, and soil color [51], thus reducing the correlation with SMC. However, the small content of these substances in SS has a small impact on reflectance. Therefore, it is difficult to establish a general SMC estimation model. In most cases, it is necessary to carry out the spectral transformation on the original soil spectral reflection data, such as reciprocal, logarithm, FOD, etc. to extract characteristic bands or parameters to obtain feature variables [52]. However, these methods have a low noise reduction function and cannot deal with data background and noise well, which directly affects the accuracy of subsequent estimation.

In this paper, the results of several traditional spectral indices for estimating SMC showed that both univariate linear regression models and multivariate PLSR models had significant errors. Therefore, it is necessary to explore the variables and methods for SMC estimation in mixed soil types.

Through correlation analysis, it can be found that the correlation between WF and SMC is significantly higher than that of the original spectra and SMC (Figure 6). It shows that the FOD spectra can eliminate some effects of background and atmosphere, but still cannot achieve satisfactory results. In this paper, the HD method was adopted. The soil spectra were converted to frequency spectra to obtain harmonic characteristic parameters based on Fourier transform theory to effectively reduce the uncertainty of spectral parameter calculation. Furthermore, harmonic parameters can better reflect soil spectral changes caused by subtle changes in soil internal components. Compared with traditional spectral parameters, harmonic characteristic parameters (remainder, amplitude, and phase) are more correlated with SMC (Figure 7). Finally, 11 harmonic characteristic parameters with high correlation (| r | > 0.7) were selected. Based on the FOD and HD, the PCA method was applied to reduce the dimensionality of data and two kinds of feature parameters were gained: WFP and WFHP.

In parameter estimation studies using empirical models, PLSR, BP, and PCR all showed good effects [16,28,30]. To explore the applicability of the two types of characteristic parameters extracted in this paper (WFP and WFHP), these three models were used for comparison of estimation. The results show that WFPH is superior to WFP in SMC estimation in these three models (Table 4 and Figure 8). When selecting the same characteristic parameters (WFP or WFHP), the effects of PLSR and BP models are significantly better than PCR. The advantage of the PLSR model is that it can strengthen the error convergence ability of the model when the sample size is not particularly sufficient, while BP is a nonlinear distribution that can better reflect SMC and is mainly good at nonlinear prediction. Soil spectra are a comprehensive reflection of various soil properties, and the selection of estimation model alone cannot effectively solve the problem of accurate estimation of SMC. Therefore, it is necessary to explore some common and stable characteristic parameters to establish a more robust and suitable SMC inversion model. The harmonic characteristic parameters constructed in this paper can transform complex signals in the time domain into simplified signals in the frequency domain, which can not only suppress or eliminate ground object background noise, but also highlight the spectral characteristics of the ground object with low order harmonic components to achieve the effect of data compression. Therefore, the SMC prediction ability of the three models (BP, PLSR, and PCR) was effectively improved. Moreover, the advantage of harmonic variables in predicting SMC also reflects that they can accurately predict different SMC, including low and high values, while WFP parameters are underestimated at high values of SMC (Figure 8).

To further check the performance of the optimal model (WFHP-BP) in this paper for SMC estimation in different soil types, the validation models for single soil types are shown in Figure 9. It can be found that the estimation accuracy of SMC is better than that of mixed soil types in both SS and LS, and neither of them shows local overestimation and underestimation. This may be because single soil types are more consistent physically or chemically and thus receive less interference from other factors. Since the BP neural network model has a nonlinear learning capability, the estimated values of SMC for different soil types did not appear to be overestimated or underestimated.

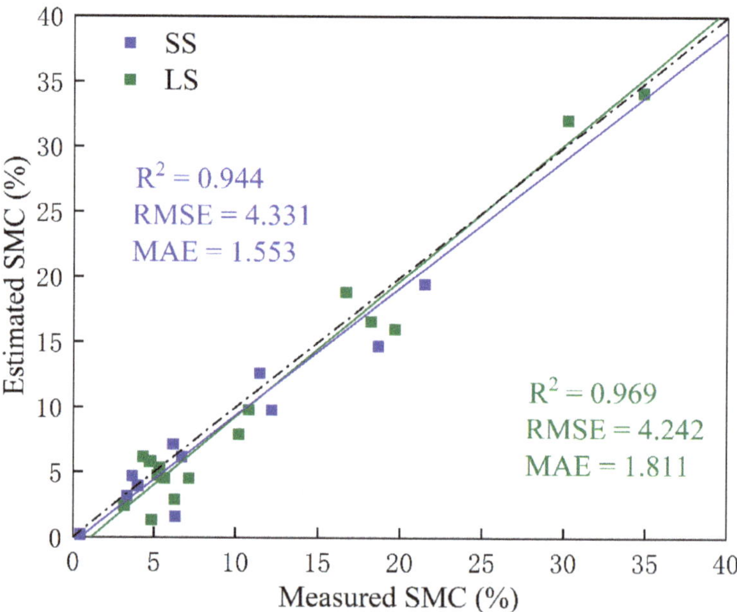

Figure 9. The comparison of measured and estimated SMC of different types of soil.

This study provided effective parameters and methods for nondestructive estimation of SMC in mixed soil types, and future research should be devoted to using satellite imagery as an alternative to ground-based measurements because of its large area, economy, time savings, and high temporal resolution, which can provide a data source for real-time field SMC mapping.

5. Conclusions

In this paper, a feature extraction method based on spectral processing technologies (WPT, FOD, and HD) and PCA was proposed, and three regression prediction methods

(PCR, PLSR, and BP) were combined to compare the accuracy and applicability of SMC estimation for mixed soil. It is observed that for SS with less impurity, the variation of spectral reflectance can well describe the difference in SMC. However, for LS and MS, the spectral reflectance cannot be directly used to predict the SMC due to the influence of organic matter content, grain size distribution, mineral composition, and soil color. After WPT and FOD transformation using the original spectral data, two sets of data can be obtained after HD: WF and WFH. Meanwhile, the PCA method was utilized to reduce the dimensionality of these two datasets to obtain two sets of characteristic parameters: WFP and WFHP. The results of three regression models (WFP-PCR, WFHP-PCR, WFP-PLSR, WFHP-PLSR, WFP-BP, and WFHP-BP) indicated that the WFHP-based models showed better performance than that of WFP-based models. Among the different regression methods, BP neural network has the highest accuracy as a result of its nonlinear prediction ability. The best prediction model is WFHP-BP (R^2 = 0.932, RMSE = 2.311, MAE = 1.834 for the validation dataset). Moreover, harmonic variables have advantages in predicting SMC values in a larger range. This study can provide a theoretical basis and technical support for establishing SMC inversion models suitable for various types and a large range of soils. Future research should focus more on the use of satellite remote sensing data and on proposing physical or chemical indicators of soils that are more suitable for SMC estimation.

Author Contributions: Conceptualization, S.L. and X.J.; methodology, X.J.; software, S.L.; validation, Q.Y.; writing—original draft preparation, X.J. and S.L.; writing—review and editing, X.L.; visualization, W.J.; funding acquisition, Q.Y. All authors have read and agreed to the published version of the manuscript.

Funding: This research was funded by the National Natural Science Foundation of China (Grant No.52078350), the National Key Research and Development Program of China (Grant No.2018YFB0505400), and the Natural Science Foundation of Shanghai (Grant No.22ZR1465700).

Institutional Review Board Statement: Not applicable.

Informed Consent Statement: Not applicable.

Data Availability Statement: Not applicable.

Conflicts of Interest: The authors declare no conflict of interest.

References

1. Ebtehaj, A.; Bras, R.L. A physically constrained inversion for high-resolution passive microwave retrieval of soil moisture and vegetation water content in L-band. *Remote Sens. Environ.* **2019**, *233*, 15. [CrossRef]
2. Bablet, A.; Viallefont-Robinet, F.; Jacquemoud, S.; Fabre, S.; Briottet, X. High-resolution mapping of in-depth soil moisture content through a laboratory experiment coupling a spectroradiometer and two hyperspectral cameras. *Remote Sens. Environ.* **2020**, *236*, 11. [CrossRef]
3. Arunrat, N.; Sereenonchai, S.; Hatano, R. Effects of fire on soil organic carbon, soil total nitrogen, and soil properties under rotational shifting cultivation in northern Thailand. *J. Environ. Manag.* **2022**, *302*, 15. [CrossRef] [PubMed]
4. Qu, W.D.; Han, G.X.; Wang, J.; Li, J.Y.; Zhao, M.L.; He, W.J.; Li, X.G.; Wei, S.Y. Short-term effects of soil moisture on soil organic carbon decomposition in a coastal wetland of the Yellow River Delta. *Hydrobiologia* **2021**, *848*, 3259–3271. [CrossRef]
5. Gibon, F.; Pellarin, T.; Roman-Cascon, C.; Alhassane, A.; Traore, S.; Kerr, Y.; Lo Seen, D.; Baron, C. Millet yield estimates in the Sahel using satellite derived soil moisture time series. *Agric. For. Meteorol.* **2018**, *262*, 100–109. [CrossRef]
6. Humphrey, V.; Berg, A.; Ciais, P.; Gentine, P.; Jung, M.; Reichstein, M.; Seneviratne, S.I.; Frankenberg, C. Soil moisture-atmosphere feedback dominates land carbon uptake variability. *Nature* **2021**, *592*, 65–69. [CrossRef]
7. Liu, L.B.; Gudmundsson, L.; Hauser, M.; Qin, D.H.; Li, S.C.; Seneviratne, S.I. Soil moisture dominates dryness stress on ecosystem production globally. *Nat. Commun.* **2020**, *11*, 9. [CrossRef]
8. Sadeghi, M.; Jones, S.B.; Philpot, W.D. A linear physically-based model for remote sensing of soil moisture using short wave infrared bands. *Remote Sens. Environ.* **2015**, *164*, 66–76. [CrossRef]
9. Zhang, Y.; Tan, K.; Wang, X.; Chen, Y. Retrieval of soil moisture content based on a modified Hapke Photometric model: A novel method applied to laboratory hyperspectral and Sentinel-2 MSI data. *Remote Sens.* **2020**, *12*, 2239. [CrossRef]
10. Luo, C.; Zhang, X.L.; Meng, X.T.; Zhu, H.W.; Ni, C.P.; Chen, M.H.; Liu, H.J. Regional mapping of soil organic matter content using multitemporal synthetic Landsat 8 images in Google Earth Engine. *Catena* **2022**, *209*, 11. [CrossRef]

11. Niyogi, D.; Jamshidi, S.; Smith, D.; Kellner, O. Evapotranspiration climatology of Indiana using in situ and remotely sensed products. *J. Appl. Meteorol. Climatol.* **2020**, *59*, 2093–2111. [CrossRef]
12. Jamshidi, S.; Zand-Parsa, S.; Niyogi, D. Assessing crop water stress index of citrus using in-situ measurements, Landsat, and Sentinel-2 data. *Int. J. Remote Sens.* **2021**, *42*, 1893–1916. [CrossRef]
13. Skakun, S.; Kalecinski, N.I.; Brown, M.G.L.; Johnson, D.M.; Vermote, E.F.; Roger, J.C.; Franch, B. Assessing within-field corn and soybean yield variability from WorldView-3, Planet, Sentinel-2, and Landsat 8 satellite imagery. *Remote Sens.* **2021**, *13*, 872. [CrossRef]
14. Eon, R.S.; Bachmann, C.M. Mapping barrier island soil moisture using a radiative transfer model of hyperspectral imagery from an unmanned aerial system. *Sci. Rep.* **2021**, *11*, 11. [CrossRef] [PubMed]
15. Muller, E.; Decamps, H. Modeling soil moisture-reflectance. *Remote Sens. Environ.* **2001**, *76*, 173–180. [CrossRef]
16. Pellegrini, E.; Rovere, N.; Zaninotti, S.; Franco, I.; De Nobili, M.; Contin, M. Artificial neural network (ANN) modelling for the estimation of soil microbial biomass in vineyard soils. *Biol. Fertil. Soils* **2021**, *57*, 145–151. [CrossRef]
17. Emamgholizadeh, S.; Mohammadi, B. New hybrid nature-based algorithm to integration support vector machine for prediction of soil cation exchange capacity. *Soft Comput.* **2021**, *25*, 13451–13464. [CrossRef]
18. Yin, Z.; Lei, T.W.; Yan, Q.H.; Chen, Z.P.; Dong, Y.Q. A near-infrared reflectance sensor for soil surface moisture measurement. *Comput. Electron. Agric.* **2013**, *99*, 101–107. [CrossRef]
19. Jiang, X.Q.; Luo, S.J.; Fang, S.H.; Cai, B.W.; Xiong, Q.; Wang, Y.Y.; Huang, X.; Liu, X.J. Remotely sensed estimation of total iron content in soil with harmonic analysis and BP neural network. *Plant Methods* **2021**, *17*, 12. [CrossRef]
20. Cheng, H.; Wang, J.; Du, Y.K.; Zhai, T.L.; Fang, Y.; Li, Z.H. Exploring the potential of canopy reflectance spectra for estimating organic carbon content of aboveground vegetation in coastal wetlands. *Int. J. Remote Sens.* **2021**, *42*, 3850–3872. [CrossRef]
21. Li, Y.; Via, B.K.; Li, Y.X. Lifting wavelet transform for Vis-NIR spectral data optimization to predict wood density. *Spectroc. Acta Pt. A-Molec. Biomolec. Spectr.* **2020**, *240*, 9. [CrossRef] [PubMed]
22. Luo, S.J.; He, Y.B.; Wang, Z.Z.; Duan, D.D.; Zhang, J.K.; Zhang, Y.T.; Zhu, Y.Q.; Yu, J.K.; Zhang, S.L.; Xu, F.; et al. Comparison of the retrieving precision of potato leaf area index derived from several vegetation indices and spectral parameters of the continuum removal method. *Eur. J. Remote Sens.* **2019**, *52*, 155–168. [CrossRef]
23. Blanco, M.; Coello, J.; Iturriaga, H.; Maspoch, S.; Pages, J. NTR calibration in non-linear systems: Different PLS approaches and artificial neural networks. *Chemometrics Intell. Lab. Syst.* **2000**, *50*, 75–82. [CrossRef]
24. Gu, X.H.; Wang, Y.C.; Sun, Q.; Yang, G.J.; Zhang, C. Hyperspectral inversion of soil organic matter content in cultivated land based on wavelet transform. *Comput. Electron. Agric.* **2019**, *167*, 7. [CrossRef]
25. Yuan, L.N.; Li, L.; Zhang, T.; Chen, L.Q.; Liu, W.Q.; Hu, S.; Yang, L.H. Modeling Soil Moisture from Multisource Data by Stepwise Multilinear Regression: An Application to the Chinese Loess Plateau. *ISPRS Int. J. Geo-Inf.* **2021**, *10*, 233. [CrossRef]
26. Xiong, J.F.; Lin, C.; Ma, R.H.; Zheng, G.H. The total P estimation with hyper-spectrum A novel insight into different P fractions. *Catena* **2020**, *187*, 11. [CrossRef]
27. Lin, N.; Liu, H.Q.; Yang, J.J.; Liu, H.L. Hyperspectral estimation of soil composition contents based on kernel principal component analysis and machine learning model. *J. Appl. Remote Sens.* **2020**, *14*, 19. [CrossRef]
28. Yang, J.C.; Wang, X.L.; Wang, R.H.; Wang, H.J. Combination of convolutional neural networks and recurrent neural networks for predicting soil properties using Vis-NIR spectroscopy. *Geoderma* **2020**, *380*, 16. [CrossRef]
29. Taghizadeh-Mehrjardi, R.; Schmidt, K.; Toomanian, N.; Heung, B.; Behrens, T.; Mosavi, A.; Band, S.S.; Amirian-Chakan, A.; Fathabadi, A.; Scholten, T. Improving the spatial prediction of soil salinity in arid regions using wavelet transformation and support vector regression models. *Geoderma* **2021**, *383*, 21. [CrossRef]
30. Shen, L.Z.; Gao, M.F.; Yan, J.W.; Li, Z.L.; Leng, P.; Yang, Q.; Duan, S.B. Hyperspectral estimation of soil organic matter content using different spectral preprocessing techniques and PLSR method. *Remote Sens.* **2020**, *12*, 1206. [CrossRef]
31. Ma, Y.; Fang, S.H.; Peng, Y.; Gong, Y.; Wang, D. Remote estimation of biomass in winter oilseed rape (*Brassica napus* L.) using canopy hyperspectral data at different growth stages. *Appl. Sci.* **2019**, *9*, 545. [CrossRef]
32. Tian, H.R.; Wang, P.X.; Tansey, K.; Zhang, S.Y.; Zhang, J.Q.; Li, H.M. An IPSO-BP neural network for estimating wheat yield using two remotely sensed variables in the Guanzhong Plain, PR China. *Comput. Electron. Agric.* **2020**, *169*, 10. [CrossRef]
33. Shi, Y.J.; Ren, C.; Yan, Z.H.; Lai, J.M. High spatial-temporal resolution estimation of ground-based global navigation satellite system interferometric reflectometry (GNSS-IR) soil moisture using the genetic algorithm back propagation (GA-BP) neural network. *ISPRS Int. J. Geo-Inf.* **2021**, *10*, 623. [CrossRef]
34. Wang, X.; An, S.; Xu, Y.Q.; Hou, H.P.; Chen, F.Y.; Yang, Y.J.; Zhang, S.L.; Liu, R. A back propagation neural network model optimized by mind evolutionary algorithm for estimating Cd, Cr, and Pb concentrations in soils using Vis-NIR diffuse reflectance spectroscopy. *Appl. Sci.* **2020**, *10*, 51. [CrossRef]
35. Liang, Y.J.; Ren, C.; Wang, H.Y.; Huang, Y.B.; Zheng, Z.T. Research on soil moisture inversion method based on GA-BP neural network model. *Int. J. Remote Sens.* **2019**, *40*, 2087–2103. [CrossRef]
36. Dong, Z.Q.; Liu, Y.; Ci, B.X.; Wen, M.; Li, M.H.; Lu, X.; Feng, X.K.; Wen, S.; Ma, F.Y. Estimation of nitrate nitrogen content in cotton petioles under drip irrigation based on wavelet neural network approach using spectral indices. *Plant Methods* **2021**, *17*, 13. [CrossRef]
37. Tao, L.L.; Wang, G.J.; Chen, X.; Li, J.; Cai, Q.K. Soil moisture retrieval using modified particle swarm optimization and back-propagation neural network. *Photogramm. Eng. Remote Sens.* **2019**, *85*, 789–798. [CrossRef]

38. Kahaer, Y.; Tashpolat, N.; Shi, Q.D.; Liu, S.H. Possibility of Zhuhai-1 hyperspectral imagery for monitoring salinized soil moisture content using fractional order differentially optimized spectral indices. *Water* **2020**, *12*, 3360. [CrossRef]
39. Li, X.; Ding, J.L. Soil moisture monitoring based on measured hyperspectral index and HSI image index. *Trans. Chin. Soc. Agric. Eng.* **2015**, *31*, 68–75. [CrossRef]
40. Zhang, X.L.; Zhang, F.; Zhang, H.W.; Li, Z.; Hai, Q.; Chen, L.H. Optimization of soil salt inversion model based on spectral transformation from hyperspectral index. *Trans. Chin. Soc. Agric. Eng.* **2018**, *34*, 110–117. [CrossRef]
41. Liu, Y.; Pan, X.Z.; Wang, C.K.; Li, Y.L.; Shi, R.J.; Li, Z.T. Prediction of saline soil moisture content based on differential spectral index: A case study of coastal saline soil. *Soils* **2016**, *48*, 381–388. [CrossRef]
42. Wu, D.H.; Fan, W.J.; Cui, Y.K.; Yan, B.Y.; Xu, X.R. Review of monitoring soil water content using hyperspectral remote sensing. *Spectrosc. Spectr. Anal.* **2010**, *30*, 3067–3071. [CrossRef]
43. Tripathi, M.; Singal, S.K. Use of principal component analysis for parameter selection for development of a novel water quality index: A case study of river Ganga India. *Ecol. Indic.* **2019**, *96*, 430–436. [CrossRef]
44. Fernandez-Delgado, M.; Cernadas, E.; Barro, S.; Amorim, D. Do we need hundreds of classifiers to solve real world classification problems? *J. Mach. Learn. Res.* **2014**, *15*, 3133–3181.
45. Meacham-Hensold, K.; Montes, C.M.; Wu, J.; Guan, K.Y.; Fu, P.; Ainsworth, E.A.; Pederson, T.; Moore, C.E.; Brown, K.L.; Raines, C.; et al. High-throughput field phenotyping using hyperspectral reflectance and partial least squares regression (PLSR) reveals genetic modifications to photosynthetic capacity. *Remote Sens. Environ.* **2019**, *231*, 10. [CrossRef]
46. Sun, W.; Huang, C.C. A carbon price prediction model based on secondary decomposition algorithm and optimized back propagation neural network. *J. Clean Prod.* **2020**, *243*, 13. [CrossRef]
47. Dominguez-Nino, J.M.; Oliver-Manera, J.; Girona, J.; Casadesus, J. Differential irrigation scheduling by an automated algorithm of water balance tuned by capacitance-type soil moisture sensors. *Agric. Water Manag.* **2020**, *228*, 11. [CrossRef]
48. Jamshidi, S.; Zand-Parsa, S.; Niyogi, D. Physiological responses of orange trees subject to regulated deficit irrigation and partial root drying. *Irrig. Sci.* **2021**, *39*, 441–455. [CrossRef]
49. Jamshidi, S.; Zand-Parsa, S.; Kamgar-Haghighi, A.A.; Shahsavar, A.R.; Niyogi, D. Evapotranspiration, crop coefficients, and physiological responses of citrus trees in semi-arid climatic conditions. *Agric. Water Manag.* **2020**, *227*, 12. [CrossRef]
50. Wu, T.H.; Yu, J.; Lu, J.X.; Zou, X.G.; Zhang, W.T. Research on inversion model of cultivated soil moisture content based on hyperspectral imaging analysis. *Agriculture* **2020**, *10*, 292. [CrossRef]
51. Jacquemoud, S.; Baret, F.; Hanocq, J.F. Modeling spectral and bidirectional soil reflectance. *Remote Sens. Environ.* **1992**, *41*, 123–132. [CrossRef]
52. Tian, A.H.; Zhao, J.S.; Tang, B.H.; Zhu, D.M.; Fu, C.B.; Xiong, H.G. Hyperspectral prediction of soil total salt content by different disturbance degree under a fractional-order differential model with differing spectral transformations. *Remote Sens.* **2021**, *13*, 4283. [CrossRef]

MDPI
St. Alban-Anlage 66
4052 Basel
Switzerland
www.mdpi.com

Agriculture Editorial Office
E-mail: agriculture@mdpi.com
www.mdpi.com/journal/agriculture

Disclaimer/Publisher's Note: The statements, opinions and data contained in all publications are solely those of the individual author(s) and contributor(s) and not of MDPI and/or the editor(s). MDPI and/or the editor(s) disclaim responsibility for any injury to people or property resulting from any ideas, methods, instructions or products referred to in the content.